# Readings in Conservation Ecology

edited by
George W. Cox
SAN DIEGO STATE UNIVERSITY

# Readings in Conservation Ecology

second edition

GOODYEAR PUBLISHING COMPANY, INC.
Pacific Palisades, California

Y-7794-4
ISBN: 0-87620-779-4
Library of Congress Catalog Card Number: 73-14412

Current printing (last digit):
10 9 8 7 6 5 4 3 2

**Library of Congress Cataloging in Publication Data**

Cox, George W comp.
Readings in conservation ecology.

Includes bibliographies.
1. Conservation of natural resources—Addresses, essays, lectures. 2. Ecology—Addresses, essays, lectures. 3. Pollution—Addresses, essays, lectures.
I. Title.
S942.C6 1973 639'.9 73-14412
ISBN 0-87620-779-4

# Editor's Introduction

The most important trend in the conservation movement today is the growing realization that conservation policy and practice must be based on sound ecological theory and must be directed toward the management of entire ecological systems. This trend has given rise to the field of conservation ecology, the aim of which is perpetuation of the economic and esthetic values of ecological systems through maintenance of their functional integrity. Recognition of the need for such an approach to conservation has developed slowly from observations of the failures and shortcomings of practices directed toward management of individual resources or individual species, and from the growing evidence that man's utilization of his environment has brought about major, yet often inconspicuous, changes in the functioning of many ecological systems. The theory and technology for this approach to conservation have been provided largely by the field of ecology, which in recent years has increasingly emphasized study of the dynamics of ecological systems. The growing acceptance of this approach is evidenced by the frequency with which papers concerned with a systems approach to diverse conservation problems are now appearing in major scientific journals.

Conservation ecology is based on recognition of the existence of functional environmental units known as ecosystems. An ecosystem is a unit consisting of groups of organisms together with their physical environment. The components of an ecosystem, both biotic and abiotic, show a high degree of interdependence as a result of processes involving the exchange of energy and specific chemical nutrients. An example of a specific ecosystem type is a lake, with the various plant, animal, and decomposer organisms representing the biotic components, and the water mass and bottom sediments representing two of the major abiotic components. Ecosystems are thus dynamic in nature, and exhibit a structure which reflects the pattern of interchange of nutrients and energy. Although some ecosystems may approach a steady-state condition in which the structure and function of the system change little through time, most ecosystems exhibit more or less rapid change in these features. Since these systems are dynamic, active management by man is usually necessary for the perpetuation of some desired condition. For ecosystems not in a steady-state condition, or for ecosystems influenced by broad environmental changes caused by man, simple protection from direct exploitation, without active management, may rapidly lead to the loss of important ecosystem values. The goal of conservation ecology is thus to develop an understanding of the functioning of world ecosystems and to devise ecologically sound techniques for their management.

Inherent in this approach to conservation is the realization that man is one of the interrelated components of most ecological systems. This relationship implies that advancing human technology, which results in the utilization of environmental resources in an increasingly intensive fashion, increases the complexity of man's interrelationships with natural ecological systems, rather than making him more independent of them. It also demands that direct consideration be given to the implications of human population behavior and the effects of human technology in an overall survey of the field of conservation ecology.

The flood of textbooks and readings collections in the past few years is evidence that courses dealing with ecology and human affairs have become standard items in the curricula of colleges and universities. However, the objectives and subject matter of these courses vary widely, and their content is rapidly changing. In some cases the course objective is simply to promote conservationist attitudes or environmental activism. In other cases it is to relate ecology, as a science, to the concerns of modern society in a manner appropriate to general education students. In still others it is to provide a basic ecological background for preprofessional students in the environmental sciences. The specific topics examined also differ widely, largely depending on the extent to which the instructor chooses to represent various interdisciplinary areas between ecology and the social sciences, physical sciences, consumer sciences, and even history, philosophy, and art. On top of this, public concern has forced rapid movement by scientific and governmental groups, leading to the continual discovery of new problem areas, as well as to a change in the information available on old ones, forcing continued change in course content.

While several different course formats may be pursued successfully, there are certain major points that deserve special emphasis. These relate to the misconceptions forming the basis of the growing "ecological backlash." This "backlash" derives from two main sources: (1) environmentalist programs aimed at superficial aspects of problems, at the expense of basic causes, and (2) environmentalist-supported programs based on inadequate analysis of the relationships involved in a particular problem situation. The first of these is reflected in the widespread feeling that the ecological movement has to do primarily with environmental "appearances," and is therefore something that only the affluent nations can afford to be concerned with. The second provides ammunition for the argument that environmentalists are misguided "do-gooders" who will, if given their way, lead us to economic disaster.

This collection of readings is aimed directly at these misconceptions. First, I have chosen studies that demonstrate that current environmental problems are indeed problems of basic ecosystem function, and problems which are becoming of importance on a global scale. Second, I have selected papers emphasizing the fact that these problems frequently result from complex ecosystem interactions that are "counter-intuitive." That is, they are problems for which simple, common sense remedies may prove inadequate or even inappropriate. The widespread recognition of these two facts is essential to continued progress in sound environmental management.

The readings in this book are organized around the concept of the ecosystem. The collection is introduced by two papers. The first, by Francis C. Evans, concisely presents the concept of the ecosystem, while the second, by Eugene P. Odum, considers in greater detail the relationships between ecosystem structure and function. In the second section, the cultural roots and basic nature of the environmental crisis are explored, and the systems approach to the study of complex environmental problems outlined. Following this, studies and discussions are organized around a variety of specific environmental problem areas. This collection is drawn together with a paper by Aldo Leopold, written about 40 years ago, but still pertinent to the problem of how the social and political attitudes necessary for the implementation of ecologically sound management of the environment must develop.

This set of readings is designed to complement a one-semester course in conservation ecology. The specific organization of such a course must, of course, depend on the philosophy of the instructor and the background of the students to whom the course is directed. At San Diego State University, where the course is available both to undergraduate majors in biology and to nonmajors who have had an introductory biology course, conservation ecology is introduced by a series of lectures on principles of ecology relating to ecosystem function. These include principles of energy flow and nutrient cycling in ecological systems, principles of population growth and regulation, and principles relating to succession and stability in communities and ecosystems. Following this introduction, specific problems, corresponding to the topics under which papers are grouped in this collection, are considered in the light of basic ecological theory.

The diversity in nature of the papers in this collection allows them to play a variable role in such a course. Some of the papers deal with basic theory of management, some summarize the state of knowledge on particular problems, and others deal with specific research studies of certain problems. A few of the papers may be most useful as substitutes for textbook coverage of certain problems. Others are probably most useful as examples of the application of an ecological approach to specific problems. One of the most important functions of many of the papers may be, however, as background reading on which class discussions of particular problems may be based. The complexity of many of these problems and the variety of ecological considerations involved in the development of appropriate management procedures may often best be explored through discussions drawing upon the ideas of many individuals.

I wish to express my gratitude to my colleagues in ecology, both at San Diego State University, and elsewhere, who have all, directly or indirectly, contributed to my interest and awareness of the field of conservation ecology. I am especially indebted to Boyd D. Collier, Charles F. Cooper, Darla G. Cox, Richard Ford, William E. Hazen, Albert W. Johnson, Herbert Melchiol, and Joy and Paul Zedler for specific suggestions on the inclusion of papers. Finally, I wish to thank the individual authors and publishers for their permission to reprint the papers included herein.

G.W.C.

# Contents

## Part IV. *Control of Undesirable Species*



## Part V. *Management of Vegetation*



## Part VI. *Human Populations*



## Part VII. *Pollutants and Ecosystems*

### Pesticides and Related Materials

**Terrestrial Ecosystems**

# I

# The Ecosystem Concept

# 1

# Ecosystem as the basic unit in ecology

*Francis C. Evans*

The term *ecosystem* was proposed by Tansley (1) as a name for the interaction system comprising living things together with their nonliving habitat. Tansley regarded the ecosystem as including "not only the organism-complex, but also the whole complex of physical factors forming what we call the environment." He thus applied the term specifically to that level of biological organization represented by such units as the community and the biome. I here suggest that it is logically appropriate and desirable to extend the application of the concept and the term to include organization levels other than that of the community.

In its fundamental aspects, an ecosystem involves the circulation, transformation, and accumulation of energy and matter through the medium of living things and their activities. Photosynthesis, decomposition, herbivority, predation, parasitism, and other symbiotic activities are among the principal biological processes responsible for the transport and storage of materials and energy, and the interactions of the organisms engaged in these activities provide the pathways of distribution. The food-chain is an example of such a pathway. In the nonliving part of the ecosystem, circulation of energy and matter is completed by such physical processes as evaporation and precipitation, erosion and deposition. The ecologist, then, is primarily concerned with the quantities of

Reprinted with permission from *Science*, 123:1127–1128 (22 June 1956).

matter and energy that pass through a given ecosystem and with the rates at which they do so. Of almost equal importance, however, are the kinds of organisms that are present in any particular ecosystem and the roles that they occupy in its structure and organization. Thus, both quantitative and qualitative aspects need to be considered in the description and comparison of ecosystems.

Ecosystems are further characterized by a multiplicity of regulatory mechanisms which, in limiting the numbers of organisms present and in influencing their physiology and behavior, control the quantities and rates of movement of both matter and energy. Processes of growth and reproduction, agencies of mortality (physical as well as biological), patterns of immigration and emigration, and habits of adaptive significance are among the more important groups of regulatory mechanisms. In the absence of such mechanisms, no ecosystem could continue to persist and maintain its identity.

The assemblage of plants and animals visualized by Tansley as an integral part of the ecosystem usually consists of numerous species, each represented by a population of individual organisms. However, each population can be regarded as an entity in its own right, interacting with its environment (which may include other organisms as well as physical features of the habitat) to form a system of lower rank that likewise involves the distribution of matter and energy. In turn, each individual animal or plant, together with its particular microenvironment, constitutes a system of still lower rank. Or we may wish to take a world view of life and look upon the biosphere with its total environment as a gigantic ecosystem. Regardless of the level on which life is examined, the ecosystem concept can appropriately be applied. The ecosystem thus stands as a basic unit of ecology, a unit that is as important to this field of natural science as the species is to taxonomy and systematics. In any given case, the particular level on which the ecosystem is being studied can be specified with a qualifying adjective—for example, community ecosystem, population ecosystem, and so forth.

All ranks of ecosystems are open systems, not closed ones. Energy and matter continually escape from them in the course of the processes of life, and they must be replaced if the system is to continue to function. The pathways of loss and replacement of matter and energy frequently connect one ecosystem with another, and therefore it is often difficult to determine the limits of a given ecosystem. This has led some ecologists to reject the ecosystem concept as unrealistic and of little use in description or analysis. One is reminded, however, of the fact that it is also difficult, if not impossible, to delimit a species from its ancestral or derivative species or from both; yet this does not destroy the value of the concept. The ecosystem concept may indeed be more useful when it is employed in relation to the community than to the population or individual, for its limits may be more easily determined on that level. Nevertheless, its application to all levels seems fully justified.

The concept of the ecosystem has been described under many names, among them those of *microcosm* (2), *naturkomplex* (3), *holocoen* (4) and *biosystem* (5). Tansley's term seems most successfully to convey its meaning and has in fact been accepted by a large number of present-day ecologists. I hope that it will eventually be adopted universally and that its application will be expanded beyond its original use to include other levels of biological organization. Recognition of the ecosystem as the basic unit in ecology would be helpful in focussing attention upon the truly fundamental aspects of this rapidly developing science.

## References

1. A. G. Tansley, *Ecology* 16, 296 (1935).
2. S. A. Forbes, *Bull. Peoria Sci. Assoc.* (1887).
3. E. Markus, *Sitzber. Naturforsch. Ges. Univ. Tartu* 32, 79 (1926).
4. K. Friederichs, *Die Grundfragen und Gesetzmässigheiten der land-und forstwirtschaftlichen zoologie*. (Parey, Berlin, 1930).
5. K. Thienemann, *Arch. Hydrobiol.* 35, 267 (1939).

# 2

# Relationships between structure and function in the ecosystem

*Eugene P. Odum*

The topic I wish to discuss with you today is: Relationships between structure and function in the ecosystem. As you know ecology is often defined as: The study of interrelationships between organisms and environment. I feel that this conventional definition is not suitable; it is too vague and too broad. Personally, I prefer to define ecology as: The study of the structure and function of ecosystems. Or we might say in a less technical way: The study of structure and function of nature.

By structure we mean: (1) The composition of the biological community including species, numbers, biomass, life history and distribution in space of populations; (2) the quantity and distribution of the abiotic (nonliving) materials such as nutrients, water, etc.; (3) the range, or gradient, of conditions of existence such as temperature, light, etc. Dividing ecological structure into these three equal divisions is, of course, arbitrary but I believe convenient for actual study of both aquatic and terrestrial situations.

By function we mean: (1) The rate of biological energy flow through the ecosystem, that is, the rates of production and the rates of respiration of the populations and the community; (2) the rate of material or nutrient cycling, that is, the biogeochemical cycles; (3) biological or ecologi-

---

Address given at the 9th Annual Meeting of the Ecological Society of Japan, April 4, 1962. Reprinted with modification from the *Japanese Journal of Ecology*, 12(3):108–118.

cal regulation including both regulation of organisms by environment (as, for example, in photoperiodism) and regulation of environment by organisms (as, for example, in nitrogen fixation by microorganisms). Again, dividing ecological function into these three divisions is arbitrary but convenient for study.

Until recently ecologists have been largely concerned with structure, or what we might call the descriptive approach. They were content to describe the conditions of existence and the standing crop of organisms and materials. In recent years equal emphasis is being placed on the functional approach as indicated by the increasing number of studies on productivity and biological regulation. Also the use of experimental methods, both in the field and in the laboratory, has increased. Today, there exists a very serious gap between the descriptive and the functional approach. It is very important that we bring together these two schools of ecology. I should like to present some suggestions for bridging this gap.

The main features of the structure of a terrestrial and an aquatic ecosystem may be illustrated by comparing an open water community, such as might be found at sea or in a large lake, with a land community such as a forest. In our discussion we shall consider these two types as models for the extremes in a gradient of communities which occur in our biosphere. Thus, such ecosystems as estuaries, marshes, shallow lakes, grasslands and agricultural croplands will have a community structure intermediate between the open water and forest types.

Both aquatic and terrestrial community types have several structural features in common. Both must have the same three necessary biological components: (1) producers or green plants capable of fixing light energy (i.e., autotrophs); (2) animals or macro-consumers which consume particulate organic matter (i.e., phagotrophs); and (3) microorganism decomposers which dissolve organic matter releasing nutrients (i.e., osmotrophs). Both ecosystems must be supplied with the same vital materials such as nitrogen, phosphorus, trace minerals, etc. Both ecosystems are regulated and limited by the same conditions of existence such as light and temperature. Finally, the arrangement of biological units in vertical space is basically the same in the two contrasting types of ecosystems. Both have two strata, an autotrophic stratum above and a heterotrophic stratum below. The photosynthetic machinery is concentrated in the upper stratum or photic zone where light is available, while the consumer-nutrient regenerating machinery is concentrated largely below the photic zone. It is important to emphasize that while the vertical extent or thickness of communities varies greatly (especially in water), light energy comes into the ecosystem on a horizontal surface basis which is everywhere the same. Thus, different ecosystems should be compared on a square meter basis, not on a cubic or volume basis.

On the other hand, aquatic and terrestrial ecosystems differ in structure in several important ways. Species composition is, of course, completely different; the roles of producers, consumers, and decomposers are carried out by taxonomically different organisms which have become adapted through evolution. Trophic structure also differs in that land plants tend to be large in size but few in number while the autotrophs of open water ecosystems (i.e., phytoplankton) are small in size but very numerous. In general, autotrophic biomass is much greater than heterotrophic biomass on land, while the reverse is often true in the sea. Perhaps the most important difference is the following: The matrix, or supporting framework, of the community is largely physical in aquatic ecosystems, but more strongly biological on land. That is to say, the community itself is important as a habitat on land, but not so important in water.

Now, we may ask: How do these similarities and differences in structure affect ecological function?

One important aspect of function is shown in Figure 2–1 which compares energy flow in an aquatic and a terrestrial ecosystem. The lower diagram is an energy flow model for a marine community; the upper diagram is a comparable model for a forest. The boxes represent the average standing crop biomass of organisms to be expected; the lined boxes are the autotrophs, the stippled boxes are the heterotrophs. Three trophic levels are shown: (1) Producers, the phytoplankton of the sea and the leaves of the forest trees; (2) primary consumers (herbivores, etc.); and (3) secondary consumers (carnivores). The pipes or flow channels represent the energy flow through the ecosystems beginning with the incoming solar energy and passing through the successive trophic levels. At each transfer a large part of the energy is dissipated in respiration and passes out of the system as heat. The amount of energy remain-

FIGURE 2–1. Energy flow models for two contrasting types of ecosystems, an open water marine ecosystem and a terrestrial forest.

Standing crop biomass (in terms of KCal/M$^2$) and trophic structure are shown by means of shaded rectangles. Energy flows in terms of KCal/M$^2$/day (average annual rate) are shown by means of the unshaded flow channels. The aquatic system is characterized by a small biomass structure (hence the habitat is largely physical) while the forest has a very large biomass structure (hence the habitat is strongly biological). In both types of systems the energy of net primary production passes along two major pathways or food chains: (1) the grazing food chain (upper sequence in the water column or vegetation), and (2) the detritus food chain (lower sequence in sediments or soil).

The marine diagram is based on work of RILEY and HARVEY, the forest diagram on the work of OVINGTON and unpublished data from research at the University of Georgia. In some cases figures are hypothetical since no complete study has yet been made of any ecosystem. Hence, the diagrams should be considered as "working models" which do not represent any one situation.

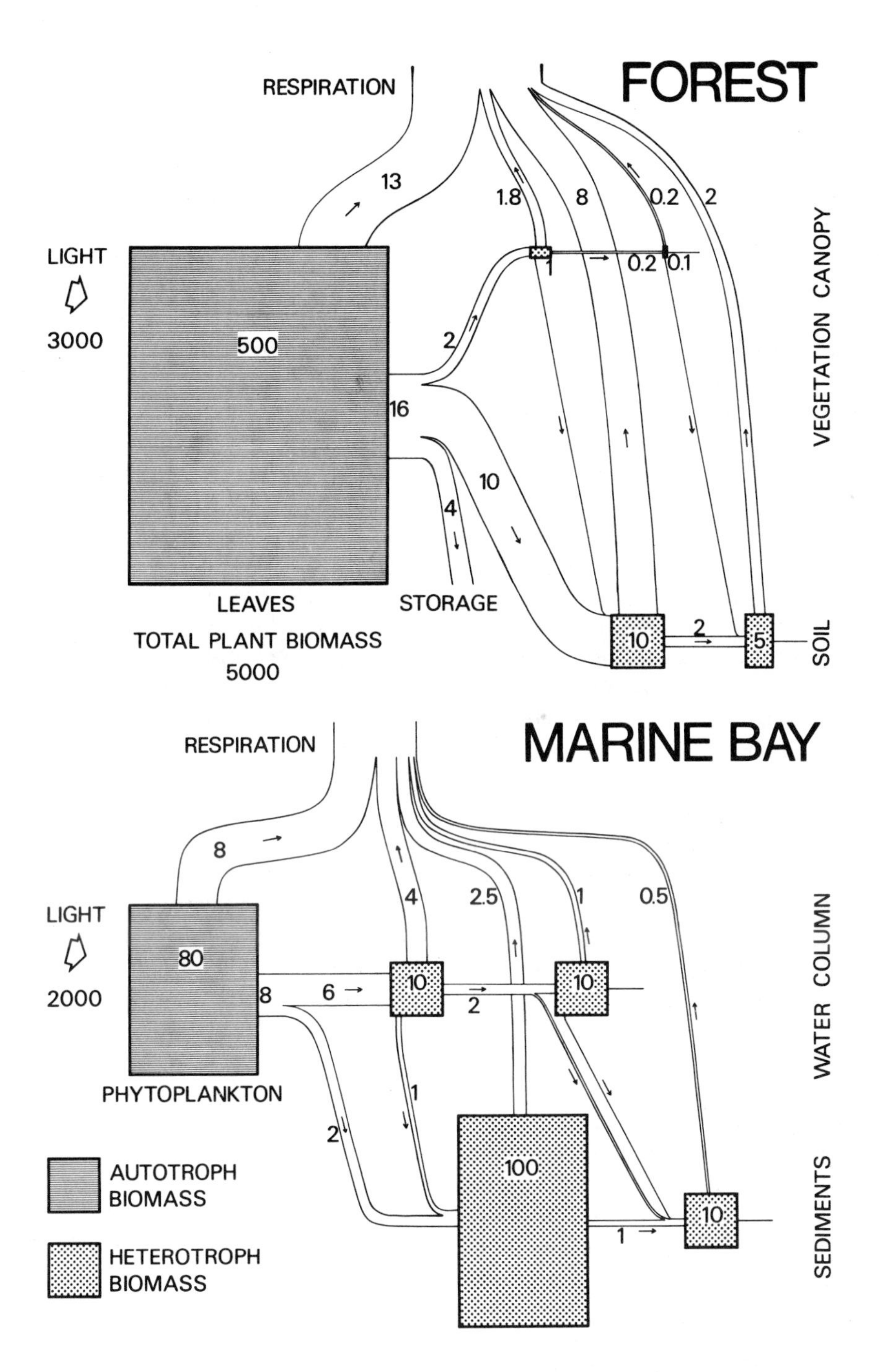


STANDING CROP BIOMASS IN KILOGRAM CALORIES / SQUARE METER

ENERGY FLOW IN KILOGRAM CALORIES / $M^2$ / DAY

ing after three steps is so small that it can be ignored insofar as the energetics of the community are concerned. However, tertiary consumers ("top carnivores") can be important as regulators; that is, predation may have an important effect on energy flow at the herbivore level.

All numbers in the diagrams are in terms of large or Kilogram Calories and square meters; standing crop is in terms of KCal/M$^2$; energy flow is in terms of KCal/M$^2$/day. The diagrams are drawn so that the areas of the boxes and the pipes are proportional to the magnitude of the standing crops and energy flows respectively. The quantities shown are a composite of measurements obtained in several different studies; some of the figures for higher trophic levels are hypothetical since complete information is not yet available for any one ecosystem. The marine community is particularly based on the work of Gordon Riley (Long Island Sound) and H.W. Harvey (English Channel), and the forest on the work of J.D. Ovington (pine forest) and unpublished data on terrestrial communities from our research group at the University of Georgia.

The autotrophic-heterotrophic stratification, which we emphasized as a universal feature of community structure, results in two basic food chains as shown in both diagrams (Fig. 2–1). The consumption of living plants by herbivores which live in the autotrophic stratum together with their predators may be considered as the *grazing food chain*. This is the classical food chain of ecology, as, for example, the phytoplankton-zooplankton-fish sequence or the grass-rabbit-fox sequence. However, a large proportion of the net production may not be consumed until dead, thus becoming the start of a rather different energy flow which we may conveniently designate as the *detritus food chain*. This energy flow takes place largely in the heterotrophic stratum. As shown in Figure 2–1 the detritus energy flow takes place chiefly in the sediments of water systems, and in the litter and soil of land systems.

Ecologists have too often overlooked the fact that the detritus food chain is the more important energy pathway in many ecosystems. As shown in Figure 2–1 a larger portion of net production is estimated to be consumed by grazers in the marine bay than in the forest; nine-tenths of the net production of the forest is estimated to be consumed as detritus (dead leaves, wood, etc.). It is not clear whether this difference is a direct or indirect result of the difference in community structure. One tentative generalization might be proposed as follows: communities of small, rapidly growing producers such as phytoplankton or grass can tolerate heavier grazing pressure than communities of large, slow-growing plants such as trees or large seaweeds.

Grazing is one of the most important practical problems facing mankind; yet we know very little about the situation in natural ecosystems. Well-ordered and stable ecosystems seem to have numerous mechanisms which prevent excessive grazing of the living plants. Sometimes, predators appear to provide the chief regulation; sometimes weather or life history characteristics (limited generation time or limited number of generations of herbivores) appear to exercise control. Unfortunately, man with his cattle, sheep and goats often fails to provide such regulation with the result that overgrazing and declining productivity are apparent in large areas of the world, especially in grasslands. A study of the division of energy flow between grazing and detritus pathways in stable natural ecosystems can provide a guide for man's utilization of grasslands, forests, the sea, etc.

The energy flow diagrams, as shown in Figure 2–1, reemphasize the difference in biomass as mentioned previously. Autotrophic biomass is very large and envelops or encloses the whole community in the forest; such extensive biological structure buffers and modifies physical factors such as temperature and moisture. In contrast, the aquatic community stands naked or exposed to the direct action of physical factors. In the marine situation the animal biomass often exceeds the plant biomass, and sessile animals (oysters, barnacles, etc.) instead of plants often provide some protection or habitat for other organisms.

Despite the large difference in relative size of standing crops in the two extreme types of ecosystems, the actual energy flow may be of the same order of magnitude if light and available nutrients are similar. In Figure 2–1 we have shown the available light (absorbed light) and the resulting net production as being somewhat lower in the marine community, but this may not always be true. Thus, 80 KCals of phytoplankton may have a net production almost as large as 5000 KCals of trees (or 500 KCals of green leaves). Therefore, productivity is not proportional to the size of the standing crop except in special cases involving annual plants (as in some agriculture). Unfortunately, many ecologists confuse productivity and standing crop. The relation between structure and function in this case depends on the size and rate of metabolism (and rate of turnover) of the organisms.

To summarize, we see that biological structure influences the pattern of energy flow, particularly the fate of net production and the relative importance of grazers and detritus consumers. However, total energy flow is less affected by structure, and is thus less variable than standing crop. A functional homeostasis has been evolved in nature despite the wide range in species structure and in biomass structure.

So far we have dealt with structure in relation to one aspect of function of the entire ecosystem. Now let us turn to structure and function at the population level and consider a second major aspect of function, namely, the cycling of nutrients. As an example I shall review the work of Dr. Edward J. Kuenzler at the University of Georgia Marine Institute on Sapelo Island. The study concerned a species of mussel of the genus *Modiolus* in the intertidal salt marshes. There are similar species of filter-feeding mollusca in the intertidal zone in all parts of the northern hemisphere.

First, we shall take a look at the salt marsh ecosystem and the distribution of the species in the marsh. The mussels live partly buried in the sediments and attached to the stems and rhizomes of the marsh grass, *Spartina alterniflora*. Individuals are grouped into colonies (clumped distribution), but the colonies are widely scattered over the marsh. Numbers average $8/M^2$ for the entire marsh and $32/M^2$ in the most favorable parts of the marsh. Biomass in terms of ash-free dry weight averages 11.5 $gms/M^2$. When the tide covers the colonies the valves partly open and the animals begin to pump large quantities of water.

Figure 2–2 illustrates the role of the mussel population in phosphorus cycling and energy flow according to Dr. Kuenzler's data. Each day the population removes a large part of the phosphorus from the water, especially the particulate fraction. Most of this does not actually pass through the body but is sedimented in the form of pseudofeces which fall on the sediments. Thus, the mussel makes large quantities of phosphorus available to microorganisms and to the autotrophs (benthic algae and marsh grass). As shown along the bottom of the diagram (Fig. 2–2) the energy flow was estimated to be about 0.15 $KCals/M^2/day$.

The most important finding of the study is summarized in the bottom line below the diagram (Fig. 2–2) which shows the ratio between flux and amount. Note that over one third of the 14 mgms of particulate phosphorus is removed from the water each day by the population, and thereby retained in the marsh. In contrast, less than one per cent of the 20 KCals of potential energy (net production estimate) available is actually utilized by the mussel population. In other words, the mussel population has a much more important effect on the community phosphorus cycle than it has on community energy flow. Or one might say that the role of the mussel in conserving nutrients in the ecosystem is more important than its role as energy transformer. In other words, the mussel population would be of comparatively little importance as food for man or animals (since population growth or production is small), but is of great importance in maintaining high primary production of the ecosystem.

To summarize, the mussel study brings out two important points: (1) It is necessary to study both energy flow and biogeochemical cycles to determine the role of a particular species in its ecosystem, (2) animals may be important in the ecosystem not only in terms of food energy, but as agents which make basic nutrients more available to auto

Finally, I think it is highly significant that the most productive ecosystems of the biosphere are those in which autotrophic and heterotrophic strata lie close together, thus insuring efficient nutrient regeneration and

ROLE OF A MOLLUSCAN POPULATION
IN NUTRIENT CYCLING AND ENERGY FLOW
IN A SALT MARSH ECOSYSTEM
SAPELO ISLAND, GEORGIA

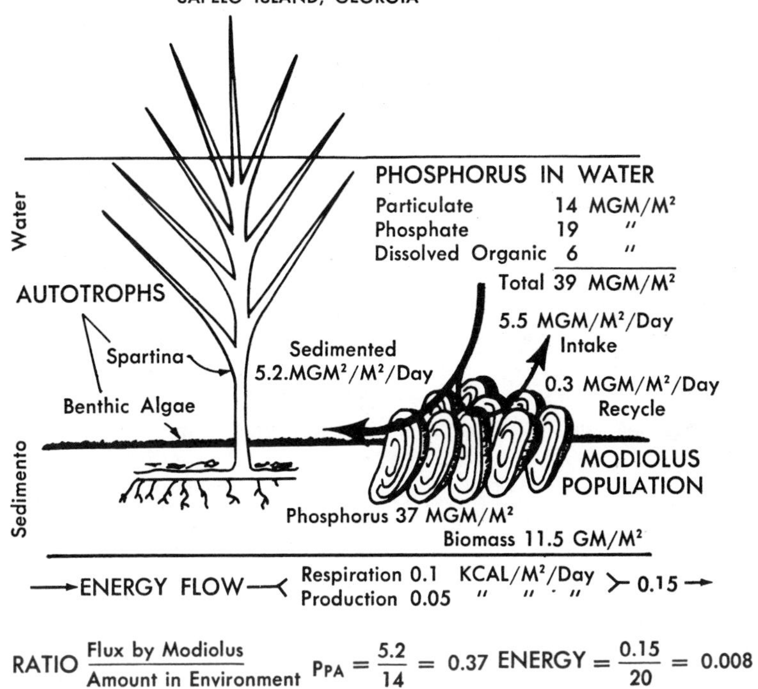


FIGURE 2–2. The effect of a population of mussels (*Modiolus*) on energy flow and the cycling of phosphorus in a salt marsh ecosystem according to the study of Dr. E. J. KUENZLER at the University of Georgia Marine Institute, Sapelo Island, Georgia, U.S.A. From the standpoint of the ecosystem as a whole, the population has a much greater effect on the cycling of phosphorus than on the transformation of energy. The study illustrates one often overlooked function of animals, that of nutrient regeneration. See text for details of the study.

recycling. Estuaries, marshes, coral reefs and rice fields are examples of such productive ecosystems.

Now let us consider the third important aspect of ecological function, that is, community regulation. Ecological succession is one of the most important processes which result from the community modifying the environment. Figure 2–3 illustrates a very simple type of ecological succession which can be demonstrated in a laboratory experiment. Yet the basic pattern shown here is the same as occurs in more complex succession of natural communities. The diagram (Fig. 2–3) was suggested to me by Dr. Ramon Margalef, hence we may call it the Margalef model of succession.

At the top of the diagram (Fig. 2–3) are a series of culture flasks containing plankton communities in different stages of succession. The graph shows changes in two aspects of structure and in one aspect of function. The first flask on the left contains an old and relatively stable community; this flask represents the climax. Diversity of species is high in the climax; species of diatoms, green flagellates, dinoflagellates and

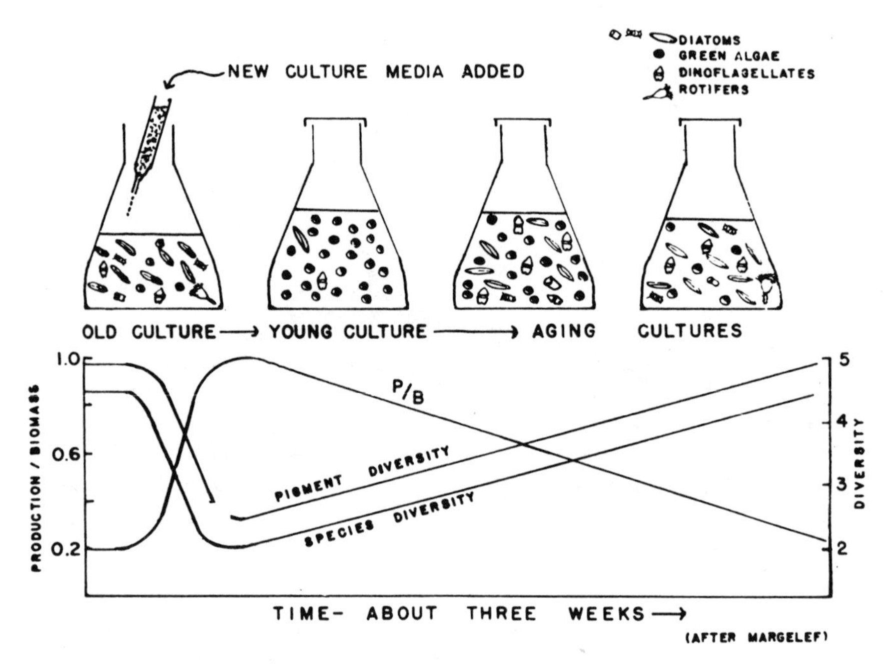


FIGURE. 2–3. The MARGALEF model of ecological succession showing a simple type of succession which can be demonstrated in laboratory cultures.

The flasks show changes in species composition occurring when succession is set in motion by the introduction of new nutrient media into an old "climax" culture. The graph shows resultant changes in two aspects of diversity and in the relation between production and biomass (P/B). See text for details of the experiment.

rotifers are shown in the diagram to illustrate the variety of plants and animals present. Biochemical diversity is also high as indicated by the ratio of yellow plant pigments (optical density at 430m$\mu$) to chlorophyll-a (optical density at 665m$\mu$). On the other hand the ratio of production to biomass (P/B in Fig. 2–3) is low in the old or climax culture, and gross production tends to equal community respiration. If we add fresh culture medium to the old culture, as shown in Figure 2–3, ecological succession is set in motion. An early stage in succession is shown in the second flask. Species diversity is low, with one or two species of phytoplankton dominant. Chlorophylls predominate so that the yellow/green ratio (O.D. 430/O.D. 665) is low, indicating low biochemical diversity. On the other hand, production now exceeds respiration so that the ratio of production to biomass becomes higher. In other words, autotrophy greatly exceeds heterotrophy in the pioneer or early succession stage. The two flasks on the right side of the diagram (Fig. 2–3) show the gradual return to the climax or steady state where autotrophy tends to balance heterotrophy.

The changes which we have just described are apparently typical of all succession regardless of environment or type of ecosystem. Although much more study is needed, it appears that differences in community structure mainly affect the time required, that is, whether the horizontal scale (X-axis in Fig 2–3) is measured in weeks, months or years. Thus, in open water ecosystems, as in cultures, the community is able to modify the physical environment to only a small extent. Consequently, succession in such ecosystems is brief, lasting perhaps for only a few weeks. In a typical marine pond or shallow marine bay a brief succession from diatoms to dinoflagellates occurs each season, or perhaps several times each season. Aquatic ecosystems characterized by strong currents or other physical forces may exhibit no ecological succession at all, since the community is not able to modify the physical environment. Changes observed in such ecosystems are the direct result of physical forces, and are not the result of biological processes; consequently, such changes are not to be classed as ecological succession.

In a forest ecosystem, on the other hand, a large biomass accumulates with time, which means that the community continues to change in species composition and continues to regulate and buffer the physical environment to a greater and greater degree. Let us refer again to Figure 2–1 which compares a forest with an aquatic ecosystem. The very large biological structure of the forest enables the community to buffer the physical environment and to change the substrate and the microclimate to a greater extent than is possible in the marine community.

Recent studies on primary succession on such sites as sand dunes or recent volcanic lava flows indicate that at least 1000 years may be required for development of the climax. Secondary succession on cutover forest land or abandoned agricultural land is more rapid, but at least 200 years may be required for development of the stable climax community. When the climate is severe as, for example, in deserts, grasslands or tundras the duration of ecological succession is short since the community cannot modify the harsh physical environment to a very large extent.

To summarize, I am suggesting that the basic pattern of functional change in ecological succession is the same in all ecosystems, but that the species composition, rate of change, and duration of succession is determined by the physical environment and the resultant community structure.

The principles of ecological succession are of the greatest importance to mankind. Man must have early successional stages as a source of food since he must have a large net primary production to harvest; in the climax community production is mostly consumed by respiration (plant and animal) so that net community production in an annual cycle may be zero. On the other hand, the stability of the climax and its ability to buffer and control physical forces (such as water and temperature) are desirable characteristics from the viewpoint of human population. The only way man can have both a productive and a stable environment is to insure that a good mixture of early and mature successional stages (i.e., "young nature" and "old nature") are maintained with interchanges of energy and materials. Excess food produced in young communities helps feed older stages which in return supply regenerated nutrients and help buffer the extremes of weather (storms, floods, etc.).

In the most stable and productive of natural situations we usually find such a combination of successional stages. For example, in continental shelf marine areas such as the inland sea of Japan the young communities of plankton feed the older, more stable communities on the rocks and on the bottom (i.e., benthic communities). The large biomass structure and diversity of the benthic communities provide not only habitat and shelter for life history stages of pelagic forms, but also provide regenerated nutrients necessary for continued productivity of the plankton.

A similar favorable situation exists in the Japanese terrestrial landscape where productive rice fields on the plains are intermingled with diverse forests on the hills and mountains. The rice fields, of course, are, ecologically speaking, "young nature" or early successional communities with very high rates of net community production which are maintained as such by the constant labor of the farmer and his machines. The forests represent older, more diverse and self-sustaining communities which have a lower net production, but do not require the constant attention of man. It is important that both ecosystems be considered together in proper relation. If the forests are destroyed merely for the sake of temporary gain in wood production, then the water and soil will wash down from the slopes and destroy the productivity of the plains. In my brief travels in Japan I have noted an unfortunate tendency in some areas to consider only the productive aspect of forests, and consequently to ignore their protective value. Complete deforestation of slopes may yield more wood for the time being but is ecologically a very dangerous procedure; also rebuilding the ecosystem is always more expensive than maintaining it in good condition. I believe ecologists should be more aggressive in bringing these principles to the attention of those charged with responsibility of national resources. Especially, ecologists need to provide good data which demonstrate the value of forests and other mature-type ecosystems in maintaining water and nutrient cycles. The value of forests should not be measured only in terms of net production.

My purpose in reviewing the three basic aspects of ecological function (that is, energy flow, nutrient cycles, and biological regulation) is to emphasize that we must study both structure and function if we are really to understand and control nature. Usually the study of function is more difficult than the study of structure; hence functional ecology has lagged behind descriptive ecology. To study function we must measure the rate of change per unit of time, and not just the situation at any one time. That is, we must measure the rate of energy flow, not just the standing crop; we must measure the rate of exchange of phosphorus, not just the amount present in the ecosystem; and we must measure the degree of regulation, not just describe it.

# II

# The Ecological Crisis and the Ecosystem Approach

# 3

# Man's impact on the biosphere

*Charles F. Cooper*

Human progress over the past millennia has largely been the result of increasing technological ability to interfere constructively in natural ecological processes. Problems of environmental quality stem mostly from the fact that this interference has unforeseen second- or third-order consequences. These consequences were unimportant when men were rare and only feebly able to disturb their surroundings, but they become of real urgency when people are abundant and each individual can command as much power as could a medieval village.

Five types of constructive interference with ecological processes are responsible, in large part, for the affluence of the modern world: simplification of ecosystems, intervention in natural biogeochemical cycles, concentration of dispersed energy, introduction of species into new environments, and induced genetic and behavioral changes in organisms. The gains to man from these five types of environmental modification tend to obscure the fact that each carries the seeds of trouble. A more complete understanding of the interrelationships among these processes of ecological change, and recognition of how man has used them through time, will aid in planning for enhancement of environmental quality in the future.

---

Originally printed in *Journal of Soil and Water Conservation*, 25:124–127 (1970). Reprinted with permission. Charles F. Cooper is Director, Center for Regional Environmental Studies, San Diego State University, San Diego, California 92115.

## Simplification of Ecosystems

The first man that deliberately scraped the weeds away from a particularly promising barley plant uncovered a fundamental ecological principle. Almost without exception, planned increase in economic yield of plant and animal products is accompanied by a decrease in richness and diversity of species. Endless rows of corn have replaced the incredible complexity of the midwestern tall-grass prairie. The oak-hickory forests of the Southeast, with their numerous tree species and rich, colorful undergrowth, are esthetically pleasing, but the lumberman prefers uniform stands of pine-forests of a single species and a nearly bare forest floor.

The whole history of agriculture and forestry is basically a history of efforts to create simple systems in which preferred crops are kept free of other species that reduce yields through competition or interference with harvest. Even modern range management is based largely on deliberate simplification of the grazed ecosystem. Grazing practices selectively favor the more productive or nutritious species over the long run. Chemical control of undesired shrubs and artificial reseeding carry the process still further.

In general, the more productive the system, the simpler it must be. The new high-yielding stiff-strawed rice varieties, for example, require more rigorous weed control than did their taller predecessors. The older varieties could outgrow competitors with relative ease, but their yield potential was limited by the ability of the stem to support the seed head. The newer varieties respond more effectively to fertilizer, but they are also more susceptible to competition from associated plants that may overtop them in the absence of cultivation.

A general conclusion of ecological theory is that in a stable natural ecosystem, the energy or "cost" required to maintain the system is high compared to the net harvestable energy. Man's strategy has been to reduce this disparity and to channel more of the incoming energy into harvestable products.

Why should this be a matter for concern? Chiefly because simple ecological systems are, in general, less stable—more subject to sudden damage from external causes—than more complex systems. To say this is not to accept the extreme position of some academic ecologists that man-simplified ecosystems cannot persist indefinitely and that stability is to be equated with diversity. Nevertheless, we have numerous examples of the rapid spread of crop pests in cultivated systems of a single species under circumstances where, if the host plants were more widely interspersed among other plant species, the pest might not reach a high enough density to affect many of the hosts. We have almost as many instances of the explosion of an unwanted species when the system is simplified by removal of the pest's natural enemies. Much of the modern strategy of pest control is, in a real sense, a substitution of human technological diversity for natural ecological diversity.

The work of Carl Huffaker and associates in the Division of Biological Control at the University of California in Berkeley provides a good example of the value of a moderate amount of ecological diversity in maintaining system stability. The eggs of the grape leafhopper (*Erythroneura elegantula*), a key pest on grapevines in California's Central Valley, are strongly parasitized by a small native wasp that would be an effective control agent if it could overwinter successfully. However, the grape leafhopper persists only as adults during the winter, and there are not suitable alternate hosts in the vineyards. When the leafhoppers reproduce in the spring, the wasp population is so low that few leafhoppers are parasitized. By the time the wasp population builds up, the leafhoppers have reached pest proportions.

Wild blackberries, however, harbor a noneconomic leafhopper (*Dikrella cruentata*) which survives the California winter in the adult, egg, and larval stages. The leafhoppers on blackberries provide a reservoir of over-winter hosts for the parasitic wasp, and when spring comes, the wasps are abundant enough to parasitize the grape leafhopper eggs effectively. Planting small patches of blackberries in the vineyards has appreciably reduced grape leafhopper infestations.

Simplification of ecosystems has other consequences also. Depletion of ground cover increases susceptibility to accelerated erosion. Elimination of habitat or food resources may cause the disappearance of birds or other wildlife of interest and value to man. And, as will be discussed below, simplification of ecosystems interacts with other processes of ecological change to generate still more far-reaching consequences.

## Biogeochemical Cycles

A major result of the industrial and agricultural revolution of the past two centuries is a significant increase in man's capability to alter natural biogeochemical cycles. Nitrogen is a good example.

Until the time of the first World War, the nitrogen needed by crops and forests was supplied almost entirely by biological processes, chiefly in animal manure, through fixation of atmospheric nitrogen by bacteria and bluegreen algae, and in rainwater. It left the soil in the protein of harvested products and as gaseous nitrogen and ammonia released by microbial activity. The cycle was tightly coupled and locally closed—little nitrogen made its way into runoff or ground water.

Economic and technological changes in the past 20 years have, to a great extent, changed the nitrogen cycle from a local to a regional or even a global phenomenon. The technology of fertilizer manufacture has made artificially fixed nitrogen readily available at low cost. Plant breeders have developed crop varieties that respond to heavier application of nitrogen than those of the past. There is thus every incentive for farmers to apply nitrogen fertilizer at rates far in excess of the rates of natural biological nitrogen supply. At the same time, increasing labor costs have made it uneconomic to disperse animal manure back on the land.

The consequences are twofold. Nitrogen fertilizer is frequently applied in quantities that exceed the capacity of the soil to retain it. As a result, nitrates leach into groundwater or into streams. Nitrate contamination of water supply is feared as a growing public health problem. Secondly, animal wastes from feedlots, broiler factories, and other highly concentrated animal operations must be disposed of. Nitrogen compounds from this disposal are finding their way into runoff water in greater amounts.

In the course of increasing the economic return from plant and animal production, we have decoupled the formerly tight nitrogen cycle and converted it into a global cycle of land, groundwater, streams, lakes, and the oceans.

Nutrients that reach lakes and streams as a result of decoupling the natural biogeochemical cycles tend to bring about further inadvertent simplification of the receiving ecosystems. Adding fertilizer to a lake or a field from external sources usually results in accelerated growth or "bloom" of a few species characteristic of the early stages of succession and decreased abundance of many species characteristic of more mature systems. Thus, as Eugene Odum has pointed out, it is inaccurate to refer to the processes occurring in lakes after the addition of man-supplied nutrients (artificial eutrophication) as equivalent to the natural "aging" of a lake.

The other side of the coin is that simplified terrestrial ecosystems are less able to hang on to nitrogen and other nutrients in the soil, whether these are of natural origin or supplied artificially. This has been demonstrated repeatedly in agricultural situations and more recently in a hardwood forest. When vegetation on the watershed of a tributary of Hubbard Brook in New Hampshire was killed, nitrogen outflow in the stream increased sixtyfold.

## Concentration of Dispersed Energy

The history of human technology is virtually a tale of progressive concentration of formerly dispersed energy, first in space and later also in time. Agriculture made possible the construction of cities, fed and fueled by surpluses from the surrounding countryside. The concentration of human energy permitted establishment of workshops, shipyards, and mines. Eventually, and most conspicuously in our own era, energy stored in ages past was concentrated into the present when coal and oil came into increasing use.

Associated with the concentration of energy, which made possible the civilization we know, was a concomitant concentration of waste products from use of that energy. These waste products of all sorts—human wastes from spatial concentration of food energy and chemical wastes from application of concentrated energy to industrial processes—are perhaps the most prominent contributors to what is now perceived as an environmental crisis. Adverse effects become increasingly critical as the waste products themselves become more concentrated. A stream that half a century ago could assimilate and biologically degrade the wastes of 10,000 people cannot today absorb the effluent from 100,000 people. Throughout the world, problems arise chiefly as biological thresholds are exceeded.

Release over a concentrated time span of fossil solar energy accumulated through eons of geologic time introduces other complications. Chief among these are increased outputs of carbon dioxide and particulate matter into the atmosphere. Carbon dioxide from burning of coal and oil and from oxidation of organic matter in cultivated soil is apparently acting to increase the mean temperature of the earth by trapping outgoing radiation. At the same time, increases in atmospheric particulates from industrial processes (including transportation) and from wind-eroded soil is reducing the transparency of the atmosphere and apparently tending to cool the earth. The relative magnitude of these two effects, even the direction of their joint impact, is unknown today but the subject of active research.

Technological concentration of energy may also simultaneously disperse formerly concentrated substances. Vincent Schaefer, who discovered the principle of artificial cloud seeding to increase precipitation or reduce hail, fears that widespread introduction of lead into the atmosphere from automobile exhaust may be increasing the natural supply of condensation nuclei for precipitation, with likely disturbance of normal rainfall patterns.

## Introduction of New Species

Most agricultural crops are economically most productive when grown under cultivation in climates somewhat different from those where they thrive naturally. Major gains in man's ability to feed himself have come from the introduction of plants and animals into environments where they are not commonly found in nature. Conversely, many of our most serious pest problems have arisen following the introduction of an insect, a fungus, or a weed plant into a new habitat.

The consequences of such introductions often depend upon interactions with other factors, such as ecosystem complexity. Perhaps the most virulent pest of a single species known to have been transferred from one continent to another within historic time is the chestnut blight. Brought to North America about 1910, this fungus within a quarter century had, for all practical purposes, rendered the American chestnut extinct as a forest tree. Grievous as was the economic and esthetic loss of this magnificent hardwood species, the effect on the forest ecosystem of Appalachia was surprisingly slight. Although chestnut was probably the most abundant single species in the mixed hardwood forest, it was associated with many other trees. When the chestnut disappeared, its place was taken, not by any one species, but by expanded dominance of several species. In contrast, less-virulent pests, including European forest insects, have occasionally devastated forests comprised predominantly of a single conifer species. Not only are there no other species readily available to fill the gaps created by the pest, but the abundance of host trees allows the pest to build its population and spread to neighboring trees.

## Genetic and Behavioral Changes

Maize from the primitive pod corn of Mesoamerica and Hereford cattle from the aurochs of Pleistocene Europe are only two of the multitude of examples of achieving productivity increases by deliberately modifying the genetic and behavioral patterns of plants and animals. Little need be said here about the accelerating efforts to improve productivity through plant and animal breeding, or about the occasional selective changes that make a formerly innocuous organism a serious pest or that make a susceptible pest resistant to standard control methods.

Less often realized is the fact that relatively minor cultural changes can transform a harmless organism into a serious pest. The effect of simplifying ecosystems in this regard has already been mentioned. As another example, in grains and other annual crops the important distinction between a severe disease or insect outbreak and a mild one is time. An unusually severe epidemic is one that peaks a week or two earlier relative to the time when the fields ripen. Many of the factors previously discussed can inadvertently bring this about.

## Treating Environmental Quality

The strategies discussed above have been used by man since he began his development toward civilization. All that has really changed is the intensity of human interference with natural processes. This is not to say that it is only in the present era that the indirect effects of these strategies have led to environmental deterioration. The productive capacity of the lands around the Mediterranean was severely depleted 2,000 and more years ago. It is not too much to say that modern western civilization is the result of the expenditure long ago of the capital represented by the productive landscape of the Mediterranean region.

But the process of environmental change is now proceeding at an accelerating rate. The effects of man's activities are local, regional, and global. They influence the productive capacity of the earth, human health and well-being, and esthetic and social values (Table 3–1).

TABLE 3–1

*Examples of nine categories of environmental change*

| *Characteristics Affected* | *Local* | *Regional* | *Global* |
|---|---|---|---|
| Productive capacity of the earth | Soil erosion | Man-induced drought | Temperature and radiation balance of the earth |
| Human health and well-being | Sulfur dioxide episodes over cities | Nitrates in groundwater | Radioactivity in atmosphere and hydrosphere |
| Esthetic values | Untreated strip mines; billboards | Extinction of birds or mammals | Loss of clear air |

Since earliest times, man has had the capability for local environmental modification. Later, after long and continued use of the landscape, he affected whole regions, such as the Mediterranean Basin. Only recently has he acquired the capability to modify the biosphere of the entire globe. This indeed is the measure of the scale of technological change. The pace of this change has, in many instances, exceeded the capacity of social institutions to adjust, leading to many of our current environmental problems.

For both political and technological reasons we are likely to continue for some time to come to treat environmental problems symptomatically. It is relatively easy to pass a law banning the use of DDT (although less easy to devise alternatives) or to set nationwide standards for sulfur content of fuel oil—standards applying as well to sparsely settled regions of sulfur-deficient rangeland as to the northeastern urban corridor. Indeed, in selling environmental action to the public, a symptomatic approach is probably the only approach that will work.

The framework presented here seems to be a useful preliminary basis for deciding on what symptomatic treatment to recommend. Analyzing the problem in terms of how it arose out of man's long-term strategy for technological development should help in modifying that strategy with the aim of coming a bit closer to optimizing human welfare.

# 4

# The cultural basis for our environmental crisis

*Lewis W. Moncrief*

One hundred years ago at almost any location in the United States, potable water was no farther away than the closest brook or stream. Today there are hardly any streams in the United States, except in a few high mountainous reaches, that can safely satisfy human thirst without chemical treatment. An oft-mentioned satisfaction in the lives of urbanites in an earlier era was a leisurely stroll in late afternoon to get a breath of fresh air in a neighborhood park or along a quiet street. Today in many of our major metropolitan areas it is difficult to find a quiet, peaceful place to take a leisurely stroll and sometimes impossible to get a breath of fresh air. These contrasts point up the dramatic changes that have occurred in the quality of our environment.

It is not my intent in this article, however, to document the existence of an environmental crisis but rather to discuss the cultural basis for such a crisis. Particular attention will be given to the institutional structures as expressions of our culture.

---

Reprinted with permission from *Science*, 170:508–512 (30 October 1970). Copyright 1970 by the American Association for the Advancement of Science. The author is associate professor and director of the Recreation Research and Planning Unit and holds a joint appointment in the Departments of Park and Recreation Resources and Resource Development at Michigan State University, East Lansing 48823. This article is based on an address given at a Man and Environment Conference at Arizona State University on April 16, 1970.

## Social Organization

In her book entitled *Social Institutions* (1), J. O. Hertzler classified all social institutions into nine functional categories: (i) economic and industrial, (ii) matrimonial and domestic, (iii) political, (iv) religious, (v) ethical, (vi) educational, (vii) communications, (viii) esthetic, and (ix) health. Institutions exist to carry on each of these functions in all cultures, regardless of their location or relative complexity. Thus, it is not surprising that one of the analytical criteria used by anthropologists in the study of various cultures is the comparison and contrast of the various social institutions as to form and relative importance (2).

A number of attempts have been made to explain attitudes and behavior that are commonly associated with one institutional function as the result of influence from a presumably independent institutional factor. The classic example of such an analysis is *The Protestant Ethic and the Spirit of Capitalism* by Max Weber (3). In this significant work Weber attributes much of the economic and industrial growth in Western Europe and North America to capitalism, which, he argued, was an economic form that developed as a result of the religious teachings of Calvin, particularly spiritual determinism.

Social scientists have been particularly active in attempting to assess the influence of religious teaching and practice and of economic motivation on other institutional forms and behavior and on each other. In this connection, L. White (4) suggested that the exploitative attitude that has prompted much of the environmental crisis in Western Europe and North America is a result of the teachings of the Judeo-Christian tradition, which conceives of man as superior to all other creation and of everything else as created for his use and enjoyment. He goes on to contend that the only way to reduce the ecologic crisis which we are now facing is to "reject the Christian axiom that nature has no reason for existence save to serve man." As with other ideas that appear to be new and novel, Professor White's observations have begun to be widely circulated and accepted in scholarly circles, as witness the article by religious writer E. B. Fiske in the *New York Times* earlier this year (5). In this article, note is taken of the fact that several prominent theologians and theological groups have accepted this basic premise that Judeo-Christian doctrine regarding man's relation to the rest of creation is at the root of the West's environmental crisis. I would suggest that the wide acceptance of such a simplistic explanation is at this point based more on fad than on fact.

Certainly, no fault can be found with White's statement that "Human ecology is deeply conditioned by beliefs about our nature and destiny—that is, by religion." However, to argue that it is the primary conditioner of human behavior toward the environment is much more than the data that he cites to support this proposition will bear. For example, White himself notes very early in his article that there is evidence for the idea that man has been dramatically altering his environment since antiquity. If this be true, and there is evidence that it is, then this mediates against the idea that the Judeo-Christian religion uniquely predisposes cultures within which it thrives to exploit their natural resources with indiscretion. White's own examples weaken his argument considerably. He points out that human intervention in the periodic flooding of the Nile River basin and the fire-drive method of hunting by prehistoric man have both probably wrought significant "unnatural" changes in man's environment. The absence of Judeo-Christian influences in these cases is obvious.

It seems tenable to affirm that the role played by religion in man-to-man and man-to-environment relationships is one of establishing a very broad system of allowable beliefs and behavior and of articulating and invoking a system of social and spiritual rewards for those who conform, and of negative sanctions for individuals or groups who approach or cross the pale of the religiously unacceptable. In other words, it defines the ball park in which the game is played, and, by the very nature of the park, some types of games cannot be played. However, the kind of game that ultimately evolves is not itself defined by the ball park. For example, where animism is practiced, it is not likely that the believers will indiscriminately destroy objects of nature because such activity would incur the danger of spiritual and social sanctions. However, the fact that another culture does not associate spiritual beings with natural objects does not mean that such a culture will invariably ruthlessly exploit its resources. It simply means that there are fewer social and psychological constraints against such action.

In the remainder of this article, I present an alternative set of hypotheses based on cultural variables which, it seems to me, are more plausible and more defensible as an explanation of the environmental crisis that is now confronting us.

No culture has been able to completely screen out the egocentric tendencies of human beings. There also exists in all cultures a status hierarchy of positions and values, with certain groups partially or totally excluded from access to these normatively desirable goals. Historically, the differences in most cultures between the "rich" and the "poor" have been great. The many very poor have often produced the wealth for the few who controlled the means of production. There may have been no alternative where scarcity of supply and unsatiated demand were economic reality. Still, the desire for a "better life" is universal; that is, the desire for higher status positions and the achievement of culturally defined desirable goals is common to all societies.

## The Experience in the Western World

In the West two significant revolutions that occurred in the 18th and 19th centuries completely redirected its political, social, and economic destiny (6). These two types of revolutions were unique to the West until very recently. The French revolution marked the beginnings of widespread democratization. In specific terms, this revolution involved a redistribution of the means of production and a reallocation of the natural and human resources that are an integral part of the production process. In effect new channels of social mobility were created, which theoretically made more wealth accessible to more people. Even though the revolution was partially perpetrated in the guise of overthrowing the control of presumably Christian institutions and of destroying the influence of God over the minds of men, still it would be superficial to argue that Christianity did not influence this revolution. After all, biblical teaching is one of the strongest of all pronouncements concerning human dignity and individual worth.

At about the same time but over a more extended period, another kind of revolution was taking place, primarily in England. As White points out very well, this phenomenon, which began with a number of technological innovations, eventually consummated a marriage with natural science and began to take on the character that it has retained until today (7). With this revolution the productive capacity of each worker was amplified by several times his potential prior to the revolution. It also became feasible to produce goods that were not previously producible on a commercial scale.

Later, with the integration of the democratic and the technological ideals, the increased wealth began to be distributed more equitably among the population. In addition, as the capital to land ratio increased in the production process and the demand grew for labor to work in the factories, large populations from the agrarian hinterlands began to concentrate in the emerging industrial cities. The stage was set for the development of the conditions that now exist in the Western world.

With growing affluence for an increasingly large segment of the population, there generally develops an increased demand for goods and services. The usual by-product of this affluence is waste from both the production and consumption processes. The disposal of that waste is further complicated by the high concentration of heavy waste producers in urban areas. Under these conditions the maxim that "Dilution is the solution to pollution" does not withstand the test of time, because the volume of such wastes is greater than the system can absorb and purify through natural means. With increasing population, increasing production, increasing urban concentrations, and increasing real median incomes for well over a hundred years, it is not surprising that our environment has taken a terrible beating in absorbing our filth and refuse.

## The American Situation

The North American colonies of England and France were quick to pick up the technical and social innovations that were taking place in their motherlands. Thus, it is not surprising that the inclination to develop an industrial and manufacturing base is observable rather early in the colonies. A strong trend toward democratization also evidenced itself very early in the struggle for nationhood. In fact, Thistlewaite notes the significance of the concept of democracy as embodied in French thought to the framers of constitutional governmen in the colonies (8, pp. 33–34, 60).

From the time of the dissolution of the Roman Empire, resource ownership in the Western world was vested primarily with the monarchy or the Roman Catholic Church, which in turn bestowed control of the land resources on vassals who pledged fealty to the sovereign. Very slowly the concept of private ownership developed during the Middle Ages in Europe, until it finally developed into the fee simple concept.

In America, however, national policy from the outset was designed to convey ownership of the land and other natural resources into the hands of the citizenry. Thomas Jefferson was perhaps more influential in crystallizing this philosophy in the new nation than anyone else. It was his conviction that an agrarian society made up of small landowners would furnish the most stable foundation for building the nation (*8*, pp. 59–68). This concept has received support up to the present and, against growing economic pressures in recent years, through government programs that have encouraged the conventional family farm. This point is clearly relevant to the subject of this article because it explains how the natural resources of the nation came to be controlled not by a few aristocrats but by many citizens. It explains how decisions that ultimately degrade the environment are made not only by corporation boards and city engineers but by millions of owners of our natural resources. This is democracy exemplified!

## Challenge of the Frontier

Perhaps the most significant interpretation of American history has been Fredrick Jackson Turner's much criticized thesis that the western frontier was the prime force in shaping our society (9). In his own words,

> If one would understand why we are today one nation, rather than a collection of isolated states, he must study this economic and social consolidation of the country.... The effect of the Indian frontier as a consolidating agent in our history is important.

He further postulated that the nation experienced a series of frontier challenges that moved across the continent in waves. These included the explorers' and traders' frontier, the Indian frontier, the cattle frontier, and three distinct agrarian frontiers. His thesis can be extended to interpret the expansionist period of our history in Panama, in Cuba, and in the Philippines as a need for a continued frontier challenge.

Turner's insights furnish a starting point for suggesting a second variable in analyzing the cultural basis of the United States' environmental crisis. As the nation began to expand westward, the settlers faced many obstacles, including a primitive transportation system, hostile Indians, and the absence of physical and social security. To many frontiersmen, particularly small farmers, many of the natural resources that are now highly valued were originally perceived more as obstacles than as assets. Forests needed to be cleared to permit farming. Marshes needed to be drained. Rivers needed to be controlled. Wildlife often represented a competitive threat in addition to being a source of food. Sod was considered a nuisance—to be burned, plowed, or otherwise destroyed to permit "desirable" use of the land.

Undoubtedly, part of this attitude was the product of perceiving these resources as inexhaustible. After all, if a section of timber was put to the torch to clear it for farming, it made little difference because there was still plenty to be had very easily. It is no coincidence that the "First Conservation Movement" began to develop about 1890. At that point settlement of the frontier was almost complete. With the passing of the frontier era of American history, it began to dawn on people that our resources were indeed exhaustible. This realization ushered in a new philosophy of our national government toward natural resources management under the guidance of Theodore Roosevelt and Gifford Pinchot. Samuel Hays (10) has characterized this movement as the appearance of a new "Gospel of Efficiency" in the management and utilization of our natural resources.

## The Present American Scene

America is the archetype of what happens when democracy, technology, urbanization, capitalistic mission, and antagonism (or apathy) toward natural environment are blended together. The present situation is characterized by three dominant features that mediate against quick solution to this impending crisis: (i) an absence of personal moral direction concerning our treatment of our natural resources, (ii) an inability on the part of our social institutions to make adjustments to this stress, and (iii) an abiding faith in technology.

The first characteristic is the absence of personal moral direction. There is moral disparity when a corporation executive can receive a prison sentence for embezzlement but be congratulated for increasing profits by ignoring pollution abatement laws. That the absolute cost to society of the second act may be infinitely greater than the first is often not even considered.

The moral principle that we are to treat others as we would want to be treated seems as appropriate a guide as it ever has been. The rarity of such teaching and the even more uncommon instance of its being practiced help to explain how one municipality can, without scruple, dump its effluent into a stream even though it may do irreparable damage to the resource and add tremendously to the cost incurred by downstream municipalities that use the same water. Such attitudes are not restricted to any one culture. There appears to be an almost universal tendency to maximize self-interests and a widespread willingness to shift production costs to society to promote individual ends.

Undoubtedly, much of this behavior is the result of ignorance. If our accounting systems were more efficient in computing the cost of such irresponsibility both to the present generation and to those who will inherit the environment we are creating, steps would undoubtedly be taken to enforce compliance with measures designed to conserve resources and protect the environment. And perhaps if the total costs were known, we might optimistically speculate that more voluntary compliance would result.

A second characteristic of our current situation involves institutional inadequacies. It has been said that "what belongs to everyone belongs to no one." This maxim seems particularly appropriate to the problem we are discussing. So much of our environment is so apparently abundant that it is considered a free commodity. Air and water are particularly good examples. Great liberties have been permitted in the use and abuse of these resources for at least two reasons. First, these resources have typically been considered of less economic value than other natural resources except when conditions of extreme scarcity impose limiting factors. Second, the right of use is more difficult to establish for resources that are not associated with a fixed location.

Government, as the institution representing the corporate interests of all its citizens, has responded to date with dozens of legislative acts and numerous court decisions which give it authority to regulate the use of natural resources. However, the decisiveness to act has thus far been generally lacking. This indecisiveness cannot be understood without noting that the simplistic models that depict the conflict as that of a few powerful special interests versus "The People" are altogether inadequate. A very large proportion of the total citizenry is implicated in environmental degradation; the responsibility ranges from that of the board and executives of a utility company who might wish to thermally pollute a river with impunity to that of the average citizen who votes against a bond issue to improve the efficiency of a municipal sanitation system in order to keep his taxes from being raised. The magnitude of irresponsibility among individuals and institutions might be characterized as falling along a continuum from highly irresponsible to indirectly responsible. With such a broad base of interests being threatened with every change in resource policy direction, it is not surprising, although regrettable, that government has been so indecisive.

A third characteristic of the present American scene is an abiding faith in technology. It is very evident that the idea that technology can overcome almost any problem is widespread in Western society. This optimism exists in the face of strong evidence that much of man's technology, when misused, has produced harmful results, particularly in the long run. The reasoning goes something like this: "After all, we have gone to the moon. All we need to do is allocate enough money and brainpower and we can solve any problem."

It is both interesting and alarming that many people view technology almost as something beyond human control. Rickover put it this way (11):

> It troubles me that we are so easily pressured by purveyors of technology into permitting so-called "progress" to alter our lives without attempting to control it—as if technology were an irrepressible force of nature to which we must meekly submit.

He goes on to add:

> It is important to maintain a humanistic attitude toward technology; to recognize clearly that since it is the product of human effort, technology can have no legitimate purpose but to serve man—man in general, not merely some men; future generations, not merely those who currently wish to gain advantage for themselves; man in the totality of his humanity, encompassing all his manifold interests and needs, not merely some one particular concern of his. When viewed humanistically, technology is seen not as an end in itself but a means to an end, the end being determined by man himself in accordance with the laws prevailing in his society.

In short, it is one thing to appreciate the value of technology; it is something else entirely to view it as our environmental savior—which will save us in spite of ourselves.

## Conclusion

The forces of democracy, technology, urbanization, increasing individual wealth, and an aggressive attitude toward nature seem to be directly related to the environmental crisis now being confronted in the Western world. The Judeo-Christian tradition has probably influenced the character of each of these forces. However, to isolate religious tradition as a cultural component and to contend that it is the "historical root of our ecological crisis" is a bold affirmation for which there is little historical or scientific support.

To assert that the primary cultural condition that has created our environmental crisis is Judeo-Christian teaching avoids several hard questions. For example: Is there less tendency for those who control the resources in non-Christian cultures to live in extravagant affluence with attendant high levels of waste and inefficient consumption? If non-Judeo-Christian cultures had the same levels of economic productivity, urbanization, and high average household incomes, is there evidence to indicate that these cultures would not exploit or disregard nature as our culture does?

If our environmental crisis is a "religious problem," why are other parts of the world experiencing in various degrees the same environmental problems that we are so well acquainted with in the Western world? It is readily observable that the science and technology that developed on a large scale first in the West have been adopted elsewhere. Judeo-Christian tradition has not been adopted as a predecessor to science and technology on a comparable scale. Thus, all White can defensibly argue is that the West developed modern science and technology *first*. This says nothing about the origin or existence of a particular ethic toward our environment.

In essence, White has proposed this simple model:

| I | | II | | III |
|---|---|---|---|---|
| Judeo-Christian tradition | → | Science and technology | → | Environmental degradation |

I have suggested hare that, at best, Judeo-Christian teaching has had only an indirect effect on the treatment of our environment. The model could be characterized as follows:

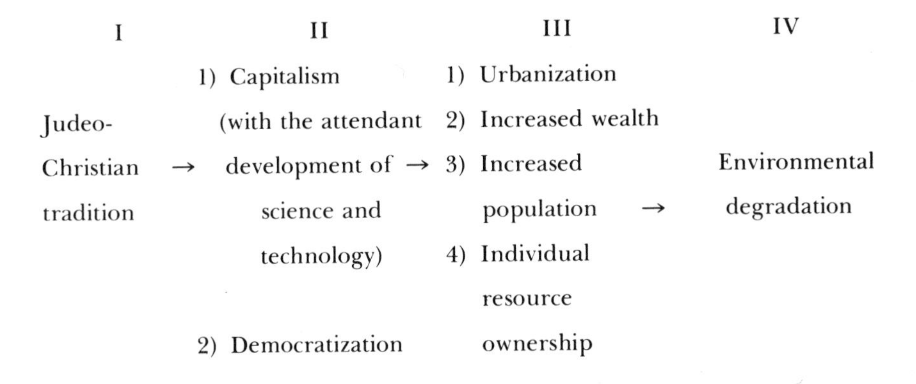

Even here, the link between Judeo-Christian tradition and the proposed dependent variables certainly has the least empirical support. One need only look at the veritable mountain of criticism of Weber's conclusions in *The Protestant Ethic and the Spirit of Capitalism* to sense the tenuous nature of this link. The second and third phases of this model are common to many parts of the world. Phase I is not.

Jean Mayer (12), the eminent food scientist, gave an appropriate conclusion about the cultural basis for our environmental crisis:

> It might be bad in China with 700 million poor people but 700 million rich Chinese would wreck China in no time. . . . It's the rich who wreck the environment . . . occupy much more space, consume more of each natural resource, disturb ecology more, litter the landscape . . . and create more pollution.

## References and Notes

1. J. O. Hertzler, *Social Institutions* (McGraw-Hill, New York, 1929), pp. 47–64.
2. L. A. White, *The Science of Culture* (Farrar, Straus & Young, New York, 1949), pp. 121–145.
3. M. Weber, *The Protestant Ethic and the Spirit of Capitalism,* translated by T. Parsons (Scribner's, New York, 1958).
4. L. White, Jr., *Science* 155, 1203 (1967).
5. E. B. Fiske, "The link between faith and ecology," *The New York Times* (4 January 1970), section 4, p. 5.
6. R. A. Nisbet, *The Sociological Tradition* (Basic Books, New York, 1966), pp. 21–44. Nisbet gives here a perceptive discourse on the social and political implications of the democratic and industrial revolutions to the Western world.
7. It should be noted that a slower and less dramatic process of democratization was evident in English history at a much earlier date than the French revolution. Thus, the concept of democracy was probably a much more pervasive influence in English than in French life. However, a rich body of philosophic literature regarding the rationale for democracy resulted from the French revolution. Its counterpart in English literature is much less conspicuous. It is an interesting aside to suggest that perhaps the industrial revolution would not have been possible except for the more broad-based ownership of the means of production that resulted from the long-standing process of democratization in England.
8. F. Thistlewaite, *The Great Experiment* (Cambridge Univ. Press, London, 1955).
9. F. J. Turner, *The Frontier in American History* (Henry Holt, New York, 1920 and 1947).
10. S. P. Hays, *Conservation and the Gospel of Efficiency* (Harvard Univ. Press, Cambridge, Mass., 1959).
11. H. G. Rickover, *Amer. Forests* 75, 13 (August 1969).
12. J. Mayer and T. G. Harris, *Psychol. Today* 3, 46 and 48 (January 1970).

# 5

# Ecosystems, systems ecology, and systems ecologists

*George M. Van Dyne*

## Ecosystems

### *Definitions*

In 1935 Tansley (106) introduced the term ecosystem, which he defined as the system resulting from the integration of all the living and nonliving factors of the environment. Webster now defines the term as a complex of ecological community and environment forming a functioning whole in nature. An ecosystem is a functional unit consisting of organisms (including man) and environmental variables of a specific area (3). Macroclimate has an overriding impact on the other components, each of which is interrelated at least indirectly (Fig. 5–1). The term "eco" implies environment; the term "system" implies an interacting, interdependent complex.

### *Trends in Ecological Research*

Although the concept of the ecosystem and many methods for studying ecosystems have been available for some time, only recently have many ecologists given more than lip service to the idea. Recently it has

---

Research performed at Oak Ridge National Laboratory and sponsored by the U. S. Atomic Energy Commission under contract with Union Carbide Corporation. Reprinted with permission and in modified form from *Oak Ridge National Laboratory Report* 3957, pp. 1–31.

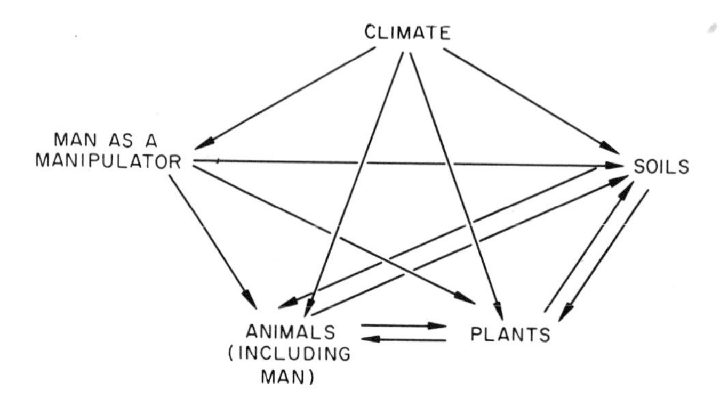


FIGURE 5–1. An ecosystem is an integrated complex of living and nonliving components. Each component is influenced by the others, with the possible exception of macroclimate. And now man is on the verge of exerting meaningful influence over macroclimate.

been suggested that the ecosystem is the rallying point for ecologists. There has been a gradual but distinct shift in emphasis in ecological studies and training from the description or inventory of ecosystems, or parts thereof, to the study of energy flow, nutrient cycles, and productivity of ecosystems. More workers are extending knowledge from the "anatomy" to the "physiology" of the environment. This requires different concepts, tools, and methods. The gradual change in emphasis from inventory to experimentation also requires more use of scientific methodology; this will be discussed below in the section, "Systems Ecology."

### *Ecosystem Components*

The controlling factors of the ecosystems are macroclimate, available organisms, and geological materials, where the last term includes parent material, relief, and ground water. Time is considered as a dimension in which the controlling factors operate, rather than as an environmental factor. The controlling factors are partially or entirely independent of each other. Each of the controlling factors is a composite of many separate elements, and each element is variable in time or space. Operationally, we may consider each controlling factor as a multiple-dimensioned matrix. Each change in a controlling agent in the ecosystem produces in time a corresponding change in the dependent elements of the ecosystem. In space and time there is a continuum of ecosystems.

Internal properties of ecosystems, such as rate of energy flow, might be considered as dependent factors which vary through time under the influence of a series of independent controlling factors. The dependent factors of the ecosystem are soil, the primary producers (vegetation), consumer organisms (herbivores and carnivores), decomposer organisms (bacteria, fungi, etc.), and microclimate. Each of these factors is dynamically dependent on the others (Fig. 5–1), and each is a product of the controlling agents operating through time.

Producers, consumers, and decomposers are not distributed at random in the abiotic part of an ecosystem. To maintain either dynamic equilibrium or ordered change in an ecosystem requires that a tremendous number of ordered interrelations exist among its dependent elements (82). To function properly ecosystems must process and store large amounts of information concerning past events, and they must possess homeostatic controls which enable them to utilize the stored information. This information may be expressed in amino acid and nucleotide sequences in genetic codes which have developed over evolutionary time, or it may be expressed in spatial or temporal patterns (20). For example, the changing patterns of plant populations and communities in secondary succession can be considered as expressions of genetically coded information. One species, population, or community is replaced by another with greater genetic potential for utilizing the resources of the changing environment.

### *Dynamics*

Ecosystem changes may be caused by fluctuations in internal population interactions or by fluctuations of the controlling factors. Such changes may be cyclical or directional (14). Directional change from less complex to more complex communities may be considered as progression or succession; directional change from more complex to less complex communities may be considered as regression or retrogression; both are shown in Figure 5–2.

Autogenic succession occurs when the controlling factors are stable and change is due to the effect of the system or some part of it on the microhabitat. Clements (15) formalized this process as migration, ecesis, competition, reaction, and stabilization. This type of primary succession produces changes which are usually gradual and continuous. Allogenic succession occurs when there is a change in the controlling factors. Most changes in the ecosystem are products of both allogenic and autogenic successions. Most macroecosystems can be said to be polygenetic and are the result of several climatic changes and erosion cycles. Purposeful alterations, such as disruption by man, in the controlling and controlled factors of the ecosystem may induce relatively permanent changes in the ecosystem.

Because ecosystems vary both temporally and spatially, and to prevent ambiguity, it is important to specify at least semiquantitative time and space scales. The importance of specifying a time scale is illustrated in Figure 5–2, where the time for primary succession (see $T_5$ for progression in Fig. 5–2) is shown as much greater than the time require-

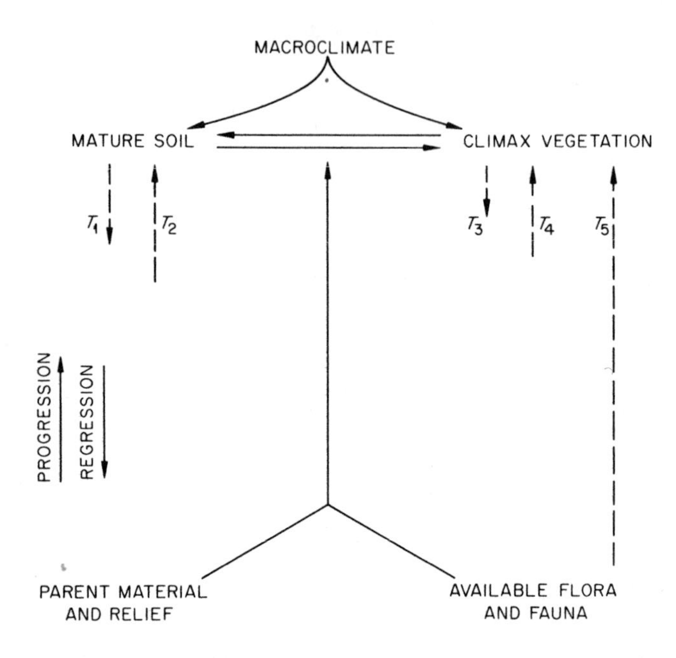


FIGURE 5–2. Ecosystems develop through time, under climatic control, from the original flora and fauna under a given set of relief and parent material conditions. A final dynamic equilibrium is reached in which there exist a mature soil and climax plant and animal populations.

ment for man to disrupt the system and alter soil or vegetation ($T_1$ and $T_3$ in Fig. 5–2). In the process of retrogression, changes take place in the vegetation more rapidly than they do in the soil. Generally, the ecosystem will recover stable state through a progressive process called secondary succession ($T_2$ and $T_4$ ). Again, the rate of progressive changes of soil properties is usually lower than that for vegetation. Recovery of the vegetation to the climax state may take an amount of time similar to that required for deterioration of the soil. Change in a given ecosystem component or property may be negligible in $T_3$ but considerable in $T_5$.

During progressive succession there is usually an increase in productivity, biomass, relative stability and regularity of populations, and diversity of species and life forms within the ecosystem (74). Finally, the ecosystem reaches a steady state or equilibrium, which is characterized by dynamic fluctuation rather than by directional change. This steady state of the ecosystem is referred to as climax (119). At climax the dependent factors are in balance with the controlling factors; the climax is an open steady state (101). A diversity of species and life forms occupies every available ecological niche at climax and, because there is a maximum number of links in the food web, the stability of the system is maximized (63). A maximum amount of the entering energy is used in maintenance of life. Fosberg (34) considers "that climax communities are those in which there is the greatest range and degree of exploitation or utilization of the available resources in the environment." There is no net output from an ecosystem in the climax state (86). Three states of ecosystems exist with regard to energy or nutrient balance: steady state or climax, positive balance or succession occurring, and negative balance or decadence and senescence (99).

There is continual interchange of matter and energy among contiguous ecosystems. This interchange or flux is an essential property of ecosystems. The fluxes in and out of an ecosystem may be difficult to measure accurately, but there is relatively less error in measuring flux in a macroecosystem than in a microecosystem, because usually the error in measurements is inversely proportional to the magnitude of the object, rate, or processes being measured. Also, the relative amount of relevant surface or area around an ecosystem decreases as its size increases; many of the measurement errors or biases occur at such interfaces because of subjective decisions in defining boundaries. Still, we may find it convenient to study microecosystems such as a sealed bottle containing nutrients, gases, organisms, and water. Essentially, this is the type of system we need to study in preparing for interplanetary travel. But even such discrete microcosms are not adiabatic with their environment, and ultimately they are dependent upon their environment for a continuing energy input.

When flux of some element in and out of a given ecosystem is negligible for a defined period of time we consider that ecosystem to be stable with regard to that element. The equilibrium is referred to as climax only if it is reached naturally. Other equilibria, or disclimaxes, can be maintained by man's intervention. Here is the essence of renewable resource management: maintaining disclimaxes at equilibrium for the benefit of man.

### *Manipulation of Ecosystems*

Man is a vital part of most major ecosystems, and there is an increasing human awareness of man's part in them and his influence on them (108) (Fig. 5–3). Traces of his pesticides probably can be found in living organisms throughout the world. Humans are both parts of and manipulators of ecosystems. Induced instability of ecosystems is an important cause of economic, political, and social disturbances throughout the world. In altering his environment in order to overcome its limitations to him, man learns that he often is faced with undesirable consequences of the environmental change (13, 38, 39). In manipulating his environment (e.g., felling forests, burning grasslands or protecting them from fire, and draining marshes), seldom has he foreseen the full consequences of his action (104).

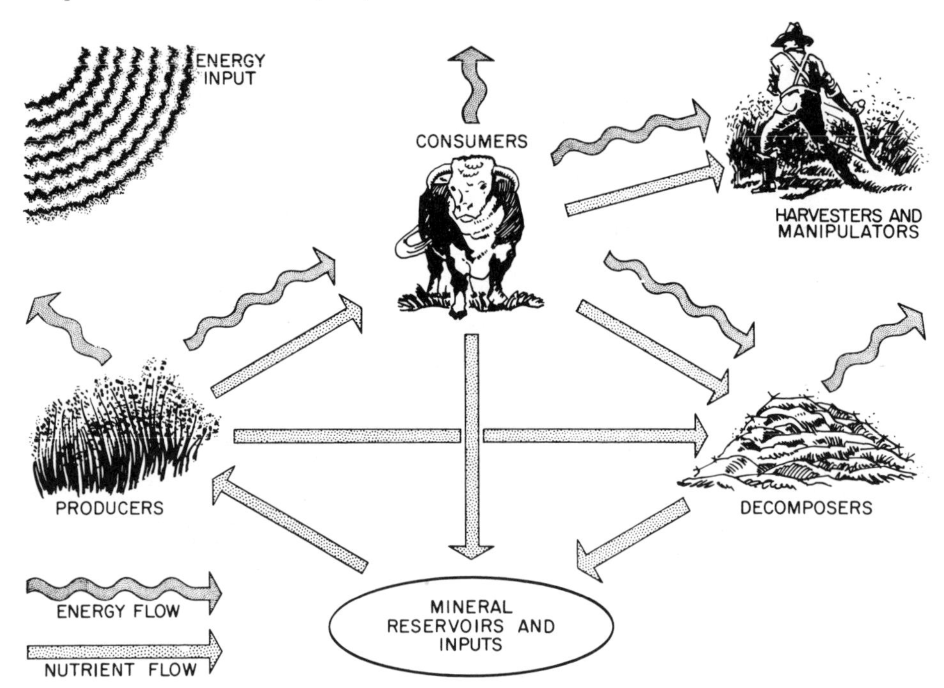


FIGURE 5–3. Man is both a spectator of, and a participant in, the functioning of ecosystems. He has manipulated ecosystems to maximize the flow of nutrients and energy to him from the producers and primary consumers. He has attempted to minimize the respiratory losses of energy from producers, consumers, and decomposers.

Most ecosystems in our country were in climax states when civilized man began to affect them, but the economy of civilized man demanded that the ecosystems produce a removable product under his domination. In order to reach this goal he disrupted the climax ecosystems, perhaps by shortening food chains or by altering the diversity of life forms of primary producers. He has altered the rate of, and amount of, nutrients cycling through the system by such means as fertilization, both in aquatic and terrestrial systems (Fig. 5–3). In some instances the fertilization has been excessive and has led to undesirable side effects, such as algal blooms caused by excesses of organic wastes. In other instances man has altered the structure of ecosystems by simplifying them and diverting the flow of energy into his food products, such as in replacing a grassland and wild animals with a wheatfield. Eventually he has produced changes in some ecosystem properties, which in some instances has led to quasi-stable levels. In other instances such changes have led to desertifi tion, such as the result of centuries of overgrazing in the Middle East.

Man has also encountered difficulties when he attempts to return ecosystems to their native state or to preserve vegetation by the development of national parks or by control of predators (104). In several instances ungulate populations have multiplied rapidly, outstripped the natural control by predators, exceeded the carrying capacity of their ranges, and severely damaged their habitat. Examples include the classical Kaibab mule deer problem (94) and the elk problem in Yellowstone National Park (64). Man himself has had a direct and profound effect on some ecosystems he has attempted to maintain in a natural state, such as in Yosemite National Park (39).

## Systems Ecology

### *Definitions*

Systems ecology can be broadly defined as the study of the development, dynamics, and disruption of ecosystems. I consider systems ecology to have two main phases—a theoretical and analytical phase and an experimental phase.

Earlier I stated that for studying function in ecology we need methods and concepts which are different from those for studying structure. Essentially, study of problems in systems ecology requires three groups of tools and processes: conceptual, mechanical, and mathematical.

## Study of Ecosystems

The tools and processes required for systems ecology are different from those for conventional phases of ecology because of the complexity of the total ecosystem as compared with a segment of it. When we consider the totality of interactions of populations with one another and with their physical environs—i.e., ecosystem ecology—we face a new degree of complexity (10). Other than some recent papers (e.g., ref. 41) only a few reasonably adequate functional analyses of natural ecosystems exist (80).

One of the major problems in systems ecology is that of analyzing and understanding interactions. Events in nature are seldom, if ever, caused by a single factor. They are due to multiple factors which are integrated by the organism or the ecosystem to produce an effect which we observe (45). To further complicate the matter, various combinations of factors and their interactions may be interpreted and integrated by the ecosystem to produce the same end result.

### Conceptual Requirements

A first conceptual requirement in systems ecology is clearer definition of problems. It is axiomatic that ambiguous use of terminology and an ambiguous statement of the problem lead to ambiguities of thought as well (19). These statements apply to many fields, but, particularly here, clear definitions are required because of the type of people systems ecologists will be and the types of people with whom they will work (discussed further below). Furthermore, in using computers, which are essential tools for systems ecologists, it is necessary to formulate the problems precisely and to clearly delineate the factors involved.

A second conceptual requirement in systems ecology is more and better use of logic and scientific and statistical methods. Essentially, we can define scientific method as the pursuit of truth as determined by logic and experimentation. In scientific method we use the approach of systematic doubt to discover what the facts really are. Experimentation is one of several tools of scientific method used to eliminate untenable theories, that is, to test hypotheses (32). Other experiments may be conducted to determine existing conditions, or to suggest hypotheses, etc. The conclusions from experiments may be criticized because the interpretation was faulty, or the original assumptions were faulty, or the experiment was poorly designed or badly executed (88). Experimental design and statistical inference are aids in testing hypotheses.

Much past ecological research has not tested a hypothesis. There is a tendency for ecologists to pass over the primary phase of analysis. The lack of understanding of what is known already (inadequate knowledge of the literature, in part) is understandable because of the volume of material to be covered (58). Glass (40) has clearly stated this dilemma— "the vastness of the scientific literature makes the search for general comprehension and perception of new relationships and possibilities every day more arduous." But inadequate examination of facts and data and inadequate formulation of hypotheses lead to uncritical selection of experiments testing poorly formulated hypotheses, and ecologists are often at fault here (51). The experimental design is, essentially, the plan or strategy of the experiment to test clearly certain hypotheses (32). Statistical methods are especially important in experimentation with ecosystems, because not all factors influencing the system can be controlled in the experiment without altering the system (29). These uncontrollable factors lead to error or "noise" in our measurements, and inferences to be made from the results of experiments should be accompanied by probability statements (32).

Eberhardt (27) has discussed many of the problems ecologists encounter in sampling, and has stressed the importance of statistical techniques in analysis of such problems. Methods of statistical inference are also useful in suggesting improvements in our mathematical models and in suggesting alterations in the design of future experiments. Some of the work initiated and developed by the late R. A. Fisher on partial correlation and regression is invaluable to us in evaluating independent and interaction effects in complex ecosystems where experimental control is neither possible nor desirable (45).

The first two conceptual needs for systems ecology, mentioned above, lead naturally to the third, the approach of modeling (Fig. 5–4), with models which are mathematical abstractions of real world situations (17, 107). In this process some real world situation is abstracted into a mathematical model or a mathematical system. Next, we apply mathematical argument to reach mathematical conclusions. The mathematical conclusions are then interpreted into their physical counterparts. In some instances we are able to proceed from the real world situation via experimentation to reach physical conclusions. In other instances, however, we cannot experiment with a situation that does not exist but may become real; examples are such situations as thermonuclear war and wide-scale environmental pollution (50). In many cases we find it too costly to exper-

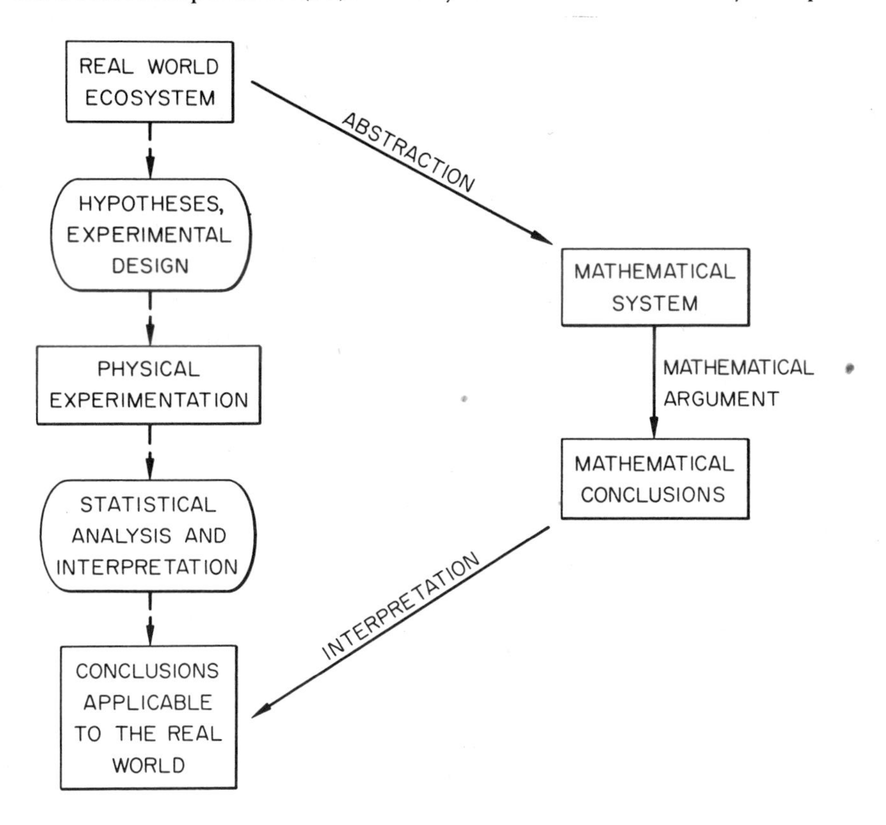


FIGURE 5–4. Two ways of experimenting with ecosystems. One involves the conventional process of formulating hypotheses, designing and conducting experiments, and analysis and interpretation of results. The second involves the abstraction of the system into a model, application of mathematical argument, and interpretation of mathematical conclusions.

iment; therefore mathematical modeling or mathematical experimentation may be especially useful.

Mathematical modeling is somewhat new to many conventional ecologists and, in part, is just as much an art as a science. To ensure that the model will be valid, the mathematical axioms must be translations of valid properties of the real world system. The application of mathematical argument gives rise to theorems which we hope can be interpreted to give new insight into our real world system. However, the value of these conclusions should, where possible, be verified by experimentation. We must then accept the conclusions or reject them and start over again. This procedure of modeling, interpretation, and verification is used in many engineering and scientific disciplines. The success of the procedure, however, depends on the existence of an adequate fund of basic knowledge about the system. This knowledge permits predictive calculations. Hollister (50) outlines some of the problems to be encountered in modeling ecological phenomena.

### *Tools for Study of Ecosystems*

The above conceptual tools should provide a framework in which to attack the complex problems of systems ecology. To implement these methods in studying ecosystems we will need both physical and mathematical tools, including digital and analog computers and electrical, mechanical, and hydraulic simulation devices, and artificial populations (44, 75, 85). The act of expressing and testing biological problems with numerical, electrical, or hydraulic analogs often reveals some unsuspected relationships and leads to new approaches in investigation. In conducting experiments in systems ecology, more refined chemical analytical equipment will be needed, such as gas chromatographs, infrared gas analyzers, and recording spectrophotometers. Physical analytical equipment required includes micro-bomb calorimeters, biotelemetric equipment, and other electronic equipment useful for rapid, nondestructive sampling and measuring of plant and animal populations and parameters under field conditions.

The importance of these chemical and physical tools is apparent when one considers the amount and variety of apparatus required to construct and maintain even the simplest aquatic ecosystems or to transplant and manipulate naturally occurring ecosystems for detailed measurements (e.g., ref. 2). A major reason for the scarcity of detailed studies of entire terrestrial ecosystems is that many ecologists are not trained to use many of the required, diversified tools. In other instances these tools may not be available to the ecologist. The systems ecologist cannot be an expert with each of these tools, but he must be aware of their applications and limitations in the the study of components and processes in ecosystems. He will need to be conversant with the specialists in other disciplines who make increasing use of these modern and complex tools.

### *Operations Research and Systems Analysis Applications*

Mathematical analysis will become increasingly important in providing advances in systems ecology. Large, fast digital computers have become available in the last 15 years and have allowed the development of special methods of analyzing and studying complex systems in industry and government. Most of these newer mathematical tools were developed in and are used primarily in two loosely defined and somewhat overlapping fields, operations research and systems analysis.

Operations research may be defined as the application of modern scientific techniques to problems involving the operation of a system looked upon as a whole (77). Included therein are any systematic, quantitative analyses aimed at improving efficiency in a situation where "efficient" is well understood (103).

Systems analysis is more difficult to define. Perhaps it can best be defined by opposites. The opposite of a systems approach is unsystematic or piecemeal consideration of problems; intuition may be taken as the opposite of analysis (46). Essentially, systems analysis is any analysis to suggest a course of action arrived at by systematically examining the objectives, costs, effectiveness, and risks of alternative policies—and designing additional ones if those examined are found to be insufficient (93).

It is easily seen that operations research and systems analysis are both alike and different. They both contain elements from mathematical, statistical, and logical disciplines. In operations research, however, there usually is an unambiguous goal to be achieved, and the operations researcher is interested in optimization. The systems analyst faces a multiplicity of goals, a highly uncertain future, a frequent predominance of qualitative elements, and an exceedingly low probability of building an accurate and satisfactory model for his total problem (103). Because of the methods and techniques he can use effectively, the systems engineer has much to offer in study of ecosystems but he will need considerable guidance. In systems ecology he will be facing a collection and coupling of "green, pink, and brown" boxes (plants, animals, and physical environment) rather than the black boxes with which he is familiar (56). The interconnections between these boxes may be known only imperfectly, and the functional significance of the boxes will need to be established.

Some of the mathematical tools to be employed and examined in systems ecology include scientific decision-making procedures, theory of games, mathematical programming, theory of random processes, and methods of handling problems of inventory, allocation, and transportation (77).

Linear, nonlinear, and dynamic programming, which are especially important to the operations researcher, already show promise in ecology (4, 112, 115) and in management of renewable resources (12, 60, 69). Mathematical programming has already been used widely in agro-ecological problems, such as crop or yield prediction (97), in formulation of least-cost rations for livestock (110), and in farm management decisions (5). Game theory has been applied to decisions in cultivated-crop agriculture (113) and appears to have potential in dealing with wildland resources. Queuing theory and network flow appear to offer much in looking at problems of flow rates in ecosystems (90). Margalef (74) has discussed and indicated some important applications of information theory in ecology. Cybernetics principles and techniques are also useful in studying biological systems (37). Simulation is another important tool in operations research, although not limited to it. Mathematical simulation models have been used to study important resource problems, such as salmon population biology (59, 95), and abstract systems (36).

### *Importance of Digital Computers*

Probably most systems ecology problems will be attacked first with deterministic models as first approximations (70, 71). However, to increase their usefulness and their realism, stochastic elements will be involved in most models or an indeterministic point of view will be taken; for example, see Leslie (65), Neyman and Scott (79), and Jenkins and Halter (53). This will require extensive use of digital computers, not only in simulation but also in analysis. Most stochastic models in ecology to date have been concerned with only one or two species rather than populations or ecosystems (6). Stochastic simulation of biological models or processes has been a useful process in some problems (109, 111). Many problems of modeling and analysis will require study and examination of the underlying statistical distributions (28). In addition to the normal distribution, other distributions which will need examination and use in systems ecology problems include the Poisson, the exponential, and the log-normal. Monte Carlo methods will be especially valuable in developing, testing, and using stochastic or probabilistic models (30, 67). Computers are essential in studying and using these statistical techniques in systems ecology. Other statistical aspects are discussed by Eberhardt (29).

Compartment model methodology, implemented with both analog and digital computers, has proven its value in theoretical studies and is beginning to be put to use in analysis and extension of real data in medical (9) and ecological fields (87, 90). Thus far, however, compartmental simulation models have been restricted to relatively simple ecological and agricultural situations, because most investigators have worked with analog systems of limited capacity (1, 35), although simulation systems have been developed for and used with digital computers in the study of renewable resources (e.g., ref. 42). Most systems ecologists will find it surprisingly easy to express many problems in the pseudoalgebraic languages, such as the many dialects of FORTRAN, used to communicate with digital computers.

An ecosystem might be depicted, as in Figure 5–5, as composed of

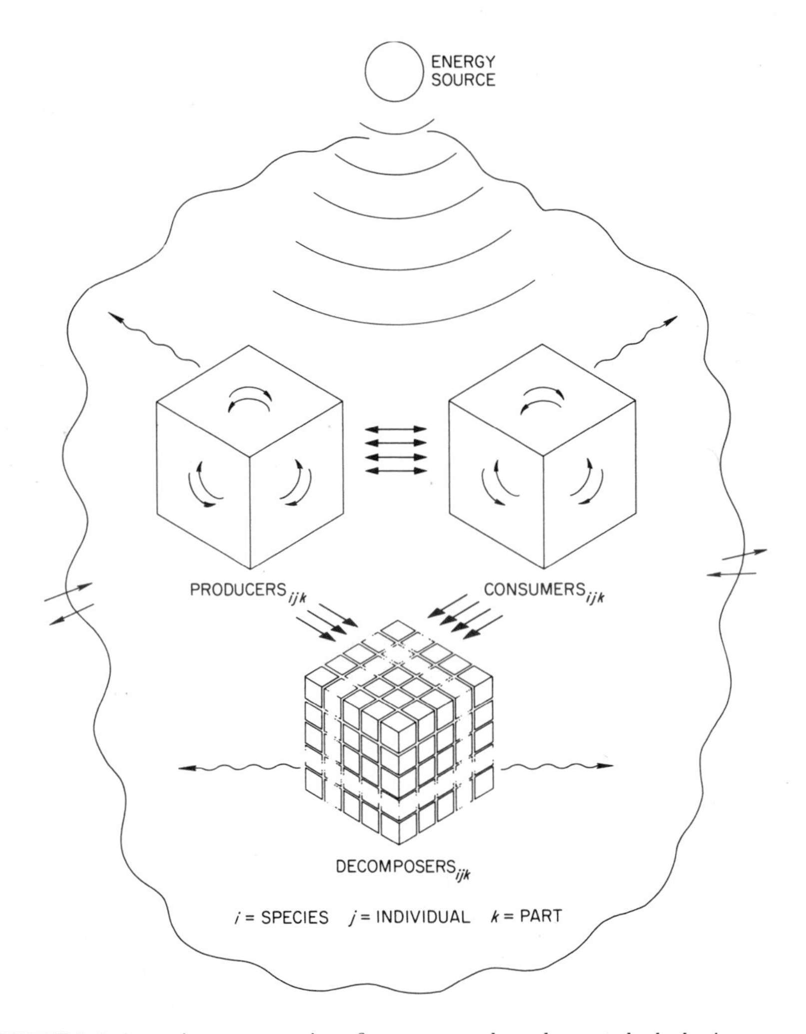


FIGURE 5–5. A matrix representation of ecosystems adapted to pseudoalgebraic computer languages. Each trophic level is represented as a three-dimensional matrix. The arrows, wavy for energy and straight for matter, represent matrices of transfer functions interconnecting parts within individuals, individuals of a species to each other and to other species, etc., on up to connections between contiguous ecosystems. The transfer functions may contain probabilistic components and may be probabilistic functions of external variables such as macroclimate.

trophic levels represented as three-dimensional matrices. PRODUCER (I,J,K), CONSUMER (I,J,K), and DECOMPOSER (I,J,K) are matrices of species, individuals, and parts. The ranges of I, J, and K in each matrix are variable and depend upon the study. Matrices of transfer functions, depicted and simplified by the arrows in Figure 5–5, are concerned with movement of matter or energy within individuals, between individuals, or between species. The latter two types of transfers may be between or within trophic levels. Also included in the figure is the fact that flux among contiguous ecosystems may be considered in matrix representation. Some of the transfer functions themselves may contain random noise and may be functions of a driving variable, such as macroclimate, acting on the system over time. Models or functions for macroclimatic influences may be constructed from actual data or may follow some prescribed hypothetical statistical distribution.

Consider the simplified case (Fig. 5–5) with only three parts per individual, three individuals per species, three species per trophic level, three trophic levels per ecosystem, and three ecosystems per problem. This leads to $3^5$ microcompartments to be accounted for in addition to the many transfer functions interrelating the compartments. Many of the transfers will be zero, but this simplified model exceeds the capacity of most analog computers even if the problems of using various random function generators with an analog computer are bypassed. This example does not indicate that analog computers will not play an important role in systems ecology, but only that they may be of limited value in many realistically complex situations. Their major role may be as teaching (and learning) tools and as components of hybrid (digital-analog) systems. The capabilities and versatility of digital computers in general are far greater than those of analog computers (62).

Maximum use of most of the above mathematical tools and others by systems ecologists depends upon access to fast digital computers with large memory capacities (115). Such access will be especially important in working with large complicated models where remote-console access to large central computers will be essential for efficient and rapid progress. Computer technology is approaching the point where the rate of debugging of programs is the limiting factor.

The role of computers in the future of systems ecology is too readily underestimated. Computers in tomorrow's technology will have larger and faster memories, remote consoles and timesharing systems. Some may accept hand-written notes and drawings, respond to human voices, and translate written words from one language into spoken words in another (96). There will be vast networks of data stations and information banks, with information transmitted by laser channels over a global network. This network will be used not only by researchers but also by engineers, lawyers, medical men, and sociologists as well as government, industry and the military. Computers could become tomorrow's reference library used by students in the university; they are already starting to revolutionize our present approaches to certain kinds of teaching. To utilize computers effectively in ecology we will have to state precisely what we know, what we do not know, and what we wish to know. Also, it will be necessary to assemble, analyze, identify, reduce, and store our ecological data and knowledge in a form retrievable by machine.

### *A Systems Approach*

Systems ecology will call for an interdisciplinary team of systems ecologists, systems analysts and operations researchers (if they can be separated), conventional ecologists, mathematicians, computer technologists, and applied ecologists, including agriculturalists and natural resource managers of various disciplines. Systems ecologists studying ecosystems will devote at least as much time to delineating the problem as they will to solving it. This gives a hint as to the nature of the work of the systems ecologist.

The physical and mathematical tools to be used by this team are impressive, even though the list in the preceding sections is only partial. It serves to show that the systems ecologist will have to have more types of specialized training than did his predecessors. That different tools and methods may be needed to solve some of today's complex ecological problems is emphasized by the fact that many important contributors to advances in ecology in recent years may not be identified as ecologists (92). This will be especially true of systems ecology in the future, even though ecologists must be generalists, and systems ecologists also will, in part, have to be generalists. Still, there are probably few, if any, authentic generalists or truly great minds who are not firmly grounded in a specialty (18). Most systems ecologists will serve their apprenticeship in basic fields. The conventional plant, animal, and aquatic ecologists will not be acceptable as systems ecologists, because they will lack the depth required in many specialties (78).

This raises the difficult question of how to train a man to be a specialist in at least one field, to be able to converse well with specialists in several fields, and yet to have a holistic or systems viewpoint.

## Systems Ecologists

### *Definitions*

The systems ecologist of tomorrow may be defined as one type of scientist who is a specialist in generalization (100). There are few if any systems ecologists today. Of today's biologists, perhaps some scientists in applied ecology can be considered systems ecologists (Fig. 5–6). In some applied ecology fields, such as forestry, it has been noted that the field is becoming so complex that more members of the profession will likely find it to their advantage to either specialize or become exceptionally well-balanced generalists (11). This trend is well established in many professional fields. In addition to the growing need for specialists in resource management, there also is a niche for the generalist (120). Undergraduate programs have been developed to train such individuals.

The applied ecologists, shown as the second group of specialists in Figure 5–6, to some degree have in their training many elements of the training of the four groups of specialists listed below them. The

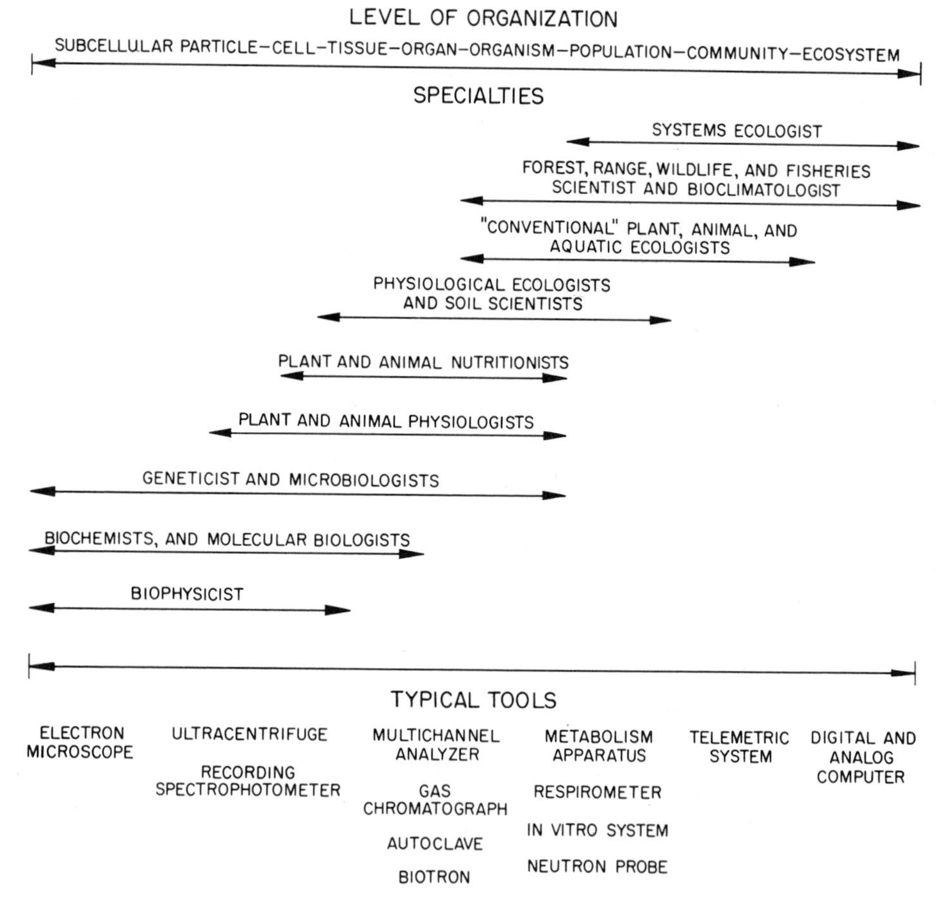


FIGURE 5–6. A schematic comparison of biologists, the level of organization of the media with which they work, and the tools they use. The double-ended arrows indicate general positions of the specialties. Tools especially, and to some extent the media, overlap widely for different specialties.

applied ecologists are closer akin to the systems ecologists than are the conventional ecologists, because in their training and in their work they usually are more cognizant of the total ecosystem and its interrelations than are many conventional ecologists.

Consider, for example, a scientist responsible for the trout population in a Rocky Mountain forest. He realizes that the trout's well-being is inextricably related to the total environment. He must consider the impact of grazing, lumbering, mining, and road-building upon the response of the watershed to uncontrolled and fluctuating precipitation. He must consider also the inherent fertility of the watershed and its impact on populations of fish and fish foods. Superimposed upon this are other factors, such as insect control by wide-scale pesticide spraying, the problems of optimum rates and places of artificial stocking of streams, of seasons and levels of bag limits, of public relations etc., ad infinitum. In contrast, for example, few plant ecologists thoroughly appreciate aquatic problems or communicate well with aquatic ecologists; few animal ecologists understand soil problems or communicate well with soil scientists.

The systems ecologist will require better mathematical, chemical, physical, and electronic training than either the applied or the conventional ecologist. Yet he must share their holistic way of thinking or approaching problems, and he must have a broad background in ecological subject matter. A lifetime may not be sufficient for any one person to prepare adequately to perform unassisted the synthesizing function, a major effort of the systems ecologist. This function requires the cooperation of specialists, and publication in each other's journals (52). No individual will be able to direct or conduct research without consulting others to obtain a complete understanding of the processes within even most fairly simple ecosystems (80). Future leaders toward this goal must have the ability to organize concepts, things, and people.

It has long been apparent to those in physical sciences that the ecologist, in the broad sense, must be an environmental specialist. For example, Jehn (52) suggested that an ecologist must simultaneously be a meteorologist, a soils physicist, a geologist, and a geographer. But because no one man can encompass all the required specialties, he must ally himself with these specialists (61). Therefore, the greatest advances to be made in systems ecology will require the effort of an interdisciplinary team. How can this be done without losing the spontaneity and originality of the individual's personality?

### *Systems Ecologists: Interdisciplinarians and Multi-disciplinarians*

That interdisciplinary teams are required to solve many physicobiological problems of national importance is becoming more and more apparent (43). For example, the understanding of pollution processes requires the cooperation of "of [systems] ecologists, physiologists, biomathematicians, microclimatologists, geneticists, microbiologists, biochemists, chemists, morphologists, and taxonomists . . ." (108). To work effectively as a member of an interdisciplinary team, the systems ecologist will need to establish a common vocabulary, an agreed-upon ideology, a set of reasonable goals, a common context for symbols, and ways for translating ideas into action (57).

The systems ecologist is one of the types of interdisciplinary scientists who should be in great demand in the near future. It has been estimated (98)

> that about ten percent of our total national effort will be going into production, development, and research based on biophysics, biomedicine, bioengineering, and related computer projects by 1970. . . . then we must hurriedly prepare to train several thousand additional students to the Master's or Ph.D. level per year in this difficult field and we must anticipate at least a doubling of our teaching and research facilities in this interdiscipline once in each three years during this decade.

Systems ecologists can and should contribute heavily to these efforts. But the increasing importance of group effort and interdisciplinary teams in the study of major, man-created environmental problems creates new paradoxes. Large-scale, expensive research activity may decrease the flexibility and freedom which are intrinsic to research. Operation of an interdisciplinary team requires unique coordination and appreciation of contributions by different skills at many levels (23). Interdisciplinary research must be reconciled with the continuing importance of distinctive contributions by highly talented and motivated individuals. Furthermore, it is historical fact that to date many major scientific achievements have been made by specialists—by scientists wearing blinders (46).

Imagination and inventiveness, like the ability to work in an interdisciplinary team, are difficult to develop by training (47). A successful systems ecologist will be one with the imagination to perceive an important problem before others do. He must have the inventiveness to devise and weigh alternatives for its solution. This emphasizes the multidisciplinary nature of systems ecologists. In order to contribute effectively in the interdisciplinary team they must have sufficient depth in more than one specialty in order to make significant contributions to the solution of the problem. Thus, systems ecologists will need both breadth and depth of interests.

The role of the systems ecologist is complex and not well defined, nor is it easy to analyze. He may be viewed by many specialists of the interdisciplinary team as an amateur, and he may be viewed by his fellow ecologists with suspicion. Both views are justified until he proves his worth to all concerned. A major problem of the systems ecologist will be to convince an ecologist that a mathematical attack is useful and to convince the mathematician that his time and methods will be productive in ecology. Is there a natural course for this convincing to take? The interdisciplinary viewpoint does not rest solely on the biologist. A team composed of biologists with no mathematical and computer training and of engineers, mathematicians, and programmers with no biological training is doomed for failure (62).

### *Some Pitfalls Facing Systems Ecologists*

The availability of such powerful tools and equipment as gas chromatographs, telemetric devices, and computers will not make the solution of new problems trivial, nor will it make systems ecology research routine. Some of these powerful tools themselves are raising important problems.

A special problem exists with computers. In general, the larger and faster a computer is, the more economical it is, even for small problems, if there is sufficient work available. But, of necessity, operating procedures of large computer centers are rigid in order to maintain output. The "people problems" of getting small problems into large computers grow disproportionately with computer size (98), so remote access to and time sharing on these big computers will be essential. Without direct access to the computer, such as provided by remote consoles, many problems can be completed more rapidly with a hand calculator, although they may require several hours' work. Even though they could be run in a few seconds on a computer, the long delay or "turn-around" time in using a computer without direct access leads to inefficiency. Our research output is best and most efficient when we are able to progress at full speed, regardless of time of day or day of week, rather than to take days or weeks to complete a problem. Remote-console access to the large computers which will be required for many ecological problems will allow concentrated work and will in every sense give rapid results.

As compared with his predecessors, the systems ecologist will still have to acquire empirical data by means of experimental or literature research, but he will need a better grasp of the biological and physical interactions in the system he is studying, and he will have to apply more ingenuity and invention to formulating and analyzing his problems in order to make significant advances. The easy problems have been solved.

Another pitfall facing the systems ecologist of tomorrow, who may be an undergraduate today, is that often he has been given equations and their coefficients and has been asked to produce numerical or graphic solutions. The problem he faces when he leaves the "ivy-covered halls" is first to design experiments correctly and then to conduct them effectively before he even obtains experimental results. Then his problem will be to infer and derive the form of the equations and to determine analytically the magnitude of the coefficients. Needless to say, this will be a much different and more difficult task than that which he faces as a student. Perhaps it will be desirable to develop "co-op" training programs wherein the student may intersperse practical experience in his undergraduate program. Graduate students in systems ecology, of course, will have numerous opportunities to test the effectiveness of their training, especially in their research work.

### Mathematical Training of Systems Ecologists

The training of systems ecologists in mathematics and computer sciences is an especially important part of their education. Watt's review (114) of the use of mathematics in population ecology gives numerous examples of mathematical methodologies and applications. Unfortunately, many ecologists receive little mathematical, statistical, or computer training, and this is only late in graduate school. An encouraging trend is the recent development of undergraduate biomathematical courses and curricula at several universities. For example, an undergraduate biologist at Colorado State University who has had college algebra can, in 12 quarter credits, complete courses in calculus and differential equations designed for biologists. Mathematical training for attacking four types of problems has been outlined for undergraduate students in biological, management, and social sciences (24).

Consider the four combinations generated by deterministic and stochastic phenomena, each with few or with many variables. Tools required in study of organized simplicity (i.e., deterministic × few variables) include the classical analytic geometry—calculus sequence, and difference and differential equations. Disorganized simplicity (i.e., stochastic × few variables) requires probability and statistics for analysis. Organized complexity requires linear algebra and many-variable advanced calculus. Study and use of complex stochastic models are needed for analysis of phenomena characterized as disorganized complexity. Computers are especially important in these last two areas, and computing practice in numerical analysis is equally important. For those systems ecologists who wish to specialize in computers in their undergraduate or graduate training, additional courses may be recommended (25).

Recently courses have been taught to ecologists which combine many systems-oriented mathematical approaches to ecological problems with practice in the use of digital computers (117) or use of both analog and digital computers (Systems Ecology, a yearlong graduate course at the University of Tennessee, which has been taught by B. C. Patten, J. S. Olson, and the author). Systems ecologists, however, should be trained in mathematics not so much for developing their skill in performing mass computations as for having the ingenuity of escaping them (47).

### Systems Ecology Research

It is considered by some that existing ecological theory often has limitations in the rapid solution of many problems (76). Also, some consider that the reliance upon analogies from physics for the solution of ecological problems has distinct limitations (102). But the field essentially is a virgin area for tomorrow's better trained and better equipped systems ecologists. The dearth of quantitative ecologists has been mentioned above, and most ecologists have been isolated and not well supported. Hopefully, we will be able to develop centers wherein "critical masses" of systems ecologists will migrate and find suitable niches.

Perhaps a routine similar to the following will be of use to the systems ecologist (7). He will often have to guess at the fundamental cause-and-effect relationships in his system and may even have to guess about the basic variables. He will then test these hypotheses by comparing the quantitative and qualitative behavior of the real world with that predicted from his model. Fortunately, he will have varied and powerful tools available for the testing of complex (and realistic) as well as simple hypotheses. But there are an infinite number of hypotheses to test about any complex system, and most of these hypotheses will be wrong. Holling (48–49) shows by example how theory and experiment can be combined in a systems analysis of predation in a way to greatly reduce the number of hypotheses to be tested. The majority of the alternatives must be excluded by means which require, not computers, but only pencils, paper, and discriminating thought. The systems ecologist will have to develop the knack of feeding on negative information and use these negative hypotheses to guide his further experimentation and theory. In instances where experimentation or measurement of parameters is impossible without disruption of the system, perturbation of the system followed by measurements may still give new insight for the definition of a model (8).

A major hurdle to overcome in systems ecology research is lack of precedent in funding detailed and integrated research on complex ecological systems. This applies both to the theoretical and analytical phase and the experimental phase. In many respects no single organization has been working in depth on complex systems ecology problems, for such work may be beyond the role, objectives, or structure of existing organizations. Universities, national laboratories, and state and federal experiment stations each have some unique resources and capabilities for studying ecosystem problems. Some advantages and disadvantages of these three types of organizations with respect to "total systems" research, based on my experience in working in these organizations, are briefly outlined as follows.

In the past, ecological research conducted at universities generally has involved one investigator, or at most a few, on a part-time basis on problems of limited extent. Extensive and intensive interdepartmental cooperation in ecological research has been the exception rather than the rule. Many sources of funds are available to these researchers, but usually in amounts insufficient to attract permanent personnel and to support long-term ecosystem research. Although the economy of graduate student use has been exploited, often there is a lack of continuity in the conduct of long-term environmental researches. By the time the student gains competence and becomes capable of independent contributions to the project, he graduates and is lost to the project. Also, universities often lack controllable research areas or have conflicting needs for them.

Applied ecologists, in state fish and game departments or in state and federal agricultural experiment stations, for instance, often have controllable research areas on which to conduct long-term ecosystem research, but continuity again is impeded because of high turnover rate of personnel. Furthermore, their research funds often are restricted to only one or a few phases of the total ecosystem problem, and their funds have become more limited in recent years. Funds from many granting agencies may not be available to them for research, special training, or foreign study.

In the past ten years considerable ecological research has been conducted at several national laboratories. Ecologists in these laboratories have available many services, tools, and consultants which the university or experiment station scientists lack. Much of the ecological work in the national laboratories, however, has been concerned with specific needs of the funding agency. Most ecological researchers in these laboratories have come directly from liberal arts departments in universities, and these laboratory staffs are divided into subject matter groups for conduct of research. A total-system, interdisciplinary approach usually has not been implemented in their research. Although these ecologists are funded comparatively well, their costs are high due to the nature of their work. Although they may have long-term control over their research areas, the number of these laboratories is limited, and some important biomes, such as grasslands, are not within the boundaries of these laboratories.

I feel that perhaps research in systems ecology could encompass the advantages held by ecological researchers of the above three categories. This would require, however, some shifts concerning funding and conduct of research, and some shifts in administrative policy of the respective agencies or institutions. The exact nature and organization of such research is uncertain, although Dubos (21) has raised some interesting questions and has made some good suggestions about similar research in environmental biology.

The long-term impact on man of fundamental, total-ecosystem research should be recognized, and the framework should be developed for extensive and intensive intercooperation of these three groups of ecologists. Analytical and experimental research on total-ecosystem complexes should be initiated as soon as possible, if man is to benefit tomorrow, because most problems of environmental magnitude require many years of study before conclusions may be reached.

The proposed International Biological Program is, in several respects, a call for systems ecology research, both experimental and theoretical. This program could provide an incentive and a means for ecologists from universities, experiment stations, and national laboratories to work cooperatively and share funds, research areas, and talent. An example follows. Other such examples of needed research on total-system problems can be found in other parts of this continent and on other continents, in both terrestrial and aquatic ecosystems.

Consider the seminatural grassland ecosystems in the Great Plains. A more complete understanding of the structure and function of these ecosystems becomes increasingly essential as these lands are called upon to provide food, water, and recreation for tomorrow's growing populations. Several state and federal agricultural experiment stations hold sizeable acreages of representative variations of grasslands, from the aspen parkland in Canada to the semidesert grasslands of Mexico. But at none of these stations is there a team equipped with suitable manpower or funds for intensive total-ecosystem research. Scientists at these experiment stations and nearby universities have accumulated considerable data and experience on and about these grassland ecosystems. There is no national laboratory in this vast area, and no university in the area can marshal many of the unique facilities, such as computing facilities and computer consultants, that are necessary for systems ecology research. Still, scientists at universities in the area can provide much necessary insight into these ecosystems, and graduate students can help conduct ecosystem research in the area. With sufficient funds and planning it should be possible to combine the special skills and resources of all these groups and bring them to bear on problems of ultimate national importance at selected locations in the Great Plains.

## Literature Cited


1. Arcus, P. L. 1963. An introduction to the use of simulation in the study of grazing management problems. *Proc. New Zealand Soc. Animal Prod.* 23:159–168.
2. Armstrong, N. E. and H. T. Odum. 1964. Photoelectric ecosystem. *Science* 143:256–258.
3. Bakuzis, E. V. 1959. Structural organization of forest ecosystems. *Proc. Minn. Acad. Sci.* 27:97–103.
4. Barea, D. J. 1963. Analisis de ecosistemas en biologia, mediante programacion lineal. *Archivos de Zootecnia* 12:252–263.
5. Barker, R. 1964. Use of linear programming in making farm management decisions. *New York Agr. Exp. Sta. Bull.* 993. 42 pp.
6. Bartlett, M. S. 1960, *Stochastic population models in ecology and epidemiology*. Methuen and Co., Ltd., London. 90 pp.
7. Bellman, R. 1961. *Mathematical experimentation and biological research*. Rand Corp. P-2300. 12 pp.
8. Berman, M. 1963. A postulate to aid in model building. *J. Theoret. Biol.* 4:229–236.
9. Berman, M. 1963. The formulation and testing of models. *Ann. New York Acad. Sci.* 108:182–194.

10. Blair, W. F. 1964. The case for ecology. *BioScience* 14:17–19.
11. Briegleb, P. A. 1965. The forester in a science-oriented society. *J. Forestry* 63:421–423.
12. Broido, A., R. J. McConnen, and W. G. O'Regan. 1965. Some operations research applications in the conservation of wildland resources. *Manage. Sci.* 11:802–814.
13. Caldwell, L. K. 1963. Environment: a new focus for public policy? *Public Administration Review* 23:132–139.
14. Churchill, E. D. and H. C. Hanson. 1958. The concept of climax in arctic and alpine vegetation. *Bot. Rev.* 24:127–191.
15. Clements, F. E. 1916. Plant succession: an analysis of the development of vegetation. *Carnegie Inst. Wash. Publ.* 242. 512 pp.
17. Coombs. C. H., H. Raiffa, and R. M. Thrall. 1954. Some views on mathematical models and measurement theory. *In:* Thrall, R. M., C. H. Coombs, and R. L. Davis (eds.). *Decision Processes.* John Wiley & Sons, Inc. New York. 332 pp.
18. Dansereau, P. 1964. The future of ecology. *BioScience* 14:20–23.
19. Davis, C. C. 1963. On questions of production and productivity in ecology. *Arch. Hydrobiol.* 59:145–161.
20. Deevey, E. S. 1964. General and historical ecology. *BioScience* 14:33–35.
21. Dubos, R. 1964. Environmental biology. *BioScience* 14:11–14.
23. Duckworth, W. E. 1962. *A guide to operational research.* Methuen and Co., Ltd. London, England. 145 pp.
24. Duren, W. L., Jr. (Chr.). 1964. Tentative recommendations for the undergraduate mathematics program of students in the biological, management and social sciences. Mathematical Association of America, Committee on the Undergraduate Program in Mathematics, 32 pp.
25. Duren, W. L. (Chr.). 1964. Recommendations on the undergraduate mathematics program for work in computing. Mathematical Association of America, Committee on the Undergraduate Program in Mathematics. 29 pp.
27. Eberhardt, L. L. 1963. Problems in ecological sampling. *Northwest Sci.* 37:144–154.
28. Eberhardt, L. L. 1965. Notes on ecological aspects of the aftermath of nuclear attack. pp. 13–25 *in:* Hollister, H. and L. L. Eberhardt. *Problems in estimating the biological consequences of nuclear war.* U.S. Atomic Energy Commission TAB-R-5.
29. Eberhardt, L. L. 1965. Notes on the analysis of natural systems. pp. 27–40 *in:* Hollister, H. and L. L. Eberhardt. *Problems in estimating the biological consequences of nuclear war.* U.S. Atomic Energy Commission TAB-R-5.
30. Elveback. L., J. P. Fox, and A. Varma. 1964. An extension of the Reed-Frost epidemic model for the study of competition between viral agents in the presence of interference. *Amer. J. Hygiene* 80:356–364.
32. Feibleman, J. L. 1960. Testing hypothesis by experiment. *Persp. Biol. and Med.* 4:91–122.
34. Fosberg, F. R. 1965. The entropy concept in ecology. pp. 157–163 *in: Symposium on Ecological Research in Humid Tropics Vegetation*, Kuching, Sarawak, July 1963.
35. Garfinkel, D., R. H. MacArthur, and R. Sack. 1964. Computer simulation and analysis of simple ecological systems. *Ann. New York Acad. Sci.* 115:943–951.
36. Garfinkel, D. and R. Sack. 1964. Digital computer simulation of an ecological system, based on a modified mass action law. *Ecology* 45:502–507.
37. George, F. H. 1965. *Cybernetics and biology.* Freeman and Co., San Francisco, California. 138 pp.
38. George, J. L. 1964. Ecological considerations in chemical control: Implications to vertebrate wildlife. *Bull. Entomol. Soc. Amer.* 10:78–83.
39. Gibbens, R. P. and H. F. Heady. 1964. The influence of modern man on the vegetation of Yosemite Valley. *Calif. Agr. Exp. Sta. Manual* 36. 44 pp.
40. Glass, B. 1964. The critical state of the critical review article. *Quart. Rev. Biol.* 39:182–185

41. Golley, F. B. 1965. Structure and function of an old-field broomsedge community. *Ecol. Monogr.* 35:113–137.
42. Gould, E. M. and W. G. O'Regan. 1965. Simulation, a step toward better forest planning. *Harvard Forest Paper 13.* 86 pp.
43. Gross, P. M. (Chr.). 1962. Report of the committee on environmental health problems. *Public Health Service Publ. 908.* 288 pp.
44. Harris, J. E. 1960. A review of the symposium: models and analogues in biology. *Symp. Soc. Exp. Biol.* 14:250–255.
45. Hasler, A. D. 1964. Experimental limnology. *BioScience* 14:36–38.
46. Hitch, C. 1955. An appreciation of systems analysis. Rand Corp. P-699. (*Symposium on Problems and Methods in Military Operations Research*, pp. 466–481.)
47. Hoag, M. W. 1956. *An introduction to systems analysis.* Rand Corp. RM-1678. 21 pp.
48. Holling, C. S. 1963. An experimental component analysis of population processes. *Mem. Entomol. Soc.* Canada. 32:22–32.
49. Holling, C. S. 1965. The functional response of predators to prey density and its role in mimicry and population regulation. *Mem. Entomol. Soc.* Canada. 45:3–60.
50. Hollister, H. 1965. Problems in estimating the biological consequences of nuclear war. pp. 1–11 *in:* Hollister, H. and L. L. Eberhardt. (same title). U.S. Atomic Energy Commission TAB-R-5.
51. Hughes, R. D. and D. Walker. 1965. Education and training in ecology. *Vestes* 8:173–178.
52. Jehn, K. H. 1950. The plant and animal environment: a frontier. *Ecology* 31: 657–658.
53. Jenkins, K. B. and A. N. Halter. 1963. A multi-stage stochastic replacement decision model. *Ore. Agr. Exp. Sta. Tech. Bull.* 67. 31 pp.
56. Jones, R. W. 1963. System theory and physiological processes: An engineer looks at physiology. *Science* 140:461–464.
57. Kennedy, J. L. 1956. A display technique for planning. Rand Corp. P-965.
58. Kramer, P. J. 1964. Strengthening the biological foundations of resource management. *Trans. N. Amer. Wildlife and Natural Resources Conf.* 29:58–68.
59. Larkin, P. A. and A. S. Hourston. 1964. A model for simulation of the population biology of Pacific salmon. *J. Fish Res. Bd. Canada* 21:1245–1265.
60. Leak, W. B. 1964. Estimating maximum allowable timber yields by linear programming. *U.S. For. Serv. Res. Paper NE-17.* 9 pp.
61. Lebrun, J. 1964. Natural balances and scientific research. *Impact of Sci. on Society* 14:19–37.
62. Ledley, R. S. 1965. *Use of computers in biology and medicine.* McGraw-Hill Book Co., Inc. New York. 965 pp.
63. Leigh, E. G. 1965. On the relation between the productivity, biomass, diversity, and stability of a community. *Proc. Nat. Acad. Sci.* 53:777–783.
64. Leopold, A. S., S. A. Cain, C. H. Cottam, I. N. Gabrielson, and T. L. Kimball. 1963. Wildlife management in the national parks. *Amer. Forests* 69:32–35, 61–63.
65. Leslie, P. H. 1958. A stochastic model for studying the properties of certain biological systems by numerical methods. *Biometrika* 45:16–31.
67. Lloyd, M. 1962. Probability and stochastic processes in ecology. *In:* H. L. Lucas (ed.) *The Cullowhee Conf. on Training in Biomath.* Institute of Statistics, N. Car. St. U., Raleigh, N.C.
69. Loucks, D. P. 1964. The development of an optimal program for sustained-yield management. *J. Forestry* 62:485–490.
70. Lucas, H. L. 1960. Theory and mathematics in grassland problems. *Proc. Intern. Grassland Cong.* 8:732–736.
71. Lucas, H. L. 1964. Stochastic elements in biological models; their sources and significance, pp. 335–383 *in:* Gurland, J. (ed.). *Stochastic models in medicine and biology.* U. Wisc. Press, Madison, Wisc. 393 pp.

74. Margalef, R. 1957. Information theory in ecology. *Mem. Real Acad. Ciencias y Artes de Barcelona* 23:373–449.
75. Margalef, R. 1962. Modelos fiscos simplificados de poblaciones de organismos. *Mem. Real Acad. Ciencias y Artes de Barcelona* 24:83–146.
76. Margalef, R. 1963. On certain unifying principles in ecology. *The Amer. Nat.* 97:357–374.
77. Miller, I. and J. E. Freund. 1965. *Probability and statistics for engineers.* Prentice-Hall Inc., Englewood Cliffs, N.J. 432 pp.
78. Miller, R. S. 1965. Summary report of the ecology study committee with recommendations for the future of ecology and the Ecological Society of America. *Bull. Ecol. Soc. Amer.* 46:61–82.
79. Neyman, J. and E. L. Scott. 1959. Stochastic models of population dynamics. *Science* 130:303–308.
80. O'Connor, F. B. 1964. Energy flow and population metabolism. *Science Prog.* 52:406–414.
82. Odum, E. P. 1963. *Ecology.* Holt, Rinehart, and Winston. New York. 152 pp.
85. Odum, H. T. 1965. An electrical network model of the rain forest ecological system. U.S. Atomic Energy Commission PRNC 67.
86. Odum, H. T. and R. C. Pinkerton. 1955. Time's speed regulator: the optimum efficiency for maximum output in physical and biological systems. *Amer. Sci.* 43:331–343.
87. Olson, J. S. 1965. Equations for cesium transfer in a *Liriodendron* forest. *Health Physics* 11:1385–1392.
88. Ostle, B. 1963. *Statistics in research.* Iowa St. Univ. Press. 2nd Ed. 585 pp.
90. Patten, B. C. 1964. *The systems approach in radiation ecology.* Oak Ridge National Laboratory Technical Memorandum 1008. 19 pp.
92. Platt, R. B., W. D. Billings, D. M. Gates, C. E. Olmsted, R. E. Shanks, and J. R. Tester. 1964. The importance of environment to life. *BioScience* 14:25–29.
93. Quade, E. S. (ed.). 1964. *Analysis for military decisions.* Rand Corp. R-387. Rand McNally & Co., Chicago.
94. Rasmussen, D. I. 1941. Biotic communities of Kaibab Plateau, Arizona. *Ecol. Monogr.* 11:229–275.
95. Royce, W. F., D. E. Bevan, J. A. Crutchfield, G. J. Paulik, and R. L. Fletcher. 1963. Salmon gear limitation in northern Washington waters. *U. Wash. Publ. in Fisheries* (N.S.) 2:1–123.
96. Sarnoff, D. 1964. The promise and challenge of the computer. *Amer. Fed. Infor. Process. Soc. Conf. Proc.* 26:3–10.
97. Schaller, W. N. and G. W. Dean, 1965. Predicting regional crop production: an application of recursive programing. *USDA Tech. Bull.* 1329. 95 pp.
98. Schmitt, O. H. and C. A. Caceres (eds.). *Electronic and computer-assisted studies of biomedical problems.* C. C. Thomas. Springfield, Illinois. 314 pp.
99. Schultz, A. M. 1961. Introduction to range management. U. California. Ditto notes. 116 pp.
100. Schultz, A. M. 1965. The ecosystem as a conceptual tool in the management of natural resources. *In:* Parsons, J. J. (ed.). Symposium on quality and quantity in natural resource management (in press), manuscript 33 pp.
101. Sears, P.B. 1963. The validity of ecological models. pp. 35–42. *In*: XVI International Congress of Zoology. Vol. 7. Science and Man Symposium—Nature, Man and Pesticides.
102. Slobodkin, L. B. 1965. On the present incompleteness of mathematical ecology. *Amer. Scientist* 53:347–357.
103. Specht, R. D. 1964. Systems analysis for the postattack environment: some reflections and suggestions. Rand Corp. RM-4030. 34 pp.

104. Stone, E. C. 1965. Preserving vegetation in parks and wilderness. *Science* 150:1261–1267.
106. Tansley, A. G. 1935. The use and abuse of vegetational concepts and terms. *Ecology* 16:284–307.
107. Thrall, R. M. 1964. Notes on mathematical models. U. Mich. Engin. Summer Conf.—Foundations and Tools for Operations Research and Management Sciences. Multilith 97 pp.
108. Tukey, J. W. *et al.* (Environmental Pollution Panel, President's Science Advisory Committee). 1965. Restoring the quality of our environment. Superintendent of Documents, U.S. Govt. Printing Office, Washington, D.C. 317 pp.
109. Turner, F. B. 1965. Uptake of fallout radionuclides by mammals and a stochastic simulation of the process. pp. 800–820 *in:* Klement, A. W. (ed.). Radioactive fallout from nuclear weapons test. *U.S. Atomic Energy Commission Symp. Ser. 5*. 953 pp.
110. van de Panne, C. and W. Popp. 1963. Minimum-cost cattle feed under probabilistic protein constraints. *Manage. Sci.* 9:405–430.
111. Van Dyne, G. M. 1965. Probabilistic estimates of range forage intake. *Proc. West. Sect. Amer. Soc. Animal Sci.* 16(LXXVII):1–6.
112. Van Dyne, G. M. 1965. Application of some operations research techniques to food chain analysis problems. *Health Physics* 11:1511–1519.
113. Walker, O. L., E. O. Heady, and J. T. Pesek. 1964. Application of game theoretic models to agricultural decision making. *Agronomy J.* 56:170–173.
114. Watt, K. E. F. 1962. Use of mathematics in population ecology. *Ann. Rev. Entom.* 7:243–260.
115. Watt, K. E. F. 1964. The use of mathematics and computers to determine optimal strategy and tactics for a given insect pest control problem. *Canad. Entomol.* 96:202–220.
117. Watt, K. E. F. 1965. An experimental graduate training program in biomathematics. *BioScience* 15:777–780.
119. Whittaker, R. H. 1953. A consideration of climax theory: the climax as a population and pattern. *Ecol. Monogr.* 23:41–78.
120. Yamber, P. A. 1964. Is there a niche for the generalist? *Trans N. Amer. Wildlife and Natural Resources Conf.* 29:352–372.

# III

# Management of Wildlife Populations

# 6

# Adaptability of animals to habitat change

*A. Starker Leopold*

All organisms possess in some measure the ability to adapt or adjust to changing environmental conditions. But the degree to which different species are capable of adjusting varies enormously. This chapter concerns the nature and extent of adaptability and demonstrates the truism that in a world undergoing constant and massive modification by man, the animals with the highest capacity for adjustment are those that persist in abundance. Specialized animals with narrow limits of adjustment are those that have become scarce or in some instances extinct.

## Relation of Ungulate Populations to Plant Successions

In any given ecosystem, there are animals that thrive best in the climax stages of plant succession and others that do better when the climax has been destroyed in some way and the vegetation is undergoing seral or subclimax stages of succession, working back toward restoration of the climax. This can be interpreted to mean that the climax animals are more specialized in their environmental needs, while the seral or successional species are more adaptable and able to take advantage of

---

transitory and unstable situations. The principle can be well illustrated by considering the status of various native ungulates in North America.

The mass of data accumulated in studies of North American deer permits us to draw some general deductions about population dynamics in these animals, particularly in relation to food supplies. The following remarks apply equally to the white-tailed deer (*Odocoileus virginianus*) and the mule deer or blacktail (*O. hemionus*). The quality and quantity of forage available to a deer population during the most critical season of the year has proven repeatedly to be the basic regulator of population level. Usually this means winter forage, but not always. In desert areas or regions of Mediterranean climate, like coastal California, summer may be the critical season. In any event, the nutritive intake of the individual deer during the critical season determines both productivity of the herd (Cheatum and Severinghaus, 1950; Taber, 1953) and mortality in the herd, whether death be caused by starvation, disease, parasites, or even to some extent by predation or accidents (Longhurst et al., 1952, and others). Average population level is a dynamic function of these two opposing variables—rate of productivity and rate of mortality. Hunting is a source of mortality artificially interposed in the formula, and although it is intercompensatory with other forms of loss (that is, hunting kill will reduce starvation losses, etc.) it is not regulated by nutrition, but by legislative fiat. However, since hunting is generally controlled in North America to remove no more than annual increment, and usually less, it cannot be construed as a primary determinant of population level in most areas. Putting all this in much simpler form, good forage ranges generally have many deer; poor ranges have few. All other influences are secondary.

Good deer ranges characteristically include stands of nutritious and palatable browse which as a rule are produced in secondary stages of plant succession (Leopold, 1950). Burned or cutover forest lands support most of the deer in the continent; some brush-invaded former grasslands are of local importance. In a few special cases, as for example that of the burro deer (*O. h. eremicus*) on the desert, sparse populations live in climax floras. But on the whole the association between deer and secondary brushlands (the connecting link being nutritional) is so general as to permit classification of deer as seral or successional species.

Assuming that range relationships are equally dominant in determining populations of other North American ungulates, and much evidence indicates that this is so, a general characterization can be made of each species, permitting classification along lines of range affinities, as has been done in Table 6–1.

This rather subjective classification requires some explanation.

### *Climax species*

The northern caribou is a classic example of an animal that depends heavily in winter on undisturbed climax vegetation of the subarctic zone. The lichens which supply much of the caribou's winter food grow either as an understory to the spruce forest or suspended from the spruce limbs. Any disturbance such as fire or grazing that depletes this particular vegetative complex lowers the carrying capacity for caribou.

Similarly bighorn sheep and mountain goats in their alpine retreats, bison on the great prairies, and musk ox on the arctic plains are adapted to feed on climax species of forbs, sedges, grasses, and a few shrubs.

In the southern reaches of the continent the two species of peccaries are generalized in their food habits, like other pigs, but the mast of oak and of many tropical fruit trees contributes heavily to their diet. Besides mast, the bulbous roots, palmettos, cacti, forbs, and grasses on which these pigs feed are on the whole characteristic of climax associations. The tapir and brocket are even more typical of climax rain forest (Leopold, 1959).

It is notable that of nine species of North American ungulates associated with climax vegetation, four are of boreal or arctic affinities, four are tropical, and only the bison and the collared peccary in part of its range occur in temperate latitudes.

TABLE 6–1

*General association of North American ungulates with climax or subclimax successional stages*

| | Biotic zone | | |
|---|---|---|---|
| | *Boreal* | *Temperate* | *Tropical* |
| 1. Associated primarily with climax forage types: | | | |
| Caribou (*Rangifer arcticus*) | x | | |
| Bighorn (*Ovis canadensis* and allied species) | x | | |
| Mountain goat (*Oreamnos americanus*) | x | | |
| Musk ox (*Ovibos moschatus*) | x | | |
| Bison (*Bison bison*) | | x | |
| Collared peccary (*Pecari tajacu*) | | x | x |
| White-lipped peccary (*Tayassu pecari*) | | | x |
| Tapir (*Tapirella bairdii*) | | | x |
| Brocket (*Mazama americana*) | | | x |
| 2. Associated primarily with subclimax forage types: | | | |
| Moose (*Alces americana*) | x | x | |
| Elk (*Cervus canadensis*) | | x | |
| White-tailed deer (*Odocoileus virginianus*) | | x | |
| Mule deer (*Odocoileus hemionus*) | | x | |
| Pronghorn antelope (*Antilocapra americana*) | | x | |

*Subclimax species*

Nearly all of the ungulates that thrive best on kinds of weeds and brush that characterize disturbed vegetative situations are native in the temperate zone. This includes the two common deer, elk, and the moose, which extends northward through the boreal zones as well. The pronghorn antelope is predominantly a weedeater (Buechner, 1950), although it may consume much sage and other browse at times. On the Great Plains, the weeds and forbs that supported antelope originally may have resulted from local overgrazing by the native bison. On the deserts of Mexico, however, the antelope almost certainly depended on climax vegetation, but this is the fringe of its continental range.

Whereas most boreal and tropical ungulates have climax affinities, the temperate-zone species thrive largely on successional vegetation. In an evolutionary sense this would suggest that these adaptive species, all highly successful today, developed in an environment subject to frequent disturbance, presumably fire. Even the bison, here classed as a climax species, would fall in this category if one accepts the prairie as a subclimax, maintained by fire (Sauer, 1950).

Recognition of successional affinities of big-game species is basic to determination of sound management policy. The subclimax species (two deer, antelope, elk, moose) fit nicely into multiple-use land programs, including logging, grazing, and controlled burning. The climax species do not. Preservation of wilderness areas, without competing or disturbing uses, is particularly important in sustaining remnants of the climax forms designated in Table 6–1.

## Plant Successions and Other Wildlife

The principle illustrated above with ungulates applies generally to wild animals. In areas heavily modified by human action, the abundant species are those adapted to take advantage of disturbed ecologic situations. Over much of the United States, the upland game species that supply most of today's recreational hunting are the bobwhite quail, cottontail rabbit, ruffed grouse, mourning dove, and the introduced ring-necked pheasant—all typical subclimax or successional species.

Game species once abundant on the continent, but now localized and scarce because of shrinkage of particular climax vegetational types on which they depended, are the prairie chicken, sharp-tailed grouse, sage hen, upland plover, and wild turkey. Extinct are the heath hen and passenger pigeon.

The case of the passenger pigeon illustrates particularly well the dilemma facing an unadaptive species. The fabled legions that "darkened the sky" were supported in large part by mast crops produced in climax stands of mature timber, especially oak, beech, and chestnut. The flocks were highly mobile and searched the eastern half of the continent for favorable feeding grounds. When a good food supply was found, millions of birds would congregate to establish one of the massive colonial nestings so well documented by Schorger (1955). With the settlement of the country, two things happened concurrently that contributed to the swift collapse of the pigeon population: uncounted millions of the birds were slaughtered in the nesting areas, and the mature timber stands that produced the mast crops were felled to make way for farms. The demise of the pigeon is traditionally blamed on the market hunters; but had there been no hunting, it is doubtful that the pigeon would have survived the depletion of its food supply. So specialized was this bird that it seemingly had no capacity to adjust to the modest, scattered food source that certainly continued after the main hardwood forests were felled. When the big pigeon flocks were reduced, the survivors simply perished without a single pair exhibiting the ability to feed and reproduce under changed conditions. Its close relative, the mourning dove, on the other hand, adjusted very well indeed to the conversion from forest to farm, and today is undoubtedly much more numerous than in primitive times, despite heavy and persistent shooting.

## The Nature of Adaptation

Precisely what is this character of "adaptability" that some animals have and others do not? What are the mechanisms by which animals adapt?

The paleontological record tells us that over the eons of time there have been enormous changes in climate and hence in habitat. With each major shift many animal species became extinct; these presumably were the unadaptable ones. Other animals persisted but evolved and were modified to meet the new conditions. One component of adaptability, therefore, is the capacity for genetic change.

At the same time, current experience offers many examples of individual animals learning new tricks of survival that contribute to longevity and hence to persistence of the population. A coyote can learn to be trap-shy; a raccoon learns to search for eggs in wood-duck boxes; mallards learn the precise hour when legal shooting ceases, which signals the exodus from a refuge to go in search of food. Some species are quick to pick up new behaviorisms; others are not. Adaptability, therefore, may include the capacity to learn.

Genetic and learned adaptations will be discussed in that order.

## Morphologic Adaptation

One manifestation of genetic adaptation to local environment is the demonstrable evidence of subspeciation in animals. Many widely distributed species are segmented into local populations that show marked and persistent differences in morphology. Some of the characters that vary and are easily observed and measured are body size, proportion of body parts, and color of plumage or pelage. The bobwhite quail (*Colinus virginianus*) is an example of a resident (nonmigratory) game bird that varies greatly from place to place. This bird occurs throughout the eastern half of North America, from New England and South Dakota, south to Chiapas in southern Mexico. Within this range, twenty-one well-differentiated races or subspecies are recognized (Aldrich, 1955). In size, the bobwhite decreases from over 200 grams in the north to slightly over 100 grams in Chiapas. Likewise there is a general north-south gradient in plumage color, the palest birds occurring in the open or arid ranges, such as the Great Plains, the darkest forms being found in the wet tropics or subtropics of southern Florida, the coast of Veracruz, and the interior valleys of Chiapas. It is presumed that each population is particularly adapted to the local habitat in which it exists. The capacity to be molded genetically by local environment doubtless underlies the bobwhite's success in occupying such an extensive range in North America.

Commenting on this general question of genetic plasticity, Grant (1963, p. 434) states:

> The great role of natural selection in the formation of races [sub-species] can be inferred from the observation that racial characteristics are often adaptive. The adaptiveness of the racial characters in many plants and animals is demonstrated by two sets of correlations. First, the different races of a species have morphological and physiological characters that are related to the distinctive features of the environment in their respective areas. Second, the same general patterns of racial variation frequently recur in a parallel form in separate species inhabiting the same range of environment.

He goes on to comment on some of the generally accepted "rules" of morphologic adaptation that have been summarized and analyzed by many other authors, including Mayr (1942, p. 90). These are,

1. *Bergmann's rule*: The smaller races of a species are found in the warmer parts of a species range, the larger races in cooler parts.
2. *Allen's rule*: Protruding body parts, such as ears, tails, bills, and other extremities, are relatively shorter in the cooler parts of the range of a species than in the warmer parts.
3. *Golger's rule*: Dark pigments (eumelanins) increase in the warm and humid parts of a range, paler phaeomelanins prevail in arid climates.

The bobwhite illustrates all of these rules of local adaptation. The same may be said of white-tailed and mule deer, the raccoon, the bobcat, hares of the genus *Lepus*, cottontails of the genus *Sylvilagus*, and many other widely distributed birds and mammals. In the case of the white-tailed deer, the size gradient is extreme: in Wisconsin an adult buck weighs well over 200 pounds, in parts of Mexico scarcely seventy pounds. The larger size and smaller ears of northern animals presumably give an advantageous ratio of body mass to exposed surface, for heat conservation. The opposite is true in warmer climates.

## Physiologic Adaptation

More difficult to measure, but perhaps even more important in fitting local populations to their environments, are the physiologic adaptations. To be successful, a population must breed at the right time of year, produce only as many young as can be cared for, be able to digest and assimilate the foods locally available, and otherwise adjust its life processes to the local scene. Migratory birds lay on fat (fuel) for their travels and require elaborate navigational machinery. Research to date has scarcely scratched the surface of this enormously complicated area of animal adaptation.

A species that has been studied in some detail and that well illustrates several facets of physiologic adaptation is the common white-crowned sparrow of the Pacific Coast (*Zonotrichia leucophrys*). There are two races of this bird, very similar in appearance, that winter together in central California; but in spring one race migrates to British Columbia to breed while the other breeds locally, on the winter range. Blanchard (1941) showed a number of differences in the life cycles of these two populations. Though living together all winter, the migrants laid on fat in spring and departed for the north; the residents did not accumulate fat but went leisurely about the business of nesting. The migrants, having less time on the breeding grounds, compressed the reproductive cycle into approximately two thirds of the time used by resident birds. Subsequent investigation by a number of workers has demonstrated that the mechanism triggering these events is changing length of day in spring, but the important differences in response reflect inherent, physiologic adaptations peculiar to the two populations.

Differences in timing of breeding are demonstrable in many other species. Black-tailed deer along the California coast fawn in May, mule deer in the Sierra Nevada in July, whitetails in northwestern Mexico in August. In each case fawning corresponds to the period of optimum plant growth—spring in California, summer rains in Mexico. Time of mating (seven months before fawning) is presumably timed by changing day length—in this case by shortening days, since the breeding occurs in fall or early winter. Ian McTaggart Cowan has kept a number of races of blacktailed deer in pens in Vancouver, and notes that the southern Alaskan and British Columbian stocks breed at almost the same time, whereas the Californian stocks have retained a response that induces antler growth, shedding of velvet, breeding, antler drop, and pelage molt a month or more in advance of the northern races kept in the same pens.

The number of young produced by a breeding population is regulated by physiologic controls. Lack (1954) presents examples of clutch size in birds varying apparently with food availability. He cites the work of Swanberg on the thick-billed nutcrackers, in which it was shown that in years when the autumn crop of nuts was below average, the birds laid only three eggs. In years of good or excellent nut crops, clutches of four eggs were normal. When the experimenter supplied nuts in winter for certain wild nutcrackers, those particular individuals had clutches of four eggs, even in years of poor mast crop. The change in number of eggs was therefore apparently a physiologic adjustment to the amount of food available. But in all cases the birds laid no fewer than three eggs, nor more than five, the limits presumably set by hereditary factors.

Clutch or litter size likewise may be a function of predator populations and the likelihood of losses of eggs or young due to predation. The mallard of continental North America lays eight to twelve eggs, and predictable losses are high. The closely related Laysan duck, on isolated Laysan Island where there are no predators, lays only three to four eggs.

Certain deep-seated physiologic differences have been detected between wild and domestic turkeys which shed some light on how the wild birds are adapted to live successfully in the woods (Leopold, 1944). In the Missouri Ozarks the native turkey persists even under highly adverse circumstances and populations respond readily to protection and management. Domestic turkeys cannot exist away from farmyards. Hybrids between the two barely hold their own in refuges under intensive management. Differential reproductive success seemed to underlie the disparities in population behavior. Time of breeding is earlier in domestic and hybrid turkeys, leading to loss of eggs and chicks in late-spring storms. Behavioral differences between hens and chicks suggested other reasons why wild birds raised more young. These differential reactions were related to size of brain and relative development of some of the endocrine organs that control behavior, suggesting a few of the components that may be involved in "local adaptations."

## Danger of Transplanting Local Races

If indeed some kinds of animals are delicately attuned to life in specific local environments, one may question the advisability of trapping and shifting these populations about in an effort to restock underpopulated ranges.

During the era 1920–40 there was a very large trade in Mexican bobwhites, imported into various midwestern states for release to augment local populations. Actual measurements of the results of this endeavor are lacking, but there seemed to be a consensus among observers that such releases never led to sustained increase in bobwhite numbers, and in fact some thought that in years following a liberation, local populations were depressed. This may well have been the case, since birds from the tropical coast of Tamaulipas (the main source of stock), and their progeny if crossed with northern birds, would not likely have been winter-hardy. In any event this program was abandoned, attesting to its failure.

Dahlbeck (1951) reported a similar failure when gray partridges from southern Europe were imported to Sweden and mixed with the hardy northern populations. A catastrophic drop in number followed. He also relates a case of shipping in Carpathian red deer stags to "improve" the stock on an island off the Scandinavian coast. The resulting hybrids apparently were unable to stand the rigors of the northern climate, and the population on the island fell to near extinction. Following these experiences, Sweden adopted regulations to prohibit import of game birds and mammals from outside the country.

## Individual Adaptability

Certain adaptive responses to a changing environment appear to be nongenetic. Some animals seem capable of internal physiological and behavioral adjustments and as a consequence can tolerate wide fluctuations in weather and other environmental factors. A classic example would be the mourning dove (*Zenaidura macroura*).

There is no more widely distributed or successful game bird in the North American continent than the dove. Its breeding range extends from the Atlantic to the Pacific and from the prairie provinces of southern Canada to Oaxaca in southern Mexico. Two weakly differentiated subspecies are recognized—an eastern and a western race. But each of these races successfully occupies a great variety of habitats. The western mourning dove, for example, breeds in the pine zone of the mountains, in the bleakest southwestern deserts, and along the tropical Mexican coast with equal success. If there are local physiologic adaptations, no one has detected them. In our present state of knowledge we must assume that the individual birds are capable of this range of adjustment.

The same can be said for some migratory birds like the mallard, which breeds from the arctic tundra to northern Baja California and from coast to coast. There are no detectable morphologic differences among North American mallards, nor is there any hint of local physiologic variation. Not only is the mallard adaptable in the sense of occupying a variety of breeding situations, but it has shown a remarkable capacity to adjust to the changes wrought in its wintering habitat. In primitive North America the mallard wintered in the natural marshes, sloughs, and backwaters and ate aquatic foods along with other ducks. Today most of these waterways are drained or otherwise made unattractive, and during the autumn much of the remaining habitat bristles with the guns of eager duck hunters. The mallard copes successfully with this situation by several adjustments in its habits. First, it feeds at night, spending the day in safety of a waterfowl refuge or on some open bay or sandbar. Secondly, it has learned to feed on the waste grain of stubble fields—wheat and corn in the midwest, rice and kafir in Texas and California. Each day with cessation of legal shooting the birds rise in great masses and fly to the stubbles for the evening repast. For a period in the 1940s shooting closed at 4 P.M. and the flight began at 4:15. When the law was changed to permit shooting till sunset, the birds adjusted their exodus to fifteen minutes after sunset, attesting to their capacity for quick reaction to circumstances. As a result, the mallard today is by far the most abundant duck in North America.

Some other species of waterfowl have learned the same tricks. The pintail and widgeon in the west, and various geese, feed on crop residues and avoid guns during the day by flocking in safe refuges. But many of the ducks have not adjusted and are steadily decreasing in number. The redhead, canvasback, wood duck, and shoveler continue to feed in the marshes and along shorelines where they are exposed to heavy shooting. These nonadaptive species require special protection and their situation will not likely improve in the future.

Another example of an adaptable species is the coyote. Originally it occurred in modest numbers through western North America, something of a hanger-on in the range of the wolf, scavenging scraps left by this lordly predator and catching such rodents as were available. In the remaining climax forests of the Mexican highlands, where wolves still occur, coyotes are scarce or absent (Leopold, 1959). But over most of the continent where the virgin flora and fauna (including wolves) have been eliminated, conditions for the coyote have been vastly improved. The scourge of rodents that came with agriculture and with over-grazing of the western ranges, plus the carcasses of domestic stock, offered a food supply much superior to that originally available. As a result, the coyote has thrived and extended its range far to the north and east. It invaded Alaska in the 1930s and currently has moved as far east as New York State. The coyote, in other words, is an example of a successional or subclimax predator that has profited from alteration in the climax biota, as much so as the deer. Because it occasionally preys on sheep and poultry, it has been the object of intensive control efforts, more so perhaps than any other carnivore in the world. Yet so adept is the coyote at learning the tricks of avoiding guns, traps, poisons, dogs, and even airplanes (from which it sometimes is shot) that it persists over nearly all of its original and adopted range, at least in modest numbers. The coyote will be among the surviving wild species long into the future.

## Evolution of Behavior

When species like the mallard and the coyote show adaptive behavior, as described above, it is difficult to say what part of this adaptation may be genetic and what part is learned. Many mallards are shot and many coyotes are trapped or poisoned. Are these the slow-witted ones? Is man, acting as a predator himself, applying a strong selective force to hunted species that may be bringing genetic changes in the survivors? If so, nothing is known of this force, but there is room for speculation.

Consider first the mallard. Much of a duck's behavior we know to be learned. The quick adjustment of the birds to a change in legal shooting hours could hardly be based on anything but experience. This quickly could become a tradition, transmitted from older experienced birds to young ones as some migratory habits are transmitted (Hochbaum, 1955). Yet over the years many individual mallards depart from this tradition and decoy into small ponds during shooting hours. They are among the missing when the breeders migrate northward in spring. Shooting, then, may be creating a new strain of mallard that tends to conform to mass behavior patterns and is less prone to make mistaken individual judgments.

In the southeastern United States, where the bobwhite has been heavily hunted for a century or more, it is generally reported that the birds have changed their habits. Old hunters claim to remember the day when a covey, flushed before a dog, would fly 100 to 200 yards and scatter in the broom-sedge or weed fields where they could be taken easily over points. Today covies tend to fly 300 to 400 yards and to seek shelter not in open fields but in dense oak thickets. Often such coveys put a "hook" on the end of their flight, turning to the right or left after entering the woods, thus being much harder to relocate. Is this change in behavior, if true, strictly learned and transmitted from adults to young? Or is there a genetic change involved as well, favoring the birds that fly far, seek woody cover, and change directions after entering the cover?

Much of the coyote's skill in avoiding peril is clearly learned. Individuals known to have escaped from a trap or to have survived a dose of strychnine become wary and are much more difficult to capture than young, inexperienced individuals. But the innate capacity for wariness may be strengthened and bolstered over the years by constant removal of the least wary individuals.

Thus, it may be that the hunted animals are evolving under a new selective force not affecting those animals that are permitted to live without persecution. In this sense, the adaptability which may be expressed as a genetic trait—or put in other words, as the ability to learn—is not a biological constant but a shifting attribute of a species.

## Summary

There are notable differences in the response of wild animals to the sweeping changes in environment brought on by man. Some species are clearly associated with and dependent upon undisturbed climax situations, and these suffer the most from environmental change. They are here designated as nonadaptive species. The list includes all the rare or endangered species and some that have become extinct.

On the other hand, other animals adjust very well to changes in vegetation and in land use, and these on the whole persist or may even increase in abundance. Included in this group are many of the common game birds and mammals that supply the bulk of the recreational hunting in North America today. There appears to be a direct correlation between the affinity with seral or subclimax biotas and adaptability in the sense of the capacity to adjust to change.

The ability to adapt seems to involve two distinct components: (1) genetic plasticity, or the capacity for segments of a population to evolve rapidly to fit local conditions; and (2) the capacity for individuals to learn new habits of survival under altered circumstances. These cannot readily be separated, since the capacity to learn is itself a genetic trait.

## Bibliography

Aldrich, J. W. 1955. Distribution of American Gallinaceous Game Birds. *U.S. Fish and Wildl. Serv.*, Wash., D.C. Circ. 34.

Blanchard, B. D. 1941. The White-crowned Sparrows (*Zonotrichia leucophrys*) of the Pacific Seaboard: Environment and Annual Cycle. *Univ. Calif. Pub. Zool.* 46:1–178.

Buechner, H. K. 1950. Life History, Ecology, and Range Use of the Pronghorn Antelope in Trans-Pecos Texas. *Amer. Midl. Nat.* 43:257–354.

Cheatum, E. L., and C. W. Severinghaus, 1950. Variations in Fertility of White-tailed Deer Related to Range Conditions. *Trans. N. Amer. Wildl. Conf.* 15:170–90.

Dahlbeck, N. 1951. [Commentary During U.N. Conf., Fish and Widl. Res.] *Proc. U.N. Sci. Conf. on Conserv. and Utiliz. of Res. Lake Success*, N.Y. Aug. 17–Sept. 6, 1949. 7:210.

Grant, V. 1963. *The Origin of Adaptations*. Columbia Univ. Press, New York and London.

Hochbaum, H. A. 1955. *Travels and Traditions of Waterfowl*. Univ. Minn. Press, Minneapolis.

Lack, D. 1954. *The Natural Regulation of Animal Numbers*. Oxford Univ. Press.

Leopold, A. S. 1944. The Nature of Heritable Wildness in Turkeys. *Condor* 46:133–97
———. 1950. Deer in relation to Plant Succession. *Trans. N. Amer. Wildl. Conf.* 15:571–80.
———. 1959. *Wildlife of Mexico: the Game Birds and Mammals*. Univ. Calif. Press, Berkeley.
Longhurst, W. M., A. S. Leopold, and R. F. Dasmann. 1952. A Survey of California Deer Herds, Their Ranges and Management Problems. *Calif. Fish and Game, Game Bul.* 6.
Mayr, E. 1942. *Systematics and the Origin of Species*. Columbia Univ. Press, New York.
Sauer, C. O. 1950. Grassland Climax, Fire, and Man. *J. Range Mgt.* 3:16–21.
Schorger, A. W. 1955. *The Passenger Pigeon: Its Natural History and Extinction*. Univ. Wisc. Press, Madison.
Taber, R. D. 1953. Studies of Black-tailed Deer Reproduction on Three Chaparral Cover Types. *Calif. Fish and Game Bul.* 39 (2):177–86.

# 7

# The dynamics of three natural populations of the deer **Odocoileus columbianus hemionus**

*Richard D. Taber*
*Raymond F. Dasmann*

## Introduction

In the course of an investigation of the ecology of the Columbian black-tailed deer [*Odocoileus hemionus columbianus* (Richardson)] in relation to chaparral management in Lake County, California, information on the dynamics of three natural populations has been obtained. These populations occupy different habitats. The original plant community and environment were common to all, but secondary modification has created three distinct range types.

The climax plant cover consists on south slopes of *chamise chaparral* and on the north slopes of *broad sclerophyll forest* as described by Cooper (1922). These associations are dominated by fire-tolerant shrubs which either sprout from the root-crown when burned or have seeds which germinate readily following heating.

On the southerly exposures the most abundant plant is chamise

---

Contribution from Federal Aid in Wildlife Restoration Project California W-31-R and the Museum of Vertebrate Zoology, University of California. Reprinted with permission from *Ecology*, 38:233–246 (1957).

(*Adenostema fasciculatum*). Other species include yerba santa (*Eriodictyon californicum*), wedgeleaf ceanothus (*Ceanothus cuneatus*), and toyon (*Photinia arbutifolia*). Occasional burning apparently took place on these slopes during prehistoric times.

The northerly exposures were largely covered, before white settlement, with a broad-sclerophyll forest, in which the dominant trees were interior live-oak (*Quercus wislizenii*), canyon oak (*Q. chrysolepis*), California laurel (*Umbellularia californica*) and madrone (*Arbutus menziesii*). Since that time (1855–65) fires have become more frequent and at present much of the north-exposure vegetation must be called mixed or mesic chaparral, consisting of broad-sclerophyl species which have been reduced to shrub form by burning. The principal constituent of this association is interior live oak (*Quercus wislizenii*). A large assemblage of other woody plants including scrub oak (*Q. dumosa*), Eastwood manzanita (*Arctostaphylos glandulosa*), and deerbrush (*Ceanothus integerrimus*) is also found on north exposures.

At present the general cycle of events is that every five to twenty years or more a fire sweeps the region bare, except for isolated patches of brush and the bare, charred trunks of the burned shrubs. Within five to ten years the sprouts and seedlings have grown up to cover the ground with a dense thicket of shrubs. Two of the range-types that were compared in this study are the two extremes of this burn-and-recover pattern: the newly burned area on which crown-sprouts and seedlings are abundant—here called the "wildfire burn"; and the area which has not been burned for at least 10 years—here called the "chaparral." The wildfire burn is admittedly a transient stage, but while it lasts it constitutes a special type of deer range, being rich in food and poor in cover.

Another modification of chaparral is possible. Certain portions may be burned deliberately in either the early spring or late summer, and the burned areas seeded to herbaceous species. After the rains, when the seedlings emerge, those of the herbaceous species compete successfully with those of the woody species. The brush-sprouts are kept hedged by deer use or reduced by re-burning. The net result is a scattering of shrubs with the intervening area occupied by herbaceous plants. When an area is managed in this way portions are left in heavy brush for cover. This range type is now called "shrubland."

The technique of managing chaparral to create shrubland is still in the experimental stage so the areas of shrubland are limited in extent. The area most intensively studied covered about 400 map acres (horizontal projection) and was followed from before burning, in 1949, until 1955. Areas which had been longer established were also studied.

Some discussion of the ecology of these chaparral cover types has been included in a previous publication (Biswell *et al.* 1952), and a fuller description of the plant cover, with special reference to use by deer, will form the body of a later report.

The three range-types under consideration may occur in close juxtaposition, and it was in such an area that the present studies were made. The area lies about five miles southwest of Lakeport, California, between the elevations of 1500 and 2500 feet. The substrate consists of Pliocene sandstones elevated in Pleistocene times (Manning and Ogle 1950). Erosion has been rapid and the soils are correspondingly thin. They are classified principally as *Maymen*, with lesser areas of *Los Gatos*. Small pockets of still other residual soils occur on the uplands, and alluvial soils are found in the stream-bottoms. The topography is moderately steep (average slope = 22 degrees), and consists of irregular drainages and ridges with rounded tops. Rainfall averages about 28 inches annually. Winter snows usually melt soon after they fall, especially on the warmer south exposures where the deer spend most of their time in winter.

The adjustment of the vegetation to the Mediterranean climate has a profound effect upon the food-regime of the deer. Most of the shrubs are evergreen, but do not grow in the winter, when it is moist enough but too cold, or in summer, when it is warm enough but too dry. Ordinarily there is some growth in crown-sprouts in late fall, but little or none in mature shrubs. The principal growth period, then, is spring. In April and May, the deer, feeding on the new shoots, gain weight rapidly. Most fawns are born in the second and third weeks of May. The last rains usually fall in May. Increasing drought dries the herbaceous plants and the shrubs become dormant in July. The moisture and crude protein levels in the browse fall steadily all summer, and the condition of the deer, especially lactating does, falls accordingly. Occasionally there is a heavy set of acorns and the deer begin to browse these directly from the low oaks in August and September, months when otherwise the forage is of low quality. In the fall come cool weather and the first rains. The timing of the first substantial rain is especially important in the shrubland, where large quantities of annual plants appear if about one inch of rain falls. Sufficient rain for germination usually falls in October, but occasionally the first heavy precipitation may be as early as August or as late as November.

The present study began in the fall of 1948 and continued through the fall of 1955. Many people aided the investigation: of the field forces of the California Department of Fish and Game, Norman Alstot, Gordon Ashcraft, Bonar Blong, Herbert Hagen and Manley Inlay; of the State Fish and Game Laboratory, John Azevedo, Art Bischoff, Oscar Brunetti and Merton Rosen. Help in the field was also provided by Richard Genelly and Gerald Geraldson of the University of California. Aid, advice, and encouragement were constantly available from H. H. Biswell, Project

Leader, A. M. Schultz, A. S. Leopold and W. M. Longhurst, of the University of California, and William Dasmann and Robert Lassen of the Department of Fish and Game. To these, and to Glen Keithley and Harold Manley on whose land much of the work was done, and to others too numerous to list, we are indebted for help.

In addition we wish to acknowledge the critical reading of the manuscript, with suggestions for improvement, by W. Leslie Robinette, J. J. Hickey, Robert F. Scott and staff members of the Museum of Vertebrate Zoology.

## Methods and Results

The basic data from which the analyses of population dynamics were derived consist of information on individual movement, population density, reproduction according to age-class, population structure, and mortality according to age and sex. These data were gathered seasonally for the three populations under study; notes on the methods involved are given below.

### *Movement*

By studying the movements of marked deer daily, seasonally and annually, it has been found that the deer in the study region are nonmigratory and that most of them occupy home ranges with diameters of about one-half mile (does) to three-quarters of a mile (bucks). Populations occupying neighboring ranges may, therefore, be considered separately (Dasmann 1953).

### *Population density*

Censuses were made at least twice a year, first by the pellet-group-count method and later by the sample-area-count method, both of which were found to be accurate when checked against populations of known density. These methods are described elsewhere (Dasmann and Taber 1955).

Table 7–1 shows population densities observed at various times on the three types of range under study. Counts in the chaparral gave a summer density of about 30 per square mile. The wildfire burn, at the same season, showed 120 the first year after the fire. The summer density dropped to 106 the second year, 52 the third and 44 the fourth. The shrubland went from 98 the year following burning to 131 the second year and then down to about 84 the fifth and sixth years at which level the population presumably stabilized.

## *Natality*

Data on ovulation rates were obtained by collecting pregnant and post-partum does. Forty-eight does over 17 months of age were taken on the three range types. Younger does were also collected and were found not to breed under our conditions (Taber 1952).

TABLE 7–1

*Deer density (individuals per map-square-mile) on three range types*

**A. Chaparral**

| | April-June | July-October | November-December |
|---|---|---|---|
| 1949 | 28 | .. | 30 |
| 1950 | 13 | 30 | 26 |
| 1951 | .. | .. | 30 |
| Average....... | 20 | .. | 29 |

**B. Wildfire burn**

| Growing season from time of burning | Early May | Mid-July | Early December |
|---|---|---|---|
| 1 | .. | 120 | 86 |
| 2 | 75 | 106 | 56 |
| 3 | 48 | 52 | 50 |
| 4 | 32 | 44 | 32 |

**C. Shrubland**

| | Early May | Mid-July | Early December |
|---|---|---|---|
| 1951 | 88 | 131 | 88 |
| 1952 | 69 | 112 | 99 |
| 1953 | 69 | 103 | .. |
| 1954 | 53 | 85* | .. |
| 1955 | 55 | 82* | .. |

* mid-June

The presence of *corpora lutea of pregnancy* is an indicator of the successful shedding, fertilization and implantation of ova, but not every *corpus luteum* represents a developing fetus. The ratio was found to be 94 fetuses per 100 corpora lutea, based on all pregnant does collected. No evidence of abortion or resorption was noted, so this six percent loss of ova must occur before fertilization or in very early pregnancy.

Does often breed first, in this region, at the age of 17 months, although on the poorest ranges some may not breed until the age of 29 or even 41 months. No yearling does (17–24 months) were collected in the poorest range type, the chaparral. Two from the shrubland had one fetus apiece and two out of three from the new wildfire burn had one fetus each. Two yearling does were taken from an older wildfire burn and neither was pregnant. These samples are so small that little confidence can be placed in them. However, some values for the contribution of yearling does to the annual fawn-increment must be assumed in the calculations which follow, so these figures for the shrubland and new wildfire burn will tentatively be accepted. In addition, since no data are available for reproduction in yearling does in the chaparral, it will be assumed that the rate of fawn production by them is 0.5 per doe.

Among adult does the samples are larger, ranging for the three range types from 10 to 16. Fawn production in the shrubland is significantly higher than that in the chaparral, the values being 1.65 and 0.77 fawns per doe respectively. The does on the new wildfire burn show an intermediate average of 1.32 fawns apiece. Fawn production is summarized in Table 7–2.

TABLE 7–2

*Fawn production by yearling and adult does on three range types*

| Range type | Yearling Does | | | Adult Does | | |
|---|---|---|---|---|---|---|
| | Number examined | Mean number of corpora lutea | Fawns produced per doe | Number examined | †Mean number of corpora lutea | Fawns produced per doe |
| Chaparral | 0 | .... | *0.50 | 11 | 0.82 (0.42 - 1.22) | 0.77 |
| Wildfire burn after one growing season | 3 | 0.66 | 0.62 | 10 | 1.40 (1.03 - 1.77) | 1.32 |
| Wildfire burn after three growing seasons | 2 | 0.0 | 0.0 | 4 | 0.75 (0.00 - 1.54) | 0.71 |
| Shrubland | 2 | 1.0 | 0.94 | 16 | 1.75 (1.51 - 1.99) | 1.65 |

* Assumed (see text).
† Values in parentheses indicate the range with a confidence limit = .05.

## *Population structure*

Population structure was determined by observing and classifying undisturbed deer at ranges consistent with accuracy, at times during which all age and sex classes were equally visible. The most favorable seasons were late July, when fawns were at heel and before hot weather caused the bucks to seek heavy cover, and early December, when the

TABLE 7–3

*Population structure according to herd composition counts on three range types*

| Month | Number of deer classified | Adult ♂ ♂ | Yearling ♂ ♂ | Adult ♀ ♀ | Yearling ♀ ♀ | Fawns |
|---|---|---|---|---|---|---|
| **A. Chaparral** | | | | | | |
| Dec. 1949 | 124 | 72 | 23 | 100 | 38 | 85 |
| Dec. 1950 | 47 | 20 | 5 | 100 | 25 | 60 |
| Average | 85 | 46 | 14 | 100 | 31 | 72 |
| **B. Wildfire burn (December counts)** | | | | | | |
| During growing season | | | | | | |
| 1 (1948 burn) counted in 1949 | 90 | 63 | 30 | 100 | 33 | 107 |
| 4 (1948 burn) counted in 1952 | 37 | 50 | 14 | 100 | 21 | 79 |
| 1 (1950 burn) | 54 | 37 | 21 | 100 | 39 | 89 |
| 2 " | 45 | 66 | 20 | 100 | 33 | 80 |
| 3 " | 90 | 48 | 7 | 100 | 27 | 34 |
| **C. Shrubland** | | | | | | |
| July, 1951 | 82 | 70 | 27 | 100 | 46 | 127 |
| Dec., 1951 | 55 | 30 | 30 | 100 | 45 | 70 |
| May, 1952 | 43 | 38 | 31 | 100 | 44 | 56 |
| July, 1952 | 70 | 48 | 9 | 100 | 30 | 117 |
| Dec., 1952 | 62 | 39 | 9 | 100 | 30 | 92 |
| May, 1953 | 43 | 32 | 9 | 100 | 14 | 41 |
| June, 1953 | 67 | 36 | 16 | 100 | 20 | 96 |
| May, 1954 | 33 | 39 | 29 | 100 | 14 | 57 |
| June, 1954 | 53 | 56 | 25 | 100 | 25 | 125 |
| May, 1955 | 34 | 46 | 31 | 100 | 8 | 77 |
| June, 1955 | 51 | 71 | 36 | 100 | 36 | 121 |

rut had subsided and the bucks had returned to their normal level of mobility but before the antlers were shed (Dasmann and Taber 1956). In shrubland, where a more intensive study was carried out, determinations were made throughout the year. The values in Table 7–3 are expressed in terms of ratios, where the number of adult does is always taken as 100.

The low density of deer and the high density of cover made herd composition counts difficult to obtain in the chaparral. Therefore use was made of the fact that when an area of chaparral is burned, and sprouts appear, as happened in 1949 and 1950, the deer whose home ranges impinge upon the burned area congregate on it to feed and are then (December) easily observed. This population may be taken to represent the population in the chaparral, so far as structure is concerned, if one correction is made. Since the home ranges of bucks are larger than those of does, a burned area attracts proportionally more bucks than does, if only those animals whose home ranges impinge upon the burned area come to it. The counts in Table 7–3(A) are corrected for

TABLE 7–4

*The classification of deer found dead on three range types*

| Sex | 0 - 3 months | 4 - 6 months | 7 - 9 months | 10 - 12 months | 13 - 24 months | 25 - 36 months | 37 months and over | Total |
|---|---|---|---|---|---|---|---|---|
| **A. Chaparral** | | | | | | | | |
| ♂ | 0 | 7 | 6 | 0 | 8 | 3 | 9 | 33 |
| ♀ | 1 | 6 | 4 | 0 | 1 | 1 | 23 | 36 |
| ? | 16 | 0 | 0 | 0 | 0 | 0 | 0 | 16 |
| | | | | | Sub-total........ | | 85 | |
| **B. Wildfire burn—one to four years after burning** | | | | | | | | |
| ♂ | 7 | 1 | 2 | 0 | 1 | 6 | 12 | 29 |
| ♀ | 1 | 4 | 1 | 0 | 1 | 1 | 11 | 19 |
| ? | 16 | 0 | 0 | 0 | 0 | 0 | 0 | 16 |
| | | | | | Sub-total........ | | 64 | |
| **C. Shrubland** | | | | | | | | |
| ♂ | 3 | 7 | 1 | 2 | 5 | 4 | 4 | 26 |
| ♀ | 1 | 5 | 0 | 0 | 0 | 4 | 18 | 28 |
| ? | 19 | 0 | 0 | 0 | 0 | 0 | 0 | 19 |
| | | | | | Sub-total........ | | 73 | |
| | | | | | Grand total........... | | | 222 |

this; i.e., adult buck counts were reduced one-third, because average buck home-range diameter is $^{3}/_{2}$ average doe home-range diameter.

## *Mortality*

Every carcass encountered was classified, if possible, according to age, sex, season of death, cause of death and range type. Occasionally special systematic searches were made for carcasses, especially along the beds of steep-walled canyons, where dead deer were most likely to accumulate. Altogether 222 carcasses were tallied; these are listed in Table 7–4. Aging was by tooth eruption and wear (Severinghaus 1949; Moreland 1952).

A carcass-count of this sort is not a true reflection of mortality in all classes; the very young deer are under-represented, because their fragile carcasses soon disintegrate. However, if this is taken into account, the carcass tally aids the study of mortality.

The principal cause of mortality was starvation, not caused by a quantitative lack of food, but rather by a seasonal drop in the quality of available forage (Taber 1956). The effects of starvation were often augmented by exposure to unfavorable weather and occasionally by disease or parasitism. Mortality from this cause was heaviest among fawns, but adults of both sexes were also affected. Principal starvation losses occurred in fall and winter. Next in importance as a mortality factor was hunting. Almost all adult buck mortality, except for old deer, was

TABLE 7–5

*Distribution by age (in per cent) of bucks killed and taken home from three range types*

| Range type | Two years old | Three years old | Four years old and older | Number in sample |
|---|---|---|---|---|
| Chaparral | 40 | 22 | 38 | 194 |
| Wildfire burn after first growing season | 36 | 22 | *42 | 59 |
| After second growing season | 63 | 21 | 15 | 52 |
| After third growing season | 37 | 37 | 25 | 83 |
| After fourth growing season | 50 | 22 | 28 | 32 |
| Shrubland | 49 | 19 | 32 | 43 |

* Old bucks are unusually vulnerable during the first hunting season after a large wildfire has removed most of the escape cover.

caused by bullet-wounds. A few deer of other classes were also affected. The loss of protected classes to hunting varies widely. It was found in the study region to be quite low. A few deer were killed in accidents or by predators, but these are relatively unimportant mortality factors.

The hunting season extends from early August to mid-September and bucks usually become legal game during their third year and remain so for the rest of their lives. The kill at this time has a profound effect on the population structure. Ths ratios existing between two-year, three-year and older bucks in the kill, which are given in Table 7–5, are affected by three factors: the age distribution of bucks in the population; the relative vulnerability of the various age classes to hunting; and the deliberate selection of the larger (hence older) bucks by the hunter.

Where hunting pressure is only moderate and escape cover is adequate, bucks over four years old are much less vulnerable than two-year-olds, with three-year-olds being intermediate. For example: in an intensively studied area of shrubland the combined buck populations for the seasons of 1951 and 1952 were 14 two-year-olds, of which 10 were killed, and 13 older bucks, of which two were killed. This difference in vulnerability is probably above the average because little drive-hunting with dogs took place. This type of hunting results in a proportionally higher kill of old bucks.

Selective hunting, involving the deliberate attempt of the hunter to bag a trophy buck, is generally not practiced. Most hunters attempt to take the first legal target that they see. If the antlers are small and inconspicuous the hunter may not recognize the deer as legal game, or if two bucks appear together, the hunter will select the larger, but these factors do not appear to affect the kill appreciably. These remarks apply to hunters on public land, and private land with public access. All the areas studied except part of the shrubland could be so classed. However, there are certain lightly hunted areas of private land where there is a definite selection of larger (hence older) bucks by the hunter. Part of the shrubland of the present study was in this category, and for that reason the percent of old bucks in the kill (Table 7–5) is believed to be higher than would ordinarily be found.

In addition to the bucks which are killed and taken home by the hunters, there are those which are shot but not found. In the study region, hunters often shoot across canyons at running deer at long range. This fact, and the heavy cover and the hard ground, which makes tracking difficult, lead to the loss of many wounded deer. Few of these recover. Intensive studies of small known populations have shown this loss to equal about 40 percent, or slightly more, of the take-home kill.

## Reconstruction of Population Dynamics

The information presented above has been used to deduce the detailed changes taking place within each population in the course of a year or, in the wildfire burn population, four years. In order that values per square mile may be readily derived, the tables of population dynamics given in this section are constructed to represent the deer population occupying 100 square miles of the range-type in question.

### *Annual population cycle: chaparral*

From data presented above it may be seen that the deer population in the chaparral is one of low density, with low reproductive rate. Since the population is stable, this low reproductive rate is matched by a relatively low mortality. Does probably do not breed as early as they do on better ranges, and they certainly do not bear as many fawns in maturity. It is possible that some individuals breed only in alternate years. This partly relieves the population of the heaviest drain on adult vitality—gestation and especially lactation, where lactation takes place at a season of a falling nutritive plane. Mortality among fawns, though high, is not as high proportionately as it is on also fully stocked but better ranges, like shrubland. Presumably this is due in some measure to the fact that most fawns are born singly. Mortality among bucks during the hunting season is lighter than in more open range because the abundance of dense escape cover in the chaparral makes hunting difficult. The general pattern is of a population with a rather low rate of replacement and a correspondingly high life expectancy from adulthood. An annual cycle is shown in Table 7–6.

TABLE 7–6

*The population dynamics, through one year, of a deer population inhabiting 100 square miles of chaparral*

| Season | 4+ yr. ♂♂ | 3 yr. ♂♂ | 2 yr. ♂♂ | 1 yr. ♂♂ | 3+ yr. ♀♀ | 2 yr. ♀♀ | 1 yr. ♀♀ | ♂Fawns | ♀Fawns | Total |
|---|---|---|---|---|---|---|---|---|---|---|
| Late May (fawn drop)...... | 343 | 121 | 190 | 239 | 1063 | 262 | 280 | 504 | 446 | 3448 |
| Early summer loss.......... | ... | ... | ... | ... | ... | ... | ... | 180 | 114 | 294 |
| July herd composition count. | 343 | 121 | 190 | 239 | 1063 | 262 | 280 | 324 | 332 | 3154 |
| Hunting season bag......... | 39 | 27 | 49 | ... | ... | ... | ... | ... | ... | 115 |
| Crippling loss.............. | 16 | 11 | 20 | ... | ... | ... | ... | ... | ... | 47 |
| Late summer and fall loss... | ... | ... | ... | 43 | 152 | 17 | 8 | 31 | 19 | 270 |
| Dec. herd composition count | 288 | 83 | 121 | 196 | 911 | 245 | 272 | 293 | 313 | 2722 |
| Late winter loss............ | 28 | ... | ... | 6 | 80 | 13 | 10 | 54 | 33 | 224 |
| Early May population...... | 260 | 83 | 121 | 190 | 831 | 232 | 262 | 239 | 280 | 2498 |
| Late May (fawn drop)...... | 343 | 121 | 190 | 239 | 1063 | 262 | 280 | 504 | 446 | 3448 |

## Annual population cycle: wildfire burn

The general trend of population density characteristic of wildfire burns is from a high point the spring following burning to lower and lower levels thereafter. Movement is the most important element in these population changes.

When the area burns the deer move ahead of the fire and are seldom directly injured. Lack of food, however, keeps them from reoccupying their home ranges, unless there happens to have been a heavy acorn crop. The brush on the warmer slopes sprouts from the root-crown about October or November, in areas burned in late summer. The wildfire burn thereupon becomes an attractive feeding area, and large numbers of deer appear on it. The deer densities observed on burned areas can be adequately explained by supposing that the deer feeding on the burned area are those whose home ranges included part of that area before it was burned. If this is true, deer are drawn to a burned area from a peripheral zone one-half mile wide for does and three-quarters of a mile wide for bucks; *i.e.*—the peripheral zone equals one home-range diameter in width. Thus if burns are small the deer density feeding upon them is high, whereas if burns are large the density is lower, because in the former case the peripheral zone is larger in proportion to the burned area and in the latter it is smaller.

Actual observations of increase in deer density following burning correspond, for the areas representative of wildfire burns, to that to be expected if a strip one-half mile in width were burned through the country. Therefore in the reconstruction of population dynamics (Table 7–7) it has been assumed that the burn consists of a strip one-half mile in width extending indefinitely across the country, and the population shown is that inhabiting 100 square miles of this burn.

In order to follow the dynamics of a deer population through four years, as has been attempted in Table 7–7, it is necessary to use some interpolated values for productivity. It has been seen (Table 7–2) that the average fawn production per doe in the new wildfire burn is about 0.62 for yearlings and 1.32 for adults. As the wildfire burn grows older the quality of the forage produced on it drops, and this is reflected in the reproductive rate. By the fourth year after burning reproduction appears, according to evidence from limited collecting and herd composition counts, to be about the same as that in the heavy brush, namely 0.5 fawns per yearling doe and 0.77 fawns per adult doe. It remains to interpolate intermediate values for productivity during the second and third years following burning. We will assume that yearling does each produce an average of 0.58 and 0.54 fawns and adult does 1.14 and 0.95 during the second and third years following burning, respectively.

Table 7–7 traces the dynamics of a deer population inhabiting a wildfire burn from the December following burning for four full years. The population, if there had been no fire, would have been about 29 per square mile. The influx of deer whose home ranges abutted on the burned area added another 53 per square mile. These deer, which bred in early November, showed a reproductive rate somewhat greater than that found in the chaparral. Presumably this was due either to a high-quality diet of new sprouts for a few weeks prior to breeding, or to the psychic stimulation of crowding in the periphery of the burn during the rut. The former seems more probable.

With the onset of cold weather there was apparently an exodus of a portion of the population from the burned area, where there was little heavy cover, to the periphery, and these did not return. There was apparently a similar movement the following winter. In both cases the population dropped but a lack of carcasses indicated that movement rather than mortality was the cause.

Mortality on new wildfire burns was found to be very low, as might be expected from the high quality of the feed there. However, as forage quality declined, mortality increased.

By the end of the fourth year the deer population in the wildfire burn had declined to about the levels in the chaparral both in density and reproduction.

### *Annual population cycle: shrubland*

The management of chaparral to create shrubland resulted in the attraction of a heavy deer population, in much the same manner as has been described for the wildfire burn. There were, however, several important points of difference between the shrubland and the wildfire burn. In the shrubland there were areas of cover closely adjacent to feeding grounds; the wildfire burn had little cover. In the shrubland there was

TABLE 7–7

*Dynamics of a deer population inhabiting 100 square miles of wildfire burn for the first four years following burning*

| Season | 4+ yr. ♂♂ | 3 yr. ♂♂ | 2 yr. ♂♂ | 1 yr. ♂♂ | 3+ yr. ♀♀ | 2 yr. ♀♀ | 1 yr. ♀♀ | ♂Fawns | ♀Fawns | Total |
|---|---|---|---|---|---|---|---|---|---|---|
| Original population (Dec.) | 406 | 83 | 99 | 196 | 740 | 344 | 362 | 276 | 395 | 2901 |
| Influx from periphery | 406 | 83 | 225 | 444 | 1247 | 580 | 610 | 695 | 1003 | 5293 |
| Early May population | 812 | 166 | 324 | 640 | 1987 | 924 | 972 | 971 | 1398 | 8194 |
| Fawn-drop | 978 | 324 | 640 | 971 | 2911 | 972 | 1398 | 2098 | 1856 | 12148 |
| Early summer loss | ... | ... | ... | ... | ... | ... | ... | 92 | 56 | 148 |
| July herd composition count | 978 | 324 | 640 | 971 | 2911 | 972 | 1398 | 2006 | 1800 | 12000 |
| Hunting bag | 126 | 66 | 108 | ... | ... | ... | ... | ... | ... | 300 |
| Crippling loss | 50 | 26 | 43 | ... | ... | ... | ... | ... | ... | 119 |
| Late summer and fall loss | ... | ... | ... | ... | ... | ... | ... | 100 | 61 | 161 |
| Exodus | 145 | 42 | 107 | 175 | 799 | 267 | 377 | 524 | 478 | 2914 |
| Dec. herd composition count | 657 | 190 | 382 | 796 | 2112 | 705 | 1021 | 1382 | 1261 | 8506 |
| Exodus | 46 | 11 | 29 | 48 | 117 | 39 | 57 | 77 | 70 | 494 |
| Late winter loss | ... | ... | ... | ... | 7 | ... | ... | 473 | 106 | 586 |
| Early May population | 611 | 179 | 353 | 748 | 1988 | 666 | 964 | 832 | 1085 | 7426 |
| Fawn-drop | 790 | 353 | 748 | 832 | 2654 | 964 | 1085 | 2176 | 1925 | 11527 |
| Early summer loss | ... | ... | ... | ... | ... | ... | ... | 589 | 437 | 1026 |
| July herd composition count | 790 | 353 | 748 | 832 | 2654 | 964 | 1085 | 1587 | 1488 | 10501 |
| Hunting bag | 66 | 92 | 277 | ... | ... | ... | ... | ... | ... | 435 |
| Crippling loss | 26 | 37 | 111 | ... | ... | ... | ... | ... | ... | 174 |
| Late summer and fall loss | ... | ... | ... | ... | ... | ... | ... | 279 | 133 | 412 |
| Exodus | 259 | 69 | 55 | 386 | 1062 | 386 | 434 | 635 | 595 | 3881 |
| Dec. herd composition count | 439 | 155 | 305 | 446 | 1592 | 578 | 651 | 673 | 760 | 5599 |
| Exodus | 114 | ... | ... | 196 | 400 | 150 | 169 | 175 | 198 | 1402 |
| Late winter loss | 35 | ... | ... | ... | 152 | ... | ... | 218 | 42 | 447 |
| Early May population | 290 | 155 | 305 | 250 | 1040 | 428 | 482 | 280 | 520 | 3750 |
| Fawn-drop | 445 | 305 | 250 | 280 | 1468 | 482 | 520 | 1067 | 943 | 5760 |
| Early summer loss | ... | ... | ... | ... | ... | ... | ... | 380 | 230 | 610 |
| July herd composition count | 445 | 305 | 250 | 280 | 1468 | 482 | 520 | 687 | 713 | 5150 |
| Hunting bag | 64 | 93 | 93 | ... | ... | ... | ... | ... | ... | 250 |
| Crippling loss | 26 | 37 | 37 | ... | ... | ... | ... | ... | ... | 100 |
| Late summer and fall loss | ... | ... | ... | ... | 18 | ... | ... | 165 | 101 | 284 |
| Dec. herd composition count | 355 | 175 | 120 | 280 | 1450 | 482 | 520 | 522 | 612 | 4516 |
| Late winter loss | 94 | ... | ... | ... | 646 | ... | ... | 336 | 204 | 1280 |
| Early May population | 261 | 175 | 120 | 280 | 804 | 482 | 520 | 186 | 408 | 3236 |
| Fawn-drop | 436 | 120 | 280 | 186 | 1286 | 520 | 408 | 822 | 727 | 4785 |
| Early summer loss | ... | ... | ... | ... | ... | ... | ... | 261 | 174 | 435 |
| July herd composition count | 436 | 120 | 280 | 186 | 1286 | 520 | 408 | 561 | 553 | 4350 |
| Hunting bag | 44 | 34 | 104 | ... | ... | ... | ... | ... | ... | 182 |
| Crippling loss | 18 | 14 | 42 | ... | ... | ... | ... | ... | ... | 74 |
| Late summer and fall loss | ... | ... | ... | 33 | 307 | 10 | 10 | 210 | 140 | 710 |
| Dec. herd composition count | 374 | 72 | 134 | 153 | 979 | 510 | 398 | 351 | 413 | 3384 |

an abundance of herbaceous forage, which the deer eat in quantity from December through March; in the wildfire burn there was little herbaceous forage. In the shrubland the browsing-pressure of the deer was sufficient to keep many shrubs hedged and within reach; the browsing pressure on the wildfire burn was not usually sufficient for this and shrubs tended to grow beyond reach. Because of these, and perhaps other factors, the shrubland continued to support a high deer density throughout the study. This density was very high during the first years due to influx and a high survival. By the fifth and sixth year following burning, however, it had stabilized at a lower, but still substantial density. It appears that this level, about 84 per square mile in July, can be maintained for some time, so it has been taken as the basis for the annual cycle of population dynamics reconstructed in Table 7–8.

The shrubland range is fully stocked and population gain is matched by loss. The reproductive rate is high. Most yearling does produce a fawn and adult does produce an average of 1.65 fawns apiece. The mortality in these fawns is high; during the first year of life 73 percent of the males and 53 percent of the females die. Among adult does mortality is also high—about 25 percent per year. Adult bucks, because of heavy hunting and the prevalence of open country, lost 62 percent of two-year-olds, 35 percent of three-year-olds and about 20 percent of older bucks to hunting, on the average.

TABLE 7–8

*Dynamics of a deer population inhabiting 100 square miles of shrubland*

| Season | 4+ yr. ♂♂ | 3 yr. ♂♂ | 2 yr. ♂♂ | 1 yr. ♂♂ | 3+ yr. ♀♀ | 2 yr. ♀♀ | 1 yr. ♀♀ | Fawns ♂♂ | Fawns ♀♀ | Total |
|---|---|---|---|---|---|---|---|---|---|---|
| Late May (fawn-drop) | 434 | 182 | 482 | 556 | 1935 | 706 | 858 | 2042 | 1810 | 9005 |
| Early summer loss | ... | ... | ... | ... | 29 | ... | ... | 350 | 224 | 603 |
| July herd composition count | 434 | 182 | 482 | 556 | 1906 | 706 | 858 | 1692 | 1586 | 8402 |
| Hunting season bag | 60 | 46 | 214 | ... | ... | ... | ... | ... | ... | 320 |
| Crippling loss | 24 | 19 | 86 | ... | ... | ... | ... | ... | ... | 129 |
| Late summer and fall loss | ... | ... | ... | 20 | 115 | 37 | 20 | 856 | 547 | 1595 |
| Dec. herd composition count | 350 | 117 | 182 | 536 | 1791 | 669 | 838 | 836 | 1039 | 6358 |
| Late winter loss | 33 | ... | ... | 54 | 414 | 111 | 132 | 280 | 181 | 1205 |
| Early May population | 317 | 117 | 182 | 482 | 1377 | 558 | 706 | 556 | 858 | 5154 |
| Late May (fawn-drop) | 434 | 182 | 482 | 556 | 1935 | 706 | 858 | 2042 | 1810 | 9005 |

Among yearling (12- to 24-month-old) deer not all the loss to the population was due to mortality. It is usual for a dispersal movement of yearlings to occur about fawning-time, when the adult does become antagonistic toward their previous offspring. In a large, homogeneous area, of uniform deer density, this movement of yearlings from a local area would be balanced by others moving into it. A small area of dense population, however, would tend to lose more than it gained; this is the situation on the 400-acre shrubland study area. Here the average loss due to emigration during the four year period 1951–55 amounted to about 6 percent of the yearling class of bucks and 38 percent of the yearling class of does. It seems probable that this higher emigration rate in yearling females is connected with the tendency toward spacing and mutual antagonism among breeding does (Dasmann and Taber 1956).

Since this measurement of emigration was made on a small area of dense population, it is large. In Table 7–8, which is an attempt to describe shrubland population dynamics for a larger area, a yearling female emigration of 15 percent is assumed.

## Summary

Population dynamics have been reconstructed for three natural populations of the Columbian black-tailed deer [*Odocoileus hemionus columbianus* (Richardson)] which differed from each other in habitat. The three habitats, chaparral, shrubland and wildfire burn, were all modifications by man of the chaparral association of the North Coast Range of California. Chaparral is a densely-growing association of shrubs; shrubland is a shrub-herb interspersion; wildfire burn is chaparral recently burned, where the shrubs are sprouting from the root-crown. These populations were compared for stability, density, structure, reproduction, mortality and movement. The principal causes of mortality were hunting (adult bucks) and starvation (other classes). The details of population gain and loss through one typical year (four years for wildfire burn) were reconstructed from these data. Some characteristics of the three populations are shown in Table 7–9.

TABLE 7–9

| | Chaparral | Shrubland | Wildfire burn (First year following burning) |
|---|---|---|---|
| Population density (Deer per square mile) | | | |
| July.......... | 33 | 84 | 112 |
| December.......... | 27 | 64 | 86 |
| Reproduction (fawns per 100 adult ♀♀)...... | 0.77 | 1.65 | 1.32 |
| Importance of movement in population gain or loss.......... | none | little (yearling does) | great (all classes) |

## References

Biswell, H. H., R. D. Taber, D. W. Hedrick, and A. M. Schultz. 1952. Management of chamise brushlands for game in the north coast region of California. *Calif. Fish and Game*, 39:453–484.

Cooper, W. S. 1922. The broad-sclerophyll vegetation of California. *Carnegie Inst. of Wash.*, Pub. 319. 14 pp.

Dasmann, R. F. 1953. Factors influencing movement of nonmigratory deer. *Western Assn. State Game and Fish Comm.*, 33:112–116.

Dasmann, R. F., and R. D. Taber. 1955. A comparison of four deer census methods. *Calif. Fish, and Game*, 41:225–228.

———. 1956. Determining structure in Columbian black-tailed deer populations. *Jour. Wildl. Mangt.*, 20:78–83.

Manning, G. A., and B. A. Ogle. 1950. The geology of the Blue Lake quadrangle. *Calif. Dept. Nat. Res., Div. Mines, Bull.* 148. 36 pp.

Moreland, R. 1952. A technique for determining age in black-tailed deer. *Western Assn. State Game and Fish Comm.*, 32:214–219.

Severinghaus, C. W. 1949. Tooth development and wear as criteria of age in white-tailed deer. *Jour. Wildl. Mangt.*, 13:195–216.

Taber, R. D. 1952. Studies of black-tailed deer reproduction on three chaparral cover types. *Calif. Fish and Game*, 39:177–186.

———. 1956. Deer nutrition and population dynamics in the North Coast Range of California. *North Amer. Wildl. Conf. Trans.*, 21:159–172.

———, and R. F. Dasmann. 1954. A sex difference in mortality in young Columbian black-tailed deer. *Jour. Wildl. Mangt.*, 18:309–315.

# 8

# History, food habits and range requirements of the woodland caribou of continental North America

*Alexander Thom Cringan*

The eastern and western woodland caribou, *Rangifer tarandus caribou* (Gmelin) and *Rangifer tarandus sylvestris* (Richardson) (Cringan, 1956) are two of the generally accepted thirteen native races of North American caribou recorded within recent times. Their populations have declined greatly since the settlement of North America by Europeans, as have the populations of all other native caribou. The purpose of this paper is to review the history of these two forms and their present status, and to present the results of a food habits and range study of the woodland caribou of the Slate Islands, Lake Superior, Ontario, which was done in 1949.

Woodland caribou formerly ranged from Prince Edward Island and Nova Scotia to western Alberta or British Columbia, south into New York, New Hampshire, Vermont and Maine in the east, and Minnesota, Wisconsin and Michigan in the Great Lakes Region, north to southern Ungava in the east and the North West Territories in the West. The

Reprinted with permission from *Trans. 22nd North American Wildlife Conference,* pp. 485–501 (1957) courtesy of Wildlife Management Institute, Washington, D. C. Field investigations were under the sponsorship of the Research Council of Ontario and the Carling Conservation Club. The author is currently affiliated with the Department of Zoology, University of Guelph, Guelph, Ontario.

morphological distinctions between these two supposed races are unclear, as is the line separating their ranges. Indeed, neither the structural differences nor the geographic ranges of races of North American caribou in general are well worked out or commonly agreed upon. For this reason, I will not further distinguish between the eastern and western woodland caribou.

## History

*Eastern Range* The woodland caribou formerly occurred throughout a large area south of the St. Lawrence River, where its range included the Gaspé Peninsula of Quebec, New Brunswick, Nova Scotia, Prince Edward Island, extreme northern New Hampshire and Vermont, and extreme northeastern New York.

Caribou permanently disappeared from Prince Edward Island between 1672 and 1873 (Adams, 1873), and from New York prior to 1800 (De Kay, 1842). They occasionally appeared in Vermont until 1840 and in New Hampshire until about 1865 (Seton, 1953). The caribou population in Maine fluctuated considerably (G. M. Allen, 1942), but decreased noticeably before 1900 (Palmer, 1938), and was exterminated about 1916 (Seton, 1953). The only caribou reliably reported in Maine since then was in 1946 (Palmer, 1949).

The woodland caribou began to decline in Nova Scotia before 1900, and disappeared from the mainland of this Province by 1915 and from Cape Breton Island by 1924 (Anderson, 1946). The population in New Brunswick was decreasing noticeably by 1915 (Anderson, 1939), but a few animals remained until about 1927.

There has been a permanent population of caribou on the Gaspé Peninsula. It was much more numerous between 1900 and 1915 than at present (Moisan, 1956a). Despite this decline, there are still about 700 woodland caribou occupying the remaining 400 square miles of winter range, suggesting that the population is still high in relation to the amount of habitat available.

In Quebec to the north of the St. Lawrence River, woodland caribou declined first in the northern part of their range, east of James Bay, late in the 19th century, and a little later in the Laurentians to the south (Low, 1896; Riis, 1938).

Reports of eastern woodland caribou in Labrador are questionable owing to possible confusion with the Ungava caribou *Rangifer tarandus caboti* (G. M. Allen).

*Central Range* In the central part of its range, the woodland caribou originally occupied the northern half of Minnesota, extreme northern

Wisconsin, the Upper Peninsula of Michigan, and all of Ontario north of the French and Mattawa Rivers. Its status in these areas appears to have changed little until after 1800. By 1850, it had disappeared from Wisconsin (Hoy, 1882), and decreased in Minnesota, Michigan, and the southeastern part of its range in Ontario.

Caribou were absent from much of their former range in Minnesota and Michigan in 1900, but still occupied most of their original range in Ontario, although in reduced numbers in the eastern half of the Province. Woodland caribou generally diminished after 1900 in this section, with the possible exception of the Patricia Portion of Ontario for which specific information was lacking until 1920 or later.

The last caribou reported from the Upper Peninsula of Michigan was in 1906, and from Michigan (Isle Royale) in 1931 (Riis, 1938). Caribou remained in the Superior National Forest, Minnesota, until 1925 (Adams, 1926) and in northwestern Minnesota until 1942 (Nelson, 1947). A caribou was seen near Manitou Rapids, Minnesota, in the winter of 1954–55 (L. Magnus, personal communication, 1955).

In Ontario, woodland caribou disappeared from 50,000 square miles of range to the east of Lake Superior and 30,000 square miles to the northwest of Lake Superior between 1900 and 1950. About 250,000 square miles of the original range of woodland caribou in Ontario remains occupied, although in many places the distribution is spotty.

*Western Range* Woodland caribou formerly occurred in all of Manitoba but the extreme southern and northern parts, the northern half of Saskatchewan, northern Alberta, the southwestern part of the North West Territories, and possibly in northeastern British Columbia. It decreased markedly in Manitoba but did not experience such range reductions as in Ontario. After reaching a low in this Province between 1930 and 1950, it now seems to be increasing (G. W. Malaher, personal communication, 1957). The history of woodland caribou in the rest of the western range is poorly recorded, but apparently the species has declined generally since settlement.

## Present Status of Woodland Caribou

The eastern and western woodland caribou are now confined for practical purposes to the Provinces of Quebec, Ontario, Manitoba, Saskatchewan and Alberta, and the North West Territories.

In Quebec, there are about 700 woodland caribou on the Shickshock Mountains in the Gaspé Peninsula, and possibly between 1,500 and 2,000 to the north of the St. Lawrence River between Anticosti Island and the Saguenay River (G. Moisan, personal communication, January 22, 1957). An additional number occur in the vicinity of James Bay and further north (Banfield, 1949). Woodland caribou are thought to be increasing in Quebec.

Woodland caribou currently occur right across northern Ontario, in the west, south to a line from Minaki to Savant Lake to the Black Bay Peninsula of Lake Superior, and in the east, south to a line from Pukaskwa on Lake Superior to Swastika. Along the southern edge of this range scattered isolated herds of caribou occur. Distribution is more continuous to the north, and along Hudson Bay, scattered voids occur. Distribution varies slightly from year to year. The total of estimated caribou populations of forest districts in Ontario was 7,200 in 1953–54 (Cringan, 1956). There are few places in Ontario where estimates exceed 1 caribou per 20 square miles over large areas (of several thousand square miles), yet there are some small areas such as the 15-square mile Slate Islands with estimated densities of between 2 and 3 woodland caribou per square mile. The woodland caribou in Ontario has undoubtedly increased since 1950, on the basis of the best information available at that time (de Vos and Peterson, 1950).

The woodland caribou is currently increasing in Manitoba, according to G. W. Malaher, Director of Game and Fisheries (personal communication, 1957), who thinks there now may be approximately 4,000 or more woodland caribou in that Province. I have no information on the current status of woodland caribou in Saskatchewan.

J. G. Stelfox (personal communication, January 11, 1957) estimated that there are between 700 and 1,000 woodland caribou in Alberta, and remarked that the population was reasonably stable.

There are a few recent records of woodland caribou in the North West Territories (Banfield, personal communication, February 1, 1957), but no estimates of populations in this area are at hand. Reports of woodland caribou in British Columbia (Banfield, 1949) are subject to some doubt, as the form is not considered to occur in that Province (J. Hatter, personal communication, January 10, 1957).

The total of these estimates is between 14,000 and 15,000 woodland caribou, to which estimates for parts of Quebec, Saskatchewan, and the North West Territories must be added. It seems probable that complete estimates for Canada would total between 20,000 and 25,000 animals. Between two-thirds and three-quarters of these are in Manitoba, Quebec, and Ontario, in which provinces a phase of population increase is being experienced.

## Factors Influencing Populations of Woodland Caribou

Emigration has been postulated as a cause of woodland caribou declines in the Maritimes (Moisan, 1956a), New England (Allen, 1942), Quebec (T. Fortin in personal communication to L. A. Richard, 1940) and

Ontario (Millais, 1907). Since confirming reports of immigration into neighboring areas are lacking, and increasing moose and deer populations were usually associated with such declines, it seems likely that in these cases caribou decreased because of changes in ecological conditions rather than as a result of emigration.

Shooting has been frequently listed as a cause of woodland caribou declines. The gregarious instincts, nomadism, and low reproductive potential of the form render it very vulnerable to overexploitation. Kills of 13 caribou in 3 days on Caribou Island, Ontario (Quaife, 1921), 120 caribou in a day in Nova Scotia (Anderson, 1939), and 2,400 in a winter on Manitoulin Island, Ontario (Blair, 1911), prove that populations were sometimes overexploited. As there was no real check on caribou hunting through law enforcement until late in the 19th century, it seems likely that shooting was formerly an important factor. Yet since then, without exception, law enforcement itself has failed to prevent the ultimate extinction of woodland caribou in areas in which the range has been drastically modified by man, such as Nova Scotia, New Brunswick, New England, Michigan, Minnesota, and certain parts of other Canadian Provinces.

There is no adequate evidence that predation, disease or parasitism could account for general declines in woodland caribou populations. An epizootic may have accompanied one decline in the Gaspé Peninsula (Moisan, 1956 a), and an infestation of warble flies (probably *Oedemagena tarandi* Linnaeus) may have been associated with a decline in central Quebec (Robertson, personal communication to L. A. Richard, 1939). Similarly, there is no adequate evidence establishing the effect of weather and climate on woodland caribou populations. Still, none of these factors should be disregarded in future investigations.

Lasting diminution and eradication of woodland caribou populations have been almost always associated with important changes in the environment. The woodland caribou seems to require extensive areas of mature coniferous or mixed-wood forest. In Nova Scotia, the final decline coincided with reduction of the areas of virgin forest to less than 2 percent of the total forested area (Fernow, 1912). The remaining woodland caribou range in the Gaspé Peninsula is being reduced through encroachment by logging, mining, and forest fires (Moisan, 1956 a). In Ontario there is a relationship between presence of caribou and remaining amount of mature forest as shown in Table 8–1.

Intergeneric competition with moose and deer has also been suggested as a factor in the decline of caribou. Since moose and deer are mammals of the early successional stages, caribou of the climax, this competition is probably unimportant. Intrageneric competition, such as between the woodland and barren-ground caribou on the latter's winter range (Clarke, 1940) logically would have a much greater effect on woodland caribou populations.

Range, and the factors affecting it, forest fires, logging, settlement and the caribou themselves, seemed likely to hold the answer to the great woodland caribou decline of 1750–1950. Therefore I selected a study of woodland caribou range and food habits as the subject of graduate research.

TABLE 8–1.

*Occurrence of woodland caribou in relation to forest cover in Ontario*

| Forest District or Wildlife Management District | Per Cent of Productive Forest in Mature Coniferous or Mixedwood and Muskeg | Status of Woodland Caribou | Square Miles Per Estimated Woodland Caribou |
|---|---|---|---|
| **Patricia West** | **67% or more** | **uncommon** | **20-25** |
| **Patricia Central** | **67% or more** | **uncommon** | **20-25** |
| **Patricia East** | **67% or more** | **scarce to uncommon** | **?** |
| **Kapuskasing** | **63.2%** | **scarce** | **40** |
| **Cochrane** | **54.5%** | **very scarce** | **300** |
| **Geraldton** | **53.2%** | **scarce** | **50** |
| **Gogama** | **48.6%** | **absent** | **....** |
| **Chapleau** | **47.6%** | **absent** | **....** |
| **White River** | **41.1%** | **scarce** | **60-70** |
| **Sioux Lookout** | **39.8%** | **uncommon** | **20** |
| **Kenora** | **31.4%** | **very scarce** | **?** |
| **Port Arthur** | **31.3%** | **very scarce** | **300** |
| **Sault Ste. Marie** | **25.2%** | **absent** | **....** |
| **Fort Frances** | **24.3%** | **absent** | **....** |
| **North Bay** | **24.0%** | **absent** | **....** |
| **Sudbury** | **20.1%** | **absent** | **....** |
| **Timiskaming** | **18.4%** | **very scarce (1 in district)** | |

## The Study Area and Methods

Field work was carried out in 1949 on the Slate Islands, Lake Superior, because the only ungulate present was the woodland caribou, the population was sufficiently isolated to be relatively unaffected by egress and ingress, and the population was sufficiently dense (between 2 and 3 caribou per square mile) for study purposes.

The Slate Islands are a group of 8 islands of 30 acres or more totaling some 15 square miles in area, situated in Lake Superior 8 miles south of the village of Jackfish. They are made up of Prekeeweenawan and Keeweenawan rocks (Parsons, 1918). The topography is rugged with high hills and steep cliffs being prominent. They are within the Superior Section of Halliday's (1937) Boreal Forest Region.

White birch, *Betula papyifera* Marsh., and balsam fir, *Abies balsamea* (L.) Mill. are the commonest trees of the islands. The flora of the islands is interesting as it includes certain relict species like *Dryas integrifolia* Vahl and *D. Drummondii* Richards.

The vegetation of the Slate Islands has been influenced by an extensive fire between 50 and 70 years ago, two periods of logging activity, the first coinciding with the fire and the second about 1930 (A. L. Parsons, personal communication, 1950) and in recent years by the actions of woodland caribou.

The only species of mammals recorded on the Slate Islands in 1949 were woodland caribou; varying hare, *Lepus americanus;* muskrat, *Ondatra zibethica;* red-backed vole, *Clethrionomys gapperi;* meadow vole, *Microtus pennsylvanicus;* colored fox, *Vulpes fulva;* short-tailed weasel, *Mustela eriminea;* and little brown bat, *Myotis lucifugus.* Evidence of a former population of beaver, *Castor canadensis,* was noted. The history of woodland caribou on the islands is poorly recorded. Parsons (1918; personal communication, 1950) did not see caribou there in 1918. Possibly the current population is of fairly recent origin.

The principal technique used to study range and food habits of the woodland caribou of the Slate Islands was a composite range study technique. Among information recorded was the following.

### General Forest Description

Subjective appraisals of forest type, height of trees, crown density, forest age and site were made on the basis of classes set forth by Seely (1949). Tree species present, aspect, slope and moisture were also recorded. Six forest types, based on two moisture regimes and three compositions, were subsequently recognized.

### Browse Analysis

Winter browsing by woodland caribou was sampled by using the Aldous system to measure availability and utilization of woody browse on 495 plots each of 1/100th of an acre in area (Aldous, 1944; Aldous and Krefting, 1946). A stratum between two and ten feet from the ground was considered to have browse available to caribou in the winter. A weighted average degree of browsing was used in analyzing the results. Individual densities were multiplied by their respective browsing values, the results totalled and divided by the sum of densities, rather than the browsing values being totalled and divided by the number of occurrences, to compute the unweighted average degree of browsing used by Aldous and Krefting.

### Tree Lichens

Species of lichens growing on branches of trees such as old man's beard, *Usnea* spp. and oakmoss, *Evernia prunastri,* were grouped together and their abundance on 1/100th acre plots and utilization subjectively estimated, using the same coefficients as in the Aldous browse analysis system. Abundance was appraised on the basis of amounts of tree lichens just beyond reach of the caribou in relation to the densest stands encountered. Utilization was judged on the basis of the difference in amounts of tree lichens within reach of caribou and beyond their reach that appeared explainable due to grazing.

## Ground Lichens

Lichens growing on the ground were tallied in three groups, as reindeer mosses (*Cladonia rangiferina, C. alpestris* and similar forms), foliose lichens (*e.g., Umbillicaria* spp., *Peltigera* spp.), and other ground lichens. Occurrence and areal density of these forms on 1/100th acre quadrats were recorded. Where the density of reindeer mosses exceeded 5 percent, degree of grazing was also estimated, again using the Aldous coefficients for utilization.

Additional information on caribou food habits was gained through observation of feeding animals and through detection of grazing of herbs and leaf-browsing of shrubs.

# Winter Food Habits of The Slate Islands Caribou

Results of the Aldous browse survey are given in Table 8–2. No species of woody forage was more than lightly utilized. As the population density of caribou was high at the time, it appears that browse is relatively unimportant in the winter diet of the Slate Islands caribou.

TABLE 8–2.

*Woodland caribou winter browse analysis Slate Islands—495 plots, 1949*

| Species | Frequency Per Cent of Plots Present | Mean Density | Weighted Average Degree Browsing | Utilization Factor | Per Cent of Woody Food Eaten | Per Cent of Browse Available |
|---|---|---|---|---|---|---|
| Mountain maple | 38 | 6.5 | 4.5 | 27.9 | 41.5 | 11.7 |
| Mountain ash | 35 | 2.1 | 4.2 | 8.7 | 13.1 | 3.7 |
| Willows | 22 | 2.1 | 3.1 | 6.6 | 9.9 | 3.8 |
| Red-osier dogwood | 26 | 3.7 | 1.5 | 5.5 | 8.2 | 6.6 |
| Highbush cranberry | 20 | 1.0 | 3.9 | 4.0 | 6.0 | 1.8 |
| Red-berried elder | 24 | 1.3 | 3.0 | 3.9 | 5.8 | 2.3 |
| Ground hemlock | 14 | 3.0 | 1.1 | 3.2 | 4.8 | 5.3 |
| Red cherry | 7.5 | 0.73 | 2.4 | 1.8 | 2.6 | 1.3 |
| White birch | 52 | 3.7 | 0.47 | 1.7 | 2.6 | 6.5 |
| Juneberry | 10 | 0.52 | 2.2 | 1.1 | 1.7 | 0.9 |
| Balsam fir | 86 | 18.1 | 0.04 | 0.72 | 1.1 | 32.3 |
| Trembling aspen | 12 | 0.69 | 0.81 | 0.56 | 0.8 | 1.2 |
| Salmonberry | 16 | 2.5 | 0.16 | 0.41 | 0.6 | 4.5 |
| Balsam poplar | 2.0 | 0.20 | 1.5 | 0.30 | 0.4 | 0.4 |
| Bush honeysuckle | 7.3 | 0.46 | 0.22 | 0.10 | 0.2 | 0.8 |
| Ribes spp. | 11 | 0.66 | 0.15 | 0.10 | 0.2 | 1.2 |
| Raspberry | 13 | 0.62 | 0.08 | 0.05 | 0.1 | 1.1 |
| White cedar | 3.4 | 0.32 | 0.16 | 0.05 | 0.1 | 0.6 |
| Buffaloberry | 1.0 | 0.10 | 0.50 | 0.05 | 0.1 | 0.2 |
| Blueberry | 0.2 | 0.01 | 5.0 | 0.05 | 0.1 | tr. |
| Rose | 2.8 | 0.14 | 0.36 | 0.05 | 0.1 | 0.3 |
| Black spruce | 18 | 1.6 | | | | 2.9 |
| Speckled alder | 12 | 4.0 | | | | 7.1 |
| Mountain alder | 1.2 | 0.16 | | | | 0.3 |
| White spruce | 15 | 1.2 | | | | 2.2 |
| Labrador tea | 1.3 | 0.28 | | | | 0.5 |
| Common juniper | 0.6 | 0.08 | | | | 0.1 |
| Tamarac | 0.4 | 0.02 | | | | tr. |
| Sweet Gale | 0.4 | 0.28 | | | | 0.5 |
| Andromeda | 0.2 | 0.01 | | | | tr. |
| Leatherleaf | 0.2 | 0.06 | | ... | | 0.1 |

Mountain maple, *Acer spicatum* Lam. was the principal local woody winter food of Slate Islands caribou at the time of the study, and was followed by mountain ash, *Pyrus americana* (Marsh.) DC, willows, *Salix* spp., red-osier dogwood, *Cornus stolonifera* Michx. and highbush cranberry, *Viburnum rafinesquiana* Shultes. These five species together contributed about 75 percent of the winter browse consumed. At least 16 other species of woody plants were also eaten in the winter.

Reindeer mosses were heavily grazed. The mean degree of grazing of reindeer mosses within dry series types, which occupied 35 percent of the island and supported 90 percent of the reindeer mosses, was 50 percent. This suggests a lower-than-actual degree of utilization, for 70 percent was the maximum recordable grazing for any one plot, yet many stands were reduced to 10 percent and less of their potential volume of reindeer mosses through the action of many years' grazing and accompanying trampling.

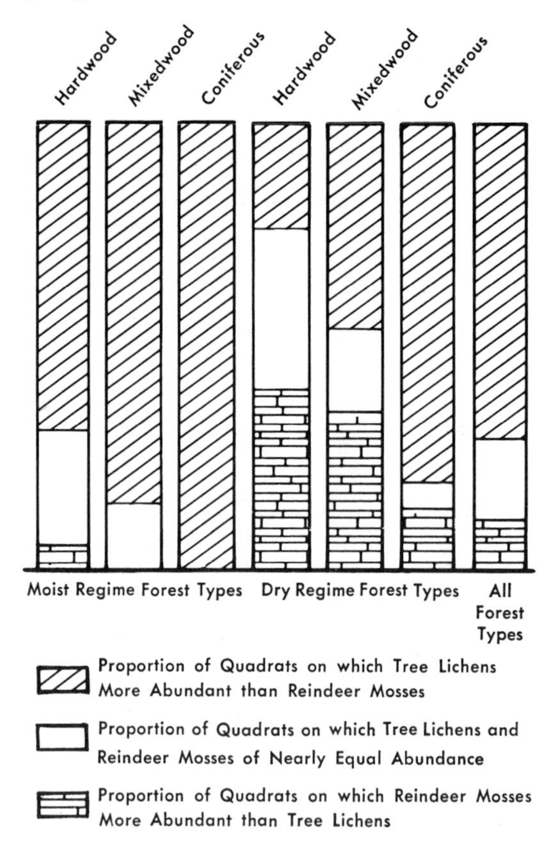


FIGURE 8–1. Relative abundance of tree lichens and reindeer mosses within principal forest types on the Slate Islands, Lake Superior, Ontario.

Utilization of tree lichens was also heavy, averaging about 41 percent out of a possible maximum of 70 percent. Utilization of tree lichens was heavy in all mixed-wood and coniferous types.

The utilization of woody foods by Slate Islands caribou was so low contrasted to that of arboreal and terrestrial lichens, that clearly the supply of lichens is critically important, while that of browse is relatively unimportant. No measure contrasting the volumes or weights of browse and lichens consumed in winter can be made from the results of this study.

It is more difficult to assess the relative importance of reindeer mosses and tree lichens. Both were heavily utilized, but as reindeer mosses grow in a nearly two-dimensional environment, and tree lichens in one which is three-dimensional, a direct contrast of abundance and utilization is not possible. To solve this problem, I examined all plot tally sheets to determine whether tree lichens or reindeer mosses were most abundant on the individual plot. If any doubt existed, they were classified as plots where the two groups were of nearly equal abundance. Figure 8–1 shows the result of this procedure. Tree lichens were more abundant than reindeer mosses in 70 percent of all plots, while reindeer mosses exceeded tree lichens on only about 10 percent of all plots! The only forest type in which reindeer mosses were generally more abundant than tree lichens was the dry hardwood type. Because of availability, it appears that tree lichens were in, 1949, much more important than reindeer mosses in the diet of woodland caribou of the Slate Islands.

## Spring and Summer Food Habits of the Slate Islands Caribou

In the spring, caribou on the Slate Islands paw for roots and shoots of herbs, and for mosses, lichens and fungi. In the summer, they eat leaves of deciduous shrubs, herbs, lichens and some aquatic plants, possibly including algae.

Foods most frequently taken in the spring included the large-leafed aster, *Aster ciliolatus* Lindl.; bunchberry, *Cornus canadensis* L.; mosses and lichens. Herbs most frequently grazed in the summer included large-leafed aster; sarsaparilla, *Aralia nudicaulis* L.; Ferns; and fireweed, *Epilobium angustifolium* L. The shrubs most frequently leaf-browsed in summer were highbush cranberry, mountain ash, mountain maple and bush honeysuckle, *Diervilla Lonicera* Mill.

## Environmental Factors Governing Production of Lichens

On the Slate Islands, reindeer mosses grow best in dry, open places. They occur more than twice as frequently in dry regime forest types than in moist regime forest types of the same composition and are more than ten times as abundant in the dry types. These relationships are shown in Table 8–3. Relationships between reindeer moss and tree crown density in dry regime forest types are shown in Figure 8–2.

TABLE 8–3.

*Lichen characteristics of principal Slate Islands forest types*

| | Forest Type | | | | | |
|---|---|---|---|---|---|---|
| | Medium Moisture Regime | | | Dry Moisture Regime | | |
| Reindeer Mosses | Hardwood | Mixedwood | Coniferous | Coniferous | Mixedwood | Hardwood |
| Per cent of plots with reindeer mosses | 31 | 38 | 27 | 67 | 79 | 74 |
| Per cent of plots where reindeer moss density 5 per cent or more | 1.5 | 0.5 | 3.3 | 35 | 41 | 38 |
| Mean density of reindeer mosses | 0.4 | 0.4 | 0.4 | 4.9 | 10 | 4.4 |
| Mean degree of grazing of reindeer mosses | ? | ? | ? | 54 | 59 | 26 |
| Per cent of islands' area occupied by type | 14.4 | 44.4 | 6.5 | 7.3 | 18.3 | 9.1 |
| Per cent of all reindeer mosses grown within type | 2.1 | 6.7 | 0.9 | 12.4 | 64.2 | 13.7 |
| **Tree Lichens** | | | | | | |
| Mean density of tree lichens—Per cent | 6.0 | 25 | 47 | 34 | 20 | 4.9 |
| Mean degree of browsing of tree lichens—Per cent | 20 | 52 | 46 | 42 | 40 | 3 |
| Per cent of all tree lichens grown within type—Per cent | 4 | 52 | 14 | 11 | 17 | 2 |

Mean density of reindeer mosses exceeded 10 percent in dry regime stands with crown densities of less than 45 percent, but was only a fifth as abundant where the tree crown density exceeded 65 percent. A vegetation survey of tundra winter range of woodland caribou in the Gaspé Peninsula showed that ground lichens occupied about 12 percent of the area (Moisan, 1956 b), a condition not unlike that in the more open dry regime forest types of the Slate Islands. The influence of the caribou themselves on the abundance of reindeer mosses could not be measured, although it was obviously great.

Abundance of tree lichens depends on forest composition, moisture conditions, tree crown density and age of the forest. Table 8–3 shows that the ratio of tree lichen densities in coniferous, mixed-wood and hardwood stands is about 7.5 : 4 : 1, and that the densities are about one-quarter greater in medium regime types than in dry regime types of the same composition. Abundance of tree lichens varies directly with age of the forest and tree crown density, the latter being shown for the six principal forest types in Figure 8–3. A further factor influencing supplies of tree lichens available to caribou is the past browsing by caribou. For this reason, availability may be very low even if excellent supplies of tree lichens are to be found in the tree tops. Both reindeer mosses and tree lichens require long periods of time to attain high densities, and both grow slowly, even when abundant.

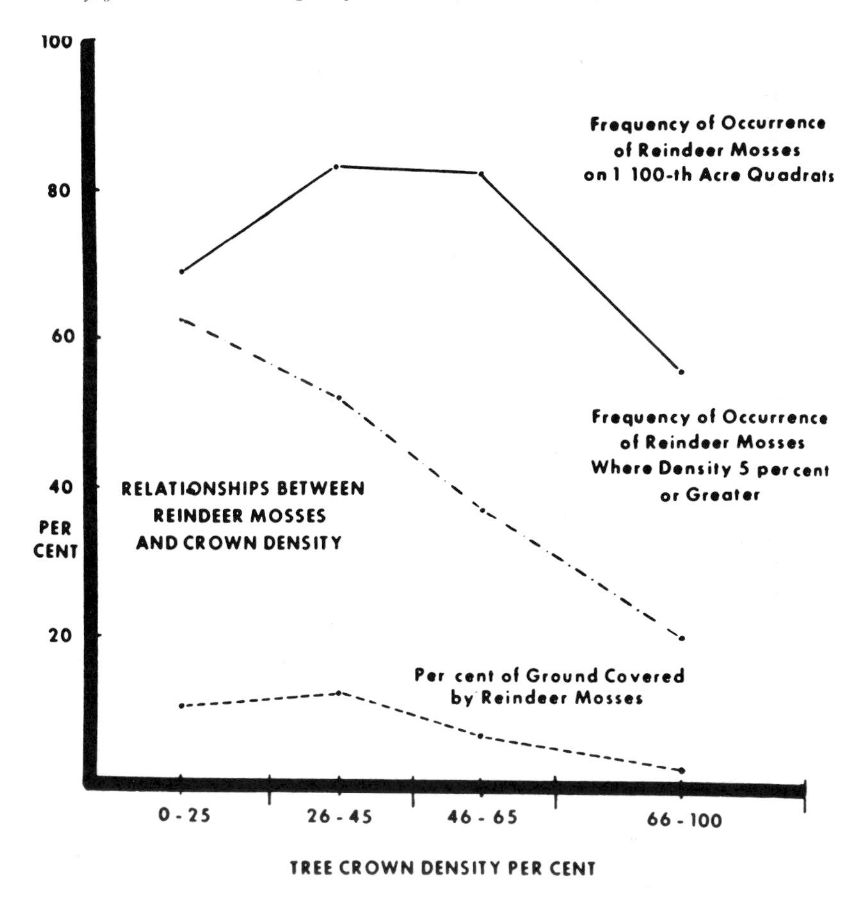


FIGURE 8–2. Relationship between frequency and areal density of reindeer mosses and tree crown density within dry regime forest types of the Slate Islands, Lake Superior, Ontario.

## Range Requirements of the Woodland Caribou

The results of the Slate Islands investigation suggest that a supply of lichens as winter food is of critical importance to woodland caribou and that tree lichens may be more important than reindeer mosses. In addition, findings point to some of the major environmental factors controlling production of lichens. The results therefore contribute to an understanding of woodland caribou range requirements.

The distribution of woodland caribou is closely related to that of the northern coniferous forest. It is generally, although not always, associated with climax stands. Climax stands of that forest are typically uneven-aged and usually coniferous or mixed in composition, with spruces, balsam fir and white birch being the principal trees.

There are some communities in this forest which have dense stands

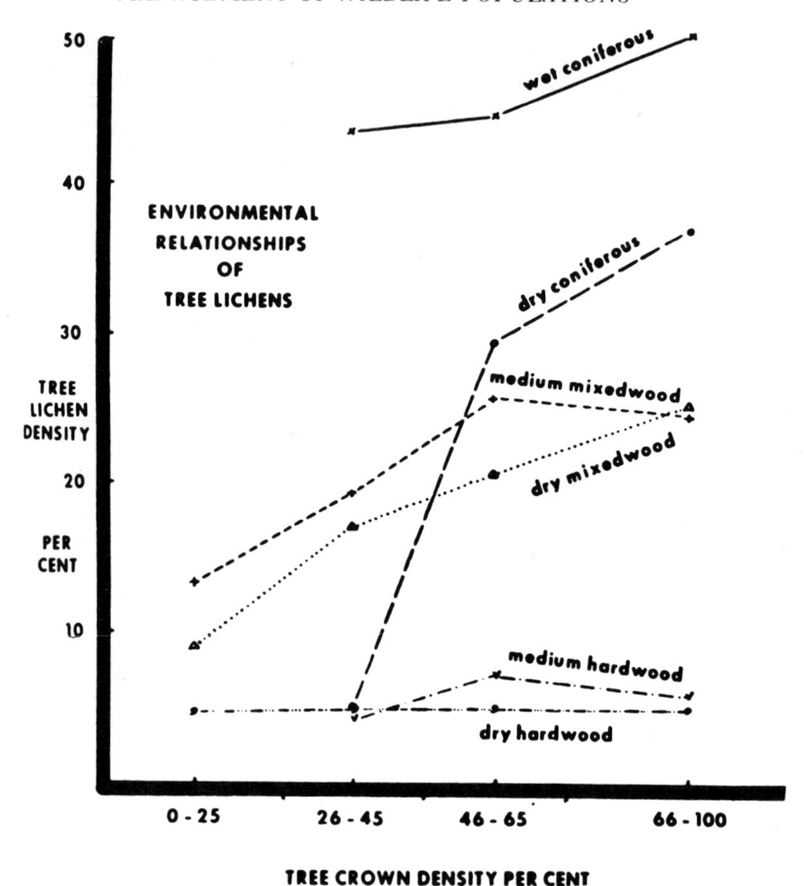


FIGURE 8–3. Relationship between mean density of tree lichens, and tree crown density within six principal forest types of the Slate Islands, Lake Superior, Ontario.

of reindeer mosses. If such communities become inhabited by woodland caribou, there is nothing to inhibit the development of a predator-prey oscillation between the caribou and its principal lichen food, the reindeer moss. *If the supply of reindeer moss is adequate to support a healthy population of woodland caribou, and does so, that supply must inevitably become reduced because of the caribou, if for no other reason, simply since the reproductive potential of caribou is much greater than the rate of increase of stands of reindeer moss.* Reindeer mosses might also be subjected to trampling by other ungulates and grazing by rodents, which would act to the detriment of woodland caribou.

There are other communities in the northern coniferous forest, which, although deficient in reindeer mosses, have fair stands of tree lichens. Because they occur as uneven-aged stands, they are fairly stable when considered over large areas. Each year, a certain number of trees mature, die, and a few years later, fall to the ground. With that, tree lichens which were previously inaccessible to woodland caribou come within their reach. Thus a mechanism for a sustained supply of essential woodland caribou foods is set up. Under these conditions, caribou would have but little effect on the supply of the principal lichens.

In full consideration of the instability of reindeer moss stands and the stability of tree lichen yields, it may be theorized that:

> Although dense stands of reindeer moss often support populations of woodland caribou, these are apt to be irruptive; the continued existence of steady populations of this animal in many parts of the northern coniferous forest is due to the peculiar tree lichen-producing characteristics of the climax stand of that forest.

If this is the basic essential of caribou range, then any factors affecting the northern coniferous forest so as to reduce total production of tree lichens or impair their sustained production is bound to be detrimental to the woodland caribou. Forest fire, logging and settlement are obviously among factors which could do so. It is reasonable to conclude that these have been important in the decline of woodland caribou. It seems unlikely that other factors which have been mentioned in connection with declines, emigration, shooting, predation, disease, parasitism and weather either could or did lead to extirpation over large areas in the absence of any basic range-altering factor.

## Future Woodland Caribou Management

The woodland caribou requires specific conditions, as provided by the typical mature northern coniferous forest. Disturbances affecting this forest influence caribou through their effects on lichens produced by the forest. Extensive forest fires create even-aged stands; centuries may pass before the uneven-aged climax stand essential to sustained caribou populations is re-established. Logging results in the cropping of trees before they have acquired dense stands of tree lichens; few lichens are produced by forests managed on short rotations for production of pulpwood. Each disturbance affects either abundance or production of tree lichens; each influences woodland caribou. These facts are of basic importance to woodland caribou management.

Forest management, forest fire protection, and harvesting of surplus animals are among the most important tools of manipulative woodland caribou management. The future of woodland caribou within the merchantably forested area of Canada is limited. As logging increases, this interesting wilderness mammal will retreat farther, as it has already done from the United States and from 80,000 square miles of former range in Ontario. However, there is a huge area of non-merchantable forest farther north in Canada which should always have sufficient mature stands to support woodland caribou.

Forest fire protection in this nonmerchantably forested area, for the primary purpose of caribou management, will probably become necessary in the future.

Even though an adequate supply of woodland caribou habitat may be assured as a result of forest management and fire protection, that habitat will be subject to damage by caribou. It is therefore necessary to prepare a harvesting plan, so that herds can be kept within the carrying capacity of their ranges.

Only a few years ago the woodland caribou was on the list of most endangered mammals, yet now it seems to be thriving, and perhaps is too numerous in certain places. If there are between 20,000 and 25,000 woodland caribou in Canada as estimated, an annual harvest of 2,500 or more animals per year would be necessary to stabilize this population. If the experience of Canadian Provinces in attempting to achieve desired harvests of moose from semi-wilderness areas serves as a guide, it will be difficult to harvest the required number of woodland caribou from wilderness areas. Failing such control, the woodland caribou will likely continue to fluctuate markedly in the wilderness portion of its range, just as it has always done.

## Literature Cited


Adams, Andrew L. 1873. *Field and forest rambles, with notes and observations on natural history of easrtern Canada.* H. S. King, London.

Adams, Charles C. 1926. The economic and social importance of animals in forestry. *Roosevelt Wildlife Bull.* 3, Syracuse, N.Y.

Aldous, Shaler, E. 1944. A deer browse survey method. *Journ. Mamm.*, 25(2):130–136.

———, and Laurits W. Krefting. 1946. The present status of moose on Isle Royale. *Trans. N. Am. Wildlife Conf.*, 11:296–308.

Allen, Glover M. 1942. Extinct and vanishing mammals of the western hemisphere. *Amer. Com. Internatl. Wildlife Protect.*, special pub. no. 11, pp. 1–620.

Anderson, Rudolph Martin. 1939 (1938). The present status and distribution of the big game mammals of Canada. *Trans. N. Am. Wildlife Conf.*, 3:390–406.

———. 1946 (1947). Catalogue of Canadian recent mammals. *Natl. Mus. Can.*, Bull. No. 102, Biol. Ser. No. 31, pp. 1–238.

Banfield, A. W. F. 1949. The present status of North American caribou. *Trans. N. Am. Wildlife Conf.*, 14:477–491.

Blair, Emma Helen. 1911. *The Indian tribes of the upper Mississippi Valley and region of the Great Lakes as described by Nicholas Perrot.* Clark, Cleveland, Vol. 1, pp. 1–372.

Clarke, C.H.D. 1940. A biological investigation of the Thelon Game Sanctuary. *Natl. Mus. Can.*, Bull. No. 96, Biol. Ser. No. 25, pp. 1–133.

Cringan, Alexander Thom. 1956. Some aspects of the biology of caribou and a study of the woodland caribou range of the slate islands, Lake Superior, Ontario. Unpublished. M. A. Thesis, University of Toronto, pp. 1–10 and 1–300.

De Kay, J. E. 1842. *Natural history of New York.* Zoology, pt. 1, Mammalia. D. Appleton and Co. and Wiley and Putnam, New York. pp. 1–146.

de Vos, Antoon and Randolph L. Peterson. 1951. A review of the status of woodland caribou (*Rangifer caribou*) in Ontario. *Journ. Mamm.*, 32(3):329–337.

Fernow, B. E. 1912. Forest conditions of Nova Scotia. *Canada Commission of Conservation*. pp. 1–93.
Halliday, W. E. D. 1937. A forest classification for Canada. *Forest Serv. Bull.* 89, Can. Dept. Mines and Resources, pp. 1–50.
Hoy. 1882. *Trans Wisc. Acad. Sci.*, 5:256.
Low, A. P. 1896. Report on explorations in the Labrador Peninsula along the Nain, Koksoak, Hamilton, Manicuagan and portions of other rivers. *Geol. Sur. Can.*, 8:70L, 86L and 318–320L.
Millias, J. G. 1907. *Newfoundland and its untrodden ways*. Longmans, Green & Co, New York. pp. 1–340.
Moisan, Gaston. 1956 a. Le caribou de Gaspé I. Histoire et distribution. *Le Naturaliste Canadien*, 83(10):225–234.
———. 1956 b. Le caribou de Gaspé II. *Le Naturaliste Canadien*. 83(11–12):262–274.
Nelson, E. C. 1947. The woodland caribou in Minnesota. *Journ. Wildlife Mgt.*, 11 (3):283–284.
Palmer, Ralph S. 1938. Late records of caribou in Maine. *Journ. Mamm.*, 19(1):37–43.
———. 1949. *Rangifer caribou* in Maine in 1946. *Journ. Mamm.*, 30(4):437–438.
Parsons, A. L. 1918. Slate Islands, Lake Superior. *Rept. Ont. Bureau Mines*, 27(1).
Quaife, Milo Milton (ed).). 1921. *Alexander Henry's travels and adventures in the years 1760–1776*. Lakeside Press, Chicago. pp. 1–340.
Riis, Paul B. 1938. Woodland caribou and–time. *Parks and recreation*. 21(10–12).
Seely, H. E. 1949. Meeting of the committee on surveys research, C.S.F.E., held in Ottawa, Feb. 4th, 1949. *Forestry Chronicle*. 25:62–65.
Seton, Ernest Thompson. 1953. *Lives of game animals*. Charles T. Branford Co., Boston, 4 vols.

# 9

# Wolf predation and ungulate populations

*Douglas H. Pimlott*

Studies of wolf (*Canis lupus*) ecology have become fairly common during the past two decades. In many instances the programs were stimulated by " . . . apprehension concerning the welfare of the big game herds," as Adolph Murie phrased it in the foreword to his classical study, "The Wolves of Mt. McKinley" (1944).

Murie's was the first of the studies that dealt intensively with the interaction of wolves and their prey. Since he completed his field work, Cowan (1947), Thompson (1952), Stenlund (1955), Mech (1966), and Shelton (1966) have also reported on the wolves of North America. Pulliainen (1965) has presented an account of the species in Finland. In addition to these published accounts other work has been in progress, some of which will be reported for the first time at this meeting.

Although virtually all the studies which have been mentioned have dealt, at least in part, with the effects of predation by wolves on the population levels of the animals on which they prey, quantitative data in many cases have been sparse and their lack has precluded a very detailed consideration of the subject. Murie (1944) obtained considerable data on predation on Dall sheep (*Ovis dalli*) and on caribou (*Rangifer rangifer*). Mech (1966) and Shelton (1966) published data on predation

---

Reprinted with permission from the *American Zoologist*, 7:267–278 (1967). The author is currently affiliated with the Department of Zoology, University of Toronto, Toronto, Canada.

on moose (*Alces alces*), and studies in Ontario (Pimlott, *et al.*, 1967) presented data on predation on white-tailed deer.

In most cases a considerable element of the problem has been, and continues to be, the difficulty that is encountered in obtaining sufficiently detailed data on the population levels of both the wolf and its principal prey species. The studies on Isle Royale, which are directed by Durward Allen and which have been conducted by Mech (1966), Shelton (1966), and Jordan (unpublished), have come the closest to laboratory studies of any big-game species which have been undertaken. They, and similar future studies, will undoubtedly provide a much firmer quantitative basis from which the principles of wolf predation will be developed.

Although we have not yet reached the stage where a broad definitive statement can be made on the role of wolf predation in controlling the populations of species on which they prey, the studies that I have mentioned have added a great deal of fresh insight on the question.

My objective in this paper is to review the state of our knowledge on wolf predation, to attempt to clarify some of the areas where thinking on the subject has not been clear and, finally, to present my preliminary thoughts on the interaction of wolves and their prey.

## Variables and Components of Predation

The literature on predation and its influence on prey populations is extensive. The great majority of the detailed studies have, however, been conducted on situations where both predator and prey were insects (Thompson, 1939) or where a vertebrate predator was preying on insect prey (Tinbergen, 1955, 1960; Holling, 1959, 1961; Morris, *et al.*, 1958; Kendeigh, 1947).

Studies of predation that have been reported make it apparent that many variable factors can influence, moderating or intensifying, the effect of predation. Leopold (1933) classified the factors into five groups: (1) the density of the prey population, (2) the density of the predator population, (3) the characteristics of the prey, *e.g.*, reactions to predators, (4) the density and the quality of alternate foods available to the predator, (5) the characteristics of the predator, *e.g.*, food preferences, efficiency of attack, and other characteristics.

Holling (1959, 1961) has developed a comprehensive theory of predation, based on his studies of small-mammal predation on the European pine sawfly; following the scheme proposed by Leopold (1933) he classified the factors into basic and subsidiary variables. The variables that are always present, predator and prey density, he referred to as universal variables. Since they are part of every predator-prey situation he considered that "... the basic components of predation will arise from these universal variables" (Holling, 1961, p. 164). The remaining variable factors (environmental characteristics, prey characteristics, and predator characteristics) are either constant or absent, so he called them subsidiary variables and the components represented by their effects as subsidiary components.

To describe the dual nature of predation he adopted terminology proposed by Soloman (1949) and used the terms "functional response" to indicate the numbers of animals consumed per predator and "numerical response" to indicate the change in the population level of the predators. The basic components of predation that he described are the functional response to prey density, the functional response to predator density, and the numerical response, which arose from the functional response and from other population processes.

Holling (1961) considers that there are two types of functional response to prey density. In one type, more prey, or hosts, are attacked as host density increases. The relationship is curvilinear and the slope of the curve decreases until the curve becomes level. In the second type, predators attack more prey as prey density increases; however, the rising phase of the curve has an S-shape. Holling documented the form of the curve by his studies, both in the laboratory and in the field, of small mammals preying on the cocoons of the European pine sawfly. However, he also stated (Holling, 1961) that the curve for functional response of vertebrate predators to the density of their prey seemed in general to be of this type. He pointed out that Leopold (1933) predicted this type of response when he suggested that vertebrates attack scarce prey by chance but develop the ability to find a greater proportion when the prey becomes abundant.

The curves of both types ultimately level off because of satiation of the predator or, if for no other reason, simply because of the time expended in finding, attacking, and killing prey. In reviewing the components of the equations which describe the two types of curve, Holling (1961, p. 170) stated,

> The two types of functional response to prey density therefore can be explained by combinations of the five components: time predator and prey are exposed, searching time, handling time (including identification, capture, and consumption), hunger, and stimulation of predator by each prey discovered. The first three are universally present and hence basic and, by themselves or in conjunction with the effects of hunger, can explain those response curves that rise with a continually decreasing slope to a plateau. If stimulation by prey discovery is added to those four components, an S-shaped response results.

The third type of response curve which Holling (1961) stated might be expected in response to prey density is a domed type which may result from a predator attacking fewer prey, when the prey are very abundant. In some cases at least this may result from the "confusion effect" described by Allee (1951) as a result of studies, by J. C. Welty, of goldfish feeding on *Daphnia*.

Holling (1961) pointed out that, in the past, studies of predation have concentrated on direct numerical responses. He cited the works of Lack (1934) and Andrewartha and Birch (1954) in which survival, fecundity, and dispersal are related to consumption of food. Holling (1961) pointed out that studies of vertebrates preying on insects (*e.g.*, Kendeigh, 1947; Morris, *et al.*, 1958; Holling, 1959) have demonstrated direct and inverse responses, as well as no response, to increasing density of prey species.

I suggest that this scheme, or structure, of predation that has been proposed by Holling (1959, 1961) is worthy of detailed consideration by students of vertebrate predation. It does a great deal to clarify this area of population dynamics that has long been a rather nebulous one. It could be valuable in guiding our thought as we seek to understand the background principles of predation by wolves on the large ungulates.

## Wolf Populations

Obtaining accurate data on the two basic variables, predator and prey densities, has proven to be the principal stumbling block to understanding the influence that wolves have on prey populations. It is mandatory, if we are to gain an understanding of the processes involved, that we continue our efforts to develop census methods that will provide accurate data at costs that are economically feasible.

The early estimates of numbers of wolves were based to a considerable extent on impressions that the individual investigators obtained as a result of their observations on the occurrence of wolves and as a result of packs reported to them by other individuals. When areas of moderate size were involved the estimates were probably quite close to the actual population. When very large areas were involved too many unknown factors entered the picture and the "estimates" could hardly warrant being called anything but guesses.

Cowan (1947) worked for three years on wolves and ungulates of the Rocky Mountain National Parks of Canada. He had the close cooperation of the wardens and made estimates of the wolf populations of Banff and Jasper Parks. In the latter, the wardens regularly patrolled the principal wolf ranges and provided him with details of their observations. The area of the park is 4200 square miles; the minimum and maximum estimates of the wolf population made by Cowan were 33 and 55 wolves (Table 9–1). Based on summer range he estimated the population density at between one wolf per 87 and one wolf per 111 square miles. He stated (Cowan, 1947, p. 150), "At the time of maximum winter compression, however, this population is present on an area that

TABLE 9–1

*Estimated densities of wolf population in North America*

| Location | Author | Area (sq. miles) | Population | Density General range | Density Winter range |
|---|---|---|---|---|---|
| N.W. Territories (Canada) | Clarke (1940) | 600,000 | 36,000 | 16+ | |
| N.W. Territories | Kelsall (1957) | 480,000 | 8,000 | 60 | |
| Mt. McKinley (Alaska) | Murie (1944) | 2,000 | 40-60 | 50± | |
| Jasper Natl. Park (Canada) | Cowan (1947) | 4,200 | 33-55 | 87-111 | 10 |
| Superior Natl. For. (Minnesota) | Stenlund (1955) | 4,100 | 240 | 16+ | |
| Isle Royale (Michigan) | Mech (1966) Shelton (1966) | 220 | 20-22 | 10± | |
| Algonquin Park (Ontario) | Pimlott, *et al.* (1967) | 1,000 | 90-110 | 10± | |

averages approximately 10 square miles per wolf."

An estimate of wolf numbers that has been quite widely quoted is the one made by Clarke (1940) for the range of the barren-ground caribou, which he estimated at 600,000 square miles. He considered that there probably was a pack (6 animals) for every 100 square miles and on this basis estimated the wolf populations at 36,000 animals. It has been suggested by both Banfield (1954) and Kelsall (1957) that the estimate was too high. Kelsall (1957) suggested, on the basis of observations made in the course of 43,624 miles of transit flying on caribou surveys, that a population of 8,000, or one wolf per 60 square miles of caribou range, would be more realistic (Table 9–1). Kelsall also pointed out that the kill of wolves (2000 to 3000 annually) made during the height of the control program would not have had any influence on a population of 36,000 but did appear to have considerable influence on the population that was present.

The relationship of the area occupied by a species to its population density can be most accurately appraised where the area occupied is essentially the same at all periods of the year. Three of the more recent studies conducted in white-tailed deer and moose ranges of eastern North America were in areas where this situation applied. In Minnesota (Stenlund, 1955), Isle Royale (Mech, 1966; Shelton, 1966), and Algonquin Park (Pimlott, *et al.*, 1967), the density of the wolf population was determined, primarily, by the use of aerial surveys during the winter (Table 9–1). Isle Royale has proven to be a particularly excellent area and the study there has provided accurate data on the density of the wolf population. On the island the population has remained at a level of approximately one wolf per 10 square miles. During this period the wolves were completely protected, a moose population of high density (approximately three per square mile) was present and, as far as could be determined, there was no movement of wolves from the island.

In Algonquin Park, the boundaries of the study area were fairly well delineated but those of the wolf ranges rarely coincided with them. Because of this, it was not possible to state with as high a degree of certainty what the relationship was between the number of wolves and the size of the study area (Pimlott, *et al.*, 1967). The work, however, was quite intensive, and extended over several years, so that the estimate of the density of the wolf population at between one wolf per 9 and one wolf per 11 square miles was considered to be very close to the actual size of the population. During the greater part of the study the wolves were protected as was the principal prey species, white-tailed deer, and the secondary prey species, moose and beaver.

Another study of a wolf population in Ontario indicated that this high density of wolves does not occur generally throughout the province. Aerial surveys in an area of 10,000 square miles of moose range, in conjunction with an experimental wolf control program, suggested a population density of between one wolf per 100 and one wolf per 200 square miles (Pimlott, *et al.*, 1961; Shannon, *et al.*, 1964).

In summary, data on wolf populations in North America indicate that densities of one wolf per 10 square miles are high, and they show that populations of a much lower density are common over very large areas.

## Food Habits of Wolves and Selection of Prey

The evidence from the studies of the food habits of wolves in Alaska (Murie, 1944), western Canada (Cowan, 1947). Wisconsin (Thompson, 1952), Minnesota (Stenlund, 1955), Isle Royale (Mech, 1966; Shelton, 1966) and from work in Algonquin Park (Pimlott, *et al.*, 1967) shows clearly that wolves are dependent to a very marked degree on large mammals for their food.

### *Summer Food*

It has been fairly generally accepted that large mammals serve as prey in winter; however, it is often stated that in summer wolves utilize small animals to a considerable degree. For example, Olson (1938, p. 329), writing about the wolf in the Superior National Forest in Minnesota, stated that, "The major portion of the food of the wolf during the summer is grouse, woodmice, meadow voles, fish, marmots, snakes, insects, and some vegetation. In fact anything that crawls, swims, or flies may be included in their diet." His conclusions have not been borne out by quantitative studies.

The greatest degree of uncertainty about the food habits of wolves in summer is for tundra areas, the range of the barren-ground caribou. Banfield (1951) stated that the observations of Farley Mowat, made near Nueltin Lake in Keewatin, N. W. T., "indicated a drastic change in diet between the denning period and the nomadic period."

He stated that there were no caribou in the vicinity of the wolves between June 17 and August 20, and during that period the wolves were observed hunting for small mammals and eating dead fish and a dead gull. Unfortunately, an intensive study of their food habits was not undertaken, and only 61 scats were examined (Kelsall, 1957). In spite of the apparent absence of caribou, 42 of these contained caribou remains, while the remains of small mammals occurred in 17. Kelsall (1957) pointed out that almost all wolf scats collected in caribou country contain caribou hair; however, he suggested that much of this may be the result of scavenging activity by the wolves.

One aspect of the uncertainty about the importance of caribou in the summer diet of wolves is caused by the comparative behavior of the two species. Barren-ground caribou are highly migratory and, it would seem, must often leave the wolves behind during the period when the pups are young and relatively immobile. Possibly under such circumstances the wolves are much more dependent on small animals, or if there are such breaks in the contact with the primary prey, they may constitute an important limiting factor to populations of tundra wolves.

It is also possible that Murie's work (1944) may provide an answer to the question that was not apparent to Mowat and other investigators. In the area of Mt. McKinley Murie found that even after the main movement of the caribou through the zone, there were usually stragglers left behind. The wolves were able to locate these animals and thus subsist on caribou long after the main herds had disappeared. It is possible that such a situation exists in caribou range much more frequently than has been realized.

In a study of the food habits of wolves in Algonquin Park during the snow-free period of the year, white-tailed deer comprised 80 percent of the food items that occurred in 1435 scats; moose comprised 8 percent and beaver 7 percent. The remaining 5 percent included snowshoe hare (*Lepus americana*), muskrat (*Ondratra zibethica*), marmot (*Marmota monax*), porcupine (*Erethizon dorsatum*), raccoon (*Procyon lotor*), and three species of mice (Pimlott, *et al.*, 1967).

In addition to Algonquin Park, studies were conducted in a number of areas in other parts of Ontario. In two of these, which lie west and north of Algonquin Park (the Pakesley area of Parry Sound Forest District and the Marten River area of the North Bay Forest District) and where the same species of prey occurred, the collections were compared with the data from Algonquin Park. For the Pakesley area (206 scats) the frequency of occurrence of the three most important species was beaver, 59 percent, deer, 27 percent, and marmot, 7 percent. For the Marten River area (226 scats) the frequency of occurrence of the three most important species was deer, 42 percent, beaver, 37 percent, and moose, 17 percent. The deer population had undergone a marked decline in the Marten River area as a result of losses during the severe winters of 1958-59 and 1959-60 and was very low the year when the study was conducted. The relatively high occurrence of deer hair in wolf scats suggested that predation on deer may have been disproportionate to their abundance in the area.

The data from Pakesley are the only ones, of which I am aware, that indicate that ungulates have comprised less than 50 percent of the summer food of wolves. However, even in this area beaver cannot be considered to be a primary food for they are unavailable three to five months of the year; wolves could not persist in the area during this period if deer and moose were not available to them.

### Selection of Prey in Summer

In Ontario the percentage of wolf scats in summer (July 1 to September 30) that contained fawn hair and calf moose hair was high. Fawn hair comprised 71 percent of the occurrences in the scats that contained deer hair, and calves 88 percent in the scats that contained moose hair (Pimlott, *et al.*, 1967). It has not been shown to what extent the frequency of occurrence of juvenile and adult remains reflects the proportion of animals in the age classes that are killed by wolves. However, Pimlott, *et al.* (1967) considered that the best assumption is that the proportion of remains in scats approaches the actual proportion in the kill; in this respect we disagreed with the conclusion of Mech (1966) that juveniles in the kill are over-represented by the occurrence of their remains in scats.

### Selection of Prey in Winter

The food habits of wolves in winter on Isle Royale and in Algonquin Park were known primarily from the remains of animals found during the aerial searches. In the latter the remains of 676 deer that were believed to have been killed by wolves were located and the mandiblles of 331 (47 percent) were collected.

The age distribution of the deer killed by wolves was not a normal one. Animals under five years of age included 42 percent of the specimens while those five years of age and older comprised 58 percent. The comparable percentages for a sample of 275 deer that were killed by cars or collected for research purposes were 87 percent and 13 percent. The comparative percentages of fawns, the age class most likely to be under-represented in the collection from wolf kills, was 17 percent and 20 percent, respectively.

The only other data on the age classes of deer killed by wolves were reported by Stenlund (1955) for the Superior National Forest. The collection (33 deer) did not show the preponderance of animals in the older age class, but the sample was too small for statistical comparisons to be valid.

The data on the kill of moose by wolves on Isle Royale show a somewhat similar trend to those from Algonquin Park. Of 80 animals examined, 50 by Mech (1966) and 30 by Shelton (1966), 22 were calves, one was a yearling, and 57 were 6 years of age (Age Class VI, Passmore, *et al.*, 1955) or older.

In Alaska, Burkholder (1959) tracked a pack of wolves from the air and reported on the ages of eight caribou and eight moose. Six of the moose that were killed were calves, one was a yearling, and one an adult of unknown age. The caribou were all adults, three of unknown age, four were between two and six, and one was over 10 years of age.

Fuller (1962) stated that the evidence from stomach samples of wolves (95) and an analysis of 63 scats, collected in Wood Buffalo Park, indicated that bison (*Bison bison*) form a staple food of wolves in both summer and winter. He found the remains of eight animals that had been killed by wolves and he observed wolves attacking bison on three occasions. Five of the bison killed were very old animals, three were calves, and three were in middle-age classes. All three of the latter animals were injured or diseased. He could not determine whether the leg of one had been broken before or during the attack by the wolves.

### Food Requirements of Wolves

The studies on Isle Royale (Mech, 1966; Shelton, 1966) have permitted an estimation to be made of the food requirements of wolves. Mech (1966) obtained data on 48 moose that were killed over a total period of 110 days. He estimated the average daily consumption at 12.3 pounds per wolf (Mech, 1966). Scrutiny of his data suggested that he overestimated the size of moose and underestimated the amount of wastage. Pimlott, *et al.* (1967) recalculated Mech's data and concluded that 10 pounds per day would be a better estimate. The suggested daily rate of consumption was 0.14 pound of food per pound of body weight (winter conditions). In their calculations they arbitrarily lowered the per diem rate to 0.12 pounds per pound for the summer (June to September) period.

## The Dynamics of Wolf Predation

The most intensive studies of vertebrate predation were conducted by the late Paul Errington (1934, 1943, 1963; Errington, *et al.*, 1940). A fundamental aspect of his theory of threshold phenomena is that vertebrate predators take a high toll of prey only when the prey are living in insecure situations, in marginal or submarginal habitats. However, he frequently referred to predation by the genus *Canis* on the ungulates and indicated that he considered that there were at least some occasions when it might be of a noncompensating nature. In his major review of the topic he stated (Errington, 1946, p. 158):

> Intercompensations in rates of gain and loss are evidently less complete in the life equations of the ungulates, however, than in the muskrats. There is vastly more reason that I can see for believing that predation can have a truly significant influence on population levels of at least some wild ungulates.
>
> Without losing sight of the fact that much more than predation or lack thereof may be involved in the great changes recorded for American deer populations of recent decades . . . , we may detect pretty strong indications of the depressive influence of predation upon the numbers of the deer.

He summed up:

> Most examples of predation upon wild ungulates showing a reasonably clear evidence of population effect have one thing in common: the predators involved had special abilities as killers—indeed were usually *Canis* spp., members of a subhuman group inferior as mammals only to man in adaptiveness and potential destructiveness to conspicuous, relatively slow-breeding forms.

The environments in which wolf predation occurs in North America are extremely variable. The rigorous arctic environment of Ellesmere Island, where the wolves prey on musk oxen and caribou, contrasts sharply with the mixed forests of the Superior National Forest in Minnesota, where the wolves prey on white-tailed deer, and with Isle Royale National Park, where the wolves prey on moose. The nature of the universal variables of predation, predator, and prey density, and the nature of the subsidiary variables are very different in the various environments. The studies that have been or are being conducted suggest that we are likely to find that the interaction of the variables of predation produce such complexities that few generalizations are possible on the influence of predation by wolves on populations of prey.

### *Influence of Wolves on Ungulates in North America*

In the case of the caribou, Murie (1944) suggested that predation was apparently an important limiting factor on a population in Alaska through predation on fawns. However, Banfield (1954) suggested that the mortality caused by wolves in the western Canadian Arctic did not exceed 5 percent of the population. Kelsall (1957) stated that an annual kill of four caribou was a likely average kill of a tundra wolf, although he estimated that it would take 14 caribou to sustain a wolf for a year. Banfield's and Kelsall's estimates, which are of approximately the same magnitude, will be subject to upward revision if future studies indicate that caribou, and particularly fawns, are primary prey of tundra wolves in summer in northern Canada, as they are in Alaska. However, if future studies confirm that small mammals are a primary component of the summer diet of wolves, then the estimates of Banfield and Kelsall may be quite realistic. Dall sheep are prey of wolves in Mt. McKinley National Park in Alaska (Murie, 1944). Although the data from scat analyses suggested that they were not as important as caribou, Murie considered that the wolves were controlling the population. The control appeared to be exercised through periodic, heavy predation on yearlings. When Murie (1961) returned to the Park in 1945, after an absence of four years, he found that the population of sheep had declined to 500 from a minimum of 1000 to 1500 in 1941. He considered that poor survival of the young, combined with the loss of old sheep that had been predominant in 1941, had caused the decline. The wolf population had also declined but, unfortunately, there was no knowledge of their role in the decline of the sheep population.

The sheep quadrupled in numbers by 1959, but the wolf population did not show any parallel increase. Murie (1961) believed that predator control operations, conducted outside the Park, were the factor that prevented an increase in wolf numbers.

Cowan (1947) gave rough estimates of the population of elk (*Cervus canadensis*) and mule deer (*Odocoileus hemionus*) in Jasper National Park. His data suggest that the ratio of wolves to the combined populations of elk and mule deer was of the order of 1:100. In addition to these two species, moose, bighorn sheep (*Ovis canadensis*), mountain goat (*Oreamnos americanus*), and caribou occurred in the area and were utilized to a lesser degree by the wolves. Their numbers would appreciably increase the ratio of predator to prey. Cowan (1947) discussed the overpopulations of ungulates that existed in the park and pointed out that, in addition to not removing the net increment of the populations, the wolves were not even removing the diseased and injured animals, which he referred to as the "cull group," from the population.

The work of Thompson (1952) in Wisconsin, though covering the ranges of only two packs of wolves, provides informative data. He showed that the two areas in which he studied wolves developed the same symptoms of an overpopulation of white-tailed deer as did areas where there were no wolves. Data on the deer population indicated that their density increased very rapidly in the late 1930's, from 10 to 30 per square mile, following extensive changes in habitat that resulted from fire and logging. The density of the wolf population was of the order of one per 35 square miles so that the ratio of wolves to deer would have been greater than 1:300.

Mech (1966) estimated the population of moose in Isle Royale, in late winter, at approximately 600 animals, and, as mentioned previously, the wolf population at 20 to 22. The ratio of wolves to moose in this case was approximately 1:30. Mech (1966) and Shelton (1966) concluded that the wolves were controlling the moose population. They estimated that the control was being accomplished by the kill of between 142 (Mech, 1966) and 150 (Shelton, 1966) moose, or approximately 25 percent of the late winter population.

The data on the deer population in Algonquin Park suggest a density of 10 to 15 per square mile, or a ratio of wolf to deer of between 1:100 and 1:150. The deer are primary prey of the wolves and predation may have been important in preventing major irruptions such as those that have occurred in many deer ranges where wolves are absent (Leopold, *et al.*, 1947). The population of deer has not been in perfect balance with the environment, however, for there have been periodic reductions caused by starvation during severe winters (Pimlott, *et al.*, 1967). The interpretation of the influence of wolves on the deer population in Algonquin Park is made difficult by the fact that wolves in the Park were subject to control by Park personnel for many years prior to the inauguration of the research program. The deer, however, were protected from hunting.

Calculations based on the data on rates of food consumption by wolves, and on the data obtained from studies of food habits of wolves in Algonquin Park, suggest that a population of a wolf per 10 square miles would require and would utilize 3.7 deer per square mile per year. This would require a deer population of a minimum density of 10 per square mile and a productivity rate of approximately 37 percent to support the wolf population (Table 9–2) (Pimlott, *et al.*, 1967).

TABLE 9–2

*Calculation of number of deer required to support a wolf population of one per 10 square miles*

| *Basic Assumptions* | | | |
|---|---|---|---|
| Size of area | | | 100 sq. miles |
| Wolf population | | | 10 |
| Gross food consumption by wolves (avg. wt. 60 lbs.) | | | |
| Oct.-May | | | 8.4 lbs./day |
| June-Sept. | | | 7.2 " " |
| Wastage | | | 20% |
| Species other than deer—winter | | | 10% |
| summer | | | 20% |
| Age-composition and weight of deer killed | | | |
| winter— | Fawns | 30% | 80 lbs. |
| | Adults | 70% | 150 lbs. |
| summer— | Fawns | 80% | 40 lbs. |
| | Adults | 20% | 150 lbs. |
| Total kill of deer—winter | 177 | | |
| summer | 190 | | |
| | 367 deer | | |
| Density of 10 deer/sq. mile, productivity of 37% is required to support 1 wolf/10 sq. miles. | | | |

## Discussion

Since Errington's review (1946) of vertebrate predation, there has been a great increase in knowledge of population dynamics of both wolves and the ungulates. The marked variation in reproductive performance that has been shown to exist among the ungulates permits considerable compensation for adverse or favorable environmental factors (*e.g.*, Cheatum and Severinghaus, 1948; Pimlott, 1959). It is conceivable that predation is a factor in triggering an increase in the reproductive rate, and, if so, it could be considered to be of a compensatory nature.

In the discussion of the selection of prey by wolves it has been shown that predation tends to be concentrated on the very young and the very old. When the old animals in a population are eliminated it probably has very little influence on the population level of the prey species, for they, like animals in submarginal habitats, would soon have died of other causes anyway. Predation then on the old animals in the population also appears to be of a compensatory nature.

A great weakness that exists in the study of wolves in summer is that there does not appear to be any way of making concrete determinations about the condition of the young that are eaten by wolves. A number of studies show conclusively (Thompson, 1952; Murie, 1944; Mech, 1966; Shelton, 1966; Pimlott *et al.*, 1967) that wolves feed heavily on the young of the year—but, what percentage of these animals was actually killed by wolves? To what extent is the feeding on young animals a scavenging activity? What percentage of those killed by wolves would have survived in the absence of wolves? Studies in a number of areas where wolves have been extirpated indicate that a significant mortality of young ungulates occurs between spring and fall. To the extent that predation by wolves removes young that would have died anyway, as in the case of adults, it is of a compensatory nature.

Although Murie's (1944) work indicated that predation fell heavily on young animals, I do not think that the full import of this fact has been realized. If a considerable portion of this predation is noncompensatory, a population of wolves of high density would exercise a considerable influence on ungulate populations. Allee, Emerson, Park, Park and Schmidt (1949) listed a series of principles that arose from their review of predation. The third is of particular interest to this discussion: "predation is frequently directed against the immature stages of the prey and as such may constitute an effective limiting factor." (p. 374).

The question of whether or not wolves constitute an effective limiting factor on ungulates, and particularly on deer, moose, and caribou, is one that has only been partially answered. In considering the population dynamics of some big-game species, deer and moose in particular, the question arises as to why intrinsic mechanisms of population control have not evolved to prevent them from increasing beyond the sustaining level of their food supply. It seems reasonable to postulate that it may be because they have had very efficient predators, and the forces of selection have kept them busy evolving ways and means not of limiting their own numbers, but of keeping abreast of mortality factors.

Contemporary biologists often have a distorted viewpoint about the interrelationships of ungulates and their predators. We live in an age when there is a great imbalance in the environments inhabited by many of the ungulates. In the case of deer and moose the environmental changes, or disturbances, have been favorable and populations are probably higher than they have ever been. Under such circumstances it is not much wonder that we have been inclined to argue that predators do not act as important limiting factors on deer and moose populations. I doubt, however, that it was a very common condition prior to intensive human impact on the environment. In other words, I consider that adaptations between many of the ungulates, particularly those of the forest, and their predators probably evolved in relatively stable environments that could not support prey populations of high density.

The history of wolves and moose on Isle Royale is an interesting example. There, as I have mentioned, in the presence of abundant food and complete protection, the wolf population stabilized at a level of one wolf per 10 square miles. In Algonquin Park the estimates indicated a population of the same magnitude; there was no significant difference between 1959 and 1964, although during most of this period the wolves were protected (Pimlott, *et al.*, 1967). These examples suggest that a wolf per 10 square miles is close to the maximum density that can be attained by a population.

The data from Isle Royale suggest that a state of equilibrium has been reached between the wolves and the moose at a ratio of approximately one wolf per 30 moose. A similar calculation, based on the data from Algonquin Park, suggests that a ratio of one wolf per 100 deer may be close to an equilibrium. On the basis of these data, and on the basis of the previous discussion of the evolution of wolf-prey population mechanisms, I suggest that wolves may not be capable of exercising absolute control of white-tailed deer at ratios that exceed 1:100. I also suggest that predation by wolves may cease to be an important limiting factor when densities of deer exceed 20 per square mile.

The fact that no animal smaller than the beaver has been shown to be the predominant food of wolves for any significant period is not surprising. Their size, and the complex social organization of the packs, are such that it would rarely be efficient for them to live on small animals. The organization of the pack is undoubtedly an adaptation which has developed because wolves prey on animals larger and often fleeter than themselves.. Such an organization would be unlikely to persist if small animals became their primary source of food. Energy relationships are undoubtedly also involved. An adult wolf may weigh between 50 and 150 pounds and it would rarely be efficient to obtain the energy to maintain this biomass by the utilization of animals that weigh a few ounces or even a few pounds, especially when these are often difficult to capture.

I suggest that energy demands alone make it very unlikely that tundra wolves regularly subsist on small animals during the summer. When the question is studied intensively it is likely that the successful rearing of a litter will, in the great majority of cases, be found to be dependent on the availability of caribou or of other large ungulates as food for the wolves.

A study of the interaction of wolves and their prey indicates that there are a number of characteristic aspects of predation that are worthy of review. They serve to sum up this discussion of the dynamics of wolf predation; a knowledge of their existence may also contribute to the further development of understanding of the underlying principles of vertebrate predation.

1. In all but one instance, intensive studies of the food habits of wolves indicate that the large ungulates are the primary prey of wolves both in summer and in winter. It remains to be demonstrated that wolves can live and raise young in areas where they must subsist on small animals.

2. The process of wolf predation does not come about simply as a result of random contacts between predator and prey, but is complicated by a process in which the ability of the prey to escape is tested. The dynamic aspects of the process have been observed in a number of areas (Murie, 1944; Crisler, 1956) and have been particularly well documented on Isle Royale (Mech, 1966; Shelton, 1966).

3. Among the ungulates, wolves prey primarily on the young-of-the-year and on animals in older age-classes. Predation is most heavy on the young during the summer but is less intensive during the winter, when old animals are vulnerable.

4. Intensive utilization of prey animals that are captured is a characteristic of wolf predation. One study (Pimlott, *et al.*, 1967) has demonstrated, however, that utilization was less complete during a winter when severe snow conditions prevailed.

## References

Allee, W. C. 1951. *Cooperation among animals*. Henry Schuman, New York.

Allee. W. C., H. E. Emerson, O. Park, T. Park, and K. P. Schmidt. 1949. *Principals of animal ecology*. W. B. Saunders Co., Philadelphia, Pa.

Andrewartha, H. G., and L. C. Birch. 1954. *The distribution and abundance of animals*, The Univ. of Chicago Press, Chicago.

Banfield, A. W. F. 1951. The barren-ground caribou. *Canad. Wildl. Serv., Dept. Res. and Dev.*

Banfield, A. W. F. 1954. Preliminary investigation of the barren-ground caribou. *Canad. Wildl. Serv., Wildl. Mgmt. Bull.* Ser. 1, No. 10B.

Burkholder, B. L. 1959. Movements and behavior of a wolf pack in Alaska. *J. Wildl. Mgmt.* 23:1–11.

Cheatum, E. L., and C. W. Severinghaus. 1950. Variations in the fertility of the white-tailed deer related to range conditions. *Trans. N. Am. Wildl. Conf.* 15:170–190.

Clarke, C. H. D. 1940. A biological investigation of the Thelon Game Sanctuary. *Natl. Mus. Canad. Bull.* No. 96, Biol. Ser. No. 25.

Cowan, I. McT. 1947. The timber wolf in the Rocky Mountain National Parks of Canada. *Canad. J. Res.* 25:139–174.

Crisler, L. 1956. Observations of wolves hunting caribou. *J. Mammal.* 37:337–346.

Errington, P. L. 1934. Vulnerability of bob-white populations to predation. *Ecol.* 15:110–127.

Errington, P. L. 1943. An analysis of mink predation upon muskrats in north-central United States. *Agric. Expt. Sta., Iowa State Coll. Res. Bull.* 320:797–924.

Errington, P. L. 1946. Predation and vertebrate populations. *Quart. Rev. Biol.* 21:144–177, 221–245.

Errington, P. L., 1963. *Muskrat populations*. Iowa State Univ. Press, Ames.

Errington, P. L., F. Hammerstrom, and F. N. Hammerstrom, Jr. 1940. The great horned owl and its prey in north-central United States. *Agric. Expt. Sta., Iowa State Coll. Res. Bull.* 277:757–850.

Fuller, W. A. 1962. The biology and management of the bison of Wood Buffalo National Park. *Canad. Wildl. Serv., Wildl. Mgmt. Bull.*, Ser. 1, No. 16.

Holling, C. AS. 1959. The components of predation as revealed by a study of small mammal predation of the European pine sawfly. *Canad. Entomol.* 91:293–320.

Holling, C. S. 1961. Principles of insect predation. *Ann. Rev. Entomol.* 6:163–182.

Kelsall, J. P. 1957. Continued barren-ground caribou studies. *Canad. Wildl. Serv., Wildl. Mgmt. Bull.*, Ser. 1, No. 12.

Kendeigh, S. C. 1947. Bird population studies in the coniferous forest biome during a spruce budworm outbreak. *Div. of Res. Ont. Dept. of Lands and Forest. Biol., Bull.* No. 1.
Lack, D. 1954. *The natural regulation of animal numbers*. Oxford Univ. Press, London.
Leopold, A. 1933. *Game management*. Charles Scribner's Sons, New York.
Leopold, A., L. K. Sowls, and D. L. Spencer. 1947. A survey of over-populated deer ranges in the United States. *J. Wildl. Mgmt.* 11:162–177.
Mech, L. D. 1966. The wolves of Isle Royale. *U.S. Natl. Park Serv.*, Fauna Ser. No. 7.
Morris, R. F., W. F. Cheshire, C. A. Miller, and D. G. Mott. 1958. Numerical response of avian and mammalian predators during a gradation of the spruce budworm. *Ecol.* 39:487–494.
Murie, A. 1944. The wolves of Mt. McKinley. *U.S. Dept. Interior, U.S. Natl. Park Serv.*, Fauna Ser. No. 5.
Murie, A. 1961. *A naturalist in Alaska*. Devin Adair Co., Ltd., New York.
Olson, S. F. 1938. A study in predatory relationships with particular reference to the wolf. *Sci. Monthly* 46:323–336.
Passmore, R. C., R. L. Peterson, and A. T. Cringan. 1955. A study of mandibular tooth wear as an index to age of moose, pp. 223–246. *In:* R. L. Peterson, North American moose. Univ. Toronto Press, Toronto, Canada.
Pimlott, D. H. 1959. Reproduction and productivity of Newfoundland moose. *J. Wildl. Mgmt.* 23:381–401.
Pimlott, D. H., J. A. Shannon, and G. B. Kolenosky. 1967. The interrelationships of wolves and deer in Algonquin Park. *Trans. Northeast Wildl. Conf.* 16 pp. mimeo.
Pimlott, D. H., J. A. Shannon, W. T. McKeown, and D. Sayers. 1961. Experimental timber wolf poisoning program in northwestern Ontario. Res. Branch, Dept. Lands and Forests, Ontario. Unpubl. 11 pp. (typed).
Pulliainen, E. 1965. Studies on the wolf in Finland. *Ann. Zool. Fenn.* 2:215–259.
Shannon, J. A., J. L. Lessard, and W. McKeown. 1964. Experimental timber wolf poisoning program in northwestern Ontario, 1963–64. Res. Branch, Dept. of Lands and Forests, Ontario. Unpubl. 4 pp. (typed).
Shelton, P.S. 1966. Ecological studies of beavers, wolves, and moose in Isle Royale National Park, Michigan. Ph.D. thesis. Purdue Univ., Lafayette, Ind. 308 pp.
Soloman, M. E. 1949. The natural control of animal populations. *J. Animal Ecol.* 18:1–35.
Stenlund, M. H. 1955. A field study of the timber wolf (*Canis lupus*) on the Superior National Forest, Minnesota. *Minn. Dept. Conserv., Tech Bull.* No. 4.
Thompson, D. Q. 1952. Travel, range, and food habits of timber wolves in Wisconsin. *J. Mammal.* 33:429–442.
Tinbergen, L. 1955. The effect of predators on the numbers of their hosts. *Vakblad voor Biologen.* 28:217–228.
Tinbergen, L. 1960. The natural control of insects in pinewoods. I. Factors influencing the intensity of predation by song birds. *Arch. Neerl. Zool.* 13:265–343.

# IV

# Control of Undesirable Species

# 10

# The integrated control concept

*Vernon M. Stern*
*Ray F. Smith*
*Robert van den Bosch*
*Kenneth S. Hagen*

All organisms are subjected to the physical and biotic pressures of the environments in which they live, and these factors, together with the genetic make-up of the species, determine their abundance and existence in any given area. Without natural control, a species which reproduces more than the parent stock could increase to infinite numbers. Man is subjected to environmental pressures just as other forms of life are, and he competes with other organisms for food and space.

Utilizing the traits that sharply differentiate him from other species, man has developed a technology permitting him to modify environments to meet his needs. Over the past several centuries, the competition has been almost completely in favor of man, as is attested by decimation of vast vertebrate populations, as well as populations of other forms of life (Thomas, 1956). But while eliminating many species, as he changed the environment of various regions to fit his needs for food and space, a number of species, particularly among the Arthropoda, became his direct competitors. Thus, when he subsisted as a huntsman or foraged for food from uncultivated sources, early man was largely content to share his subsistence and habitat with the lower organisms. Today, by

Reprinted with permission from *Hilgardia*, 29(2):81–101 (1959). Published by the California Agricultural Experiment Station, University of California, Berkeley, California.

contrast, as his population continues to increase (Hertzler, 1956) and his civilization to advance, he numbers his arthropod enemies in the thousands of species (Sabrosky, 1952).

The increase to pest status of a particular species may be the result of a single factor or a combination of factors. In the last century, the most significant factors have been the following:

First, by changing or manipulating the environment, man has created conditions that permit certain species to increase their population densities (Ullyett, 1951). The rise of the Colorado potato beetle, *Leptinotarsa decemlineata* (Say) to pest status occurred in this manner (Fig. 10–1). When the potato, as well as other solanaceous plants, was brought under widespread cultivation in the United States, a change favorable to the beetle occurred in the environment, which enabled it to become very quickly an important pest (Trouvelot, 1936). Similarly, when alfalfa, *Medicago sativa* L., was introduced into California about 1850, the alfalfa butterfly, *Colias philodice eurytheme*, Boisduval, which had previously occurred in low numbers on native legumes, found a widespread and favorable new host plant in its environment, and it subsequently became an economic pest (Smith and Allen, 1954).

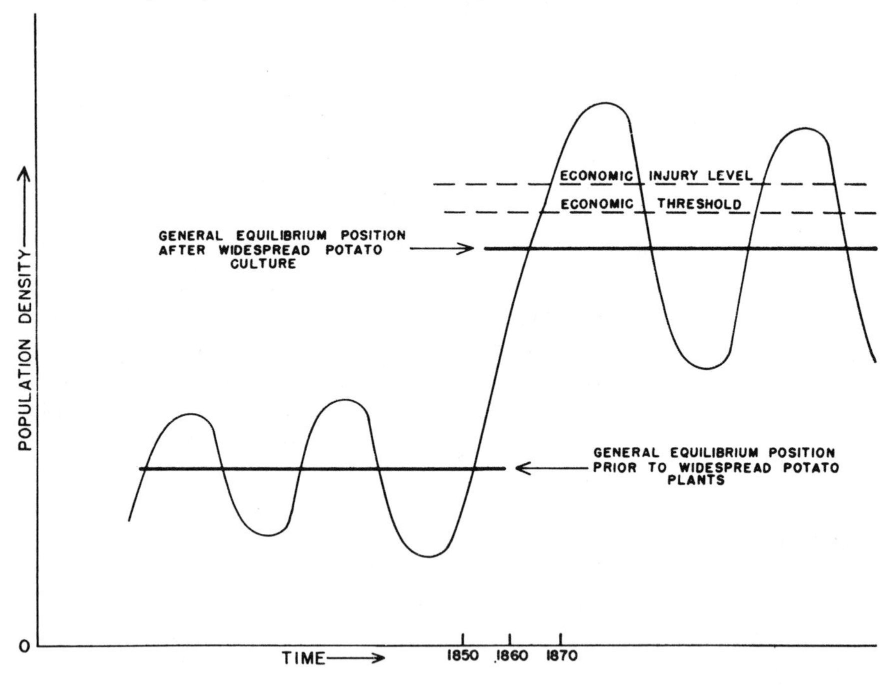


FIGURE 10–1. Schematic graph of the change in general equilibrium position of the Colorado potato beetle, *Leptinotarsa decemlineata* following the development of widespread potato culture in the United States. For a discussion of the significance of economic-injury levels and economic thresholds in relation to the general equilibrium position, see p. 89.

A second way in which arthropods have risen to pest status has been through their transportation across geographical barriers while leaving their specific predators, parasites, and diseases behind (Smith, 1959). The increase in importance through such transportation is illustrated by the cottony cushion scale, *Icerya purchase* (Maskell) (Fig. 10–2). This scale insect was introduced into California from Australia on acacia in 1868. Within the following two decades, it increased in abundance to the point where it threatened economic disaster to the entire citrus industry in California. Fortunately the timely importation and establishment of two of its natural enemies, *Rodolia cardinalis* (Mulsant) and *Cryptochaetum iceryae* (Williston), resulted in the complete suppression of *I. purchasi* as a citrus pest (Doutt, 1958).

The cottony cushion scale again achieved the status of a major pest when the widespread use of DDT on citrus in the San Joaquin Valley eliminated the vedalia (Ewart and DeBach, 1947).

A third cause for the increasing number of pest arthropods has been the establishment of progressively lower economic thresholds (see p. 143 for definition and discussion). This can be illustrated by lygus bugs (*Lygus*

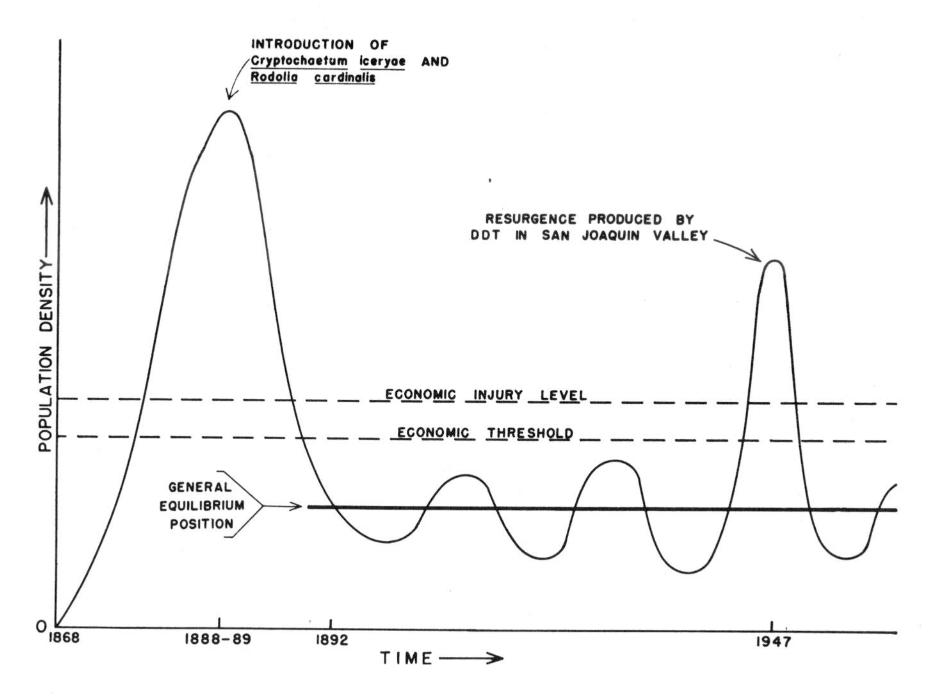


FIGURE 10–2. Schematic graph of the fluctuations in population density of the cottony cushion scale, *Icerya purchasi*, on citrus from the time of its introduction into California in 1868. Following the successful introduction of two of its natural enemies in 1888 this scale was reduced to noneconomic status except for a local resurgence produced by DDT treatments.

spp.) on lima beans. Not too many years ago the blotches caused by lygus bugs feeding on an occasional lima bean were of little concern, and lygus bugs were considered a minor pest on this crop. However, with the emphasis on product appearance in the frozen-food industry, a demand was created for a near-perfect bean. For this reason, economic-injury thresholds were established and lygus bugs are now considered serious pests of lima beans.

In the face of this increased number of arthropod pests man has made remarkable advances in their control, and economic entomology has become a complex technical field. Of major importance have been new developments in pesticide chemistry and application.

The discovery of the insecticidal properties of DDT, and its spectacularly successful application to arthropod-borne disease and agricultural pest problems, spurred research in chlorinated hydrocarbon chemistry and stimulated the development of other organic pesticides. On a national scale, the experiment stations, state and governmental agencies, and commercial companies, all searching for new or better answers to old insect-pest problems, eagerly accepted the new chemicals. Within a short period many became an integral part of public health and agricultural pest-control programs. Without question, the rapid and widespread adoption of organic insecticides brought incalculable benefits to mankind, but it has now become apparent that this was not an unmixed blessing. Through the widespread and sometimes indiscriminate use of pesticides, the components and intricate relations of crop environments have been drastically altered, and as a result a number of serious problems have arisen (Wigglesworth, 1945; Michelbacher, 1954; Pickett, 1949; Pickett and Patterson, 1953; Solomon, 1953; and others). Among these new problems and old ones which have been aggravated are:

1. Arthropod resistance to insecticides. This phenomenon relating to the genetic plasticity of the arthropods has been reviewed by Metcalf (1955), Hoskins and Gordon (1956), Crow (1957), Brown (1959), and others. In many cases, resistance is already drastic enough to have eliminated certain insecticides from important pest-control programs. There are today in excess of 70 demonstrated cases of arthropod resistance. Actually, a much larger number of pest species exist which are developing resistance or have already done so, but there has not been time to evaluate these cases.

2. Secondary outbreaks of arthropods other than those against which control was originally directed (Massee, 1954; DeBach and Bartlett, 1951; Ripper, 1956; and others). These outbreaks usually result from the interference of the insecticide with biological control (Lord, 1947; Bartlett and Ortega, 1952; Michelbacher, 1954; Michelbacher and Hitchcock, 1958; and others). This may also occur through the effect of the insecticide on the plant, which, in turn, affects the development of the second-

ary pest (Fleschner and Scriven, 1957). An example is the increase in mites on plants growing in soil receiving certain chemical treatments (Klostermeyer and Rasmussen, 1953).

3. The rapid resurgence of treated species necessitating repetitious insecticide applications (Holloway and Young, 1943; Bovey, 1955; Schneider, 1955; Stern and van den Bosch, 1959; and others). These flarebacks occur from individuals surviving treatment or from individuals migrating into the treated areas, where they can reproduce unhindered because their natural enemies have been eliminated.

4. The toxic insecticide residues on food and forage crops (Brown, 1951; Linsley, 1956). This problem may result from two sources. First, untimely applications or accidental increases in dosage may result in residues above the tolerance limits. Second, the first three problems mentioned above are interrelated and by aggravating one another may lead to excessive treatment and a residue problem. For example, where the level of resistance is increasing, it requires either more frequent applications or higher insecticide dosages to control the pest, or both. This increased insecticide program may in turn have a drastic effect on the ecosystem, which frequently results in outbreaks of secondary pests or rapid resurgence of the resistant pest for which control was originally intended. Often, under such conditions, where insects threaten the crop or marketability of a crop close to harvest, the grower is faced with the problem of suffering a severe monetary loss or of making an insecticide application closer to harvest than is ordinarily permissible. In many instances, the end result is a residue far above the accepted tolerance limit at harvest time.

5. Hazards to insecticide handlers and to persons, livestock, and wildlife subjected to contamination by drift (Hayes, 1954; Petty, 1957; Upholt, 1955).

6. Legal complications from suits and other actions pertaining to the above problem.

Unquestionably, some of these problems have arisen from our limited knowledge of biological science; others are the result of a narrow approach to insect control. Few studies have included basic investigations on the effects the chemicals might have on other components of the ecosystems to which the pests belong. The entomologist may recognize the desirability of a thorough investigation of these aspects, but because of the need for immediate answers to pressing problems and because of other pressures, he does not have the necessary time. In other instances because fundamental knowledge is lacking, the investigator may be unaware of the intricate nature of the biotic complex with which he is dealing, and of the destructive potential that many chemicals in use today have on the environment of the pests. Finally, and most unfortunately, there are workers who are highly skeptical that biotic factors are of any consequence in the control of pest population densities and thus choose to ignore any approach to pest control other than the use of chemicals.

Whatever the reasons for our increased pest problems, it is becoming more and more evident that an integrated approach, utilizing both biological and chemical control, must be developed in many of our pest problems if we are to rectify the mistakes of the past and avoid similar ones in the future (DeBach, 1951, 1958*a*; Pickett, Putnam, and Roux, 1958; Ripper, 1944; Huffaker and Kennett, 1956; Wille, 1951; Michelbacher and Middlekauff, 1950; and others).

## Terminology

To clarify the discussion in other parts of this paper, some definitions and explanations of terms are here given:

*Biological control. The action of parasites, predators, or pathogens on a host or prey population which produces a lower general equilibrium position than would prevail in the absence of these agents.* Biological control is a part of natural control (*q.v.*) and in many cases it may be the key mechanism governing the population levels within the framework set by the environment. If the host or prey population is a pest species, biological control may or or may not result in economic control. Biological control may apply to any species whether it is a pest or not, and regardless of whether or not man deliberately introduces, manipulates, or modifies the biological-control agents.

*Biotic insecticide. A biotic mortality agent applied to suppress a local insect pest population temporarily.* The effects of the agent usually do not persist and they are similar to those resulting from the use of a chemical insecticide in that they do not produce a permanent change in the general equilibrium position. A polyhedrosis virus applied as a spray to control the alfalfa caterpillar is a typical example of a biotic insecticide. Preparations of microörganisms used in this manner are sometimes referred to as *microbial insecticides.* Predators, such as lady beetles, or parasites, when they are released in large numbers, can also act, in some instances, as biotic insecticides.

*Biotic reduction. Deaths or other losses to the population* (e.g., *dispersal, reduced fecundity) caused or induced by biotic elements of the environment in a given period of time.*

*Economic control. The reduction or maintenance of a pest density below the economic-injury level* (q.v.).

*Economic-injury level. The lowest population density that will cease economic damage.* Economic damage is the amount of injury which will justify the cost of artificial control measures; consequently, the economic-injury level may vary from area to area, season to season, or with man's changing scale of economic values.

*Economic threshold. The density at which control measures should be determined to prevent an increasing pest population from reaching the economic-injury level.* The economic threshold is lower than the economic-injury level to permit sufficient time for the initiation of control measures and for these measures to take effect before the population reaches the economic-injury level.

*Ecosystem. The interacting system comprised of all the living organisms of an area and their nonliving environment.* The size of area must be extensive enough to permit the paths and rates of exchange of matter and energy which are characteristic of any ecosystem.

*General equilibrium position. The average density of a population over a period of time (usually lengthy) in the absence of permanent environmental change.* The size of the area involved and the length of the period of time will vary with the species under consideration. Temporary artificial modifications of the environment may produce a temporary alteration of the general equilibrium position (*i.e.*, a temporary equilibrium).

*Governing mechanism. The actions of environmental factors, collectively or singly, which so intensify as the population density increases and relax as this density falls that population increase beyond a characteristic high level is prevented and decrease to extinction is made unlikely.* The governing mechanisms operate within the framework or potential set by the other environmental elements.

*Integrated control. Applied pest control which combines and integrates biological and chemical control.* Chemical control is used as necessary and in a manner which is least disruptive to biological control. Integrated control may make use of naturally occurring biological control as well as biological control effected by manipulated or introduced biotic agents.

*Microbial control. Biological control that is effected by microörganisms (including viruses).*

*Natural control. The maintenance of a more or less fluctuating population density within certain definable upper and lower limits over a period of time by the combined actions of abiotic and biotic elements of the environment.* Natural control involves all aspects of the environment, not just those immediate or direct factors producing premature mortality, retarded development, or reduced fecundity; but remote or indirect factors as well. For most situations, governing mechanisms (*q.v.*) are present and determine the population levels within the framework or potential set by the other environmental elements. In the case of a pest population, natural control may or may not be sufficient to provide economic control.

*Natural reduction. Deaths or other losses to the population caused by naturally existing abiotic and biotic elements of the environment in a given period of time.*

*Population. A group of individuals of the same species that occupies a given area.* A population must have at least a minimum size and occupy an area containing all its ecological requisites to display fully such characteristics as growth, dispersion, fluctuation, turnover, dispersal, genetic variability, and continuity in time. The minimum population and the requisites in occupied area will vary from species to species.

*Population dispersion. The pattern of spacing shown by members of a population within its occupied habitat and the total area over which the given population may be spread.*

*Selective insecticide. An insecticide which while killing the pest individuals spares much or most of the other fauna, including beneficial species, either through differential toxic action or through the manner in which the insecticide is utilized (formulation, dosage, timing, etc.).*

*Supervised insect control. Control of insects and related organisms supervised by qualified entomologists and based on conclusions reached from periodically measured population densities of pests and beneficial species.* Ideally, supervised control is based on a sound knowledge of the ecology of the organisms involved and projected future population trends of pests and natural enemies.

*Temporary equilibrium position. The average density of a population over a large area temporarily modified by a procedure such as continued use of insecticides.* The modified average density of the population will revert to the previous or normal density level when the modifying agent is removed or expended (*cf.* "general equilibrium position").

## The Nature and Working Principles of Biological and Chemical Control

### *Biological Control*

Biological controls are part of natural control which governs the population density of pest species. On the other hand, with certain exceptions, chemical controls involve only immediate and temporary decimation of localized populations and do not contribute to permanent density regulation. This distinction is not always clearly made, and biological control is often thought of as being similar in its action to chemical control. Perhaps one reason for the misunderstanding is that in spectacular instances a biotic agent may act in the manner of a chemical in eliminating a local pest population. For example, this may occur when weather conditions are favorable and disease pathogens eliminate a localized pest population. Parasites and predators may sometimes act in a similar manner. However, the important prevailing characteristic of biological control is one of permanent population-density regulation. Usually these governing mechanisms occur over such a large area and are so subtle or intricate in their action that they are not easily observed and recorded; thus they tend to be overlooked.

A principal phase of applied biological control is the importation and establishment of natural enemies of pests that accidentally gain entry into new geographical regions. These new pests frequently escape the natural enemies that help to regulate their densities in the areas to which they are indigenous (Elton, 1958). Under satisfactory conditions in the new environment, the pest may flourish and reach damaging abundance. As a countermeasure, the natural enemies are obtained from the native home of the pest and transplanted into the new environment to increase the biotic resistance of the environment to the pest.

Biological control is thus utilized to permanently increase environmental resistance to an introduced pest. The hope is that the introduced enemies will lower the general equilibrium position of the pest sufficiently to maintain it permanently below the economic threshold. Most often the introduction of a biotic agent is not so spectacular, and it is an exception when the general equilibrium position of the introduced pest is lowered sufficiently to prevent its occasionally or even commonly reaching economic abundance at certain times or places (Clausen, 1956; Simmonds, 1956). This, of course, is precisely the status of a native pest which, though attacked by a complex of parasites and predators, still has a general equilibrium position high enough to permit it occasionally to cause damage of greater or lesser severity. Thus, in any geographic area the governing mechanisms in the environment are constantly at work to counteract the inherent natality of plant and animal pest species. In terms of crop protection, these regulating factors actually keep thousands of potentially harmful arthropod species permanently below economic thresholds. Moreover, these environmental pressures tend to localize the outbreaks of those forms which on occasion are capable of rising above economic thresholds. A biological control agent is self-perpetuating and capable of response to fluctuations in the population density of the pest it attacks. Biological controls, whether imported or native, are permanent characteristics of a given environment.

### *Chemical Control*

Chemical control of an arthropod pest is employed to reduce populations of pest species which rise to dangerous levels when the environmental pressures are inadequate. When chemicals are used, the damage from the pest species must be sufficiently great to cover not only the cost of the insecticidal treatment but also the possible deleterious effects, such as the harmful influence of the chemical on the ecosystem. On some occasions, the pest outbreak may cover a wide area; in other instances, damaging numbers occur in very restricted locations. These outbreaks occur during the season favorable to the pest, with the relaxed environmental pressures occurring some time before the outbreak. Chemical control is only needed at those times and places where natural control is inadequate. Chemical control should act as a complement to the biological control.

An insecticide must always be manipulated by man, who adds it to a restricted segment of the pest's environment to decimate a localized pest population. Because chemical insecticides are nonreproductive, have no searching capacity, and are nonpersistent, they constitute short-term, restricted pressures. They cannot permanently change the general equilibrium position of the pest population nor can they restrain an increase in abundance of the pest without repeated applications. Therefore, they must be added to the environment at varying intervals of time.

In certain pest-control programs, the insecticide is applied over extensive geographical areas. In some areas, after application, the pest population density may be far below the economic threshold and below its general equilibrium position; but since the insecticide is not a permanent part of the environment, the pest may return to a high level when the effects of the insecticide are gone.

The effectiveness of a chemical insecticide or a biotic insecticide is measured in percent of kill or in percent of clean fruits, uninjured cotton bolls, and so forth, in the area of application. Such applications have little influence on the pest in adjoining areas except as localized population depressants. In general, this contrasts sharply with the role of the permanent biotic mortality agent, whose effectiveness is best measured by its influence on the general equilibrium position of the pest species over an entire geographical region or a long period of time.

## Economic Thresholds and the General Equilibrium Position

Chemical control should be used only when the economic threshold is reached and when the natural mortality factors present in the environment are not capable of preventing the pest population from reaching the economic-injury level. The economic-injury level is a slightly greater density than the economic threshold. This difference in densities provides a margin of safety for the time that elapses between the detection of the threatening infestation and the actual application of an insecticide. The economic threshold and the economic-injury level of a pest species can vary depending upon the crop, season, area, and desire of man; the general equilibrium position, on the other hand, barring "permanent" changes in the environment, is a fixed population level (Griffiths, 1951; Strickland, 1954).

A species population is plastic and is undergoing constant change within the limits imposed upon it by its genetic constitution and the characteristics of its environment. Typical fluctuations in population and dispersion are shown in Figure 10–3. The population dispersions shown at the three points in time A, B, and C are not static but rather are instantaneous phases of a continuously changing dispersion.

Thus at point A, when the population is of greatest numerical abundance, it also has its widest distributional range (as depicted by the maximum diameter of the base of the model), and is of maximum economic status (as depicted by the number and magnitude of the blackened pinnacles representing penetrations of the economic threshold). At point B, on the other hand, when the species population is at its lowest numerical abundance, it is also most restricted in geographical range and is of only minor economic status. Point C represents an intermediate condition between points A and B.

In order to determine the relative economic importance of pest species, both the economic threshold and general equilibrium position of the pests must be considered. It is the general equilibrium position and its relation to the economic threshold, in conjunction with the frequency and amplitude of fluctuations about the general equilibrium position, that determine the severity of a particular pest problem.

In the absence of permanent modifications in the composition of the environment, the density of a species tends to fluctuate about the general equilibrium position as changes occur in the biotic and physical

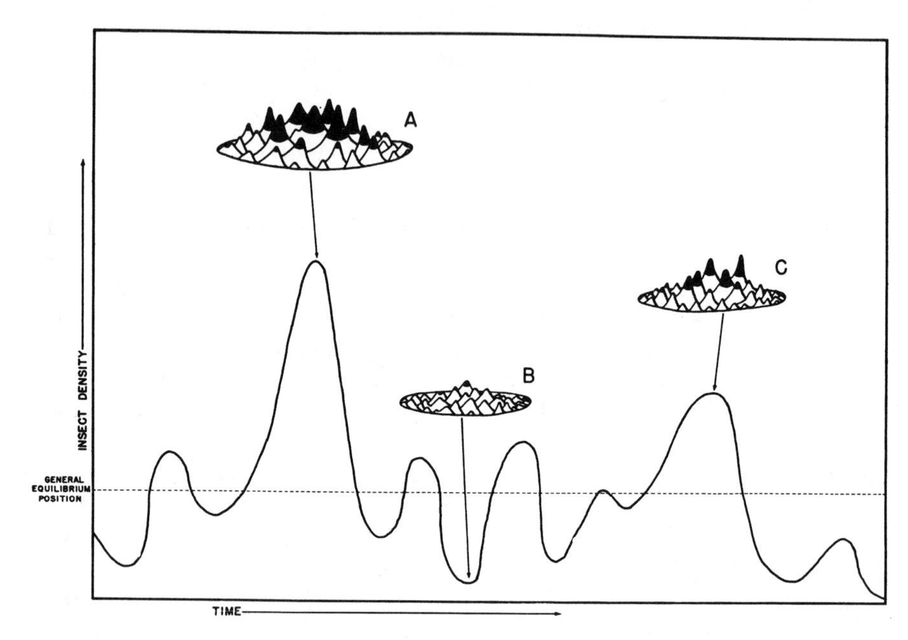


FIGURE 10–3. Schematic graph of the population trend and population dispersion of a pest species over a long period of time. The solid line depicts the fluctuations in the population density with time. The broken line depicts the general equilibrium position. The population dispersion is indicated at the specific times *A*, *B*, and *C*. The basal area of these models reflects the distributional range, the height indicates population density. Population densities above the economic threshold are black.

components of the environment. As the population density increases, the density-governing factors respond with greater and greater intensity to check the increase; as the population density decreases, these factors relax in their effects. The general equilibrium position is thus determined by the interaction of the species population, these density-governing factors, and the other natural factors of the environment. A permanent alteration of any factor of the environment—either physical or biotic—or the introduction of new factors may alter the general equilibrium position.

The economic threshold of a pest species can be at any level above or below the general equilibrium position or it can be at the same level. Some phytophagous species may utilize our crops as a food source but even at their highest attainable density are of little or no significance to man (Fig. 10–4, *A*). Such species can be found associated with nearly every crop of commercial concern.

Another group of arthropods rarely exceeds the economic thresholds and these consequently are occasional pests. Only at their highest population density will chemical control be necessary (Fig. 10–4, *B*).

When the general equilibrium position is close to the economic

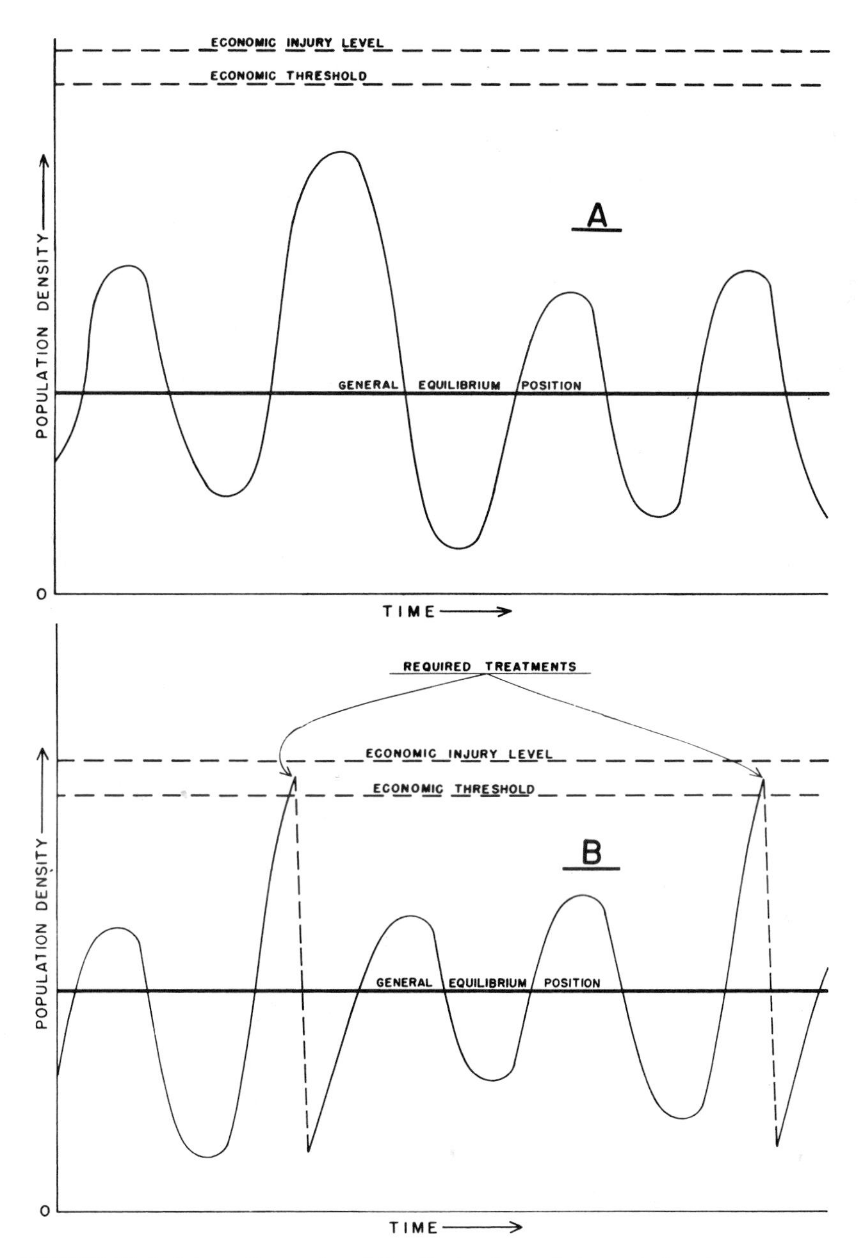


FIGURE 10–4. Schematic graphs of the fluctuations of theoretical arthropod populations in relation to their general equilibrium position, economic thresholds, and economic-injury levels. *A*, Noneconomic population whose general equilibrium position and highest fluctuations are below the economic threshold, *e.g.*, *Aphis medicaginis* Koch on alfalfa in California. *B*, Occasional pest whose general equilibrium position is below the economic threshold but whose highest population fluctuations exceed the economic threshold, *e.g.*, *Grapholitha molesta* Busck on peaches in California.

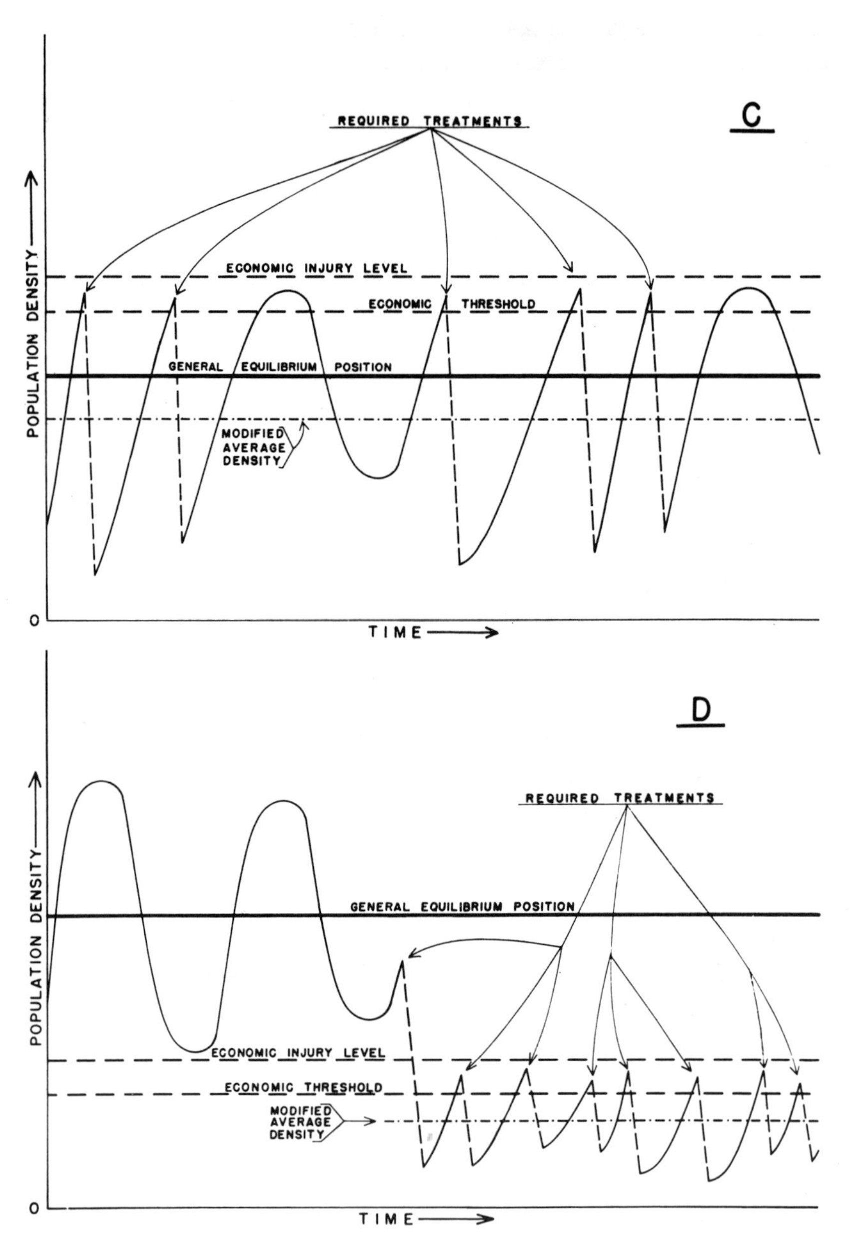


FIGURE 10–4 (Cont.). *C*, Perennial pest whose general equilibrium position is below the economic threshold but whose population fluctuations frequently exceed the economic threshold, *e.g.*, *Lygus* spp. on alfalfa seed in the western United States. *D*, Severe pest whose general equilibrium position is above the economic threshold and for which frequent and often widespread use of insecticides is required to prevent economic damage, *e.g.*, *Musca domestica* in Grade A milking sheds.

threshold, the population density will reach the threshold frequently (Fig. 10–4, *C*). In some cases, the general equilibrium position and the economic threshold are at essentially the same level. Thus, each time the population fluctuates up to the level of the general equilibrium position insecticidal treatment is necessary. In such species the frequency of chemical treatments is determined by the fluctuation rate about the general equilibrium position, which in some cases necessitates almost continuous treatment.

Finally, there are pest species in which the economic threshold lies below the general equilibrium position; these constitute the most severe pest problems in entomology (Fig. 10–4, *D*). The economic threshold may be lower than the level of the lowest population depression caused by the physical and biotic factors of the environment, e.g., many insect vectors of viruses. In such cases, particularly where human health is concerned, there is a widespread and almost constant need for chemical control. This produces conditions favorable for development of insecticide resistance and other problems associated with heavy treatments.

One solution to pest problems and particularly those in this last category is to change the environment permanently so that the general equilibrium position will be lowered. For example, this might be accomplished through the introduction of a new biological control agent or through the permanent modification of a large portion of a required habitat. This has been done in certain areas with malaria-vector mosquitoes and similar pests by the draining of swamps and the destruction of other favorable habitats. Such methods may completely eliminate the species from some areas.

Environmental changes unfavorable to the pest may also be made through the use of plants and animals resistant to the pest species. This control method may involve three different aspects—tolerance, preference, and antibiosis (Painter, 1951). If tolerance alone is involved, the general equilibrium position may not be changed but the economic threshold is raised. Where preference or antibiosis is involved, the ability of the pest to reproduce upon the host is reduced, so that the general equilibrium position is lowered.

The lack of a sound measure of economic thresholds, in many cases, has been a major stumbling block to the development of integrated pest-control programs. Our changing economy, variations in natural governing mechanisms from one geographical area to another, differences in consumer demands, and the the complexity of measuring the total effect of insects on yield and quality often make the assessment of economic damage extremely difficult. Yet the economic threshold and the economic-injury level must be determined reasonably and realistically before integrated pest control can develop to its fullest. Success in any well-balanced pest-control program is dependent on the aim of holding insect populations below experimentally established economic levels rather than attempting to eliminate all the insects.

## The Integration of Biological and Chemical Control

Biological control and chemical control are not necessarily alternative methods; in many cases they may be complementary, and, with adequate understanding, can be made to augment one another. One reason for the apparent incompatibility of biological and chemical control is our failure to recognize that the control of arthropod populations is a complex ecological problem. This leads to the error of imposing insecticides on the ecosystem, rather than fitting them into it. It is short-sighted to develop a chemical control program for the elimination of one insect pest and ignore the impact of that program on the other arthropods, both beneficial and harmful, in the ecosystem. On the other hand, this approach is no worse than the other extreme which would eliminate chemicals to preserve the biological control even in the face of serious economic damage. For we must recognize that modern agriculture could not exist without the use of insecticides. The evidence that biological and chemical control can be integrated is mounting. It has come from many sources involving many kinds of pests in various situations: see Ullyett (1947), Pickett and Patterson (1953), Ripper (1956), Huffaker and Kennett (1956), DeBach (1958*a*), Stern and van den Bosch (1959), and many others.

In approaching an integrated control program, we must realize that man has developed huge monocultures, he has eliminated forests and grasslands, selected special strains of plants and animals, moved them about, and in other ways altered the natural control that had developed over thousands upon thousands of years. We could not return to those original conditions if it were desirable. We may, however, utilize some of the mechanisms that existed before man's modifications to establish new balances in our favor.

### *Recognition of the Ecosystem*

To establish new, favorable balances, it is first necessary to recognize the "oneness" of any environment, natural or man-made. The populations of plants and animals (including man) and the nonliving environment together make up an integrated unit, the ecosystem. If an attempt is made to reduce the population level of one kind of animal (for example, a pest insect) by chemical treatment, modification of cultural practices, or by other means, other parts of the ecosystem will be affected as well. For this reason, the production of a given food or fiber must be considered in its entirety. This includes simultaneous consideration of insects, diseases, plant nutrition, plant physiology, and plant resistance, as well as the economics of the crops (Forbes, 1880; Ullyett, 1947; Pickett, 1949; DeBach, 1951; Solomon, 1953; Pickett and Patterson, 1953; Glen, 1954; Michelbacher, 1954; Huffaker and Kennett, 1956; Simmonds, 1956; Balch, 1958; Decker, 1958; and others).

In most agricultural ecosystems, some potentially harmful organisms are continually held at subeconomic levels by natural controlling forces. In others, the pests are held below economic levels only part of the time. A pest species may be under satisfactory biological control over a large area or a long period of time, but not in all individual fields or during all periods. In a single field or orchard, or during a portion of a year, the pest population may rise to economic levels, while elsewhere or at other times the pest may be subeconomic. It is in such situations that integrated control programs are especially important. These intermittently destructive populations must be reduced in a manner that permits the biological control which prevailed before or prevails elsewhere to take over again. If a chemical treatment destroys the biotic agents without eradicating the pest, then repeated treatments may become necessary.

### *Population Sampling and Prediction*

The sampling methods utilized by most research investigators for experimental plots are usually too time-consuming and tedious to be of practical value in establishing pest population levels in commercial crops. Special index methods are needed that are rapid and simple to use. Ideally, these should be of such nature that they can be easily utilized by the person examining the crop. But in many cases the grower is not able to evaluate all situations because of the difficulties and complexities involved in determining the status of some pest populations at the times of the year when they must be controlled. Then qualified entomologists will be required to evaluate the populations (Ripper, 1958).

One answer to this problem has been the development of supervised control in California, Arkansas, Arizona, and elsewhere. In a supervised control program the farmer, or a group of farmers, contracts with a professional entomologist who determines the status of the insect populations. On the basis of his population counts, other conditions peculiar to each situation, and his knowledge of the ecology of the pests and their biological controls, the entomologist makes predictions as to the course of the population trends and advises as to when controls should be applied and what kind. For instance, in the case of the alfalfa caterpillar, *Colias philodice eurytheme*, when economic thresholds are reached, the recommended procedure may involve immediate cutting of the hay crop without treatment, application of the polyhedrosis virus (Steinhaus and Thompson, 1949; Thompson and Steinhaus, 1950), or treatment with an insecticide. The course to be taken depends on the characteristics of the particular infestation (Smith and Allen, 1954).

Wherever possible, knowledge must be developed so that we can predict the times when occasional pests will be present in outbreak numbers. This will eliminate unnecessary and environment-disturbing "insurance" treatments. When this is not possible, the treatments can be timed according to the actual pest population levels, as is now done with many field-crop pests.

With those crops that do not yet have fixed chemical control schedules, every effort should be made to plan programs dependent upon pest population levels and to avoid dependence upon insurance and prophylactic treatments. If this is not done there is real danger that on these crops, too, pest-control problems will become increasingly complex.

### Augmentation of Natural Enemies

In some situations, the development of integrated control requires the augmentation of the natural-enemy complex (DeBach, 1958*a*). The introduction of additional natural enemies is usually the simplest and best solution. This may not be possible or effective with some pest species, and methods of overcoming the inefficiency of the natural-enemy complex must be sought. This can be done by periodic colonization of parasite or predators (Doutt and Hagen, 1950; DeBach, Landi, and White, 1955), artificial inoculation of the host at times of low density (Smith and DeBach, 1953; Huffaker and Kennett, 1956), modification of the environment, or selective breeding of parasites and predators.

The modification of the environment may involve changes in irrigation, introduction of a covercrop, or development of greater plant heterogeneity. Refuges for beneficial forms can be produced by strip treatments with chemicals (DeBach, 1958 *a*) or by the development of uncultivated and untreated areas (Grison and Biliotti, 1953). Modifications of the environment may also involve the control of ants or other organisms which curtail parasite and predator activity (Flanders, 1945; DeBach, Fleschner, and Dietrick, 1951). The selective breeding of parasites and predators may be directed toward increased or modified tolerance ranges to physical conditions (DeBach, 1958 *b*) or insecticides (Robertson, 1957).

Where prophylactic treatments are proved to be necessary for a perennial pest, selective materials must be developed and utilized to foster biological control both of other pests and of the pest of direct concern at other times.

### *Selective Insecticides*

Chemical control programs are limited by the nature of the available materials. In the past, nonselective insecticides applied for one insect in a pest complex often have eliminated the biotic factors holding other pests in check. More recently, we have had available a greater variety of materials, some of them selective in their action (Ripper, 1956).

The selective use of insecticides may be accomplished in at least four ways. First, the insecticide itself may be selective in its toxicological action. Narrow-range toxicants may be utilized to reduce a pest of concern and at the same time spare the beneficial forms (Ripper, 1944; Ripper, Greenslade, and Hartley, 1951). A particular material may be selective in one situation and not in another; or it may be selective at low dosages but not at high dosages. Furthermore, the manner of application (Ripper, 1956) and especially the type of carrier and residue deposit may produce differential effects on the insect complex (Flanders, 1941; Holloway and Young, 1943; DeBach and Bartlett, 1951).

Second, we can produce a selective action on a pest-parasite complex by treating only those areas where the pest-parasite ratio is unfavorable. This method is one of the bases of supervised control of the alfalfa caterpillar in California (Smith and Allen, 1954). Population levels of both the host caterpillar and its parasite, *Apanteles medicaginis* (Meusebeck), are determined at appropriate intervals in all fields. A prediction of possible damage is made on the basis of these population levels, and only those infestations which are potentially damaging are treated. In this way, on an area-wide basis, the balance is shifted in favor of the parasites, even though parasite adults and parasite larvae within the host caterpillars are often destroyed in the treated fields. The success of such programs will depend on the exact nature of the local problem and the quality of supervision. The rates of dispersal of parasites, predators, and pests are complicating factors.

Third, proper timing of insecticides can produce a selective action on the pest and natural-enemy complex (Ewart and DeBach, 1947; Michelbacher and Middlekauff, 1950; Bartlett and Ewart, 1951; Jeppson, Jesser, and Complin, 1953; Massee, 1954; DeBach, 1955). In such situations, an intimate knowledge of the behavior patterns of the pests and their natural enemies is required.

Fourth, nonselective materials with short residual action may be used if the beneficial forms can survive in a resistant stage or in an untreated reservoir area. Stern and van den Bosch (1959) have demonstrated that parasites of the spotted alfalfa aphid can survive nonselective treatments if they are in the more resistant pupal stage. DeBach (1958 *a*) reports successful integration of biological and chemical control of purple scale on citrus where alternate pairs of tree rows were sprayed at 6-month intervals with a nonselective oil treatment.

For some pests a disease pathogen may be used as a selective insecticide (Steinhaus, 1954; 1956). For example, under supervised control in the Dos Palos area of California, the polyhedrosis virus affecting the alfalfa caterpillar has been used successfully either alone or in combination with a selective insecticide to avoid the use of a nonselective treatment. More recently, interest has developed in the commercial use of virulent strains of *Bacillus thuringiensis* (Berliner), for the control of certain truck- and field-crop pests in California. The use of disease pathogens as selective insecticides is in its infancy, but can be expected to increase in importance with additional research (Steinhaus, 1951, 1957).

The ideal selective material is not one that eliminates all individuals of the pest species while leaving all of the natural enemies. Use of such a material would force the predators and parasites to leave the treated area or starve (Clausen, 1936; Flanders, 1940). The ideal material is one that shifts the balance back in favor of the natural enemies (Boyce, 1936; Ripper, 1944; Wigglesworth, 1945).

### *Future of Integrated Control*

If our knowledge were adequate today to outline an ideal integrated control program for a crop now utilizing an intensive fixed spray program, it would not be possible to switch to such a program immediately. The effects of the previous treatments may last several years. In some instances, effective biological control no longer exists and would have to be reëstablished. This may be a slow process (DeBach, 1951; DeBach and Bartlett, 1951; Pickett and Patterson, 1953; and others).

It should be emphasized also that the development of integrated control is not a panacea that can be applied blindly to all situations, for it will not work if biotic mortality agents are inadequate or if low economic thresholds preclude utilizing biological control (Barnes, 1959). However, it has worked so well in some appropriate situations that there can be no doubt as to its enormous advantages and its promise for the future.

## Literature Cited

Balch, R. E. 1958. Control of forest insects. *Ann. Rev. Ent.* 3:449–68.

Barnes, M. M. 1959. Deciduous fruit insects and their control. *Ann. Rev. Ent.* 4:343–62.

Bartlett, B., and W. H. Ewart. 1951. Effect of parathion on parasites of *Coccus hesperidum*. *Jour. Econ. Ent.* 44:344–347.

Bartlett, B. R., and J. C. Ortega. 1952. Relations between natural enemies and DDT-induced increases in frosted scale and other pests of walnuts. *Jour. Econ. Ent.* 45(5):783–85.

Bovey, Paul. 1955. Les actions secondaires des traitements antiparasitaires sur les populations d'insectes et d'acariens nuisibles. *Schweiz. Landw. Monatsh.* 33(9/10):369–79.

Boyce, A. M. 1936. The citrus red mite. *Paratetranychus citri* McG. in California, and its control. *Jour. Econ. Ent.* 29(1):125–30.

Brown, A. W. A. 1951. *Insect control by chemicals*. John Wiley & Sons, Inc., New York, N.Y. 817 pp.

———. 1959. Spread of insecticide resistance. *Adv. Pest Control Res.* 2:351–414. Interscience Publishers, Inc., New York, N.Y.

Clausen, C. P. 1936. Insect parasitism and biological control. *Ent. Soc. Amer. Ann.* 29:201–23.

———. 1956. Biological control of insect pests in the continental United States. *U.S. Dept. Agr. Tech. Bul.* 1139. 151 pp.

Crow, J. F. 1957. Genetics of insect resistance to chemicals. *Ann. Rev. Ent.* 2:227–46.

DeBach, P. 1951. The necessity for an ecological approach to pest control on citrus in California. *Jour. Econ. Ent.* 44(4):443–47.

———. 1955. Validity of the insecticidal check method as a measure of the effectiveness of natural enemies of diaspine scale insects *Jour. Econ. Ent.* 48(5):584–88.

———. 1958a. Application of ecological information to control of citrus pests in California. *Tenth Internatl. Congr. Ent. Proc.* 3:185–97.

———. 1958b. Selective breeding to improve adaptations of parasitic insects. *Tenth Internatl. Congr. Ent. Proc.* 4:759–67.

DeBach, P., and B. Bartlett. 1951. Effects of insecticides on biological control of insect pests of citrus. *Jour. Econ. Ent.* 44(3):372–83.

DeBach, P., C. A. Fleschner, and E. J. Dietrick. 1951. A biological check method for evaluating the effectiveness of entomophagous insects. *Jour. Econ. Ent.* 44(5):763–66.

DeBach, P., J. H. Landi, and E. B. White. 1955. Biological control of red scale. *California Citrog.* 40:254, 271–72, 274–75.

Decker, G. C. 1958. Don't let the insects rule. *Agr. Food Chem.* 6(2):98–103.

Doutt, R. L. 1958. Vice, virtue and the vedalia. *Ent. Soc. Amer. Bul.* 4(4):119–23.

Doutt, R. L., and K. S. Hagen. 1950. Biological control measures applied against *Pseudococcus maritimus* on pears. *Jour. Econ. Ent.* 43(1):94–96.

Elton, Charles S. 1958. *The ecology of invasions by animals and plants*. 181 pp. Methuen & Co., Ltd., London; John Wiley & Sons, Inc., New York, N.Y.

Ewart, W. H., and P. DeBach. 1947. DDT for control of citrus thrips and citricola scale. *California Citrog.* 32:242–45.

Flanders, S. E. 1940. Environmental resistance to the establishment of parasitic Hymenoptera. *Ent. Soc. Amer. Ann.*, 33:245–53.

———. 1941. Dust as inhibiting factor in the reproduction of insects. *Jour. Econ. Ent.* 34(3):470–1.

———. 1945. Coincident infestations of *Aonidiella atrana* and *Coccus hesperidum*, a result of ant activity. *Jour. Econ. Ent.* 38:711–12.

Fleschner, C. A., and G. T. Scriven. 1957. Effects of soil-type and DDT on ovipositional response of *Chrysopa californica* (Coq.) on lemon trees. *Jour. Econ. Ent.* 50(2):221–22.

Forbes, S. A. 1880. On some interactions of organisms. *Illinois State Lab. Nat. Hist. Bul.* 3:3–17.

Glen, R. 1954. Factors that affect insect abundance. *Jour. Econ. Ent.* 47:398–405.

Griffiths, J. T. 1951. Possibilities for better citrus insect control through the study of the ecological effects of spray programs. *Jour. Econ. Ent.* 44:464–68.

Grison, P., and E. Biliotti. 1953. La signification agricole des "stations refuges" pour la faune entomologique. *Acad. d'Agr. de France Compt. Rend.* 39(2):106–9.

Hayes, W. J. 1954. Agricultural chemicals and public health. *U.S. Public Health Reports* 69(10):893–98.

Hertzler, J. O. 1956. *The crisis in world population*. Univ. Nebraska Press, Lincoln. 279 pp.

Holloway, J. K. and T. Roy Young, Jr. 1943. The influence of fungicidal sprays on entomogenous fungi and on the purple scale in Florida. *Jour. Econ. Ent.* 36(3): 453–457.

Hoskins, W. M., and H. T. Gordon. 1956. Arthropod resistance to chemicals. *Ann. Rev. Ent.* 1:89–148.

Huffaker, C. B., and C. E. Kennett. 1956. Experimental studies on predation: Predation and cyclamen-mite populations on strawberries in California. *Hilgardia* 26(4):191–222.

Jeppson, L. R., M. J. Jesser, and J. O. Complin. 1953. Timing of treatments for control of citrus red mite on orange trees in coastal districts of California. *Jour. Econ. Ent.* 46:10–14.

Klostermeyer, E. C., and W. B. Rasmussen. 1953. The effect of soil insecticide treatments on mite populations and damage. *Jour. Econ. Ent.* 46:910–12.

Linsley, E. G., (ed.). 1956. Evaluation of certain acaricides and insecticides for effectiveness, residues, and influence on crop flavor. *Hilgardia* 26(1):1–106.

Lord, F. T. 1947. The influence of spray programs on the fauna of apple orchards in Nova Scotia: II. Oystershell scale, *Lepidosaphis ulmi* (L.) *Canad. Ent.* 79:196–209.

Massee, A. M. 1954. Problems arising from the use of insecticides: effect on the balance of animal populations. *6th Commonwealth Entomol. Conf. Rept.* pp. 53–57. London, England.

Metcalf, R. L. 1955. Physiological basis for insect resistance. *Physiol. Rev.* 35:197–232.

Michelbacher, A. E. 1954. Natural control of insect pests. *Jour. Econ. Ent.* 47(1):192–94.

Michelbacher, A. E., and S. Hitchcock. 1958. Induced increase of soft scales on walnut. *Jour. Econ. Ent.* 51(4):427–31.

Michelbacher, A. E., and W. W. Middlekauff. 1950. Control of the melon aphid in northern California. *Jour. Econ. Ent.* 43(4):444–47.

Painter, R. H. 1951. *Insect resistance in crop plants.* xi + 520 pp. Illus. Macmillan Company, New York, N.Y.

Petty, C. S. 1957. Organic phosphate insecticide poisoning; an agricultural occupational hazard. *Louisiana St. Med. Soc. Jour.* 109(5):158–64.

Pickett, A. D. 1949. A critique on insect chemical control methods. *Canad. Ent.* 81(3):67–76.

Pickett, A. D., and N. A. Patterson. 1953. The influence of spray programs on the fauna of apple orchards in Nova Scotia. IV. A Review. *Canad. Ent.* 85(12):472–78.

Pickett, A. D., W. L. Putnam, and E. J. Le Roux. 1958. Progress in harmonizing biological and chemical control of orchard pests in eastern Canada. *Tenth Internatl. Congr. Ent. Proc.* 3:169–74.

Ripper, W. E. 1944. Biological control as a supplement to chemical control of insect pests. *Nature* 153:448–52.

———. 1956. Effect of pesticides on balance of arthropod populations. *Ann. Rev. Ent.* 1:403–38.

———. 1958. The place of contracting organizations and professional supervision in the application of pest control methods. *Tenth Internatl. Congr. Ent. Proc.* 3:93–97.

Ripper, W. E., R. M. Greenslade, and G. S. Hartley. 1951. Selective insecticides and biological control. *Jour. Econ. Ent.* 44(4):448–59.

Robertson, J. G. 1957. Changes in resistance to DDT in *Macrocentrus ancylivorus Rohw. Canad. Jour. Zool.* 35(5):629–633.

Sabrosky, C. W. 1952. How many insects are there? pp. 1–7 *in:* Insects, 1952 Yearbook of Agriculture. 780 pp. U.S. Dept. Agr., Washington, D.C.

Schneider, F. 1955. Beziehungen zwischen nützlingen und chemischer Schädlingsbekämpfung. *Deutsche Gesell. f. Angw. Ent. Verhandl.* (13[te] Mitglied versamm.). 1955: 18–29.

Simmonds, F. J. 1956. The present status of biological control. *Canada. Ent.* 88(9):553–63.

Smith, H. S., and P. DeBach. 1953. Artificial infestation of plants with pest insects as an aid in biological control. *Seventh Pacific Sci. Congr. Zoology* (1949) Proc. 4:255–59.

Smith, Ray F. 1959. The spread of the spotted alfalfa aphid, *Therioaphis maculata* (Buckton), in California. *Hilgardia* 28(21):647–91.

Smith, R. F., and W. W. Allen. 1954. Insect control and the balance of nature. *Sci. Amer.* 190(6):38–42.

Solomon, M. E. 1953. Insect population balance and chemical control of pests. *Chem. and Indus.* 1953:1143–47.

Steinhaus, E. A. 1951. Possible use of *Bacillus thuringiensis* Berliner as an aid in the biological control of the alfalfa caterpillar. *Hilgardia* 20(18):359–81.

———. 1954. The effects of disease on insect populations. *Hilgardia* 23(9):197–261.

———. 1956. Potentialities for microbial control of insects. *Agr. Food Chem.* 4(8):676–80.

———. 1957. Concerning the harmlessness of insect pathogens and the standardization of microbial control products. *Jour. Econ. Ent.* 50(6):715–20.

Steinhaus, E. A., and C. G. Thompson. 1949. Preliminary field tests using a polyhedrosis virus in the control of the alfalfa caterpillar. *Jour. Econ. Ent.* 42(2):301–5.

Stern, V. M., and R. van den Bosch. 1959. Field experiments on the effects of insecticides. *Hilgardia* 29(2):103–30.

Strickland, A. H. 1954. The assessment of insect pest density in relation to crop losses. *6th Commonwealth Ent. Conf. Rept.* pp. 78–83. London, England.

Thomas, W. L. Jr. (ed.). 1956. *Man's role in changing the face of the earth.* 1193 pp. Univ. Chicago Press, Chicago, Ill.

Thompson, C. G., and E. A. Steinhaus. 1950. Further tests using a polyhedrosis virus to control the alfalfa caterpillar. *Hilgardia* 19(14):411–45.

Trouvelot, B. 1936. Le doryphore de la pomme de terre (*Leptinotarsa decemlineata* Say) en Amerique de Nord. *Ann. Epiphyt.* (n.s.) 1:227–336.

Ullyett, G. C. 1947. Mortality factors in population of *Plutella maculipennis* (Tineidae:Lep.) and their relation to the problem of control. *South African Dept. Agr. and Forestry Ent. Mem.* 2(6):77–202.

———. 1951. Insects, man and the environment. *Jour. Econ. Ent.* 44(4):459–64.

Upholt, W. M. 1955. Evaluating hazards in pesticides use. *Agr. Food Chem.* 3(12):1000–6.

Wille, J. E. 1951. Biological control of certain cotton insects and the application of new organic insecticides in Peru. *Jour. Econ. Ent.* 44:13–18.

Wigglesworth, V. B. 1945. DDT and the balance of nature. *Atlantic Monthly* 176(6):107–13.

# 11

# Life against life—nature's pest control scheme

*C. B. Huffaker*

Two vast and penetrating developments meeting concurrently in recent human events have served to catapult a formerly little recognized method of pest control to a position of eminence, both in the minds of the lay public and the pursuits of research scholars. The method of control referred to is "biological control," in which man utilizes one kind of organism against another. Biological control as a scientific and applied endeavor is relatively recent; the name itself was first used in 1919 by Professor H. S. Smith of the University of California.

The explosive increase in the world's human population, with consequent need to feed an ever-increasing number of hungry mouths, and the concomitant, and largely resultant, pollution of the environment in which man must live and produce the things he needs, are forcing him to search for means of solving the first problem without intensifying the latter.

Rachel Carson's "Silent Spring" was perhaps only the timely spark that crystallized in the public mind and even in the halls of Congress a consciousness of the fact that some of our common pests can be controlled by using their own enemies against them—that we are not obligated to rely solely on toxic chemicals in order to produce the food and fiber needed in today's world.

In any event, support for research on biological methods of pest

---

Huffaker, C. B. 1970. Life against life—nature's pest control scheme. Environmental Research, *3:* 162–175.

control very quickly surpassed that of any previous period. Prior to the book's publication, only two or three states had effective research programs in this area, and the program of the Federal government was feebly supported and largely unproductive. Today, in many of the states some such research is conducted, and the Federal government has greatly expanded its program. Various sources of special awards or grants-in-aid have become available. The total number of scientific papers in this area published in 1955 throughout the world was about 240, but this number increased rapidly within a few years before and following "Silent Spring," exceeding 1,400 by 1963 (Simon, 1967).

The International Biological Program (I.B.P.) has recognized the need for expansion in biological control research. It is sponsoring a concerted worldwide effort, but here we get ahead of the story. We must first tell you what "biological control" is all about, about its strategy, and why it may be desirable over the use of chemicals in the suppression of mites, lice, or weeds, for example.

Man's strategy in pest control would ideally be one of complete eradication of those pests that are his unmitigated enemies. Achievement of this state was forecast by some entomologists when DDT first arrived on the scene. Among other difficulties, as we are now aware, however, organisms may counterattack, developing resistance to such toxicants. More recently, it has been suggested that we may eventually be able to eradicate from producing areas all forms of life other than the ones involved in production (Beirne, 1967). Such a view tends to ignore the vast complexity of nature, of environmental variability, and of the forces tending to sustain life, even under great adversity.

A more realistic strategy would seem to be one simply of control of pest species at noneconomic densities. With this goal, man has available a vast arsenal of the tools for such a war. He can use the species' own enemies against them—that is, biological control—first simply by stopping the measures now used that harm them, where feasible, by introducing new and effective ones from other lands, by artificially augmenting their numbers, and by creating more favorable circumstances for them in crop environments.

For years, economic entomologists appeared to feel that ". . .salvation lies at the end of a spray nozzle," as Professor Paul DeBach of the University of California writes in the book, "Biological Control of Insect Pests and Weeds." In their view, the chemist could always have ready another pill when the last one no longer sufficed. Yet, a more enlightened economic entomology is at hand. Thus, advocates of a holistic, integrated pest control look at use of pesticides as a supplemental measure, to be used in ways that will least upset natural controls and, perhaps, make them more effective.

Chemical controls interfere to greater or less degree with use of natural enemies, but some chemicals are less harmful than others. In a pest management approach, the use of well-chosen pesticides, selected for their specific action on the target pest and for their low toxicity to natural enemies of the pest (and to those of other potential pests), offers much promise. While all forms of control are utilized, emphasis is usually on methods favoring natural enemies, for in areas highly favorable to pests the enemies are the principal element of natural control. Where suitable ones are lacking, introductions are suggested. Thus, use of "enemies," whether man-made and unnatural ones (sterilized males or genetic weaklings)[1] or simply nature's own enemies (parasitic, predatory, and pathogenic species), can be more satisfactory than use of broad-spectrum toxicants, which harm the beneficial forms as well as the pests and, by destroying the enemies, tend to intensify the specific problem or cause new ones, sometimes on a vast scale.

## Use of Biological Control

Biological control, where effective, is cheap, usually persistent without need for recurrent expense, entails no significant genetic counterattack in the pests, usually does not occasion the rise to pest status of forms normally innocuous, does not add to the ever-growing problem of man's pollution of his environment, and is not attendant with the serious toxic hazards to the workers using the method, to consumers of the products, or to our cherished and declining wildlife. Moreover, because of the expense in the use of other methods it is often the only method available in underdeveloped countries.

The most successful biological control has involved importation of parasitic, predatory, and pathogenic enemies from foreign lands—usually, but not invariably, from the pest's original home area. Such enemies are thus introduced to redress the imbalance in its affairs engendered by its new-found escape from those enemies.

While such cases present the most promising prospects, comparable control of endemic potential pests by their endemic enemies is evident all around us. One can demonstrate such by the simple procedure of excluding the enemies by hand removal (Fleschner, 1958; Huffaker and Kennett, 1953), chemical removal (DeBach, 1946; Huffaker and Kennett, 1956), or by use of colonies of ants (DeBach et al., 1951), for example, that may interfere with their action. In fact, this is exactly what has often been done, sometimes on a vast scale, when man has applied broad-spectrum pesticides. The notorious outbreaks in spider mites, scales, and caterpillars, for example, following World War II have been widely con-

[1] While such deranged males or freaks are certainly not *natural* enemies of the target pest, they accomplish what a natural enemy would accomplish, but in a different way—i.e., their mating with the normal individuals results in a reduction in population progeny.

sidered as largely due to the destruction of their enemies by the newly introduced chlorinated hydrocarbon and organophosphorous insecticides.

The first major biological control effort virtually saved the citrus industry in California at the turn of the century. Australia experienced two examples wherein biological control virtually saved from economic ruin and desolation, vast agricultural areas in Queensland and other dry-land regions, these being the results of introduction of a moth for control of prickly pear and of myxomatosis disease for control of rabbit plagues. One of the main supporting crops of the Hawaiian economy, sugar cane, has been saved through use of biological control introductions.

Until recently, biological control has had scant support in view of the striking results obtained from the support it has had. It has been estimated (DeBach, 1964) that research in biological control in California has returned far more per dollar invested than has insecticide research. On a worldwide basis, of the 66 cases of complete biological control of pest insects achieved by introduced enemies, a quite disproportionate number involved introductions in Hawaii, California, and certain British Commonwealth islands or possessions, in that order. This ranking corresponds in general to the support and effort put into biological control by competent, enthusiastic researchers. While properly oriented, ecologically based work is essential, the key to greater use of biological control is much greater support, and more and more effort at introductions and supporting programs by dedicated, well-trained specialists who understand its potentials and the ecosystem philosophy essential to the perseverance required in giving it a real chance.

Successful biological control in California, involving introduced enemies, include the infamous cottony cushion scale, red scale, olive scale, yellow scale, nigra scale, black scale, San Jose scale, spotted alfalfa aphid, pea aphid, citrophilus mealybug, western grapeleaf skeletonizer, alfalfa weevil, klamath weed, and tansy ragwort. Substantial or complete control has been experienced in each case.

On a worldwide basis, about 250 cases of partial to complete success are on record, with about 1/3 of these being so complete that the species now present no problem at all, and the economic impact in many instances has been momentous. Yet, it is also true that the discovery and introduction of new enemy species is being largely neglected, and there are many major pests on which little or no work has been conducted (DeBach, 1964).

One of the most intriguing stories in biological control has just recently been revealed by the work of Professor R. L. Doutt in California which highlights the sort of rewards that may be had from a concerted and thorough effort. This involves control of a *native* species, the grape leafhopper, which was for many years the principal grape pest in the extensive vineyards of the Great Central Valley. It was noted that in certain locations, the growers never had to use pesticides to control this species whereas most vineyards were under heavy treatment programs and yet the problem was ever-present.

It was found that a tiny egg parasite, *Anagrus epos*, parasitized the eggs of the leafhopper in these untreated vineyards, and very few young leafhoppers developed. These vineyards were all located along streams. However, this leafhopper does not overwinter in the egg stage, and the parasites could not be found overwintering in the vineyards as adult wasps. Studies in the native riverine haunts, where the leafhopper's only original host—wild grape—grows, revealed that the parasite does overwinter there but as immature parasites in the eggs of still another leafhopper, *Dikrella cruentata*, that is not found on grape but on evergreen native blackberries that grow in association with the wild grapes. Planting of very small patches of evergreen blackberries near vineyards located miles away from riverine sources of native blackberries, permits survival of the parasite on its alternate winter host. In the spring they move readily, up to three miles, into surrounding vineyards lacking them. Thus, the outlines are clear for a very promising and highly rewarding scheme for control of this serious pest in California.

The story of control of prickly pear, *Opuntia stricta*, in Australia holds rather less of the intriguing subleties often encountered in biological control, but the immensity of the result nearly dwarfs imagination. These spiny desert cacti were so favored by climate that by the mid-1920's they had figuratively devoured some 50 million acres of agricultural lands in Queensland alone. To save the economy of a vast region, specialists were sent to desert areas of southwestern United States and South America to obtain natural enemies of the cactus.

Some 50 species of *Opuntia*-eating insects were introduced from western U. S. A., Mexico, and Argentina. One of these, a moth, *Cactoblastis cactorum*, which is native to Argentina where it attacks close relatives of *Opuntia stricta* (a native of North America) readily became established. Figure 11–1 is a photographic documentation of an event that happened on a very vast scale as the result of the action of this moth. Within three years after introduction of the moth, dense, impenetrable stands of the weed began to disappear. In summing up the results of this program, Dodd (1959) stated: "The bold statement that (in Queensland alone) some 50,000,000 acres have been brought from virtual uselessness to successful settlement cannot convey the real picture of the transformation scene, from stagnation under the stranglehold of the pest to thriving townships, prosperous farms growing a variety of agricultural crops, herds of dairy and beef cattle, and flocks of sheep, and the developmental improvements associated with this reclamation." Here then is an example of an insect which, acting as a regulating agent, accomplished

FIGURE 11–1. Destruction by the introduced moth, *Cactoblastis cactorum*, of dense prickly pear in belar scrub country, Chinchilla, Queensland, Australia (after Dodd, 1940): *Upper view*, taken in October, 1926, the prickly pear in its virgin state. *Middle view*, three years later, in October, 1929, showing the characteristic destruction resulting from the feeding on the fleshy pads of pear by larvae of *Cactoblastis*. *Lower view*, taken in December, 1931, after trees had been cut and burnt off and the land put back into use, showing a prolific growth of Rhodesgrass.

a feat totally beyond the economic capacity of a nation to manage in any other way.

As another example, California red scale, *Aonidiella aurantii*, has probably been the subject of the most intensive, long-standing biological control effort ever attempted. This diaspine scale insect is a formidable pest in citrus orchards of California. Explorations in many parts of the world over the past 70 years have resulted in importation and colonization of at least 30 species of parasites and predators of which eight or more have become established (the accepted number depending on whether certain complexes are recognized as single species or as several distinct species).

Dr. Paul DeBach and associates of the Citrus Experiment Station at Riverside have intensively studied the results from some of these importations. They and their fellow staff members at Berkeley also developed three distinct methods of evaluation of the role of such parasites in the control of their hosts' populations: (1) The insecticidal check-method, discussed later relative to the olive scale, is the most generally useful. (2) The biological check-method consists in measuring the effectiveness of a parasite on its host through comparing the host densities in ant-present and ant-free trees. Ants interfere with the controlling action of the parasites only in some situations. Hence, the method is of limited application. (3) The removal of natural enemies from plants by hand as a "check-method" was used by Fleschner (1958) and by Huffaker and Kennett (1953).

DeBach has also recently established a number of apparently distinct species of *Aphytis* for use against California red scale. He finds them to be ecologically very different in their capacity to control the scale, and some of the more recently introduced ones are replacing their previously entrenched close relatives in some environments. Effectiveness of these parasites is, as expected, different in diverse climatic areas of Southern California.

In Figure 11–2 a comparison is shown of red scale populations under three different conditions in the same orchard. The results were followed by DeBach for six years. In one plot normal activity of enemies was fostered by ant-control operations. In plot 2 no ant control was practiced, and the interference by the ants resulted in a marked increase in red scale densities. In plot 3 natural enemy action was more fully inhibited by applications of DDT. The DDT-treated trees developed heavy infestations of scale and other pests (the latter complicated comparisons) and since the tree was endangered, this test was shifted to another tree (Plot 3a). The new DDT-treated plot responded in the same way. These test results are representative of occurrences in many field situations.

The Agricultural Research Service of the U.S. Department of Agriculture has recently engaged in an intriguing program of biological control in a quite different environment—i.e., control of a noxious weed

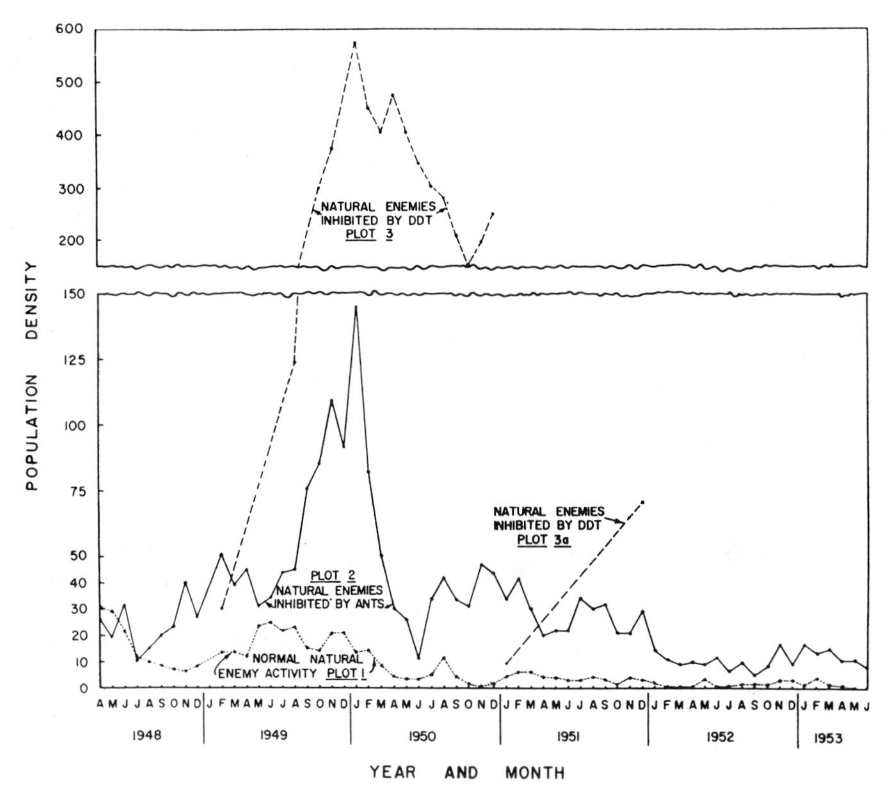


FIGURE 11–2. Different degrees of natural control as demonstrated by population density trends of the California red scale on lemon trees in adjacent plots under conditions of experimentally varied natural enemy effectiveness: Plot 1—excellent natural control under conditions of normal natural enemy effectiveness; Plot 2—fair natural control with natural enemy activity retarded by ant interference; Plots 3 and 3a—very poor natural control with natural enemy activity greatly retarded by the toxic effects of DDT sprays. Neither ant interference nor DDT residue had any appreciable effects on red scale populations except indirectly through adverse effects on natural enemies (after DeBach, 1958).

that displaces other water plants and clogs waterways (Fig. 11–3). A flea beetle, *Agasicles* sp., introduced from Argentina is clearing certain areas of this weedy aquatic (Maddox, 1967; Hawkes *et al.*, 1967).

I now wish to discuss three cases of technically complete biological control that I have myself conducted in California. These involve a weed, a scale insect, and a phytophagous mite.

## *The Klamath-weed Story*

Klamath weed (*Hypericum perforatum*), known more generally as St. Johnswort, is a serious weed in temperate regions of the world having moderate rainfall at least during the cold season. Although it is a toxic plant and may cause losses in weight and death of livestock if eaten

FIGURE 11–3. Control of alligatorweed, *Alternanthera phylloxeroides*, by a flea beetle, *Agasicles* sp. Top—Release site in Florida prior to attack by beetle; Bottom—Same site after attack by beetle. (After Hawkes *et al.*, 1967.)

in quantity, the greatest source of losses is due to the weed's displacement of desirable forage on the range. Prior to initiation of biological control in 1946, this weed infested approximately 5,000,000 acres of rangelands in the western United States. In some counties of California, banks were refusing loans for purchase of such lands, and chemical control measures were more expensive than the value of the properties. The striking success of this program occasioned the erection of a monument to the responsible beetle by a grateful ranching industry.

A number of insects have been introduced into various parts of the world for control of this pest, the pioneer example being in Australia. However, due to the unique climatic pattern in California (and to a lesser degree in adjacent western states), control in California has been phenomenal while in Australia and Canada the program has not been marked with the same high degree of achievement (Huffaker, 1967).

In this program repeated censuses, life-table data, experimental methods, extensive general observations, and use of photographs of "before" and "after" situations were employed (e.g., Huffaker, 1967).

Holloway and Huffaker (1952) reviewed the general results covering the first five years of this program, and Huffaker and Kennett (1959) presented in detail a ten-year record of changes in range vegetation associated with the control of this weed by the principal insect, a chrysomelid beetle, *Chrysolina quadrigemina.* The authors pointed out that excellent control of the weed by this species, in contrast to that achieved by its close relative, *Chrysolina hyperici*, is due to the close synchronization in the life history and general ecology of the superior beetle, with both the host weed and the climatic pattern prevailing. The longer, more severe dry season in California, contrasted to the conditions in Canada and Australia, makes the plant more susceptible to control by a given intensity and duration of damage. Comparing the two introduced beetles, in California *C. hyperici* does not respond as readily to the appearance of the fall rains, and consequently its reproductive phase is prolonged far into the early summer period. At that time the soil becomes too hard and dry for pupation. Consequently, *C. hyperici* is only partly successful, whereas *C. quadrigemina*, which escapes this adversity through earlier completion of development, is a strikingly successful control agent.

The leaf beetle, *C. quadrigemina*, was introduced in February, 1946. Within three years small areas at the locations of original plantings of colonies showed marked declines in weed densities. Figure 11–4 shows a sequence in development of the control experienced at the original Humboldt County site near Blocksburg (California). The upper view shows the field after the portion faced by the observer was already dead from attack of the beetles, while the remainder of the field (foreground) was still an undisturbed solid stand of the weed in full bloom. The middle and lower views show the same field two years, and 17 years later after the weed had been brought and maintained under complete control and a valuable grass cover had returned. At this site and at many others in coastal counties the climax bunchgrass, *Danthonia californica*, was an important element in range recovery.

FIGURE 11–4. St. Johnswort or Klamath weed control by *Chrysolina quadrigemina* at Blocksburg, California: (A) Photo in 1948: foreground shows weed in heavy flower while remainder of field has just been killed by the beetle; (B) shows same location in 1950 when heavy cover of grass had developed: (C) photo in 1966 showing the degree of control that has persisted since 1949. [This story was repeated all over California.[ (Upper photos by J. K. Holloway, lower one by Junji Hamai.)

The very high degree of control (Fig. 11–4) shows that this beetle is adept at locating isolated patches and individual plants of this weed. The fact that in local "escape" situations now, and as it did in the original heavy densities, it quickly builds up in numbers and then destroys the weed, proves its reproductive response capabilities. This numerical response dwarfs any lessening in individual or functional, i.e., behavioral response, at high host or beetle densities.

The fact that this species, *C. quadrigemina*, has essentially displaced three other species otherwise capable of survival in many open ranges again testifies to its high efficiency and searching power. These other species cannot compete with it, nor do they interfere with it, in practice, contrary to the thesis of Turnbull and Chant (1961), Zwölfer (1963), and Watt (1965).

### *Control of Olive Scale*

The diaspine scale, *Parlatoria oleae*, was first discovered in California in 1934 near Fresno. This insect attacks over 200 different host plants, including olive, ash, walnut, all the common deciduous fruits of the rose family, and a host of common shrubs valuable as landscaping ornamentals. The insect spread rapidly in the Central Valley, and in the 1940's it constituted perhaps the most destructive scale insect that had threatened the crops of California since the early years following establishment of the San Jose scale, some forty years previously. Orchard, park, and dooryard trees became encrusted with the scales, and death of limbs and branches was characteristic. At the time of introduction of the first natural enemies, no really effective chemical means of control was at hand, although oil treatments offered some relief.

Concentrated studies on the effects of the introduction have been conducted for 17 years (e.g., Huffaker *et al.*, 1962; Huffaker and Kennett, 1966; Kennett et al., 1966), for 10 years using life table data, and for the entire period, repeated censuses, experimental (check) methods, and "before" and "after" photographs.

Again, the degree of control by the parasitic wasps, *Aphytis maculicornis* and *Coccophagoides utilis*, compares favorably with any insecticidal program (Fig. 11–5). Densities are so low that from 1,000 to 15,000 leaves and subtending twig segments are required in a sample to obtain one female scale surviving parasitism. For our purpose, a sample 50 times this size must be carefully examined. The extremely efficient searching power of the parasites is indicated by the fact that even at such extremely low densities, 70 to 95 per cent of the female hosts are commonly parasitized.

These two parasites supplement one another, giving a much better control than either can accomplish alone. This is again contrary to concepts of Turnbull and Chant (1961), Zwölfer (1963), and Watt (1965). Indeed, *C. utilis* simply fills the vacant niche left by *A. maculicornis* in the period of summer activity on the spring generation of scales. This

FIGURE 11–5. Biological control of olive parlatoria scale. (Photos by Frank Skinner.) Top: Trees on right, clean, vigorous, scale under control by parasites (untreated with DDT). Trees on left, scaly, dying back, parasite control of scale inhibited by DDT treatment; Middle: Typical infestation on DDT treated trees; Bottom: Typically clean fruits, stems, and foliage on untreated, biological control trees.

is the area of its supplementing value. Although it often kills as many scales of the other (fall) generation as does *A. maculicornis*, this part of its activity is largely meaningless (dispensable), for *A. maculicornis* would attack most of those attacked by *C. utilis* if they were not already attacked (Huffaker and Kennett, 1966).

Again, the two efficient introduced parasites have so reduced the scale that four other enemy species formerly common, even abundant at times in high density situations, have been virtually eliminated as faunal elements in olive groves. Again, these less efficient enemies did not "interfere" with the efficient ones causing the host to be more abundant than would be true if they weren't present; they were simply eliminated in the competition, a pattern also observed by DeBach (1966) relative to red scale on citrus, and one we would expect on logical grounds.

### *The Cyclamen Mite*

Time will permit only brief treatment of the third case, cyclamen mite, *Steneotarsonemus pallidus* (Banks), but, again, good commercial control of this most serious pest of strawberries in California was shown to be practicable by stocking the fields early in their growth with two species of predatory mites, *Amblyseius cucumeris* (Ouds.) and *Amblyseius aurescens* A-H, using as a source those on clippings from older fields. Some 70 pairs of plots were employed over a period of 6 years, using check methods (predator-present vs. predator-absent plots), combined with frequent censuses and following of results independently by small locales. Again, other predators, such as the anthocorids, staphylinids, predatory thrips, and *Metaseiulus occidentalis* (Nesbitt), which are more efficient enemies of tetranychids but do feed on cyclamen mites when the latter are abundant, were seldom found feeding on them at the very low densities at which the prey was maintained by the two principal predators. They seemed never to interfere with the latter's efficiency.

In conclusion, we feel that given a greatly increased support for training of specialists, conduct of research and foreign importations, biological control of the classical type offers real opportunities for control of many of our pest problems, thus alleviating to that extent the ever-increasing problem of environmental pollution.

## References

Beirne, B. P. (1967). The future of integrated controls. *Mushi Suppl.* **39**, 127–130.

DeBach, P. (1946). An insecticidal check method for measuring the efficacy of entomophagous insects. *J. Econ. Entomol.* **39**, 695–697.

DeBach, P. (1958). The role of weather and entomophagous species in the natural control of insect populations. *J. Econ. Entomol.* **51**, 474–484.

DeBach, P. (Ed.) (1964). *Biological control of insect pests and weeds*. Reinhold, New York, 844 pp.

DeBach, P. (1966). The competitive displacement and coexistence principles. *Ann. Rev. Entomol.* **11**, 183–212.

DeBach, P., Fleschner, C. A., and Dietrick, E. J. (1951). A biological check method for evaluating the effectiveness of entomophagous insects. *J. Econ. Entomol.* **44**, 763–766.

DeBach, P., Fleschner, C. A., and Dietrick, E. J. (1953). Natural control of California red scale in untreated citrus orchards in southern California. *Proc. Seventh Pacific Sci. Congr. 1949 (Zoology)* **IV**, 236–248.

Dodd, A. P. (1940). "The biological campaign against prickly pear." Commonwealth Prickly Pear Board, Brisbane, Australia. pp. 177.

Dodd, A. P. (1959). "The biological control of prickly pear in Australia." In *Biogeography and Ecology in Australia.* (Monographiae Biologicae Vol. **8**, 565–577).

Fleschner, C. A. (1958). Field approach to population studies of tetranychid mites on citrus and avocado in California. *Proc. 10th Intern. Congr. Entomol.* **2**, 669–674.

Hawkes, R. B., Andres, L. A., and Anderson, W. H. (1967). Release and progress of an introduced flea beetle, *Agasicles* n.sp., to control alligator-weed. *J. Econ. Entomol.* **60**, 1476–1477.

Holloway, J. K., and Huffaker, C. B. (1952). "Insects to control a weed." In *Insects, the Yearbook of Agriculture*, 1952, pp. 135–140.

Huffaker, C. B. (1967). A comparison of the status of biological control of St. Johnswort in California and Australia. (*Proc. XI Pacific Sci. Congr.* Symp. No. 28); *Mushi Suppl.* **39**, 51–73.

Huffaker, C. B., and Kennett, C. E. (1953). Developments towards biological control of cyclamen mite on strawberries in California. *J. Econ. Entomol.* **46**, 802–812.

Huffaker, C. B., and Kennett, C. E. (1956). Experimental studies on predation: (I) Predation and cyclamen mite populations on strawberries in California. *Hilgardia* **26**, 191–222.

Huffaker, C. B., and Kennett, C. E. (1966). Studies of two parasites of olive scale, *Parlatoria oleae* (Colvée). IV. Biological control of *Parlatoria oleae* (Colvée) through the compensatory action of two introduced parasites. *Hilgardia* **37**, 283–335.

Huffaker, C. B., Kennett, C. E., and Finney, G. L. (1962). Biological control of olive scale, *Parlatoria oleae* (Colvée), in California by imported *Aphytis maculicornis* (Masi) (Hymenoptera: Aphelinidae). *Hilgardia* **32**, 541–636.

Kennett, C. E., Huffaker, C. B., and Finney, G. L. (1966). Studies of two parasites of olive scale. *Parlatoria oleae* (Colvée). III. The role of an autoparasitic aphelinid, *Coccophagoides utilis* Doutt, in the control of *Parlatoria oleae* (Colvée). *Hilgardia* **37**, 255–282.

Maddox, D. M. (1967). Bionomics of an alligator-weed flea beetle, *Agasicles* sp. in Argentina. *Ann. Entomol. Soc. Am.* **61**, 1299–1305.

Simon, H. R. (1967). Lassen sich Entwicklungslinien der biologischen Schädlingsbekämpfung an einer Bibliographie aufzeigen? *Entomophaga* **12**, 81–84.

Turnbull, A. L., and Chant, D. A. (1961). The practice and theory of biological control of insects in Cadada. *Can. J. Zool.* **39**, 697–753.

Watt, K. E. F. (1965). Community stability and the strategy of biological control. *Can. Entomol.* **97**, 887–895.

Zwölfer, H. (1963). The structure of the parasite complexes of some lepidoptera. *Z. Angew. Entomol.* **51**, 346–357.

# 12

# Eradication of the screwworm fly

*Alfred H. Baumhover*

Screwworms (Diptera, Calliphoridae, *Cochliomyia hominivorax* [Coquerell]) are important obligate parasites of various terrestrial mammals, including man. Distribution is restricted during winter months to tropical and subtropical areas of North and South America; but during other periods, the screwworm fly may disperse 500 miles or more from its normal overwintering area.

## *Life Cycle*

When held at 80 F (27 C), the adult flies mate at 2 to 3 days of age, and at 7 days of age, females oviposit shingle-like masses of 200 to 500 eggs. After an incubation period of 16 hours, larvae emerge and immediately start feeding in compact pockets, consuming primarily live flesh and wound fluids. In four to nine days after hatching, the larvae attain a body weight of 70 to 120 mg, drop from the wound, and burrow into the soil to pupate; adults emerge from the soil within eight days.

---

Reprinted from *The Journal of The American Medical Association* April 18, 1966, Vol. 196, pp. 240–248. 

From the Entomology Research Division, US Department of Agriculture, Oxford, NC.

Read as part of the Symposium on Insects and Disease during a joint meeting of the Section on Dermatology with the AMA Committee on Cutaneous Health and Cosmetics and the Entomology Research Division, US Department of Agriculture, at the 114th annual convention of the American Medical Association, New York, June 23, 1965.

### Economic Importance

Although the screwworm is primarily a pest of domestic animals and wildlife, man is highly susceptible to attack when he lives in primitive and unsanitary conditions. As its Latin name indicates, early records of the insect indicate that it infested man. With man as with other animals, breaks in the skin as well as fetid odors emanating from body openings invite oviposition. During 1935, when a severe screwworm outbreak caused 1,200,000 known cases in domestic animals in southern United States, 55 cases were recorded in man, with twice this number of cases being estimated (1) as unrecorded.

In recent years before control was effected, domestic livestock losses have been estimated at $20 million annually in southeastern United States and $50 to $100 million in the Southwest. These losses occurred in spite of availability of effective wound treatments. During the late 1800's, livestock producers, lacking effective remedies, were forced occasionally to restock because of heavy losses inflicted by the screwworm.

## Development of the Sterile-Male Technique

In 1937, Dr. E. F. Knipling (2) first proposed eradication of the screwworm through release of sexually sterile males. Surveys indicated that normal populations were quite low, which suggested the feasibility of rearing and releasing sufficient sterile males to compete with native males for mating opportunities and thus to suppress reproduction. Early attempts at chemical sterilization failed; however, in 1951 Bushland and Hopkins (3) successfully sterilized screwworms with 5 kr of x-radiation or gamma radiation ($^{60}$Co) administered to 5-day-old pupae.

When sterile and fertile males were caged with fertile females, the numbers of sterile and fertile eggs produced were almost in the same proportion as the numbers of sterile and fertile males. Preliminary field tests on Sanibel Island, Fla, in 1952 and 1953 demonstrated that the laboratory-reared sterilized males could compete for mating opportunities to effectively reduce native populations.

## Eradication of the Screwworm From Curaçao

Since Sanibel Island lies only a few miles from the mainland, an attempt to effect eradication in this area was considered impractical because of the threat of invasion from untreated areas. The island of Curaçao, 40 miles north of Venezuela in the Netherlands Antilles, was thus chosen for the initial eradication effort (4). This 170-square-mile island swarmed with goats and had a large population of screwworms. Severity of the screwworm problem in the goat population prompted a local veterinary officer, B. A. Bitter, to request recommendations for control of screwworms from the US Department of Agriculture. After the author completed a survey, the governor of Curaçao and the USDA administrator of the Agricultural Research Service agreed to use the island as a test area.

Screwworms were reared and sterilized at the USDA Entomology Research Division laboratory at Orlando, Fla, and air-shipped to Curaçao as pupae nearing eclosion. The pupae were placed in small paper bags, and the flies emerged there. Prior to release, the flies were fed honey through a screened portion of the bag. Releases were made weekly by dropping flies from light aircraft flying lines 1 mile apart. Egg masses were collected from host animals at 11 locations throughout the island to determine population levels and effectiveness of the sterile males in inducing sterility.

Preliminary releases indicated that 800 sterile flies per square mile would be required weekly to effectively reduce the native population. Both sexes were released since no convenient method was available for separation. The results (Table 12–1) closely followed theoretical calculations. Total egg masses declined from 49 during the first week to 0 during the ninth week, and egg-mass sterility rose from 60% to 100%. Although it was impossible to record the passing of the last fertile, mated native female, it appeared that eradication had been approached in three to four generations. Releases were continued an additional three months, and careful surveillance of host animals verified eradication of the screwworm.

TABLE 12–1

Screwworm Egg-Mass Collections*

| Week | No. of Egg Masses: Total | No. of Egg Masses: Sterile | % Sterile |
|---|---|---|---|
| 1 | 49 | 34 | 69 |
| 2 | 55 | 38 | 69 |
| 3 | 53 | 36 | 68 |
| 4 | 47 | 37 | 79 |
| 5 | 49 | 42 | 86 |
| 6 | 26 | 23 | 88 |
| 7 | 10 | 10 | 100 |
| 8 | 12 | 12 | 100 |
| 9 | 0 | 0 | ... |

*Collections made on Curaçao, Netherlands Antilles, during an experiment to eradicate the screwworm through the release of sexually sterile males.

## Mass Production, Sterilization, Release, and Field Evaluation

Successful eradication of the screwworm from Curaçao greatly increased interest in this approach for eradication of the screwworm from southeastern United States. As a result, research efforts were devoted to the development of improved rearing, sterilization, release, and field evaluation techniques needed to efficiently combat the screwworm in a 50,000-square-mile area.

During 1955, additional rearing facilities were constructed near Bithlo, Fla, in a relatively uninhabited area 20 miles east of Orlando. An isolated area was selected to avoid complaints previously encountered because of the foul odors associated with screwworm rearing. Initial goals were to establish a vigorous Florida strain of screwworms, develop mass-production techniques, decrease operation costs, and solve the odor problem. A production goal of 1 million flies per week, or six times the Curaçao production, was scheduled. Screwworms were collected from 12 locations in Florida and from 1 in southern Georgia. Single males were held with 15 to 25 females for 12 days to permit selection of progeny from long-lived parents and sexually active fathers. In five generations, mating frequency increased from 5 to 17 for the males, and longevity of both sexes was greatly increased. Females normally mate only once during their lifetimes.

In the initial trial, the production goal of 1 million was obtained by expanding operations based on procedural modifications adopted for Curaçao. Odor was greatly reduced by circulating conditioned air from the rearing cabinets through charcoal filters. Subsequent changes, described later in this section, were developed for sustained production of 2 million screwworm flies weekly. These insects were used for a 2,000-square-mile field test southeast of Orlando, Fla. This test was planned by personnel of the Entomology Research Division, USDA, in cooperation with officials of the Florida Livestock Board and the Animal Disease Eradication Division, USDA, who were to administer the state-federal program in southeastern United States (5). Releases were begun May 1, 1957, at the rate of 1,000 flies per square mile when screwworm infestations had reached outbreak proportions, and when almost every newborn calf in the test area required one or more treatments for navel infestations. Progress in suppressing the screwworm population was not as rapid as anticipated (6); however, weekly egg-mass collections declined from 41 per pen during May and June to only 3 per pen during August. Sterility ranged from 4% to 41% during May and June and reached 70% during July. Since the area was not isolated from untreated screwworm populations, eradication could not be demonstrated. Many of the fertile egg masses collected during the low population level in August likely were from migrant females mated outside of the test area. However, since the level of sterility obtained during July equaled that during the successful eradication attempt on Curaçao, the experiment was considered highly successful, and realistic planning for eradication of the screwworm from southeastern United States was begun.

### *Mass Production*

Screwworms were first reared successfully on artificial medium in 1936 by R. C. Bushland (7). Prior to this time they were reared in small numbers in living animals. The artificial diet originally consisted of milk, citrated calf blood, lean ground beef, and formalin (to deter decomposition). Tubs containing a thin layer of medium were held in a warm room, and at six to seven days after hatching, larvae crawled from the tubs to pupate in moist sand. Later, milk was deleted from the diet, and a mixture of 2 parts water, 2 parts lean ground beef, 1 part citrated blood, and 0.24% formalin was used. This basic formula, except for substitution of horsemeat for beef and minor adjustments in water and formalin content, was used successfully throughout the Curaçao campaign. Subsequent attempts by Graham and Dudley to use various animal and plant protein substitutes were unsuccessful in reducing costs of the rearing medium without reducing larva size. However, whale meat, which cost only half as much as horsemeat, was found acceptable (8). Graham and Dudley also found that large vats, heated from below, were superior to smaller containers held in heated cabinets or rooms.

Oviposition of screwworm flies for small-scale rearing was readily induced by transferring 10 to 15 flies from a cage measuring 12 × 12 × 20 inches to a small vial containing a few grams of lean meat and holding the flies in an incubator at 95 F (35 C). Egg production was greatly facilitated through the development of a large screened cage measuring 3½ × 3½ × 6 feet that was draped with paper toweling to provide adequate resting space for 50,000 flies (equivalent to 70 of the smaller cages). It was necessary to hold these large cages in darkness to avoid high mortality caused by the flies crowding into areas where light intensity was greatest. Oviposition was attained in a heated tray measuring 5½ × 1 feet that was 1¼ inches deep and contained a mixture of meat and an oviposition stimulant produced by incubating albumin from blood or meat juices. A wooden grate, placed over the vat, provided the dry area the flies preferred for oviposition, and a series of 7½ w incandescent lamps attracted the flies to the oviposition stimulant. After the four hours allowed for egg laying, oviposition was considered complete and the cage was rolled into a room held at 32 F (0 C) to inactivate the flies and allow easy removal of the oviposition device without allowing flies to escape. The eggs, cemented tightly together, were scraped from the tray and grid and weighed into 6-gm lots (120,000 eggs). Each cage of flies produced up to 6 million eggs.

Hatching began 12 hours after eggs were placed in the incubation chambers, since eggs laid at the beginning of the oviposition period were by this time 16 hours old. The hatching eggs were then placed on a starting medium in which blood plasma was substituted for whole blood. A pan 16 × 26 × 4 inches with medium ½ inch deep supported the larvae for 30 hours. Larvae were then transferred from the high-humidity starting room to the main floor of the plant onto 4 × 5-foot vats 1½ inches deep. Depth of the medium was increased from ½ to 1 inch as the larvae grew in size. Thermostatically controlled heating cables beneath the vats maintained the medium temperature near 95 F (35 C). Larvae fed across the vats in a compact mass, and unused medium and excreta were periodically removed and replaced with fresh medium. Growth of larvae was complete in four to six days, and during this time the vats were positioned over collecting funnels which directed the emerging larvae to sand-filled trays moving on an endless belt on the lower floor of the building. Trays were transferred to a circular monorail, and, eight hours later, when pupation had begun, the trays were emptied onto a mechanical sifter. Sand and pupae were separated, with the larvae and prepupae dropping onto a moving, endless screen belt. Larvae crawled through the belt to be returned to the pupation monorail. Pupae which remained on the belt were collected at the far end of the separator to be placed in screened trays for transfer to a holding room held at 80 F (27 C) and 70% to 80% relative humidity. Separation of the pupae at 8-hour intervals and holding them at a constant temperature were essential to obtain insects of uniform age, to regulate the age of pupae irradiated, and to schedule releases accurately.

### *Sterilization Procedures*

A deep-therapy x-ray unit and an 11-curie $^{60}$Co unit were used for small-scale studies during the early field trials in Florida. However, a 70-curie $^{60}$Co unit (9) was obtained for the Curaçao test, and later a single 480-curie $^{60}$Co unit was obtained for the 2,000-square-mile test in Florida. This manually operated unit provided the prototype for the six automated units (10) acquired for mass sterilization. The six units, ranging from 450 to 660 curies each, contained a total of 3,600 curies of $^{60}$Co. Eight $^{60}$Co strips 2 inches wide, 1 inch thick, and 13 inches long, encapsulated in stainless steel, were arranged in a cylinder to accommodate the 5-inch irradiation canister. This provided a usable field with an initial-dose rate varying from 788 to 910 roentgens/min. In order to completely sterilize the screwworms, 8,000 r ± 10% was delivered to pupae at 5.2 to 5.7 days of age.

Irradiation canisters were loaded with 2 liters of pupae (18,000 insects) and attached to a carriage which automatically inserted the canisters in the irradiation chamber and removed them after the required treatment.

### Release Procedures

After irradiation, canisters of pupae were conveyed to the packaging room. Here the pupae were automatically packaged at the rate of 440 pupae per carton (4½ × 5½ × 2 inches) by using equipment commonly employed in the food-processing industry. Cardboard inserts in the carton provided the resting area the flies required after they eclosed. Packaged pupae were held at 80 F (27 C) and 70% to 80% relative humidity for eclosion. Releases were made from the local distribution center. Cartons were also transported by refrigerated van to four other release centers throughout the eradication area. When inclement weather delayed release, insects were held at 60 F (16 C) to avoid loss from extended confinement and starvation. Although flies released on Curaçao, a low rainfall area, were fed a honey solution prior to release, feeding was not considered essential in southeastern United States since abundant rainfall and flowering plants assured availability of food.

As soon as eclosion neared completion, cartons were loaded into small, single-engine aircraft modified to hold 1,200 cartons. An employee riding in a reversed seat placed the cartons in an automatic device calibrated to dispense them at the required rate. During one week, cartons were released over the entire area in parallel lines 2 miles apart, and during the following week the lines were shifted by 1 mile so that complete coverage on a 1-mile basis was attained every two weeks.

### Field Evaluation

Livestock inspectors were employed to visit ranches weekly to inspect animals for myiasis and to enlist rancher cooperation in following screwworm-prevention measures and in submitting specimens for identification. Since other dipterous larvae of minor importance are associated with wound myiasis, positive identification was essential, particularly in areas free of screwworms, to avoid waste of released flies. Information obtained from the inspectors and livestock owners was tabulated and analyzed in the Survey Data Center at Sebring to schedule release rates appropriate to areas requiring treatment.

## Eradication Program in Southeastern United States

After the successful field trial in 1957 near Orlando, Fla, the Florida state legislature appropriated $3 million to share the cost of a joint state-federal program. A planning committee, consisting of entomologists, veterinarians, and an engineer, convened to draw up plans for construction and operation of a mass-rearing facility. A former US Air Force base near Sebring, Fla, was selected as the plant site because of its central location in southern Florida and the availability of a large hangar and other buildings. The Florida Livestock Board began renovation of the hangar early in 1958, with completion scheduled for July 1958.

In the meantime, rearing and release activities were reinstated at Orlando, Fla, to train personnel for operation of the planned mass-rearing plant at Sebring. Production at the level of 2 million per week was begun in January 1958. However, the winter of 1957-1958 proved to be one of the coldest on record, and it appeared that screwworms would overwinter only below a line running from Tampa eastward to Vero Beach, or in an area only half as large as the average overwintering area bordered on the north by Jacksonville, Fla. Since it was impossible to complete the Sebring plant in time to ward off the northward movement expected in April, production at Bithlo was increased to 14 million screwworm flies weekly until the Sebring plant reached full production in August 1958.

During April 1958, sterile screwworms were released at a weekly rate of 200 per square mile throughout peninsular Florida from Miami northward to Jacksonville. This treatment covered most of the overwintering area and the area to the north within 100 miles of the northernmost-occurring case of infestation with screwworms. However, because sporadic breakthroughs of native flies occurred, sterile flies were released as far north as southern Georgia to maintain the 100-mile barrier. Although it was temporarily necessary to abandon treatment of the overwintering populations in southern Florida, the applied treatment prevented establishment of screwworms beyond the overwintering area. Only 31 of 864 infestations confirmed in Florida during the entire program occurred more than 50 miles north of the overwintering line, only 3 of these after June 1958.

In early September 1958, screwworms occurred near Montgomery, Ala; presumably they were brought in with a shipment of animals from the West. Since rapid spread of this infestation might have produced an overwintering population if a mild winter occurred, the infestation warranted treatment. Infestation was confined to a three-county area, and with the aid of cold weather, no further activity was evident after

November 1958. During this period, as many as 80 million flies were released weekly over 85,000 miles.

During the campaign, special releases were made in several areas in which screwworms persisted. Except for Broward County, Florida, infestations in these intensively treated areas were readily brought under control. A special task force of entomologists and inspectors was assigned to Broward County in December 1958. After intensified releases of sterile flies, the final infestation was recorded on Feb 19, 1959. Then followed a four-month period free of screwworms, which ended on June 17, 1959, when a single case of myiasis occurred within 10 miles of the Sebring plant. Whether this case resulted from the escape of fertile screwworm flies from the rearing facility or from a shipment of infested animals from the West was not established. Releases were discontinued on 10,000 square miles in southern Florida in late July 1959 and in the entire Southeast on Nov 14, 1959. Except for several infestations traced to the movement of infested animals during the period from 1960 to 1962, a screwworm problem has been nonexistent in the Southeast since early in 1959.

During the two-year campaign, 3.7 billion screwworm pupae were produced, and 6.3 million lb of horsemeat and whale meat were used (11). Twenty light aircraft were used to release flies over a maximum area of 85,000 square miles, and peak employment, including plant personnel, fly distributors, field inspectors, and clerical and administrative help, totaled 500. However, for a research cost of only $250,000 and an eradication-program cost of $10 million, ranchers in the Southeast have experienced $140 million in savings since inception of the program in 1958.

## Eradication Program in Southwestern United States

Initially screwworm eradication in southwestern United States was considered impractical because of the lack of isolation from infested areas, as is afforded in the Southeast. It was realized that screwworms would not respect international boundaries and that a program in the Southwest must involve a constant battle against reinfestations from Mexico.

Representatives of the livestock industry in Texas visited the Sebring (Fla) plant and were briefed on the operation in the Southeast. They were apprised of the greater challenge that would face those who operated a program in the Southwest compared with one in the Southeast, but they were willing to take the gamble. They returned home to raise $3 million from livestock owners to match $3 million from the State of Texas. An additional $6 million in federal funds brought the total to the $12 million required for a three-year program. In phase 1 eradication from the overwintering area in Texas would be attempted. If successful, phase 2 would test the feasibility of a sterile-fly barrier in preventing reinfestations from Mexico.

## Expansion of Research

In anticipation of the screwworm eradication program in the Southwest, research activities were expanded by the USDA, Entomology Research Division at Kerrville, Tex, in 1961. Research personnel were later transferred to Mission, Tex, which became headquarters for the program.

Although the eradication program was eminently successful in the Southeast, several important problems in production, irradiation, and release remained unsolved. In addition, directors of the program had to contend with the vastness of the southwestern area, an area subject to infiltration by screwworms from Mexico (12). They also had to reckon with the differing ecological conditions, and possibly the incompatibility of the Florida strain with native screwworm populations in the Southwest and Mexico. These problems could not be solved from the experience gained in the Southeast.

### *Desiccation of Pupae*

Pupal mortality at Sebring continued to be a problem during the entire program and eclosion averaged only 75%. This problem was resolved in a study which demonstrated that larvae and prepupae were highly susceptible to desiccation during the first 48 hours after the larvae had completed feeding (13). In the Sebring plant, pupation occurred under ambient temperature and humidity conditions, and the pupae were transferred later to a temperature and humidity controlled chamber for storage prior to irradiation. Unfortunately, the insects were not adequately protected from desiccation during the critical period, particularly when overcrowding in the pupation trays resulted in large numbers of pupae appearing on the surface of the sand. At the southwestern screwworm eradication plant in Mission, larvae are regularly transferred within a few minutes of crawl-off to a chamber with the temperature and humidity controlled, where they pupate. Since this protection was provided, eclosion of pupae has been 95% or better. Development of a sawdust pupation medium by A. J. Graham, in charge of the program's methods development unit, undoubtedly has contributed to the improved emergence, since early formed pupae are not forced to the surface by constant movement of larvae, as occurs when the heavier sand is used.

### *Handling Flies for Release*

During screwworm releases in the southeastern campaign, only minor losses of sterilized flies occurred. However, eclosion in the release cartons usually required several days, and timing of releases involved a compromise between premature releases in which pupae were destroyed by concussion and delayed releases which increased adult mortality owing to starvation. In the high rainfall area of Florida, little concern was given to survival if the adults were reasonably active prior to release, since water and flowering plants were readily available. However, in the semidesert areas of the Southwest the adults must carry water and food reserves needed for survival until they arrived in a favorable habitat.

Studies of diurnal periodicity showed that screwworms eclose primarily from daylight until noon. Pupae in a uniform age group failing to eclose the first day did not emerge until the following morning. Since differences in biological age were less than 24 hours, a reduction in temperature from the normal 80 F (27 C) to 55 F (13 C) for 12 hours resulted in a saving of water and food reserves of emerged flies, without delaying time of eclosion for the remaining pupae. Further studies (14) indicated that in spite of the availability of food and water, adult reserves were depleted in less than four days and that under conditions of high temperature and drought, releases should be made promptly after fly emergence. In contrast, it was determined that during cool weather males should be held until they are sexually mature to assure prompt competitiveness with native males, particularly in newly treated areas. More than a week is required for development of sexual maturity at 60 F (16 C) compared with two days at 80 F (27 C).

### *Irradiation Studies*

Early irradiation studies (15) showed that a sterilizing dose of x-radiation or gamma radiation could be administered to screwworms without seriously affecting longevity or male competitiveness under laboratory conditions; however, efforts were continued to more closely define the optimum dose and to reduce the 15% loss in survival recorded at Sebring, Fla. Development of a test for sexual aggressiveness (16) greatly facilitated testing for any factor that might affect male vigor. Previously, the tedious and time-consuming procedure of maintaining individual records of egg hatch was required to measure a male's ability to mate. However, the need for such records was eliminated when it was discovered that male vigor and female mortality under close confinement were inversely related. Females weakened by frequent attacks from the males were finally subdued and failed to survive a second mating. By using the sexual-aggressiveness test, which merely involved obtaining female mortality data, it was clear that up to 12 kr could be given to male screwworm pupae 6 or 7 days of age with minimum loss in sexual vigor, but at 5 days of age sexual vigor was greatly reduced by the same radiation exposure.

Sterility data obtained in the quality control section at Sebring, Fla, occasionally showed a disturbing degree of fertility in the irradiated females. Exposure times were thus arbitrarily increased until most of the radiation units were delivering 12 kr, or twice the dose normally required to sterilize the flies. Tests conducted during the final stages of the Southeast program (17) demonstrated that reduced sterility resulted from anoxia in the closed radiation canisters. Oxygen consumption by the pupae rises sharply after 5 days of age, and a mere 15-minute delay between loading and irradiating 7-day-old pupae resulted in complete anoxia, so that the dose required under these conditions was double that required when an oxygen deficit did not exist. Use of screened canisters and positive control of the interval between loading and treatment eliminated this problem.

Prior to the southwest screwworm eradication program, limited funds, personnel, and time precluded the gathering of basic information essential to an understanding of the effects of radiation on the screwworm. However, in the since expanded research program the services of a geneticist and a cytologist have been available. The relationship between the stage of oogenesis (18) or spermatogenesis and susceptibility to treatment with gamma radiation was thoroughly explored.

### Strategic Release of Flies

Ecological studies conducted during the 1930's in Texas showed that during hot, dry periods, native screwworm flies concentrated along watercourses where food and water were much more abundant than on the open range. These studies, continued by Hightower and Alley (19), provided a sound basis for the recommendation that priority be given to watercourse treatment in lieu of the uniform grid used in the Southeast.

At the beginning of the southwest screwworm eradication program, in 1962, only 20 million screwworms weekly were available, and an attempt to contain the overwintering populations in the southern tier of counties was unsuccessful. Screwworms dispersed throughout Texas and into Oklahoma by midsummer. Failure to control this population was variously attributed to (1) overwintering farther north than expected, (2) movement of infested animals, (3) failure of the released males to compete for mating opportunities, or (4) long-range dispersal of native screwworms. Earlier studies had shown that screwworms normally move northward in the spring at the rate of only 30 miles per week (20), and most critics did not subscribe to the long-range dispersal theory. However, in the spring of 1963 when populations of screwworms again were spreading at the rate of 100 miles per week, E. F. Knipling suggested that individual native screwworms may disperse 200 miles or more. Hightower and Alley (21) obtained field data to support this theory. Irradiated, plant-reared females were marked and released near Austin, Tex, on the Colorado River, and a single female was recovered in the most distantly located trap 180 miles upstream. Further recoveries were made at shorter distances from the point of release. Rapid screwworm dispersal observed during these studies did not necessarily refute the earlier studies, since release of large numbers of sterile flies may create the same flushing effect that is observed among other animals when populations are high. Since the odds for long-distance recoveries would be astronomical if dispersal were random, further evidence was obtained to show that screwworms orient to watercourses. Based on this and study of the progressive spread of isolated infestations, plans were made to extend the sterile-fly barrier 200 – 300 miles into Mexico.

### *Evaluation of Strains for Release*

Research planning in support of screwworm eradication in the Southwest originally included a study of various strains of screwworms in Texas and Mexico to provide a sound basis for selecting a strain more suitable for release than the Florida strain. Laboratory tests showed that the Florida strain survived as well as two Texas strains held in the laboratory under simulated Florida and Texas conditions, respectively. Early inauguration of the southwest screwworm eradication program precluded field comparisons of the various strains, and release of the Florida strain continued. Subsequent evaluations of egg-mass collection in the spring of 1962 indicated that the Florida males were reducing effective mating as expected.

### *Genetic Markers*

Difficulties encountered in mass-marking released screwworms for identification in ecological studies or to separate released from native populations suggested need for development of a genetic marker for adults. In an intensive program, numerous markers (22) were isolated, including body and eye color variations, wing-venation anomalies, and gross morphological variations. However, as geneticists point out, most mutations are deleterious, and most of the mutants were not acceptable for release. Genetic variations affecting flight, vision, etc, obviously are not likely to produce usable markers. Limited releases of mutants screened for acceptable behavior, vigor, and longevity in the laboratory were not successful in the field until recently, when a wing variant performed well, as based on recovery of released sterile females.

The belated discovery of a screwworm infestation near the Sebring (Fla) plant prompted the search for a genetic marker for larvae. In order to avoid the accidental dispersal of live screwworms, larval specimens collected in the field usually are submitted for identification in a killing agent. In this situation, an adult marker would have no value in determining whether plant escapees were responsible for infestations. As in the search for adult genetic markers, many interesting anomalies were observed among larvae, but few have been established in pure culture. Fortunately, it was discovered that among laboratory-reared specimens of the Florida strain of screwworms a high incidence of complete banding of spines occurs on the anterior of the 11th segment, in contrast to incomplete banding in wild strains. Appearance of this character in a field collection would strongly indicate origin from the rearing facility.

### *Attractants*

Decomposing liver has been used as a screwworm attractant since the beginning of screwworm research. Efforts at Orlando, Fla, prior to inauguration of the southeastern eradication program were unsuccessful in developing a more efficient attractant. Decomposing liver is nonspecific in effect and attracts thousands of other insects for each female screwworm attracted. In addition, male response to this material is negligible. Recent progress with sex pheromones of other species that are used for attracting the opposite sex has prompted consideration of this approach in screwworm control. Extracts of material collected by L. A. Fletcher in a cold trap from cages of virgin males induce responses from females. However, efforts to collect a material that elicits responses from male screwworm flies have been unsuccessful. Other materials have shown promise with the Florida strain in the laboratory, but field tests have yielded inconsistent results.

### *Chemosterilants*

Chemosterilants, chemicals which include antimetabolites, and radiomimetic compounds, which produce effects similar to x-radiation or gamma radiation, have been investigated (23) to solve the problem of reduced vigor previously noted after treatment with radiation and to provide an effective approach to sterilization of screwworm populations in the field. Pending development of an efficient and specific screwworm attractant, a chemosterilant combined with an attractant may afford an effective method for establishing a screwworm barrier, particularly in Panama where the area involved would be of relatively small size. Chemosterilants may present the same disadvantages inherent in conventional iinsecticides, i.e., toxic residues and possibility of insect resistance.

## Methods Development

During the southeast eradication program, efforts were made by the methods development unit to increase rearing efficiency and to reduce overall program costs. These efforts, continued by a nucleus of key program personnel stationed at Kerrville, Tex, to maintain a standby colony for combating outbreaks that might occur in Florida, resulted in further improvements. Major changes in plant design were developed by C. N. Husman, supervisory equipment specialist, who, with the help of entomologists, mechanized mass production of screwworms at Sebring.

### *Pupa-Larva Separation*

The Sebring pupa-larva separator allowed the larvae to crawl through an endless screen-mesh belt for recycling and conveyed the pupae to a collection container. This unit, however, failed to work effectively when larval size varied significantly. Large larvae nearing pupation had difficulty in crawling through the mesh and became mixed with the pupae. Some of these found their way out of the pupation tray and produced flies that emerged in the plant to present a security hazard. Others pupated later in the trays and thus were irradiated and released prematurely before eclosion occurred. Small pupae either were caught in the screen mesh and crushed by the terminal pulley or fell through the screen and were recycled with the larvae. This situation presented a security hazard due to emergence of flies in the pupal storage chamber or in the radiation canister, where the flies could crawl out of the effective radiation field. These problems were effectively solved by replacing the screen-mesh belt with a solid, ribbed belt illuminated with fluorescent lights. The larvae, negatively phototropic, readily crawled to either side of the belt, where they could be collected and recycled, and the pupae were collected at the end without loss.

### *Larva Collection*

Larval collection at Sebring involved eight huge funnels on the lower floor of the building. The funnels were constructed of heavy-gauge aluminum sheets riveted every few inches. However, when the funnel surface was moist, larvae clung to the funnel instead of falling into the collection trays. Larvae congregating in the funnel were able to force their way through the seams and spilled onto the ground floor. This difficulty was remedied by removing larvae from the funnels every hour with compressed air and cleaning the funnels as often as the production schedule permitted. However, the number of larvae per tray varied considerably with the flow from the funnels so that larvae had to be continually transferred from one tray to another to obtain optimum loading rates. A water conveyor with sluices below the floor level, developed to transport the larvae to a water larva separator where the larvae were measured into pupation trays, took care of this difficulty. This major development in handling the larvae, now in use at Mission, Tex, not only eliminated the need for a two-story plant and the difficulties encountered in the huge funnels, but also enabled one man to collect and accurately measure larvae for placement in the pupation trays.

### *Irradiation Equipment*

At Sebring, involved security measures were required to assure irradiation of each canister of pupae leaving the plant, since canisters, inadvertently, could be transferred directly from the passthrough to the packaging-room conveyor. The Sebring radiation units were thus remodeled to drop the irradiated canister onto the packaging line conveyor instead of returning it to the operator. Thus operator error was essentially eliminated, and the possibility of a mechanical error is extremely remote. Should a canister drop onto the conveyor prior to irradiation, "in" and "out" time records maintained by both the radiation and packaging attendants indicate that it should be intercepted. This latter procedure, of course, is subject to human error, as at Sebring.

### *Larva Rearing*

Meat costs for rearing at Sebring averaged $304 per million pupae and totaled over $1 million for the entire program. A. J. Graham used fillers such as oat and cottonseed hulls and substituted fresh fish for whale meat to reduce costs to $200 or less per million pupae. At Mission, meat costs have been reduced further through the use of pork lungs and nutria, a small fur-bearing animal dressed and ground whole. A major change in diet currently being tested on a large scale consists of using a fluid medium containing spray-dried blood and other dehydrated materials mixed with water; the dehydrated materials tested include fish flour, powdered milk, and cottage cheese. The fluid medium is dispensed onto a layer of cotton which supports the larvae and allows them to feed without drowning. The fluid is removed and replaced periodically with a vacuum device. The chief advantages of this diet are that it eliminates the need for mass cold-storage facilities, increases the capacity per vat since larvae occupy entire vats instead of only areas containing unused medium, and reduces cost of the medium to $100 or less per million pupae. However, efforts are continuing to increase larva size, which has been 15% less than when larva are on the meat diets.

### Current Status of the Southwestern Program

In February 1962, the southwestern screwworm eradication program was inaugurated in a state of urgency to capitalize on weather that was unfavorable to the screwworm, as occurred in the Southeast in 1958. A severe winter brought repeated and severe cold fronts deep into Texas and limited screwworm activity to a few counties at the southern tip of Texas. Facilities at Kerrville, Tex, were expanded to rear 20 million flies weekly, while the Mission plant was rushed to completion. Although sterile flies were released in the known infested areas in Texas and farther north in an attempt to control the native populations, northward movement of the flies continued unabated throughout Texas and into Oklahoma. Over 50,000 cases of infestation with screwworm were recorded during 1962. Releases were confined to central and southern Texas to include primarily the normal overwintering area. Although much of the infested area in Texas could not be treated, surveys indicated an overall reduction of 75% in case incidence.

During the spring of 1963, it was apparent that the previous year's releases had adequately coped with the potential overwintering population; however, in spite of sterile-fly releases in and beyond the overwintering area, the spring buildup and dispersal began to follow the same pattern as in 1962. Since the major spread represented only scattered cases beyond the primary treatment area, which extended as far north as Austin, Tex, the decision was made to treat these isolated infestations weekly for three weeks at the rate of 100,000 flies per premise. In addition, major watercourses in and beyond the primary grid zone were treated at 2,000 flies per linear mile. Although screwworms again spread through much of Texas and into Oklahoma, hundreds of isolated infestations were brought under control, and only 5,000 cases of infestation with screwworm were recorded in Texas in 1963.

As mentioned earlier, both field research and the pattern of screwworm infestations indicated that a barrier at least 200 miles into Mexico would be required to reduce further screwworm incidence in Texas and other southwestern states. Release requirements to combat screwworms in Texas and New Mexico during 1962 left only enough flies for release 25 to 50 miles into northeastern Mexico. However, during the winter of 1963-1964 releases were extended 200 miles into strategic areas throughout northern Mexico (24), only 223 cases were recorded in Texas during 1964.

During 1965, Arizona and California, along with Texas, Louisiana, Arkansas, Oklahoma, and New Mexico, were included in the program to reduce this multimillion-dollar pest to a noneconomic level in the entire United States and much of Mexico. During the present year, 1965, those conducting the eradication program have met their most severe challenge. Screwworm activity has been potentiated by a mild winter and abundant rainfall. However, extension of the sterile-fly barrier 300 miles into strategic areas of Mexico has limited the number of cases in Texas to less than 300 through August.

Currently, control of the screwworm involves more than 300,000 square miles. Eradication to the Bay of Tehuantepec in Mexico would reduce the area to 50,000 square miles and eradication to the Isthmus of Panama to less than 20,000. From a long-range standpoint, vast savings in program costs would obviously result if the reproductive potential of the screwworm were reduced to a minimum in North America. In addition, the livestock industry in Mexico would be spared an annual loss of $160 million, which could be added to savings presently enjoyed by the livestock industry in the United States.

From 1960 to 1964, screwworm research costs were $750,000, and from 1962 to 1964, eradication costs were approximately $15 million. Personnel of the livestock industry have estimated that they have been spared a cost of about $275 million, but other more conservative estimates place the savings at half this amount. In 1963, fly production peaked at 150 million and averaged 114 million weekly. However, in 1964, with improvements in release strategy, an average of only 90 million flies was required weekly to cover the expanded area. As of Jan 1, 1965, over 12 billion sterile flies had been released in the southwestern United States and northern Mexico. It is a great tribute to the program and the research personnel that improvements in operation have allowed extension of the screwworm barrier to a depth of 200 to 300 miles at an annual cost of only $5 million compared with $4 million originally budgeted for a 100-mile barrier.

Operation of the southwestern eradication program has greatly reduced the threat of reinfestation in the Southeast. In spite of a costly quarantine line operated along the Mississippi River during and following the southeastern program, 6 cases were recorded from 1960 to 1962 in Florida and 17 in Georgia, with many more in Alabama and Mississippi. Since 1963, with an effective program operating in the Southwest, none of the southeastern states have reported a confirmed case of screwworm infestation.

## Impact of the Sterile-Male Technique

Success of the sterile-male technique in controlling or eradicating screwworms has prompted investigations of the possibility of using this technique for control of other species of insect pests in the United States and elsewhere (25). A partial list of these pest species includes melon fly (*Dacus cucurbitae* [Coquillett], oriental, Mediterranean, Mexican, and olive fruit flies (*Dacus dorsalis* [Hendel], *Ceratitis capitata* [Wiedemann], *Anastrepha ludens* [Loew], and *Dacus oleae* [Gmelin]), pink bollworm (*Pectinophora gossypiella* [Saunders]), boll weevil (*Anthonomus grandis* [Boheman]), sugarcane borer (*Diatraea saccharalis* [F.]), European cornborer (*Ostrinia nubilalis* [Hübner]), gypsy moth (*Porthetria dispar* [L.]), mosquitos, housefly (*Musca domestica* [L.]), tsetse fly (*Glossina morsitans* [Westwood]), and codling moth (*Carpocapsa pomonella* [L.]). Among the successes already achieved are eradication of the melon fly from Rota and the oriental fruit fly from Guam in the Marianas Islands in the Pacific.

Use of the sterile-male technique has required detailed studies of biology and ecology to the target species. Many insects are difficult to rear efficiently under laboratory conditions. Not only requirements for food, temperature, light, and humidity must be known for immature and adult stages, but the conditions for optimum mating, oviposition, and longevity of the adults also must be known. Diseases, parasites, and predators must be controlled. Before sterile insects are released, population densities, congregation sites, and seasonal fluctuations must be known.

We are especially fortunate that the sterile-male technique can replace the use of insecticides for large-scale suppression or eradication of specific insects. Insecticides may eventually become ineffective when insects develop resistance to them, or their continuous application may be hazardous because of contamination of our food and environment. Although tremendous economic gains have been achieved with the sterile-male technique applied against screwworms, its significance extends far beyond its economic value. As Dr. Knipling has stated, "It suggests a principle of insect control that might be applied to other major insect problems in the United States and other parts of the world. The success of the operation under extremely difficult circumstances in a vast area is truly phenomenal when it is considered that we are dealing with an insect that has the capacity to migrate in the range of 300 miles." As the population grows and the competition between man and insects for food and space increases, eradication of major insect pests may become the rule rather than the exception. As we obtain greater knowledge of insect physiology, behavior, communication, and response to environment, we surely will develop other new approaches to insect control.

## References

1. Dove, W.E.: Myiasis of Man, *J Econ Entom* **30:**29–39 (Feb) 1937.
2. Knipling, E.F.: Possibilities of Insect Control or Eradication Through the Use of Sexually Sterile Males, *J Econ Entom* **48:**459–462 (Aug) 1955.
3. Bushland, R.C., and Hopkins, D.E.: Sterilization of Screw-Worm Flies with X-Rays and Gamma-Rays, *J Econ Entom* **46:**648–656 (Aug) 1953.
4. Baumhover, A.H., et al: Screw-Worm Control Through Release of Sterilized Flies, *J Econ Entom* **48:**462–466 (Aug) 1955.
5. Baumhover, A.H.: Florida Screw-Worm Control Program, *Vet Med* **53:**216–219 (April) 1958.
6. Baumhover, A.H., et al: Field Observations on the Effects of Releasing Sterile Screw-Worms in Florida, *J Econ Entom* **52:** 1202–1206 (Dec) 1959.
7. Melvin, R., and Bushland, R.C.: The Nutritional Requirements of Screw-Worm Larvae, *J Econ Entom* **33:**850–852 (Dec) 1940.
8. Graham, A.J., and Dudley, F.H.: Culture Methods for Mass Rearing of Screw-Worm Larvae, *J Econ Entom* **52:**1006–1008 (Oct) 1959.
9. Darden, E.B.; Maeyens, E.; and Bushland, R.C.: A Gamma-Ray Source for Sterilizing Insects, *Nucleonics* **12:**60–62 (Oct) 1954.
10. Jefferson, M.E.: Irradiated Males Eliminate Screw-Worm Flies, *Nucleonics* **18:**74–76 (Feb) 1960.
11. Smith, C.L.: Mass Production of Screw-Worms (*Callitroga hominivorax*) for the Eradication Program in the Southeastern States, *J Econ Entom* **53:**1110–1116 (Dec) 1960.
12. Knipling, E.F.: The Eradication of the Screw-Worm Fly, *Sci Amer* **203:**54–61 (Oct) 1960.
13. Baumhover, A.H.: Susceptibility of Screw-Worm Larvae and Prepupae to Desiccation, *J Econ Entom* **56:**473–475 (Aug) 1963.
14. Baumhover, A.H., and Spates, G.E.: Artificial Selection of Adult Screw-Worms for Extended Survival Without Food and Water, *J Econ Entom* **58:**645–649 (Aug) 1965.
15. Bushland, R.C., and Hopkins, D.E.: Experiments With Screw-Worm Flies Sterilized by X-Rays, *J Econ Entom* **44:**725–731 (Oct) 1951.
16. Baumhover, A.H.: Sexual Aggressiveness of Male Screw-Worm Flies Measured by Effect on Female Mortality, *J Econ Entom* **58:**544–548 (June) 1965.
17. Baumhover, A.H.: Influence of Aeration During Gamma Irradiation of Screw-Worm Pupae, *J Econ Entom* **56:**628–631 (Oct) 1963.
18. LaChance, L.E., and Bruns, S.B.: Oogenesis and Radiosensitivity in *Cochliomyia hominivorax* (Diptera: Calliphoridae), *Biol Bull* **124:**65–83 (Feb) 1963.
19. Hightower, B.G., and Alley, D.A.: Local Distribution of Released Laboratory-Reared Screw-Worm Flies in Relation to Water Sources, *J Econ Entom* **56:**798–802 (Dec) 1963.
20. Barret, N.L., Jr.: Natural Dispersion of *Cochliomyia americana, J Econ Entom* **30:**873–876 (Dec) 1937.
21. Hightower, B.G.; Adams, A.L.; and Alley, D.A.: Dispersal of Released Irradiated Laboratory-Reared Screw-worm Flies, *J Econ Entom* **58:**373–374 (April) 1965.
22. LaChance, L.E., and Hopkins, D.E.: Mutations in the Screw-Worm Fly, *J Econ Entom* **55:**733–737 (Oct) 1962.
23. Crystal, M.M.: The Induction of Sexual Sterility in the Screw-Worm Fly by Antimetabolites and Alkylating Agents, *J Econ Entom* **56:**468–473 (Aug) 1963.
24. Hightower, B.G., et al: Seasonal Abundance of Screw-Worms in Northern Mexico, *J Econ Entom*, to be published.
25. Bushland, R.C.: "Insect Eradication by Release of Sterilized Males," in *Large Radiation Sources in Industry*, London: Her Majesty's Stationery Office, 1960, pp 273–290.

# V

# Management of Vegetation

# 13

# Preserving vegetation in parks and wilderness

*Edward C. Stone*

Federal efforts to preserve natural vegetation go back to 1872, when Yellowstone National Park was carved out of the public domain; state efforts go back to 1885, when the New York Adirondack Forest Preserve was established (1). All efforts, however, have been largely unsuccessful because of a failure to appreciate fully that vegetation is a living, dynamic complex and cannot be preserved in the sense in which a building or an archeological site can be preserved. Even the most uniform vegetation is a mosaic created by local variations in the environment and by prior events such as fire, drought, and insect infestation. When a mature plant dies, hundreds of seedlings spring up to take its place, some or all of which may be of different species. Which seedlings survive, and for how long, depends upon their relative growth potential, what effect the dead plant had on its environment before it died, and what kind of environment resulted when it died. Vegetation can only be preserved by controlling the complicated successional forces that have created it and that, if unchecked, will in turn destroy it.

The very efforts made to preserve a natural system of vegetation may bring on unplanned and undesired changes in it. That steps taken to preserve animal wildlife may have this effect is well known to the general public. By 1930 there were overpopulations of elk and bison

Reprinted with permission from *Science*, 150(3701):1261–1267 (3 December 1965).

in Yellowstone National Park, of mule deer in Zion National Park, and of deer and elk in Rocky Mountain National park, all brought about by control of predators in and around the parks (2). Recognition of the problem led to a reconsideration of these practices, and today, although hampered by a lack of basic data and a restrictive budget, specialists in wildlife preservation are employed in the national parks to plan and apply sounder regulatory methods. While not so dramatic and not so widely publicized as imbalances in wildlife populations, drastic changes in the composition of many of the plant communities in the national parks have occurred during the last 50 years under fire-protection policies and heavy concentrations of use. In a number of cases these changes have progressed so far that even the once dominant plants in a wide variety of plant communities have been replaced, and now trees and shrubs occupy slopes and meadows once clothed in grass and sedge (3).

There are two federal agencies largely responsible for the management of national wildlands, each by charter concerned with conservation of this resource but each with different primary objectives. The Forest Service was organized in 1905 within the Department of Agriculture to manage the forest reserves—later renamed national forests—to secure favorable watershed conditions and to furnish a continuous supply of timber. Shortly thereafter, however, the Forest Service recognized that recreation was an important use of these areas compatible with its other uses, and began developing the recreational facilities that now serve 125 million visitors a year. Some 15 years later it recognized the need for wilderness reserves, and by appropriate administrative action over the next several years set aside almost 12 million acres for this purpose. Subsequently both of these administrative decisions have been sanctioned by congressional action (4).

The Park Service was organized in 1916 within the Department of the Interior to bring together under one administrative head a number of independent national parks formerly administered by several federal departments. It was specifically charged with preserving on these lands plant and animal life, and geological and archeological features of national value, for the enjoyment of the public. Vegetation preservation thus constitutes a minor part of the Forest Service's responsibilities but a major part of the Park Service's responsibilities.

The Forest Service has moved ahead rapidly in meeting its responsibilities as watershed and timber manager and purveyor of recreation facilities. It has been able to do so for a number of reasons. From its inception it was able to staff its key administrative posts with men trained for the job of managing forests for watershed and timber; it could draw upon a wealth of European experience, and it had an excellent research staff engaged in developing workable silvicultural techniques, based upon sound understanding of ecology, for use by its foresters operating in the field. In its minor role as vegetation-preservation manager of 12 million acres of wilderness, the Forest Service has yet to do much of anything.

The Park Service has moved slowly in its major role, that of preservation manager, although it has successfully operated the land under its control for the enjoyment of the public. When established, the Park Service, unlike the Forest Service, had no ready source of professional help to which it could turn. There was no such thing as a vegetation specialist versed in preservation management—that is, a vegetation-preservation manager. Furthermore, administrators could not rely upon European experience for guidance, because there was none. Nor could they turn to a research staff for developing the necessary management techniques, because there was none. To make matters even more difficult, they were forced almost from the beginning to fight a rear-guard action with private companies and government agencies that wanted to open park lands for mining, hunting, water impoundment behind massive dams, logging, and other commercial activities. Thus administrative energies and funds were all but exhausted in maintaining existing park boundaries; and the problem of preserving a variety of undescribed ecosystems, in which changes at the time were not well advanced or readily apparent to the untrained eye, was largely solved as far as the administrator was concerned once an efficient fire-control system had been established, livestock excluded, and insect epidemics brought under control. This does not imply that the Park Service has failed to attract competent biologists to its professional staff and that there has been no effort to stem the successional tide; this is not true. Characteristically, however, individuals in the Park Service who have been trained as biologists have been called upon more for protective and interpretive service than for specialized management of the vegetation complex, because overall park policies have not until recently included the concept of vegetation management except in the narrow aspects of fire, insect, and disease control.

In 1963, public attention was drawn to this state of affairs in the national parks by a report of a committee appointed by the National Academy of Sciences at the request of Secretary of the Interior Udall to aid in "the planning and organizing of an expanded research program of natural history research by the National Park Service." The committee presented "the pressing need for research in the national parks by citing specific examples in which degradation or deterioration has occurred because research on which proper management operations should have been based was not carried out in time; because the results of research known to operational management were not implemented; or because the research staff was not consulted before action was taken" (5). The report stresses the need to develop, by means of extensive research, the ecological basis for managing the preservation of both plant and animal life in the ecosystems involved. "It is inconceivable," the authors remark, "that property so unique and valuable as the national parks, used by such a large number of people, and regarded internationally as one of the finest examples of our national spirit should not be provided adequately with competent research scientists in natural history as elementary insurance for the preservation and best use of the parks." This is an excellent, long overdue report, exhaustive in its treatment of research needs, but it did not in my opinion focus strongly enough on the need for professional vegetation specialists at the operational level, specialists who will be responsible for carrying out the manipulative steps recommended by the research staff, much as the Forest Service professional foresters carry out the cutting practices recommended by its research staff.

The various state agencies that administer park and forest preserves with preservation as the stated objective have come into existence at different times in response to different pressures and differ widely in the composition of their administrative staffs. All engage in protection of some kind, but none, as far as I can ascertain, is involved in manipulative procedures to preserve the integrity of specific types of vegetation. Some ecological research is under way, but again there are no professional vegetation specialists available to carry out the manipulative program that will come out of this research.

## The Objectives of Preservation

Since vegetation is never static, preservation must consist, in effect, of managing change. Consequently, it is necessary to determine exactly what the objective is, and thereby to determine how much change, and what kind, can be tolerated.

One of the most common objectives is to keep park lands green or, in the arid West, green and golden. Fire protection is considered the principal means to this end. The fact that vegetation protected from fire may change completely in a relatively short period has rarely been considered, because administrators and the public have not appreciated that this can happen.

Probably the next most common objective is the preservation of certain favored dominant species within the vegetative complex. When this is the objective, the fact that certain successional stages may be fast disappearing and that the overall vegetative structure may be changing within rather broad limits is usually ignored, as long as the dominants remain dominant and the general appearance of the landscape is not altered. Change that does not interfere with the effective display of the dominant

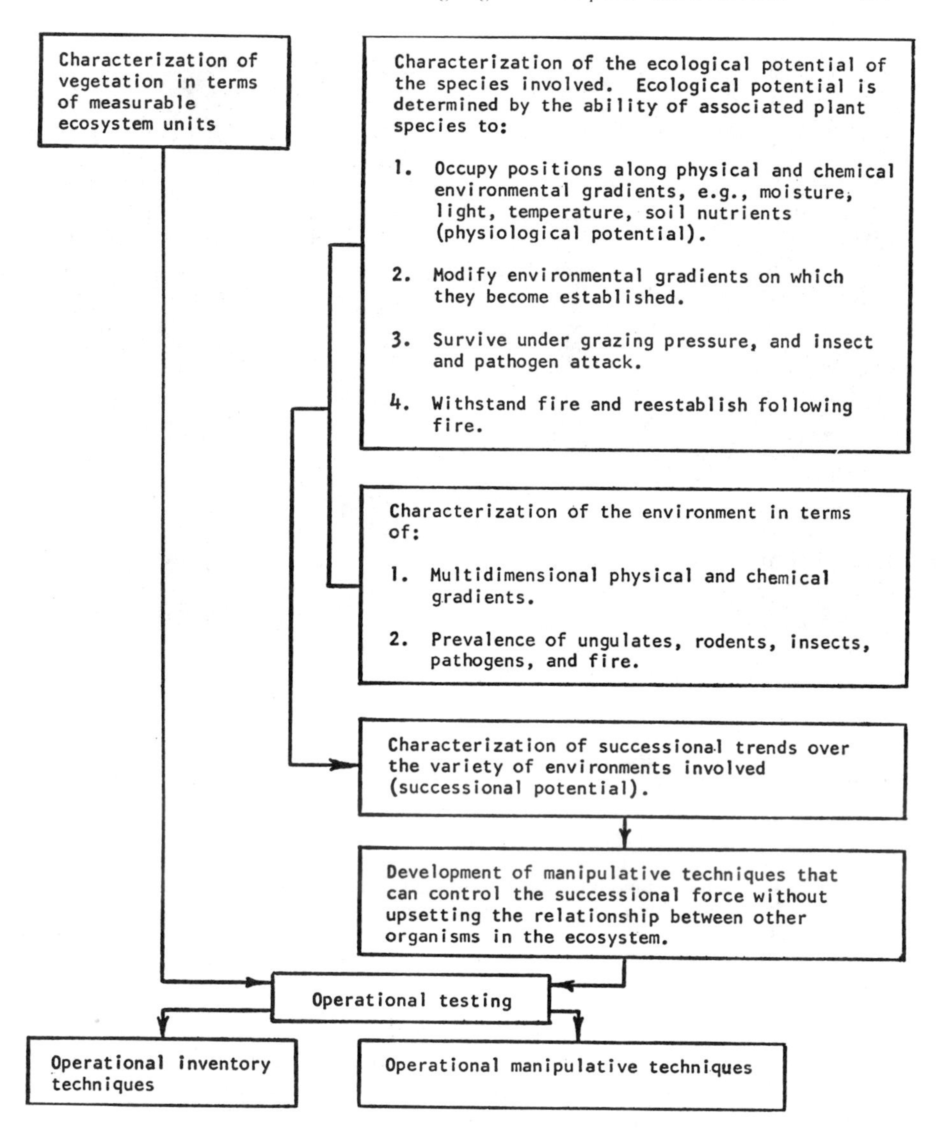


FIGURE 13–1. Information required to develop manipulative techniques for vegetation preservation.

species has, in general, been acceptable. Preservation of the redwood tree, for example, has been the sole objective in the world-famous string of redwood parks that extends from south of San Francisco to the Oregon border. Consequently, change among associated species has been ignored.

A less common but still popular objective is the preservation of par-

FIGURE 13–2. Coast redwood is physiologically attuned to periodic inundation and is provoked by the accompanying deposition of silt to regenerate a new, possibly more active, feeder-root system. Large amounts of organic matter incorporated in the silt can be fatal, however, and the vegetation-preservation manager must be ready to test for this condition and remove these silt deposits quickly if necessary. What steps he will need to take once upstream flood-control measures have altered this peculiar environment are not yet clear.

ticular successional stages. Most often this objective has reflected a desire to preserve a piece of virgin forest or native grassland and generally has involved a *climax* phase of vegetation, that is, a condition of dynamic equilibrium in which species composition remains more or less constant. On occasion, however, there has been a strong desire to preserve a particular *subclimax* phase, such as Douglas fir on the Olympic Peninsula, red pine in the Lake States, and Caribbean pine in the Everglades, where successional change may be proceeding rapidly. In support of this type of preservation, a committee appointed by Secretary Udall to review wildlife-management practices in the national parks has recently recommended that certain successional stages be re-created and maintained in order to present "vignettes" of early America to the park visitor (6). The ease with which various successional sequences can be maintained varies with the area and the type of vegetation involved. Consequently, preservation may not be as cheaply achieved in one area as in another. Provided he is aware of this, the park manager can select appropriate areas and achieve his objectives with a minimum effort.

A closely associated objective is that of slowing succession. Supporters of this objective argue that it is futile to try to stop succession, that if it can be slowed sufficiently a vegetation mosaic containing most of the successional stages could be maintained, and that such a mosaic is what we should strive for in the national parks. Certainly the degree of preservation desired is always an important consideration. Succession can often be slowed for only a few cents per acre, while costs of stopping succession can run to several hundred dollars per acre.

Today, many wilderness supporters argue that we should leave large

FIGURE 13–3. Fire and browsing by deer (see trees in left foreground) have been important in the creation and maintenance of grass prairies in the redwood-Douglas fir forest along the north coast of California. The vegetation-preservation manager can use fire but is not dependent upon it to maintain this open, park-like intermingling of grassy glades and trees. A variety of selective herbicides–none of which enter the biotic food chain–is available and can be used effectively in conjunction with either spot burning or mowing.

areas of vegetation alone to change as they may; that man should keep his hands off and let nature run its course, unimpeded by controls against fire, insect, and disease. When pressed on the point of fire control, however, proponents of this policy have usually agreed that some fire control is reasonable, provided it does not interfere with the occurrence of natural fire. Accordingly, lightning fires would be allowed to run unchecked, and if the aboriginal arsonist were alive today he would not be discouraged because he would be a part of the natural environment. Paradoxically, fire started by a careless camper would be dealt with vigorously.

On most of the areas that might be affected by such a program, succession is extremely slow and, because of extensive areas of exposed rock, wildfires soon burn themselves out. With sufficient control of human use these areas will change little in the next 100 or even 200 years, and this is probably what most proponents of a hands-off policy visualize. There are other areas, however, where the understory is now very dense and highly inflammable throughout much of the year and not in the condition that prevailed when the areas were set aside. Uncontrolled wildfire would be catastrophic. Thus, in the absence of fire control, vegetation in one area would be maintained more or less as it is today for many years to come, while in another area it might be violently changed within the next few years.

## Compatibility of Objectives

Many preservationists consider management per se to be an unwarranted interference with nature by man. This need not be true. Management consists merely of those actions that are necessary to achieve one or more objectives, whatever they may be, even if the objective is "no management." Management dealing with vegetation may be intensive or extensive, depending upon the objectives, but unless the objectives are thoroughly outlined, effective management is impossible. Because vegetation preservation may be only one of several objectives, all must be carefully considered together to determine whether they are compatible. Intensive public use may be compatible with a general policy of keeping certain park lands green, but may be incompatible with a specific policy of preserving dominant species or particular stages in a successional sequence. Probably incompatible objectives are much more widespread in current efforts to preserve vegetation than is generally recognized, because change in vegetation can proceed for many years without detection by the public or even by the park administrator responsible for its preservation.

## Research on Vegetation Preservation

The National Academy of Sciences Advisory Committee, in reviewing the research program of the National Park Service, "was shocked to learn that for the year 1962 the research staff (including the chief naturalist and fieldmen in natural history) was limited to ten people and that the Service budget for national history research was $28,000—about the cost of one camp-ground comfort station" (5). If we consider only the magnitude of the research job required to support a realistic vegetation-preservation program, it is easy to understand why the committee was shocked. A million-dollar annual budget and a staff of several hundred scientists, with several times as many supporting personnel, are needed. The Yellowstone National Park staff, for example, has indicated (5) that the research required to support its vegetation-preservation program would entail an analysis of the current climatic trends; a detailed soil survey; and analyses of the vegetative mosaic and the factors creating it; successional patterns in the various biotic communities; the interrelationships of plants and animals, particularly dominant species like ungulates; variations of current ecological conditions from the original; the factors that have caused these deviations; the practicability of re-creating original ecological conditions where ecological damage or deterioration, for instance, soil loss, has occurred; and the direct effect of visitors on important natural features. Thus, dealing effectively with the problem of vegetation preservation in this one park, only one of 31 national parks, will require a dozen or more scientists—climatologists, pedologists, and ecologists of various specializations—with a supporting staff of perhaps a hundred or more.

The Forest Service has not as yet committed its research staff to studying the overall problems of vegetation preservation, but much of what its silviculture and range-management researchers have discovered over the last 35 years is directly applicable. Both the silviculturist and the range manager, like the vegetation-preservation manager, are interested in successional processes and their control. The distinction lies in the end products desired and the tools that can be used to obtain them. The silviculturist is interested in the amount and quality of timber produced. He selects trees to this end and in the process completely alters the structure of the vegetation units involved. His imprint in the form of skid trails, neatly sawn stumps, and extraction roads is everywhere apparent. The range manager is interested in the weight and quality of beef or mutton that vegetation produces. He uses his animals, through rotational grazing schemes of various kinds, to control plant succession. Except for the presence of domestic animals, fences, and occasional scars of a disc harrow on a reseeded, overgrazed range, his mark is not apparent and the structure of the vegetation units involved may not be greatly altered.

What is most needed to get a full-scale vegetation-preservation research program underway by the Forest Service is an administrative decision to do so. Only a few shifts in research emphasis at key points within the present research program, along with a relatively modest augmentation of the basic research staff, are needed. Once embarked upon such a program, the Forest Service soon would be able to develop suitable operational techniques for preserving vegetation on the 12 million acres of wilderness that are its responsibility.

All the state parks involved in vegetation preservation need research support, but few are receiving it. The State Division of Beaches and Parks in California, for example, has an annual support budget of $10,000,000 and is responsible for vegetation preservation on more than 600,000 acres. The size of the individual parks varies from a few hundred to several thousand acres, and the type of vegetation to be preserved varies through cactus and scrub on the Mohave Desert, oak woodland in the Central Valley, mixed conifers in the Sierra, and redwoods along the North Coast. Its annual research budget amounts to only $28,000. A third of this is being spent on a crash program to develop recommendations for preserving redwood groves along the Eel River which are subjected to periodic flooding and to more than 500,000 visitors annually. The rest is being spent by the interpretive-services section. Nothing is being spent on research to determine how the variety of vegetation types that occur in the other parks in the state should be maintained.

The type of information required to develop manipulative techniques for the preservation of vegetation is summarized in Figure 13–1. Some of this information, obtained through the efforts of university-based scientists, their graduate students, and Forest Service researchers, is already available for certain types of vegetation. Rarely, however, will this information be complete from a vegetation-preservation viewpoint, partly because preservation has not generally been the objective of the studies, but mostly because studies of plant succession in this country have closely followed an approach developed by the well-known ecologist F. E. Clements (7). Clements was convinced that successional sequences, which involve changes with time, can be determined by observing changes in vegetative patterns in space, and through his persuasive pen he was able to convince others that this was feasible. The major difficulty encountered in this approach has been that the only method of evaluating the accuracy of a researcher's estimate of successional trends, that is, his first approximation, has been to wait and see.

Where short-lived grasses and herbs have been involved and reestablishment of a dynamic equilibrium following disturbance has been rapid, the Clementsian approach has been reasonably effective. Researchers have been able to modify their first approximation through several subsequent approximations, improving the accuracy of their estimate each time. This is the approach that largely has been used by range-management researchers. Recently, however, they also have experienced its limitations and have begun to turn to detailed environmental analyses and growth-performance studies under controlled environments.

FIGURE 13–4. Big cone spruce, shown against the sky line, is not a fire-resistant species but has survived in the chaparral of Southern California because of natural firebreaks created by shallow soils and the regular occurrence of widespread fires in the past. Today these natural firebreaks are overgrown and no longer offer protection. The vegetation-preservation manager must reestablish these firebreaks if stands of big cone spruce are to be preserved.

Where long-lived plants have been involved and reestablishment of a dynamic equilibrium following disturbance has been slow and will not be reached for another hundred years or so, first approximations based on Clements' approach are often little better than educated guesses. In dealing with vegetation of this type, a more sophisticated analysis of the ecological potential and the relative magnitude of the environmental factors involved will be required (Fig. 13–1). Studies of comparative growth performance of associated species in controlled environments along the lines suggested by Hellmers (8), and environmental-gradient analyses along the lines suggested by Whittaker (9), Bakuzis (10), and Waring and Major (11), are essential.

FIGURE 13–5. Top aerial photograph was taken shortly before the grass-covered slopes in the right foreground were purchased by a local park district and cattle were excluded. Bottom aerial photograph was taken early in 1965–35 years later–and shows the extent to which brush has replaced grass within the park boundaries. If the grass cover is to be preserved, the vegetation-preservation manager must remove the brush and take steps to keep it out. Fire as a tool has been ruled out by local smog-control officials. Introducing cattle is difficult because of the number of youth groups using the area. Bulldozers, herbicides, and mowing machines appear to be the alternative tools available.

## Specialists in Vegetation-Preservation Management

The National Academy of Sciences Advisory Committee (5) points out briefly that the Park Service has applied research in a piecemeal fashion and has "failed to insure the implementation of the results of research in operational management." The committee concludes with the comment that "Reports and recommendations on the subject will remain futile unless and until the National Park Service itself becomes research-minded and is prepared to support research and to apply its findings." That the "implementation of the results of research" calls for experts on the management of vegetation has as yet not been recognized. Even preservation-oriented conservationists, who are the backbone of the leading conservation groups in this country, have been slow to perceive this. Many of them still regard vegetation much as they do their own gardens and are quick to suggest how a particular vegetative cover can best be preserved, whether it be in the local nature reserve, in a state park dominated by 1000-year-old redwood trees, or in an untrammeled wilderness.

Obviously the decision as to what should be preserved cannot be left entirely to the specialist. The concerned public, although amateurs in vegetation preservation, must be heard and heeded. But at the same time a realistic assessment must be made of what can be achieved at costs commensurate with public interest, and this depends upon a knowledge of various alternatives and the relative cost and feasibility of achieving each one. Only the vegetation specialist can furnish this kind of information.

The vegetation-preservation specialist must be trained in management, must possess a thorough knowledge of ecology, must be experienced in assessing the relative growth potential of each species in the vegetative mosaic, must be experienced in the use of various manipulative techniques, and must understand research methods. Today there are few men so qualified. There is an impressive number of competent plant ecologists scattered throughout related professions who are oriented toward management, but there are relatively few who have had experience in a detailed assessment of the environmental complex, and even fewer who have had experience in manipulative techniques.

The vegetation-preservation specialist will not replace the research ecologist and to a large extent will be dependent upon him. He must be competent to understand research, to evaluate research findings in terms of his management function, and to translate research into manipulative techniques particularly suited to the specific vegetation he must manage. These manipulative techniques must be based on an understanding of the ecology of the vegetation in question; if such information is not available and ecologists are not employed to develop it, the preservation specialist will be forced to forego his primary responsibility and to spend his time collecting basic ecological data.

Because success in the field of vegetation preservation requires several—usually many—years to evaluate, the vegetation-preservation specialist often will operate in an atmosphere in which unsubstantiated opinions are forcefully urged. Many fire enthusiasts, for example, are convinced that fire protection should be curtailed, and do not recognize that merely because fire control has led to some undesired effects it does not necessarily follow that fire control should be abandoned or prescribed burning introduced. Involved is the whole process of recognizing the management objective, evaluating the ecological forces in play, identifying the conditions which must be achieved to develop the desired vegetation response, and, finally, evaluating all the possible ways of moving toward those conditions economically and with a minimum of unwanted side effects. In all of this the vegetation-preservation specialist will need a fine sense of perspective.

Little can be accomplished in the field of vegetation-preservation management until a source of competently trained specialists has been developed—and perhaps not until considerable numbers of these specialists have infiltrated the various responsible administrative bodies. How can we develop such a source? At the moment I can see only one solution: Ask those universities that have strong programs both in ecology and in land management, for example, those with forestry and range-management curricula, to take on the job. It should be possible to train these specialists by means of a 2-year graduate program, provided it is preceded by an undergraduate degree with a proper emphasis on basic biology and is followed by an appropriate period of apprenticeship. Several universities could readily meet this challenge, provided financial support were assured. The question that remains to be answered is: How soon will the universities that have staffs capable of carrying out this graduate program be asked to join in creating this new profession?

## References

1. Yellowstone National Park Establishment Act, 17 *Stat.* 32 (1872); S. W. Allen, *An Introduction to American Forestry* (McGraw-Hill, New York, ed. 2, 1938).
2. J. Ise, *Our National Park Policy* (Johns Hopkins Univ. Press, Baltimore, 1961).
3. R. P. Gibbens and H. F. Heady, "The influence of modern man on the vegetation of Yosemite Valley," *Calif. Agr. Exp. Sta. Manual No. 36* (1964); Univ. of Calif. Wildland Research Center, *Outdoor Recreation Resources Rev. Comm. Study Rept. No. 3* (Government Printing Office, Washington, D.C., 1962).
4. Multiple Use Sustained Yield Act, 86 *Stat.* 517 (1960); the National Wilderness Preservation System Act, 78 *Stat.* 890 (1964).

5. National Academy of Sciences–National Research Council, *A Report by the Advisory Committee to the National Park Service on Research* (Washington, D.C., 1963).
6. A. S. Leopold, S. A. Cain, I. N. Gabrielson, C. M. Cottam, and T. L. Kimball, *Living Wilderness* 83, 11 (1963).
7. F. E. Clements, "Plant succession," *Carnegie Inst. Wash. Publ. No. 242* (1916).
8. H. Hellmers and W. P. Sundahl, *Nature* 184, 1247 (1959).
9. R. H. Whittaker, "Vegetation of the Great Smoky Mountains," *Ecol. Monographs* 26, 1 (1956).
10. E. V. Bakuzis, thesis, Univ. of Minnesota (1959).
11. R. Waring and J. Major, "Some vegetation of the California coastal redwood region in relation to gradients of moisture, nutrients, light, and temperature," *Ecol. Monographs* 34, 167 (1964).

# 14

# Interaction between animals, vegetation, and fire in Southern Rhodesia

*A. S. Boughey*

On the wide savannas of the great tableland of eastern and southern Africa, as Huxley (1962) has described, there evolved during the Tertiary the most striking climax community to persist into modern times. Attempts are now being made to preserve at least parts of this unique ecosystem by the institution of a number of large game reserves scattered throughout the area.

Huxley refers to the scientific work already in progress on a number of different aspects of this game conservation, but it is apparent from the studies which he and others have described, that comparatively little attention has yet been given to the vegetation. Nor has it yet been generally appreciated that the wide diversity of game animals is not only reflected in, but also is dependent largely upon, an equal diversity of habitats.

The Wankie National Park, in which the investigations described here were undertaken, is situated in the south-west of Southern Rhodesia (see map). In its northern and tourist section, where additional permanent sources of drinking water have been made available, this game

Reprinted with permission from *Ohio Journal of Science*, 63:193–209 (1963). The author is with the Department of Population and Environmental Biology, University of California, Irvine, California 92664.

reserve provides a unique opportunity for studies on the interaction between artificially high populations of the larger game animals and the natural vegetation.

The Park lies at about 3,500 ft altitude and is on the Kalahari Sands formation. This is an aeolian deposit up to 100 ft in depth, superimposed on Karroo sandstones and volcanics, extending northwards from Bechuanaland to the Congo River, and in southern Rhodesia reaching eastwards from the Victoria Falls to near Bulawayo. The Kalahari Sands were formed in the Late Tertiary, but were disturbed and re-deposited

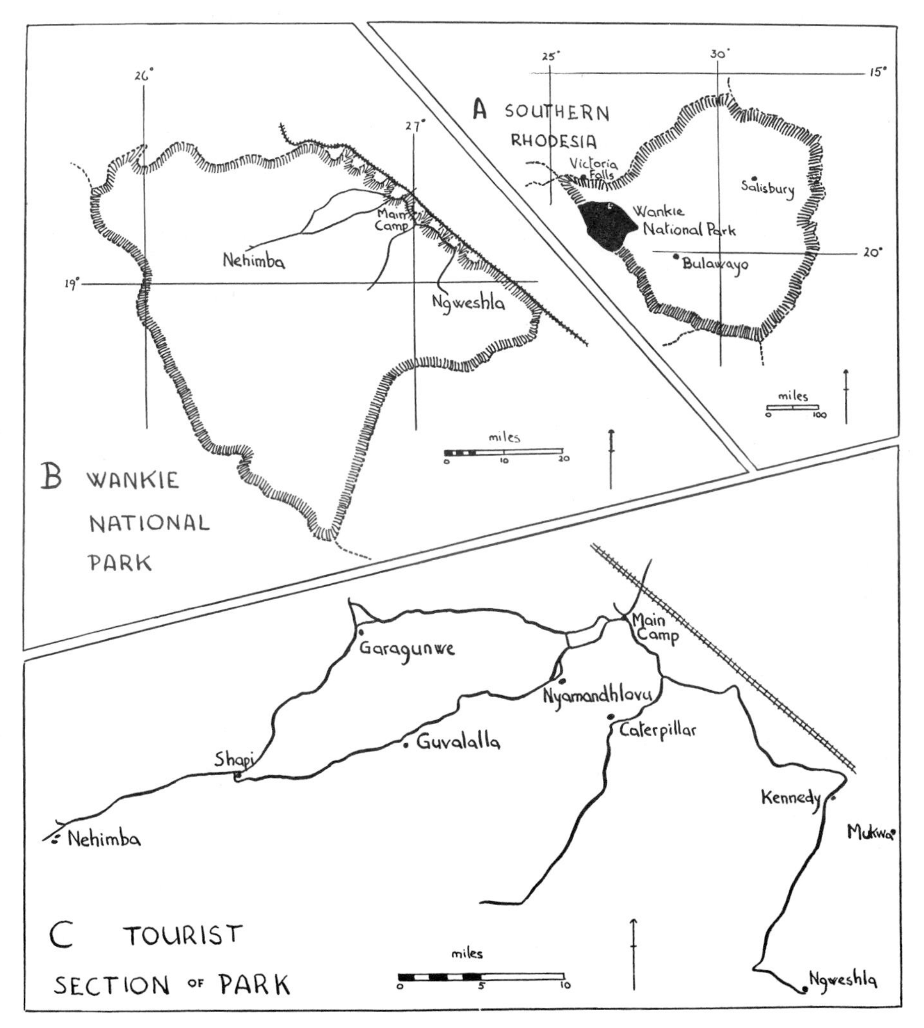


Map showing the location of: A, the Wankie National Park in southern Rhodesia; B, the position of the area of the Park investigated; C, the distribution of the various pans, drinking holes, and roads in the tourist section of the Park.

during the Pleistocene, in some instances being known to have covered human occupation sites, before the sand surface was again consolidated (Clark, 1959).

Summarized meteorological data for the area are given in Table 14–1. The annual rainfall recorded over the last 14 years varies from 19 to 37 inches, averaging approximately 24 inches. Although subject to high tropical temperatures in summer, frosts may occur in winter, more especially in June and July.

The Park was formally taken over for its present purpose in 1928 (Gale, 1963). The land around what is now the Main Camp had previously been farmed for some years on an extensive basis by Europeans, and other sections hunted over from time to time by both Europeans and Africans. It would appear that, while game concentrations have undoubtedly been present in the region for a very long time, protection from hunting, and, more especially, the provision of further permanent drinking supplies, have led to great increases in the numbers of game animals using the northern section, more particularly those of elephant, wildebeeste, zebra, and buffalo.

As described in another paper (Boughey, 1961), the undisturbed vegetation of the region in which the Park lies can be defined in terms of a catena type in which the better-drained flat or gently sloping land is covered by teak woodland, with a great abundance of Teak (*Baikiaea plurijuga* Harms). A timber company had operating rights for the extraction of Teak in the area before 1928. This timber exploitation could partly or entirely explain the absence of tall Teak trees of an *emergent* form, and account for the fact that the vegetation here is of a *woodland*

TABLE 14–1

*Summarized meteorological data from screen situated at 3,519 ft altitude, 18° 45' S, 26° 56' E, adjacent to main camp Wankie National Park**

1) *Amount of Annual Rainfall (July through June)*

| Season | Total inches per annum | Number of rainy days | Season | Total inches per annum | Number of rainy days |
|---|---|---|---|---|---|
| 1948/49 | 28.0 | 59 | 1955/56 | 21.9 | 78 |
| 1949/50 | 19.8 | 79 | 1956/57 | 18.2 | 68 |
| 1950/51 | 17.0 | 62 | 1957/58 | 39.5 | 88 |
| 1951/52 | 37.0 | 76 | 1958/59 | 27.9 | 80 |
| 1952/53 | 29.0 | 83 | 1959/60 | 25.8 | 58 |
| 1953/54 | 25.2 | 73 | 1960/61 | 28.2 | 85 |
| 1954/55 | 34.5 | 87 | 1961/62 | 19.2 | 72 |

2) *Average of mean monthly temperature 1951—1960.*

July (coldest month) 57.1 F (absolute min. 19F)
October (hottest month) 76.8F (absolute max. 102F)

3) *Average of mean monthly relative humidity 1951–1960.*
September (driest month) 37% (at 2PM 23%)
February (wettest month ) 77% (at 6AM 94%)

*All figures are by courtesy of the Federal Metereorological Department.

rather than a dry deciduous forest type. The effect of fire also has to be taken into consideration. According to Mitchell (1961a), a drastic reduction in the numbers of game which once inhabited the teak forests of the Kalahari Sands is responsible for the almost complete disappearance of those forests; the animals prevented the accumulation of sufficient material from the usually sparse herbaceous cover to sustain a grass fire. Mitchell's general explanation cannot however be applied to account for the degradation of the teak forest to woodland in the northern part of the Wankie Park, for it would seem that animal populations here have always been fairly high. As will be discussed later, it must also be borne in mind that grass fires always have occurred over the elephant damaged Teak areas even before there was significant interference by man.

The catena pattern of plant communities in the northern section of the Park is illustrated diagramatically in figure 14–1. Degraded teak forests in the form of woodland occupy the upper sections of the gentler slopes and rises of the deeper sands, while the poorer drained shallower sands carry a mixed savanna woodland in which tree species such as *Lonchocarpus capassa* (Rolfe), *Ziziphus mucronata* (Willd)., *Acacia galpinii Burtt-Davy*, *A. giraffae* (Willd.), *Combretum inberbe* (Wawra), and *C.*

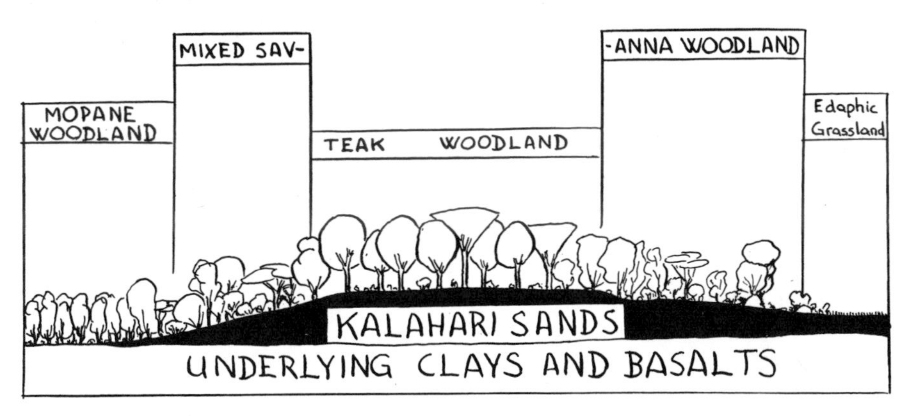


FIGURE 14–1. Diagramatic scheme illustrating the catena pattern of plant communities in the northern section of the Wankie National Park, and showing the relative positions of teak woodland, mixed savanna woodland, mopane woodland, and edaphic grasslands (vleis).

*hereroense* (Schinz) are characteristic dominants. The clay depressions bare of Kalahari Sands which occur more especially west and north of Shapi Pan, and occupy the base of the catena pattern, develop what is called locally mopane woodland. This is a savanna woodland, closing locally to a woodland, composed almost entirely as a dominant of the Mopane tree, *Colophospermym mopane* (Kirk ex Benth.) Kirk ex J. Leonard.

The area of the northern section of the Park with a catena pattern formed principally of the three main plant communities, teak woodland, mixed savanna woodland, and mopane woodland, which was examined in the course of these studies, lies between the Main Camp and Nehimba Pan about 50 miles to the west, and the Main Camp and Ngweshla Pan 35 miles to the south. The work was carried out mainly in the course of visits to the area in January, July, and December 1960, and May and July, 1961. This area of the reserve was selected because of the high animal populations in the drought months of May to October, when drinking water is scarce or absent elsewhere in the park, and the animals have to congregate near permanent drinking holes. It was considered that under these extreme conditions ecological interactions would be heightened and the more readily identifiable. A secondary consideration was that this particular area is provided with a network of roads, so essential for access in the rainy season.

The procedure followed in the preliminary investigations described here was highly subjective. What was judged to be the least disturbed vegetation in each of the three plant communities recognized was located and examined. An attempt was then made to identify the progressive steps in its degradation and to record the game animals especially associated with each successive stage.

Vegetation was recorded in sample belt transects, 20 ft wide and 20, 40, or 60 ft, long according to the relative density of woody growth, laid down in representative portions of each community. The position, identity, height, and canopy spread of all woody plants in the transect were recorded. The profile diagrams and ground plans reproduced here as text figures are prepared from these data, as are the diagrams of hypothetical succession schemes.

The terms used for physiognomic types of vegetation are those recommended by the Yangambi ecological conference (C.S.A. 1956). Figures for distance, altitude, space, and height have been left in the original units of miles and feet in which they had to be measured, rather than converted to metric units.

## Successional Relationships—Teak Woodland

### *Undisturbed Teak Woodland*

Small areas of teak woodland, about 5 to 20 acres in extent, still occur in the area investigated, especially in the western and better-drained part; rather larger copses occur along the first portion of the road leading south to Ngweshla Pan. This teak woodland is composed of stands, often nearly pure, of Teak trees between 40 and 50 ft high (fig 14–2). The commonest tree associates in these Teak stands are *Guiborta coleosperma* (Benth.) J. Léonard, and *Croton zambesicum* (Muell.

Arg.), with on the rises *Mundulea sericea* (Wild.) Greenway and undescribed form of *Pterocarpus angolensis* Dc. Shrubby associates are *Baphia obovata* Schinz, *Diplorhynchus condylocarpon* (Muell. Arg.) Pichon subsp. *mossambicensis* (Benth.) Duvign., and *Bauhinia mendonçae* Torre & Hillcoat.

Such teak woodlands are visited throughout the year by older elephants either singly or in small parties, and seasonally from about June through November, by large breeding herds commonly up to 100 to

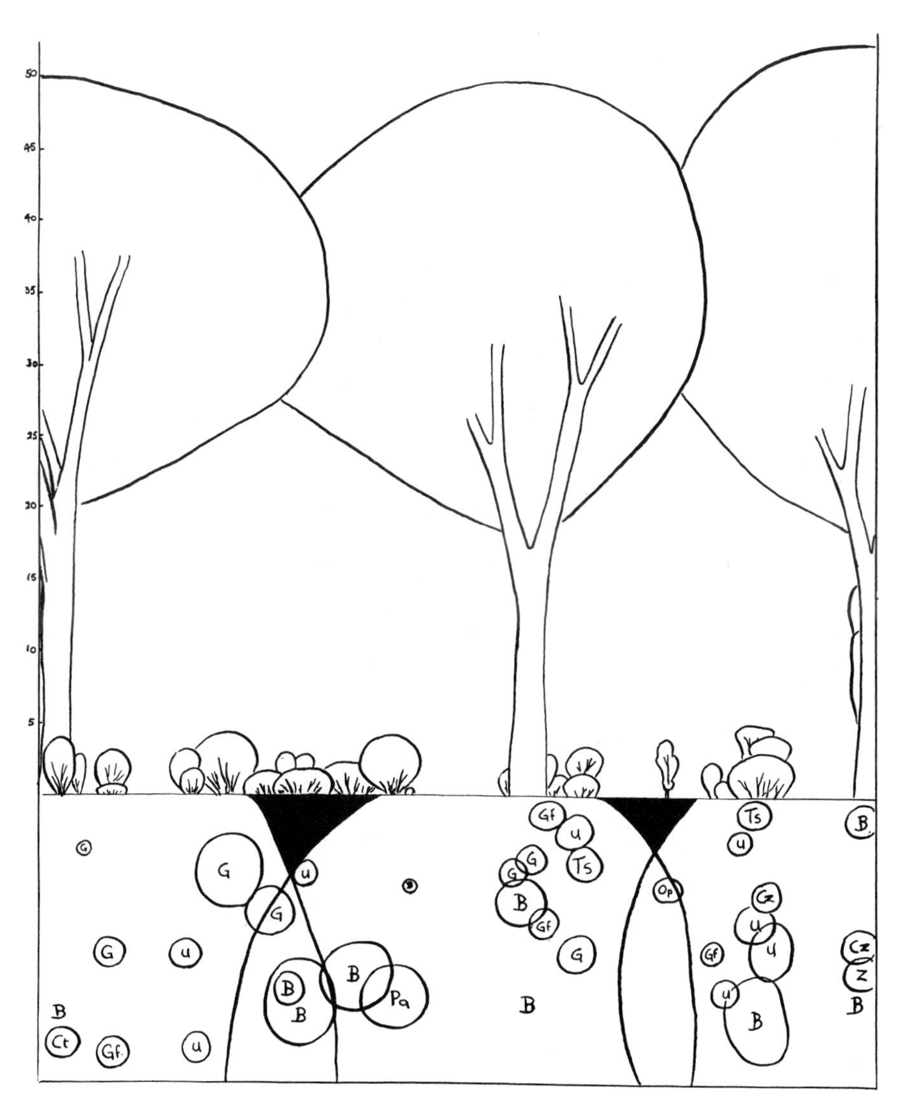


FIGURE 14–2. Profile diagram and ground plan of a portion of teak woodland, prepared from a transect taken some nine miles from Main Camp along the Guvalalla road.

FIGURE 14–3. Profile diagram and ground plan of *Baikiaea–Terminalia* woodland, ravaged and opened up by elephants, prepared from a transect taken some 13 miles from Main Camp along the Guvalalla road.

KEY TO THE SYMBOLS FOR PLANT SPECIES EMPLOYED IN FIGURES 1–5

The scale shown on all profile diagrams is the same, figures indicating the height in feet. Each ground plan is 20 ft wide, and is composed of one, two or three 20 ft squares as the case may be. The area on the ground plan covered with a mechanical tint indicates in each case the extent of ground not directly shaded from above by woody plants, and is therefore some indication of the amount of grass cover in the community illustrated.

A —*Acacia galpinii*
Ag —*Acacia giraffae*
As —*Asparagus* sp.
At —*Acacia tortilis*
B —*Baikiaea plurijuga*
Bp —*Baphia obovata*
Bx —*Bauhinia mendoncae*
Bu —*Burkea africana*
Ch —*Combretum hereroense*
Ct —*Canthium* sp.
Cz —*Croton zambesicum*
D —*Dichrostachys cinerea*
Di —*Diplorhynchus condylocarpon*
Ea —*Erythrophlelum africanum*
G —*Grewia bicolor*
Gf —*Grewia flavescens*
Mc —*Maytenus cymosus*
Op —*Ochna pulchra*
Pa —*Peltophorum africanum*
R —*Rhus tenuinvervis*
S —seedling of *Terminalia sericea*
Ts —*Terminalia sericea*
U —indet. sp.
Z —*Ziziphus mucronata*

200 head strong. Few trees in the woodlands are without some scars of these visits. The most significant damage caused by the elephants at this stage is the stripping of bark from Teak trees in large sheets. While many trees successfully recover from this damage, those which have been completely girdled by the bark stripping eventually die. It is considered by the Park Wardens that the elephants do not strip off Teak bark to eat, indeed they have never been observed to do so, but are merely whiling away the time whilst sheltering in the woodland from the heat of the day, or resting at night.

### *Baikiaea—Terminalia Woodland*

Gaps created in the teak woodlands by the death and eventual fall of ringed trees, and by the odd tree pushed over by elephants, are colonized by a number of invading tree species. Of these *Terminalia sericea* (aggregate) is by far the most abundant, with many Teak saplings, a little *Burkea africana* Hook., some *Combretum* sp., and an increase in *Baphia obovata* and *Bauhinia mendonçae* in the shrub layer (fig. 14–3). These younger and smaller invaders appear to be left relatively undisturbed for a time, while the elephants persist in the destruction of the remaining older and larger teak trees, but sooner or later they turn their attention to the young growth in the gap, and commence to break off the tops of the *Terminalia* and other tree species.

### *Burkea—Terminalia Savanna Woodland*

By the time only a few old relict Teak trees survive, the original woodland has been opened up considerably, and the community is more properly now called savanna woodland. In these more open conditions there is a proportionally increased representation of the tree *Burkea africana*, and more dense stands of grass are established.

The foliage in an undisturbed Teak woodland is not only mostly unreachable, it is also apparently unpalatable. The presence of more plentiful grass and of palatable young tree browse within reach in the *Burkea—Terminalia* savanna woodland may have two direct and important consequences. For the first time large game animals in addition to elephant begin to feed extensively in the area; sable antelope, kudu, and less importantly gemsbok, were commonly observed here. The relative intensity and the time of year of the grazing, as opposed to browsing, which these animals do is of great ecological significance. If the animals mostly browse in the rainy season, and graze during the dry season, the grass will be favoured. Should the reverse occur, the woody vegetation will increase at the expense of the grass. In the first circumstance there may be an ample supply of flammable material remaining in October. If a grass fire is then started accidentally by some human agency,

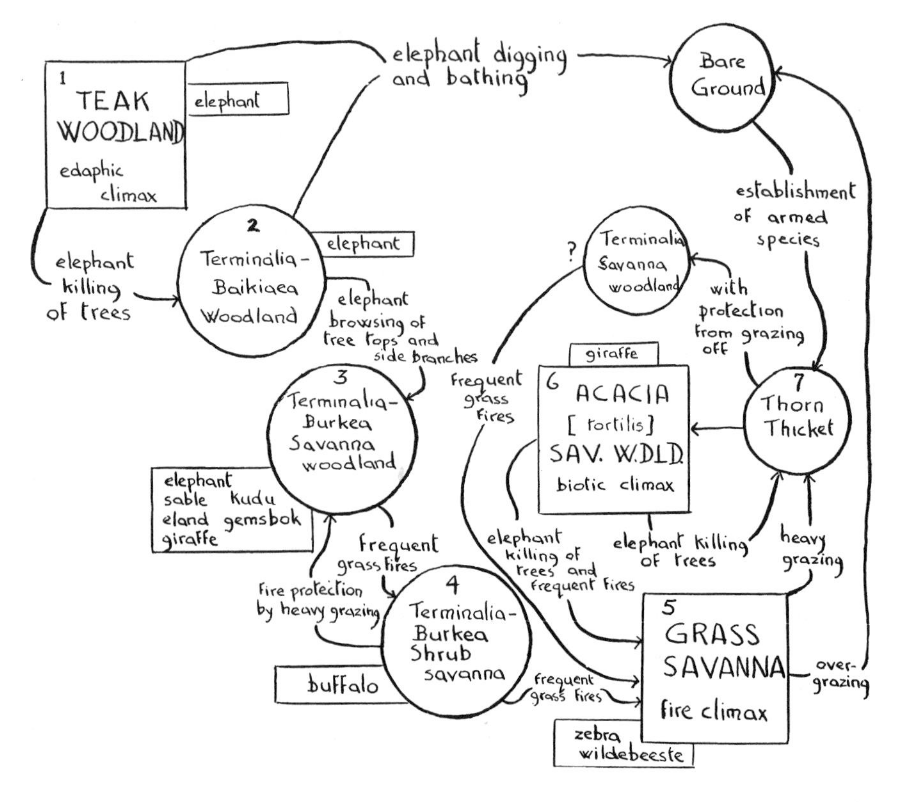


FIGURE 14–4. Diagramatic hypothetical scheme showing the various secondary communities derived from the teak woodland vegetation type on Kalahari Sands in the northern section of the Wankie National Park, the major ecological factors which operate in this ecosystem, and the larger game animals associated with particular plant communities.

as frequently occurs in this area, most tree seedlings will perish. Although the thick bark of mature trees in this community usually insures their being fire resistant, saplings have not yet developed this and may be burnt to the ground, generally sprouting again from the burnt collars during the next rainy season.

Elephants are liable to grub for edible roots such as those of *Bauhinia fassoglossis* (Kotschy ex Schweinf.) and perhaps to dig for salt or water, in any of the types of woodland or savanna woodland so far discussed. At the beginning of the rainy season in November, the holes which they have dug fill with water and may be used as mud baths by the elephants. The vegetation around such holes is destroyed, leaving patches of bare ground. Thickets of thorny species, in particular *Dichrostachys cinerea* (Wight and Arn.) and *Acacia* sp. develop on these bare sites.

Elephants commonly browse on this *Burkea—Terminalia* savanna

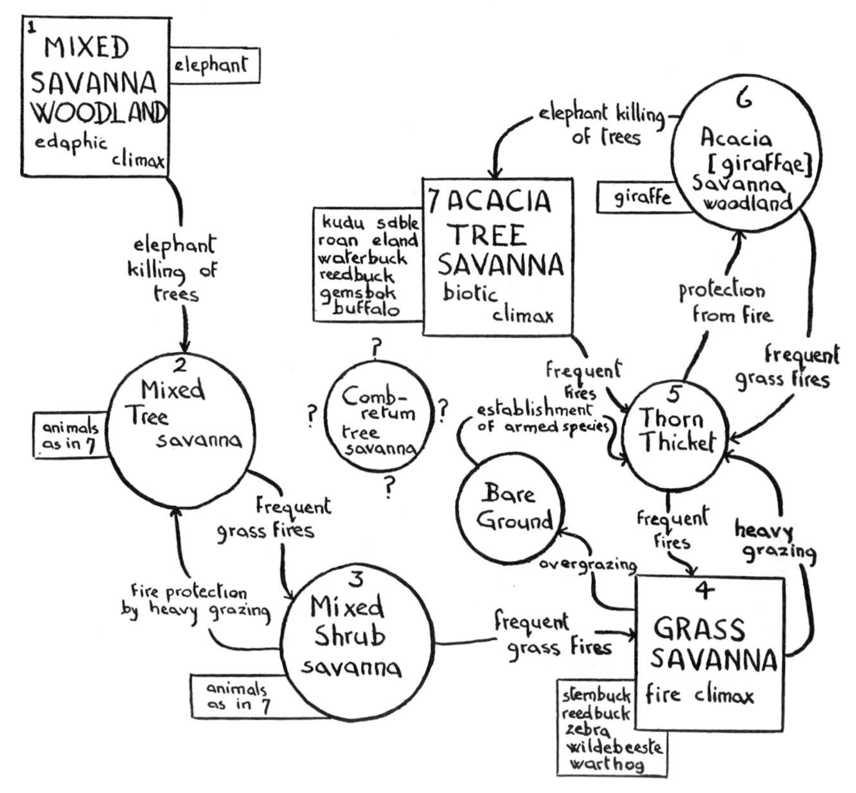


FIGURE 14–5. Diagramatic hypothetical scheme showing the various secondary communities derived from the mixed savanna woodland type of vegetation on Kalahari Sands in the northern section of the Wankie National Park, the major ecological factors which operate in this ecosystem, and the larger game animals commonly associated with particular plant communities.

woodland and often "top" the trees at a height convenient for their reach, about 4 ft. Numerous giraffe also browse the tops of the trees, but seemingly without any significant ecological effect.

### *Burkea—Terminalia Shrub Savanna*

As already noted, grass fires in the *Burkea—Terminalia* savanna woodland will kill most tree seedlings and burn tree saplings to the ground. From the latter a sprout growth develops, which may again be burned to the ground by a fire one or two years later. Ultimately all the mature trees, which have probably mostly been able to withstand the frequent fires but are relatively short-lived die out, leaving only a shrub savanna of sprout growth.

This is the stage which buffalo most seem to favor; they graze the grass between the shrub clumps and trample the area heavily. Such a grazing pressure may be maintained that fires do not occur for several years, and the sprout growths may get 3 or 4 ft high, but usually a fire eventually comes along and burns them down again.

### Grass Savanna

Continued trampling of shrub savanna by buffalo combined with only moderate grazing pressure permits the survival of sufficient flammable material to provide for late (October, November) fires annually. This in a few years results in the complete suppression of all woody growth. The resultant grasslands come to support large herds of wildebeeste and zebra as well as smaller animals such as ostrich and stembuck.

A good example of such a grassland in course of formation in the northern section of the Park is to be seen at Nehimba Pan, which lies in the middle of a grass savanna about one mile across. Nehimba Pan provides a permanent water supply, due to seepage into it of water from the surrounding area, and it must have been visited by large concentrations of animals in the dry season before the game reserve was founded. Under present conditions something like two thousand buffalo come down to water between noon and dusk every day during the height of the dry season. Despite this large animal population using the drinking holes, Nehimba Pan plain does not appear to be heavily grazed, nor are large herds of wildebeeste or zebra in occupation of or visiting the area. As a consequence the flammable grass material which persists into the dry season each year is considerable, and the plain appears to burn annually. This frequency of burning is sufficient to keep in check the re-growth from such sprout clumps as have survived from the tree cover of the original savannas and teak woodland.

Not all grass savanna in this northern section of the park has been derived in this manner from the breakdown of teak woodland and savanna. Many of the smaller pans are surrounded by an edaphic grassland, or *vlei* as such vegetation is called locally; these grasslands are discussed later.

### Acacia Savanna Woodland

With heavy grazing of grass savanna by wildebeeste and zebra herds, and the consequent disappearance or at least the infrequency of grass fires, various armed woody species such as *Acacia giraffae*, *A. galpinii*, *A. tortilis* (Forsk.) Hayne, *Dichrostachys cinerea*, *Dalbergia melanoxylon* (Guill.& Perr.), and *Maytenus cymosa* (Soland.) Exell, become established. Presumably these thorny species are able to invade such areas because their spiny shoots largely escape being grazed off. At least the young shoots are not bitten off wholesale, whereas the unprotected seedlings of unarmed tree species tend to be grazed right off.

As they mature, these *Acacia* savanna woodlands seem to be more frequently visited by giraffe than other plant communities in the area; giraffe browsing nevertheless would not appear significantly to affect the ecological succession described here.

Grazing pressure is relaxed in these developing *Acacia* savannas presumably because of their thorny nature; perhaps spines are liable to penetrate the feet of animals as well as to discourage them from grazing. In more open communities grass fires may therefore once more become frequent, and this vegetation can be reduced again to a grass savanna. The thorny tree species of this savanna woodland do not appear to have the vigorous sprouting abilities of the broad-leaved tree species of the original teak woodland, so that a shrub savanna is not formed as an intermediate degraded community.

The best examples of this biotic climax *Acacia* savanna woodland in which the dominant tree was *Acacia tortilis*, were seen in the area some 12 miles from the Main Camp along the road passing Caterpillar Pan. The low hill-tops here, which must once have been covered with teak forest, now carry mature *Acacia* savanna woodland. It has been shown experimentally (Boughey, 1963) that seedlings of *Acacia tortilis heteracantha* (Burch.) Brenan cannot germinate in the shade of the parent trees. If the trees in this instance, which appear to belong to *A. tortilis spirocarpa* (Hochst. ex A. Rich.) Brenan, behave similarly, these *Acacia* savanna woodlands will be unable to maintain themselves for more than one generation on a given site unless seedling reproduction again becomes established as the stand deteriorates. Certainly no seedlings or saplings of the dominant species were observed in the savanna woodlands.

The animal populations which must have been responsible for the break-down of the original teak woodland in this area beyond Caterpillar Pan no longer appear to be present in the same concentrations. Perhaps this once favoured area was deserted by the game when artificial drinking holes were established along the tourist roads a little to the north. There is some suggestion that in these circumstances seedlings of *Terminalia sericea* will invade the area and become established. Indeed young almost pure *T. sericea* savanna woodlands are very commonly to be seen in the area associated with *Acacia* savanna woodlands. It is possible that this *Terminalia* community takes over when the latter become overmature.

### *Thicket*

Ground laid bare by over-grazing or by animal digging in any of the plant communities discussed is liable to be invaded by armed thicket-forming species: the most aggressive of these is *Dichrostachys cinerea.* It seems likely that the armed tree species cited as forming the principal dominants of *Acacia* savanna woodland become established first as a low-growing more or less open thicket. The longer-lived tree species such as *Acacia tortilis* eventually emerge above the shorter-lived, and in this area shrubby, species like *Dichrostachys cinerea.* As already stated, young plants of the broad-leaved tree species are protected from grazing-off when they become established in these thorny thickets and savannas, and such species as *Terminalia sericea* frequently constitute an important element of such vegetation.

## Successional Scheme—Teak Woodland

The successional interrelationships of these several plant communities which arise with the degradation of teak woodland by animals and by fire, and the points at which these major ecological factors operate, are expressed diagramatically in figure 14–4. It is to be noted that with a continuing rise in the number of game animals in this area, many of the trends are at present unidirectional: this point will be discussed later.

## Successional Relationships—Mixed Savanna Woodland

### *Undisturbed Mixed Savanna Woodland*

As is illustrated in figure 14–1, areas of the northern section of the park which occupy depressions in the Kalahari Sands, where drainage is poorer but not so seasonally impeded as to exclude tree growth altogether, could be expected to be covered with mixed savanna woodland. The principal tree dominants in this community are *Acacia giraffae*, *A. galpinii*, *Lonchocarpus capassa*, *Combretum imberbe*, *C. hereroense*, *Peltophorum africanum* (Sond.), *Rhus tenuinervis* (Engl.), with most frequently on termite mounds, *Ziziphus mucronata* (Willd.) and *Diospyros mespiliformis* (Hochst. ex A.DC). Around Mukwa Pan, in the east of the section, the palm *Hyphaene crinita* Gaertn. is associated with these last two species and largely confined to termite mounds.

In the northern section of the park however, because of the proximity of this type of vegetation to drinking holes, in no area has it remained even relatively undisturbed. As in the case of teak woodland, it is again elephants which are the main destructive agents. They strip the bark more particularly from the *Acacia giraffae* trees, in this instance apparently finding some nutritional need satisfied by it. Again trees are commonly killed by girdling, and whole trees may be pushed over and wrecked. Mixed savanna woodland therefore does not exist in the area of the park under consideration although it must have been the starting point for the successions described below, and it must exist elsewhere in the park.

### *Mixed Tree Savanna*

Thinning out of the trees in a mixed savanna woodland by elephants produces a tree savanna, which is now the most conspicuous vegetation surrounding the pans and water-holes on Kalahari Sands in this northern section.

Digging by elephants may clear the ground, and thorn thickets usually develop on such bare areas as in the teak woodland succession. A number of other animals may also contribute to the laying bare of the ground, especially wart-hogs which closely graze the grass margins of pans during the rains, and grub there for edible roots in the dry season.

In some areas the mixed tree savanna may be replaced by a tree savanna or a savanna woodland in which the principal tree dominant is *Combretum hereroense.* In many instances, around the margins of the pans an invasion of broad-leaved species, and more particularly of *Terminalia sericea* occurs. A particularly striking example of this broad-leaved invasion was found at Mukwa Pan. Here *Terminalia sericea* and *Burkea africana* are colonizing the flat land between the termite mounds which was originally treeless, supposedly because of the former degree of seasonal water-logging in the area.

It would seem that as a pan providing a permanent source of drinking water develops, the removal of soil particles with the drinking water causes it to deepen until it has a significant effect on the local water-table. Weir (1961) in a paper on the evolution of drinking holes in the Wankie National Park, has remarked on the quantities of mud as well as water that drinking animals remove. He did not however continue the evolution of the pan to its logical end, when so much sand has been removed with drinking water that the hole is so deep the water table of the surrounding soil has lowered beyond the point when it can be restored to the same level each year by rain falling on the surface of the surrounding soil. The hole then ceases to have available water in the dry season, and being no longer visited, presumably rapidly fills up with blown or washed soil.

### *Mixed Shrub Savanna*

As in the teak successions the occurrence of frequent grass fires at this point will prevent the survival of seedlings and will burn saplings down to sprout growth, resulting in the development of a shrub savanna formed mostly from the same species as the two previous communities. There is also a *Combretum hereroense* shrub savanna, and this is particularly conspicuous around several of the pans near the Main Camp.

### *Grass Savanna*

Around a pan which dries out in the dry season, grazing pressure will not be sufficient to remove the vigorous grass growth which appears on these comparatively well-watered sites. Annual grass fires are then likely to occur, and these will gradually exhaust and eliminate the woody plants of the mixed shrub savanna, leaving a grass savanna remaining. Where drinking water is available all the year round however, the grass is grazed too short to provide flammable material. The woody species of the shrub savanna then persist and are supplemented by a further invasion of tree species.

Grass savanna resulting from degradation of the mixed savanna woodland is similar in appearance to that derived from the teak woodland. Its specific composition, which is not considered in this paper, is however different, and in addition to herds of wildebeeste and zebra, it is frequented by animals which do not go too far from water like wart-hog, ostrich, reedbuck, and stembuck.

### *Thorn Thicket*

Where grazing activity around permanent water-holes is sufficiently intense to prevent the persistence of sufficient flammable material to support a grass fire, the woody tree species of the mixed savannas increase greatly, producing a thicket growth. The commonest species are *Acacia giraffae, A. galpinii, A. tortilis, Dichrostachys cinerea, Maytenus cymosus*, and *Dalbergia melanoxylon*, the same group of pioneer species in fact which are present in the thicket and savanna woodland of the teak woodland succession. All these species, as noted previously, are armed, and while this does not prevent their leaves being eaten, it is usually sufficient to ensure that whole shoots are not bitten off. The plants when merely defoliated can then grow a new set of leaves in the rainy season, when animals are more dispersed and the grazing pressure lighter.

### *Acacia Savanna Woodland*

Although the pioneer thicket species are the same in both the teak woodland and the mixed savanna successions, the savanna woodland which emerges from the thicket appears to be formed of a different balance of species, and more particularly, of *Acacia* species. In the teak series it was *Acacia tortilis* which emerged as a single dominant; on these more poorly-drained Kalahari Sands of the mixed savanna succession, it is *A. giraffae* which assumes this dominant position. Such *A. giraffae* savanna woodlands are very conspicuous in the area round Kennedy; they are also very prominent adjacent to the Main Camp. In both areas these savanna woodlands may owe their survival to a reduced elephant pressure, for elephant herds devastate these savanna woodlands to which they may partly be attracted by their fondness for *A. giraffae* pods. As mentioned earlier, elephants will strip the bark from these acacias, wreaking incredible damage to the savanna woodland. Unless elephants avoid a particular savanna woodland because of lack of convenient drinking supplies, or dislike of the proximity of humans, it therefore has little chance of survival, and in fact in the northern area of the park, most of such communities have been broken down to tree savannas.

## Successional Scheme—Mixed Savanna Woodland

The probable interrelationships of these various secondary communities derive from mixed savanna woodland, and the principal ecological influences which affect them, are expressed diagramatically in figure 14–5.

## Successional Relationships—Mopane Woodland

### *Mopane Woodland*

This woodland, more correctly termed savanna woodland (Boughey, 1961), develops on the clay depressions which occur to the north and west of Shapi Pan, in the northern area of the Park under consideration. Undisturbed mopane savanna woodland here is comparatively rare, and only one example was found along the road between Shapi Pan and Nehimba Pan, although the road from Shapi Pan to Garagunwe Pan passes through some half-dozen such undisturbed mopane woodlands.

Such woodlands are inhabited especially and typically by herds of impala, which seem to require the browse provided by Mopane leaves when the predominantly annual grasses of this vegetation type have been entirely grazed off towards the close of the dry season. Other animals, particularly herds of sable antelope, were occasionally seen in mopane woodland, and herds of elephant are common in the dry season, although the clay is too soft for them to penetrate in the wet season. Elephants browse on the Mopane trees usually by pulling off side branches or tops, but occasionally they will push over a whole tree.

### *Mopane Shrub Savanna and Grassland*

Around the pans and water-holes which are very common in the necessarily low-lying Mopane areas (fig. 14–1), trees are ravaged by elephant and other game, until they are little more than poles about 4 ft high, with stubby side shoots bristling out all round. Should such an area be left sufficiently lightly grazed to provide material for a grass fire, these stubby Mopane trees will be burned down to the ground and will sprout later. This is the typical appearance of Mopane in areas of high animal populations. This kind of shrub savanna is extremely widespread in the northern sections of the Kruger National Park, and also in remote areas of the Rhodesian low veld between this park and Wankie. It is sometimes maintained that frost-damage rather than grass fires are responsible for holding the Mopane in sprout form.

Repeated fires in the mopane shrub savanna will so weaken the Mopane stools that the plants eventually die out, leaving a grass savanna. Over-grazed areas of this grass savanna do not however appear to be invaded by armed woody species, as in the two previous successions. Possibly the seasonal water-logging in the rainy season which such plants would have to endure is sufficient barrier to invading species from the Kalahari Sands areas.

## Discussion

It is concluded from these preliminary studies that most of the vegetation of the Wankie National Park, insofar as the northern areas studied are concerned, is composed of secondary communities. These secondary communities, whose nature is determined primarily by the intensity of biotic or fire factors, appear mostly to be much more suitable for game occupation than the undisturbed edaphic communities of the region. Each secondary community appears to have associated with it a particular group of game animals, making up a complex ecosystem.

The two predominant variable ecological factors in the ecosystem are elephant damage and burning by grass fires. The operation of both these factors and therefore the nature of the balance within the ecosystem, are clearly controllable, and can be brought within a game management policy. For the present the only control on elephant numbers exercised is an attempt to disperse them over a wider area in the dry season by the provision of additional permanent drinking holes. For nearly thirty years grass savannas adjoining the tourist roads near the Main Camp in the northern areas of the park described here were burnt annually, to provide an early crop of fresh grass which would bring the animals down where the tourists could see them from their cars. Latterly, with insufficient flammable material, it has only been possible to burn every other year. As grazing animals control the amount of such material remaining, control of species such as wildebeeste and zebra, which tend to occupy the same grasslands continuously throughout the year, offers scope for additional management practice.

The purely subjective conclusions of this preliminary study can be verified in two ways, by the laying down of permanent observation areas, and by statistical sampling. The first of these methods has already been adopted, and a permanent transect was laid down in 1960. Also a map of the principal vegetation types which he recognized in one small part of this area has been prepared by Mitchell (1961b). Over the years such records will provide an unchallengeable account of change. The statistical approach however offers promise of more immediate information.

The demonstrable existence of the various plant communities described in this paper may be determined by using as a criterion the degree of association between plant species. It is also possible to assess the relative extent of the various communities arranged on this basis. This statistical investigation has already been carried out in this northern area of the park, and the results will be described in a further paper. It should also be possible to correlate animal presence with plant species, so that the whole picture of the ecosystem is obtained.

More detailed information is required both as to the operation of the elephant damage factor and the influence of grass fires. The possible existence of these factors and of the secondary communities described in the less densely populated areas of the park must be investigated. It is certain that many of the changes recorded as unidirectional in the hypothetical schemes illustrated in figures 14–4 and 14–5 will be found to proceed also in the opposite direction in areas where game populations are smaller and more scattered.

The spread of plant species is obviously much dependent on animal activity. Elephants consume large quantities of *Acacia giraffae* pods, and seeds of this and numerous other flowering plant species may be recovered undamaged from elephant dung. In the rainy season dung beetles rapidly bury any fresh dung, so that great quantities of seed are in this manner inadvertently planted in an excellent seed-bed.

Until more detailed information of this nature can be obtained, management policies in this and other game parks can at best be based on *ad hoc* measures, decided by devoted officers forced to formulate some definite policy with only their long experience to guide them in their decisions. It is fortunate that this experience coupled with sound judgement has so far prevented disastrous upset in this finely balanced ecosystem.

## Acknowledgments

I wish to record my grateful thanks to Mr. T. E. Davison and Mr. B. Austen, the past and present Senior Wardens of the Wankie National Park respectively, for many discussions on the habits of game animals and for the excellent facilities and great encouragement given for work inside the park, for which thanks are also due to the Director of National Parks, Mr. R. E. Stewart.

## References

Boughey, A. S. 1961. The vegetation types of Southern Rhodesia. *Proc. Rhod. Sci. Assoc.*, 49':54–98.

———. 1963. The significance of *Acacia tortilis* woodlands in the Rhodesian Low Veld. (in press).

Clark, J. D. 1959. Carbon 14 chronology in Africa south of the Sahara. (MS) Read to the Fourth Pan-African Congress on Prehistory 1959.

C.S.A. 1956. Phyto-geography. *Commission Techn. Co-operation in Africa*, publ. 22.

Gale, W. D. 1963. How Southern Rhodesia's National Parks began. Africa Calls No. 18, March/April:7–15.

Huxley, J. 1962. Eastern Africa: the ecological base. *Endeavour* 21:98–107.

Mitchell, B. L. 1961a. Ecological aspects of game control measures in Africa. *Kirkia* 1:120–128.

———. 1961b. Some notes on the vegetation of a portion of the Wankie National Park. *Kirkia* 2:200—209.

Weir, J. S. 1961. A possible course of evolution of animal drinking holes (pans) and reflected changes in their biology. *Proc. 1st Fed. Sci. Congr.* 1960:301–305.

# VI

# Human Populations

# 15

# World population growth: an international dilemma

*Harold F. Dorn*

During all but the most recent years of the centuries of his existence man must have lived, reproduced, and died as other animals do. His increase in number was governed by the three great regulators of the increase of all species of plants and animals—predators, disease, and starvation—or, in terms more applicable to human populations—war, pestilence, and famine. One of the most significant developments for the future of mankind during the first half of the 20th century has been his increasing ability to control pestilence and famine. Although he has not freed himself entirely from the force of these two regulators of population increase, he has gained sufficient control of them so that they no longer effectively govern his increase in number.

Simultaneously he has developed methods of increasing the effectiveness of war as a regulator of population increase, to the extent that he almost certainly could quickly wipe out a large proportion, if not all, of the human race. At the same time he has learned how to separate sexual gratification from reproduction by means of contraception and telegenesis (that is, reproduction by artificial insemination, particularly with spermatozoa preserved for relatively long periods of time), so that he can regulate population increase by voluntary control of fertility. Truly it can be said that man has the knowledge and the power to direct, at least in part, the course of his evolution.

---

Reprinted with permission from *Science*, 135(3500):283–290 (26 January 1962). 

This newly gained knowledge and power has not freed man from the inexorable effect of the biological laws that govern all living organisms. The evolutionary process has endowed most species with a reproductive potential that, unchecked, would overpopulate the entire globe within a few generations. It has been estimated that the tapeworm, *Taenia*, may lay 120,000 eggs per day; an adult cod can lay as many as 4 million eggs per year; a frog may produce 10,000 eggs per spawning. Human ovaries are thought to contain approximately 200,000 ova at puberty, while a single ejaculation of human semen may contain 200 million spermatozoa.

This excessive reproductive potential is kept in check for species other than man by interspecies competition in the struggle for existence, by disease, and by limitation of the available food supply. The fact that man has learned how to control, to a large extent, the operation of these biological checks upon unrestrained increase in number has not freed him from the necessity of substituting for them less harsh but equally effective checks. The demonstration of his ability to do this cannot be long delayed.

Only fragmentary data are available to indicate the past rate of growth of the population of the world. Even today, the number of inhabitants is known only approximately. Regular censuses of populations did not exist prior to 1800, although registers were maintained for small population groups prior to that time. As late as a century ago, around 1860, only about one-fifth of the estimated population of the world was covered by a census enumeration once in a 10-year period (*1*). The commonly accepted estimates of the population of the world prior to 1800 are only informed guesses. Nevertheless, it is possible to piece together a consistent series of estimates of the world's population during the past two centuries, supplemented by a few rough guesses of the number of persons alive at selected earlier periods. The most generally accepted estimates are presented in Figure 15–1.

These reveal a spectacular spurt during recent decades in the increase of the world's population that must be unparalleled during the preceding millennia of human existence. Furthermore, the rate of increase shows no sign of diminishing (Table 15–1). The period of time required for the population of the world to double has sharply decreased during the past three centuries and now is about 35 years.

Only a very rough approximation can be made of the length of time required for the population of the world to reach one-quarter of a billion persons, the estimated number at the beginning of the Christian era. The present subgroups of *Homo sapiens* may have existed for as long as 100,000 years. The exact date is not necessary, since for present purposes the evidence is sufficient to indicate that probably 50,000 to 100,000 years were required for *Homo sapiens* to increase in number until

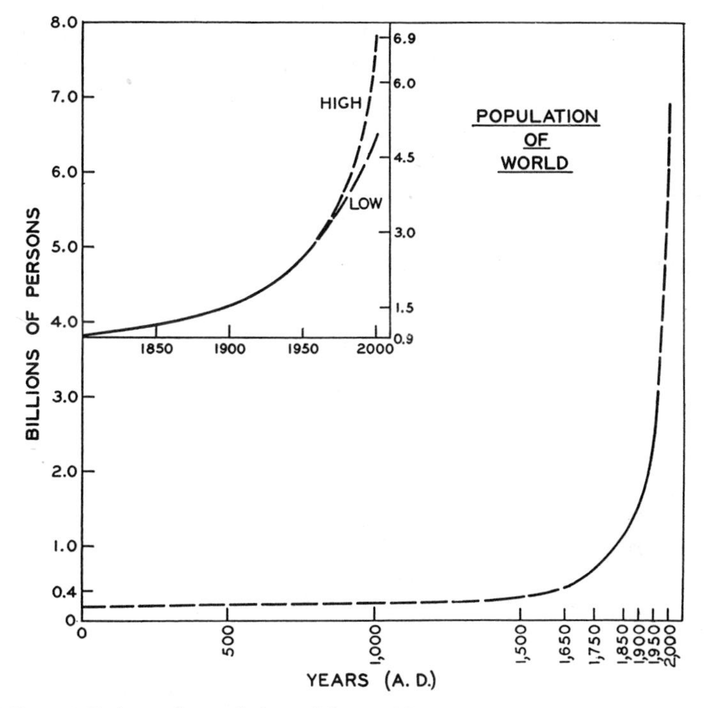


FIGURE 15–1. Estimated population of the world, A.D. 1 to A.D. 2000.

he reached a global total of one-quarter of a billion persons. This number was reached approximately 2000 years ago.

By 1620, the year the Pilgrims landed on Plymouth Rock, the population of the world had doubled in number. Two hundred years later, shortly before the Civil War, another 500 million persons had been added. Since that time, additional half billions of persons have been

TABLE 15–1

*The number of years required to double the population of the world. [From United Nations data (9, 14)]*

| Year (A.D.) | Population (billions) | Number of years to double |
|---|---|---|
| 1 | 0.25 (?) | 1650 (?) |
| 1650 | 0.50 | 200 |
| 1850 | 1.1 | 80 |
| 1930 | 2.0 | 45 |
| 1975 | 4.0 | 35 |
| 2010 | 8.0* | ? |

*A projection of United Nations estimates.

* A projection of United Nations estimates.

added during increasingly shorter intervals of time. The sixth half billion, just added, required slightly less than 11 years, as compared to 200 years for the second half billion. The present rate of growth implies that only 6 to 7 years will be required to add the eighth half billion to the world's population. The change in rate of growth just described has taken place since the first settlers came to New England.

## Implications

The accelerating rate of increase in the growth of the population of the world has come about so unobtrusively that most persons are unaware of its implications. There is a small group who are so aroused by this indifference that, like modern Paul Reveres, they attempt to awaken the public with cries of "the population bomb!" or "the population explosion!"

These persons are called alarmists by those who counter with the assertion that similar warnings, such as "standing-room only" and "mankind at the crossroads," have been issued periodically since Malthus wrote his essay on population, about 200 years ago. Nevertheless, says this group, the level of living and the health of the average person has continued to improve, and there is no reason to believe that advances in technology will not be able to make possible a slowly rising level of living for an increasing world population for the indefinite future. Furthermore, the rate of population increase almost certainly will slow down as the standard of education and living rises and as urbanization increases.

A third group of persons has attempted to estimate the maximum population that could be supported by the world's physical resources provided existing technological knowledge is fully utilized. Many of these calculations have been based on estimates of the quantity of food that could be produced and a hypothetical average daily calorie consumption per person.

As might be expected, the range of the various estimates of the maximum world population that could be supported without a lowering of the present level of living is very wide. One of the lowest, 2.8 billion, made by Pearson and Harper in 1945 on the assumption of an Asiatic standard of consumption, already has been surpassed (*2*). Several others, ranging from 5 to 7 billion, almost certainly will be exceeded by the end of this century. Perhaps the most carefully prepared estimate as well as the largest—that of 50 billions, prepared by Harrison Brown—would be reached in about 150 years if the present rate of growth should continue (*3*).

I believe it is worthwhile to prepare estimates of the maximum population that can be supported and to revise these as new information becomes available, even though most of the estimates made in the past already have been, or soon will be, demonstrated to be incorrect (in most instances too small), since this constitutes a rational effort to comprehend the implications of the increase in population. At the same time it should be recognized that estimates of the world's carrying capacity made in this manner are rather unrealistic and are primarily useful only as very general guidelines.

In the first place, these calculations have assumed that the earth's resources and skills are a single reservoir available to all. In reality this is untrue. The U.S. government attempts to restrict production of certain agricultural crops by paying farmers not to grow them. Simultaneously, in Asia and Africa, large numbers of persons are inadequately fed and poorly clothed. Except in a very general sense there is no *world* population problem; there are population problems varying in nature and degree among the several nations of the world. No single solution is applicable to all.

Since the world is not a single political unity, the increases in production actually achieved during any period of time tend to be considerably less than those theoretically possible. Knowledge, technical skill, and capital are concentrated in areas with the highest level of living, whereas the most rapid increase in population is taking place in areas where such skills and capital are relatively scarce or practically non-existent.

Just as the world is not a single unit from the point of view of needs and the availability of resources, skills and knowledge to meet these needs, so it also is not a single unit with respect to population increase. Due to political barriers that now exist throughout the entire world, overpopulation, however defined, will become a serious problem in specific countries long before it would be a world problem if there were no barriers to population redistribution. I shall return to this point later, after discussing briefly existing forecasts or projections of the total population of the world.

Most demographers believe that, under present conditions, the future population of areas such as countries or continents, or even of the entire world, cannot be predicted for more than a few decades with even a moderate degree of certainty. This represents a marked change from the view held by many only 30 years ago.

In 1930 a prominent demographer wrote, "The population of the United States ten, twenty, even fifty years hence, can be predicted with a greater degree of assurance than any other economic or social fact, provided the immigration laws are unchanged" (*4*). Nineteen years later, a well-known economist replied that "it is disheartening to have to assert that the best population forecasts deserve little credence even for 5 years ahead, and none at all for 20–50 years ahead."(*5*).

Although both of these statements represent rather extreme views, they do indicate the change that has taken place during the past two decades in the attitude toward the reliability of population forecasts. Some of the reasons for this have been discussed in detail elsewhere and will not be repeated here (*6*).

It will be sufficient to point out that knowledge of methods of voluntarily controlling fertility now is so widespread, especially among persons of European ancestry, that sharp changes in the spacing, as well as in the number, of children born during the reproductive period may occur in a relatively short period of time. Furthermore, the birth rate may increase as well as decrease.

## Forecasting Population Growth

The two principal methods that have been used in recent years to make population forecasts are (i) the extrapolation of mathematical curves fitted to the past trend of population increase and (ii) the projection of the population by the "component" or "analytical" method, based on specific hypotheses concerning the future trend in fertility, mortality, and migration.

The most frequently used mathematical function has been the logistic curve which was originally suggested by Verhulst in 1838 but which remained unnoticed until it was rediscovered by Pearl and Reed about 40 years ago (*7*). At first it was thought by some demographers that the logistic curve represented a rational law of population change. However, it has proved to be as unreliable as other methods of preparing population forecasts and is no longer regarded as having any unique value for estimating future population trends.

A recent illustration of the use of mathematical functions to project the future world population is the forecast prepared by von Foerster, Mora, and Amiot (*8*). In view of the comments that subsequently were published in this journal, an extensive discussion of this article does not seem to be required. It will be sufficient to point out that this forecast probably will set a record, for the entire class of forecasts prepared by the use of mathematical functions, for the short length of time required to demonstrate its unreliability.

The method of projecting or forecasting population growth most frequently used by demographers, whenever the necessary data are available, is the "component" or "analytical" method. Separate estimates are prepared of the future trend of fertility, mortality, and migration. From the total population as distributed by age and sex on a specified date, the future population that would result from the hypothetical combination of fertility, mortality, and migration is computed. Usually, several estimates of the future population are prepared in order to include what the authors believe to be the most likely range of values.

Such estimates generally are claimed by their authors to be not forecasts of the most probable future population but merely indications of the population that would result from the hypothetical assumptions concerning the future trend in fertility, mortality, and migration. However, the projections of fertility, mortality, and migration usually are chosen to include what the authors believe will be the range of likely possibilities. This objective is achieved by making "high," "medium," and "low" assumptions concerning the future trend in population growth. Following the practice of most of the authors of such estimates, I shall refer to these numbers as population projections.

The most authoritative projections of the population of the world are those made by the United Nations (*9, 10*) (Table 15–2). Even though

TABLE 15–2

*Estimated population of the world for* A.D. *1900, 1950, 1975, and 2000.* [*From United Nations data* (9), *rounded to three significant digits*]

| Area | Estimated population (millions) | | Projected future population (millions) | | | |
|---|---|---|---|---|---|---|
| | | | Low assumptions | | High assumptions | |
| | 1900 | 1950 | 1975 | 2000 | 1975 | 2000 |
| World | 1550 | 2500 | 3590 | 4880 | 3860 | 6900 |
| Africa | 120 | 199 | 295 | 420 | 331 | 663 |
| North America | 81 | 168 | 232 | 274 | 240 | 326 |
| Latin America | 63 | 163 | 282 | 445 | 304 | 651 |
| Asia | 857 | 1380 | 2040 | 2890 | 2210 | 4250 |
| Europe including U.S.S.R. | 423 | 574 | 724 | 824 | 751 | 987 |
| Oceania | 6 | 13 | 20 | 27 | 21 | 30 |

the most recent of these projections were published in 1958, only 3 years ago, it now seems likely that the population of the world will exceed the high projection before the year 2000. By the end of 1961 the world's population at least equaled the high projection for that date.

Although the United Nations' projections appear to be too conservative in that even the highest will be an underestimate of the population only 40 years from now, some of the numerical increases in population implied by these projections will create problems that may be beyond the ability of the nations involved to solve. For example, the estimated increase in the population of Asia from A.D. 1950 to 2000 will be roughly equal to the population of the entire world in 1958! The population of Latin America 40 years hence may very likely be four times that in 1950. The absolute increase in population in Latin America during the last half of the century may equal the total increase in the population of *Homo sapiens* during all the millennia from his origin until about 1650, when the first colonists were settling New England.

Increases in population of this magnitude stagger the imagination. Present trends indicate that they may be succeeded by even larger increases during comparable periods of time. The increase in the rate of growth of the world's population, shown by the data in Table 15–1, is still continuing. This rate is now estimated to be about 2 percent per year, sufficient to double the world's population every 35 years. It requires only very simple arithmetic to show that a continuation of this rate of growth for even 10 or 15 decades would result in an increase in population that would make the globe resemble an anthill.

But as was pointed out above, the world is not a single unit economically, politically, or demographically. Long before the population of the entire world reaches a size that could not be supported at current levels of living, the increase in population in specific nations and regions will give rise to problems that will affect the health and welfare of the rest of the world. The events of the past few years have graphically demonstrated the rapidity with which the political and economic problems of even a small and weak nation can directly affect the welfare of the largest and most powerful nations. Rather than speculate about the maximum population the world can support and the length of time before this number will be reached, it will be more instructive to examine the demographic changes that are taking place in different regions of the world and to comment briefly on their implications.

## Decline in Mortality

The major cause of the recent spurt in population increase is a world-wide decline in mortality. Although the birth rate increased in some countries—for example, the United States—during and after World War II, such increases have not been sufficiently widespread to account for more than a small part of the increase in the total population of the world. Moreover, the increase in population prior to World War II occurred in spite of a widespread decline in the birth rate among persons of European origin.

Accurate statistics do not exist, but the best available estimates suggest that the expectation of life at birth in Greece, Rome, Egypt, and the Eastern Mediterranean region probably did not exceed 30 years at the beginning of the Christian era. By 1900 it had increased to about 40 to 50 years in North America and in most countries of northwestern Europe. At present, it has reached 68 to 70 years in many of these countries.

By 1940, only a small minority of the world's population had achieved an expectation of life at birth comparable to that of the population of North America and northwest Europe. Most of the population of the world had an expectation of life no greater than that which prevailed in western Europe during the Middle Ages. Within the past two decades, the possibility of achieving a 20th-century death rate has been

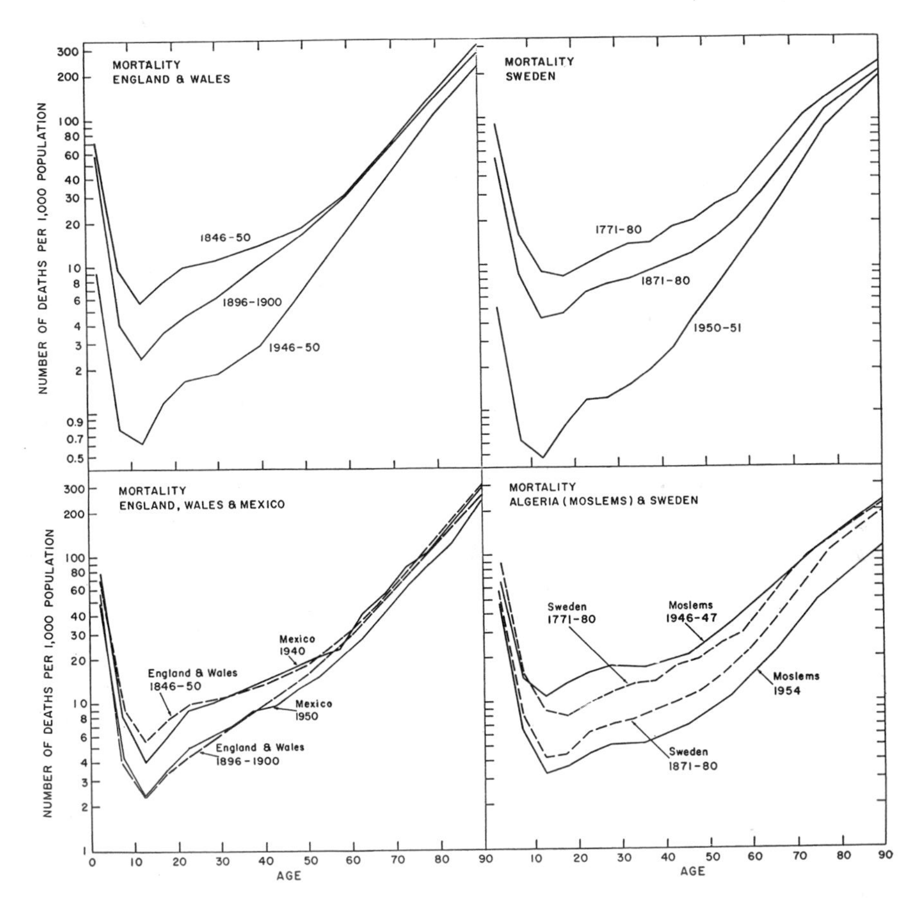


FIGURE 15–2. Age-specific death rates per 1000 per year for Sweden, England and Wales, Mexico, and the Moslem population of Algeria for various time periods from 1771 to 1954.

opened to these masses of the world's population. An indication of the result can be seen from the data in Figure 15–2.

In 1940, the death rate in Mexico was similar to that in England and Wales nearly 100 years earlier. It decreased as much during the following decade as did the death rate in England and Wales during the 50-year period from 1850 to 1900.

In 1946–47 the death rate of the Moslem population of Algeria was higher than that of the population of Sweden in the period 1771–80, the earliest date for which reliable mortality statistics are available for an entire nation. During the following 8 years, the drop in the death rate in Algeria considerably exceeded that in Sweden during the century from 1771 to 1871 (*11*).

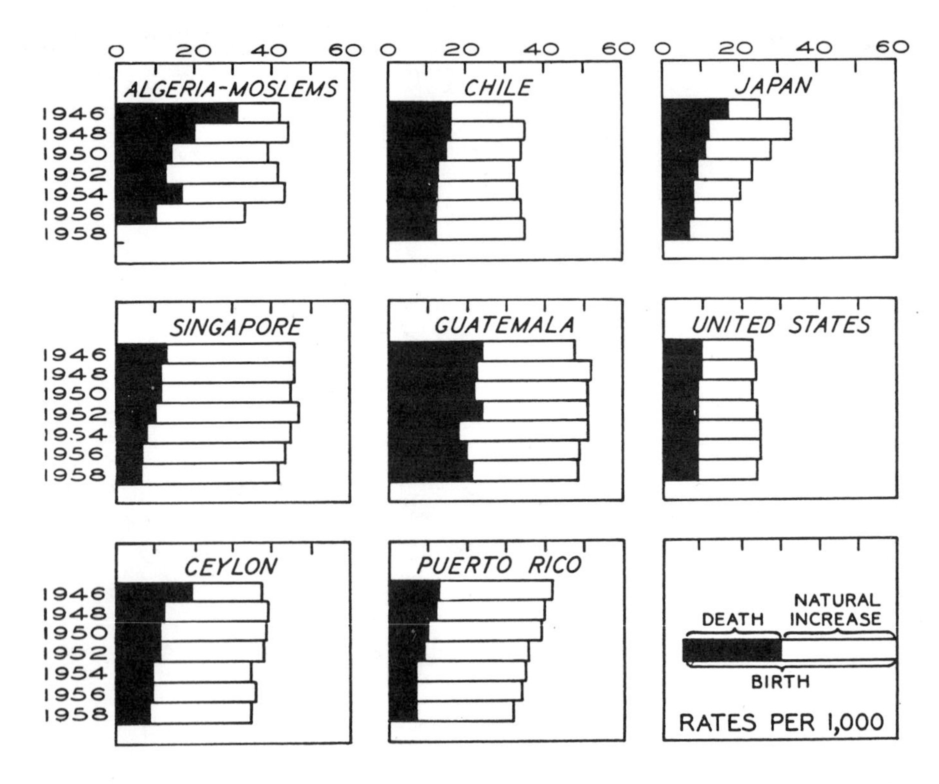


FIGURE 15–3. Birth rate, death rate, and rate of natural increase per 1000 for selected countries for the period 1946–58.

The precipitous decline in mortality in Mexico and in the Moslem population of Algeria is illustrative of what has taken place during the past 15 years in Latin America, Africa, and Asia, where nearly three out of every four persons in the world now live. Throughout most of this area the birth rate has changed very little, remaining near a level of 40 per 1000 per year, as can be seen from Figure 15–3, which shows the birth rate, death rate, and rate of natural increase for selected countries.

Even in countries such as Puerto Rico and Japan where the birth rate has declined substantially, the rate of natural increase has changed very little, owing to the sharp decrease in mortality. A more typical situation is represented by Singapore, Ceylon, Guatemala, and Chile, where the crude rate of natural increase has risen. There has been a general tendency for death rates to decline universally and for high birth rates to remain high, with the result that those countries with the highest rates of increase are experiencing an acceleration in their rates of growth.

## Regional Levels

The absolute level of fertility and mortality and the effect of changes in them upon the increase of population in different regions of the world can be only approximately indicated. The United Nations estimates that only about 33 percent of the deaths and 42 percent of the births that occur in the world are registered (*12*). The percentage registered ranges from about 8 to 10 percent in tropical and southern Africa and Eastern Asia to 98 to 100 percent in North America and Europe. Nevertheless, the statistical staff of the United Nations, by a judicious combination of the available fragmentary data, has been able to prepare estimates of fertility and mortality for different regions of the world that are generally accepted as a reasonably correct representation of the actual but unknown figures. The estimated birth rate, death rate, and crude rate of natural increase (the birth rate minus the death rate) for eight regions of the world for the period 1954–58 are shown in Figure 15–4.

The birth rates of the countries of Africa, Asia, Middle America, and South America average nearly 40 per 1000 and probably are as high as they were 500 to 1000 years ago. In the rest of the world—Europe, North America, Oceania, and the Soviet Union—the birth rate is slightly more than half as high, or about 20 to 25 per 1000. The

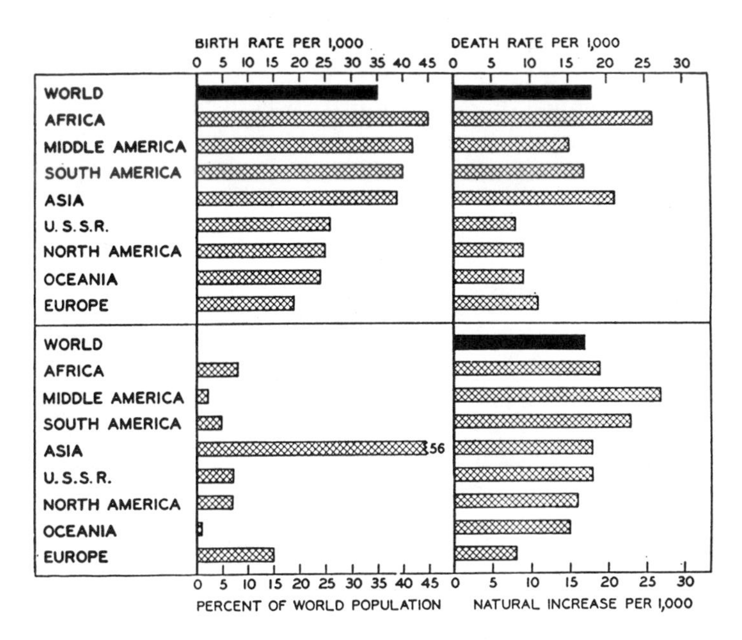


FIGURE 15–4. Percentage of the 1958 world population, birth rate, death rate, and rate of natural increase, per 1000, for the period 1954–58 for various regions of the world.

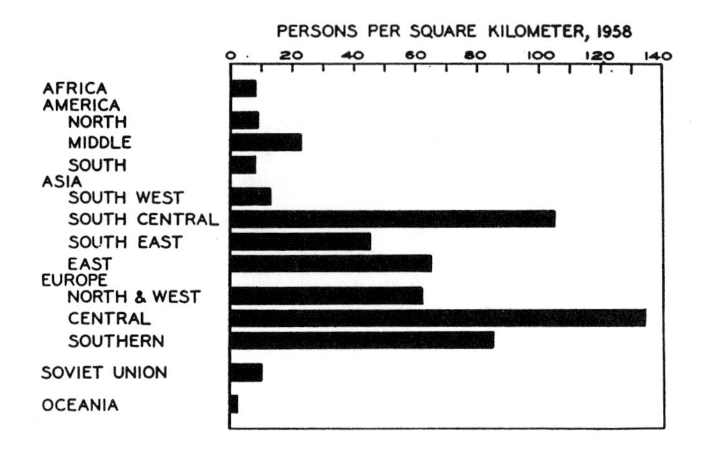


FIGURE 15–5. Number of persons per square kilometer in various regions of the world in 1958.

death rate for the former regions, although still definitely higher, is rapidly approaching that for people of European origin, with the result that the highest rates of natural increase are found in the regions with the highest birth rates. The most rapid rate of population growth at present is taking place in Middle and South America, where the population will double about every 26 years if the present rate continues.

These regional differences in fertility and mortality are intensifying the existing imbalance of population with land area and natural resources. No matter how this imbalance is measured, that it exists is readily apparent. Two rather crude measures are presented in Figures 15–4 and 15–5, which show the percentage distribution of the world's population living in each region and the number of persons per square kilometer.

An important effect of the decline in mortality rates often is overlooked—namely, the increase in effective fertility. An estimated 97 out of every 100 newborn white females subject to the mortality rates prevailing in the United States during 1950 would survive to age 20, slightly past the beginning of the usual childbearing age, and 91 would survive to the end of the childbearing period (Fig. 15–6). These estimates are more than 3 and 11 times, respectively, the corresponding estimated proportions for white females that survived to these ages about four centuries ago.

In contrast, about 70 percent of the newborn females in Guatemala would survive to age 20, and only half would live to the end of the childbearing period if subject to the death rates prevailing in that country in 1950. If the death rate in Guatemala should fall to the level of that in the United States in 1950—a realistic possibility—the number of new-

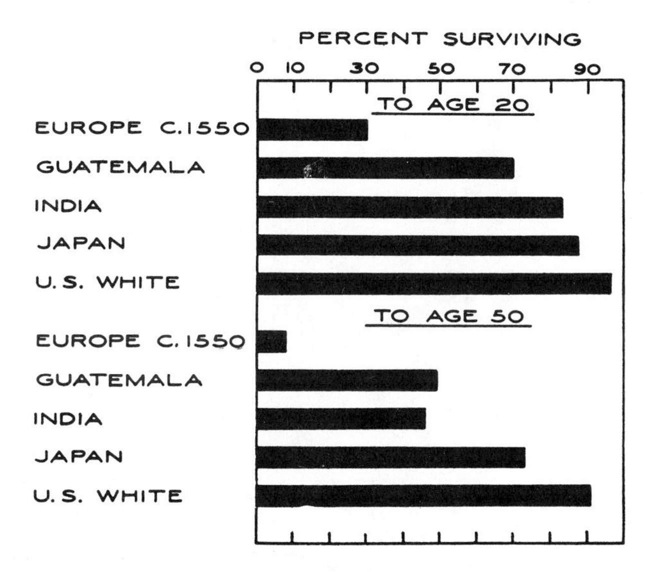


FIGURE 15–6. Percentage of newborn females who would survive to the end of the reproductive period according to mortality rates in Europe around A.D. 1500 and in selected countries around 1950.

born females who would survive to the beginning of the childbearing period would increase by 36 percent; the number surviving to the end of the childbearing period would increase by 85 percent. A corresponding decrease in the birth rate would be required to prevent this increase in survivorship from resulting in a rapid acceleration in the existing rate of population growth, which already is excessive. In other words, this decrease in the death rate would require a decrease in the birth rate of more than 40 percent merely to maintain the status quo.

As can be seen from Figure 15–3, the birth rate in countries with high fertility has shown little or no tendency to decrease in recent years. Japan is the exception. There, the birth rate dropped by 46 percent from 1948 to 1958—an amount more than enough to counterbalance the decrease in the death rate, with the result that there was a decrease in the absolute number of births. As yet there is very little evidence that other countries with a correspondingly high birth rate are likely to duplicate this in the near future.

Another effect of a rapid rate of natural increase is demonstrated by Figure 15–7. About 43 percent of the Moslem population of Algeria is under 15 years of age; the corresponding percentage in Sweden is 24, or slightly more than half this number. Percentages in the neighborhood of 40 percent are characteristic of the populations of the countries of Africa, Latin America, and Asia.

This high proportion of young people constitutes a huge fertility

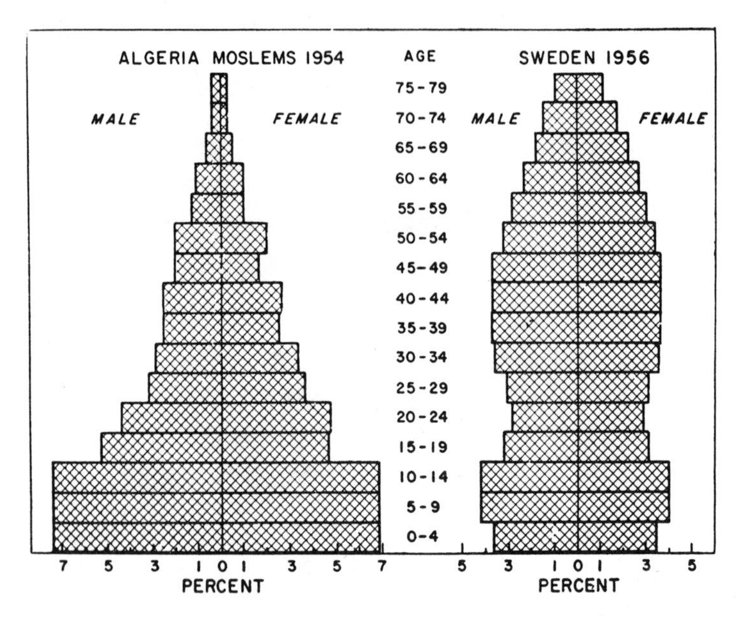


FIGURE 15–7. Percentage distribution by age of the population of Sweden in 1956 and the Moslem population of Algeria in 1954.

potential for 30 years into the future that can be counterbalanced only by a sharp decline in the birth rate, gives rise to serious educational problems, and causes a heavy drain on the capital formation that is necessary to improve the level of living of the entire population. A graphic illustration of this may be found in the recently published 5-year plan for India for 1961–66, which estimates that it will be necessary to provide educational facilities and teachers for 20 million additional children during this 5-year period (*13*).

## Historical Pattern in Western Europe

Some persons, although agreeing that the current rate of increase of the majority of the world's population cannot continue indefinitely without giving rise to grave political, social, and economic problems, point out that a similar situation existed in northwestern and central Europe during the 18th and 19th centuries. Increasing industrialization and urbanization, coupled with a rising standard of living, led to a decline in the birth rate, with a consequent drop in the rate of increase of the population. Why should not the rest of the world follow this pattern?

There is small likelihood that the two-thirds of the world's population which has not yet passed through the demographic revolution from high fertility and mortality rates to low fertility and mortality rates can repeat the history of western European peoples prior to the development of serious political and economic problems. A brief review of the circumstances that led to the virtual domination of the world at the end of the 19th century by persons of European origin will indicate some of the reasons for this opinion.

Around A.D. 1500 the population of Europe probably did not exceed 100 million persons (perhaps 15 to 20 percent of the population of the world) and occupied about 7 percent of the land area of the earth. Four hundred years later, around 1900, the descendants of this population numbered nearly 550 million, constituted about one-third of the world's population, and occupied or controlled five-sixths of the land area of the world. They had seized and peopled two great continents, North and South America, and one smaller continent, Australia, with its adjacent islands; had partially peopled and entirely controlled a third great continent, Africa; and dominated southern Asia and the neighboring islands.

The English-, French-, and Spanish-speaking peoples were the leaders in this expansion, with lesser roles being played by the Dutch and Portuguese. The Belgians and Germans participated only toward the end of this period of expansion. Among these, the English-speaking people held the dominant position at the end of the era, around 1900.

The number of English-speaking persons around 1500, at the start of this period of expansion, is not known, but it probably did not exceed 4 or 5 million. By 1900 these people numbered about 129 million and occupied and controlled one-third of the land area of the earth and, with the non-English-speaking inhabitants of this territory, made up some 30 percent of the population of the world.

This period was characterized by an unprecedented increase in population, a several-fold expansion of the land base for this population, and a hitherto undreamed of multiplication of capital in the form of precious metals, goods, and commodities. Most important of all, the augmentation in capital and usable land took place more rapidly than the growth in population.

A situation equally favorable for a rapid improvement in the level of living associated with a sharp increase in population does not appear likely to arise for the people who now inhabit Latin America, Africa, and Asia. The last great frontier of the world has been closed. Although there are many thinly populated areas in the world, their existence is testimony to the fact that, until now, these have been regarded as undesirable living places. The expansion of population to the remaining open areas would require large expenditures of capital for irrigation, drainage, transportation facilities, control of insects and parasites, and other purposes—capital that the rapidly increasing populations which need these areas do not possess.

In addition, this land is not freely available for settlement. The entire land surface of the world is crisscrossed by national boundaries. International migration now is controlled by political considerations; for the majority of the population of the world, migration, both in and out of a country, is restricted.

The horn of plenty, formerly filled with free natural resources, has been emptied. No rapid accumulation of capital in the form of precious metals, goods, and commodities, such as characterized the great 400-year boom enjoyed by the peoples of western-European origin, is possible for the people of Africa, Asia, and Latin America.

Last, but not least, is the sheer arithmetic of the current increase in population. The number of persons in the world is so large that even a small rate of natural increase will result in an almost astronomical increment over a period of time of infinitesimal duration compared to the duration of the past history of the human race. As was pointed out above, continuation of the present rate of increase would result in a population of 50 billion persons in another 150 years. A population of this magnitude is so foreign to our experience that it is difficult to comprehend its implications.

Just as Thomas Malthus, at the end of the 18th century, could not foresee the effect upon the peoples of western Europe of the exploration of the last great frontier of this earth, so we today cannot clearly foresee the final effect of an unprecedented rapid increase of population within closed frontiers. What seems to be least uncertain in a future full of uncertainty is that the demographic history of the next 400 years will not be like that of the past 400 years.

## World Problem

The results of human reproduction are no longer solely the concern of the two individuals involved, or of the larger family, or even of the nation of which they are citizens. A stage has been reached in the demographic development of the world when the rate of human reproduction in any part of the globe may directly or indirectly affect the health and welfare of the rest of the human race. It is in this sense that there is a world population problem.

One or two illustrations may make this point more clear. During the past decade, six out of every ten persons added to the population of the world live in Asia; another two out of every ten live in Latin America and Africa. It seems inevitable that the breaking up of the world domination by northwest Europeans and their descendants, which already is well advanced, will continue, and that the center of power and influence will shift toward the demographic center of the world.

The present distribution of population increase enhances the existing imbalance between the distribution of the world's population and the distribution of wealth, available and utilized resources, and the use of nonhuman energy. Probably for the first time in human history there is a universal aspiration for a rapid improvement in the standard of living and a growing impatience with conditions that appear to stand in the way of its attainment. Millions of persons in Asia, Africa, and Latin America now are aware of the standards of living enjoyed by Europeans and North Americans. They are demanding the opportunity to attain the same standard, and they resist the idea that they must be permanently content with less.

A continuation of the present high rate of human multiplication will act as a brake on the already painfully slow improvement in the level of living, thus increasing political unrest and possibly bringing about eventual changes in government. As recent events have graphically demonstrated, such political changes may greatly affect the welfare of even the wealthiest nations.

The capital and technological skills that many of the nations of Africa, Asia, and Latin America require to produce enough food for a rapidly growing population and simultaneously to perceptibly raise per capita income exceed their existing national resources and ability. An immediate supply of capital in the amounts required is available only from the wealthier nations. The principle of public support for social welfare plans is now widely accepted in national affairs. The desirability of extending this principle to the international level for the primary purpose of supporting the economic development of the less advanced nations has not yet been generally accepted by the wealthier and more advanced countries. Even if this principle should be accepted, it is not as yet clear how long the wealthier nations would be willing to support the uncontrolled breeding of the populations receiving this assistance. The general acceptance for a foreign aid program of the extent required by the countries with a rapidly growing population will only postpone for a few decades the inevitable reckoning with the results of uncontrolled human multiplication.

The future may witness a dramatic increase in man's ability to control his environment, provided he rapidly develops cultural substitutes for those harsh but effective governors of his high reproductive potential—disease and famine—that he has so recently learned to control. Man has been able to modify or control many natural phenomena, but he has not yet discovered how to evade the consequences of biological laws. No species has ever been able to multiply without limit. There are two biological checks upon a rapid increase in number—a high mortality and a low fertility. Unlike other biological organisms, man can choose which of these checks shall be applied, but one of them must be. Whether man can use his scientific knowledge to guide his future evolution more wisely than the blind forces of nature, only the future can reveal. The answer will not be long postponed.

## References and Notes

1. *Demographic Yearbook* (United Nations, New York, 1955), p. 1.
2. F. A. Pearson and F. A. Harper, *The World's Hunger* (Cornell Univ. Press, Ithaca, N.Y., 1945).
3. H. Brown, *The Challenge of Man's Future* (Viking, New York, 1954).
4. O. E. Baker, "Population trends in relation to land utilization," *Proc. Intern. Conf. Agr. Economists, 2nd Conf.* (1930), p. 284.
5. J. S. Davis, *J. Farm Economics* (Nov. 1949).
6. H. F. Dorn, *J. Am. Statist. Assoc.* 45, 311 (1950).
7. R. Pearl and L. J. Reed, *Proc. Natl. Acad. Sci. U.S.* 6, 275 (1920).
8. H. von Foerster, P. M. Mora, and L. W. Amiot, *Science* 132, 1291 (1960).
9. "The future growth of world population," *U.N. Publ. No. ST/SOA/Ser. A/28* (1958).
10. "The past and future growth of world population—a long-range view," *U.N. Population Bull. No. 1* (1951), pp. 1–12.
11. Although registration of deaths among the Moslem population of Algeria is incomplete, it is believed that the general impression conveyed by Figure 15–2 is essentially correct.
12. *Demographic Yearbook* (United Nations, New York, 1956), p. 14.
13. New York *Times* (5 Aug. 1961).
14. "The determinants and consequences of population trends," *U.N. Publ. No. ST/SOA/Ser. A/17* (1953).

# 16

# The world outlook for conventional agriculture

*Lester R. Brown*

The problem of obtaining enough food has plagued man since his beginnings. Despite the innumerable scientific advances of the 20th century, the problem becomes increasingly serious. Accelerating rates of population growth, on the one hand, and the continuing reduction in the area of new land that can be put under the plow, on the other, are postponing a satisfactory solution to this problem for at least another decade and perhaps much longer.

Conventional agriculture now provides an adequate and assured supply of food for one-third of the human race. But assuring an adequate supply of food for the remaining two-thirds, in parts of the world where population is increasing at the rate of 1 million weekly, poses one of the most nearly insoluble problems confronting man.

## Dimensions of the Problem

Two major forces are responsible for expanding food needs: population growth and rising per capita incomes.

Populations in many developing countries are increasing at the rate of 3 percent or more per year. In some instances the rate of increase appears to be approaching the biological maximum. Populations growing

Reprinted with permission from *Science*, 158:604–611 (3 November 1967). 

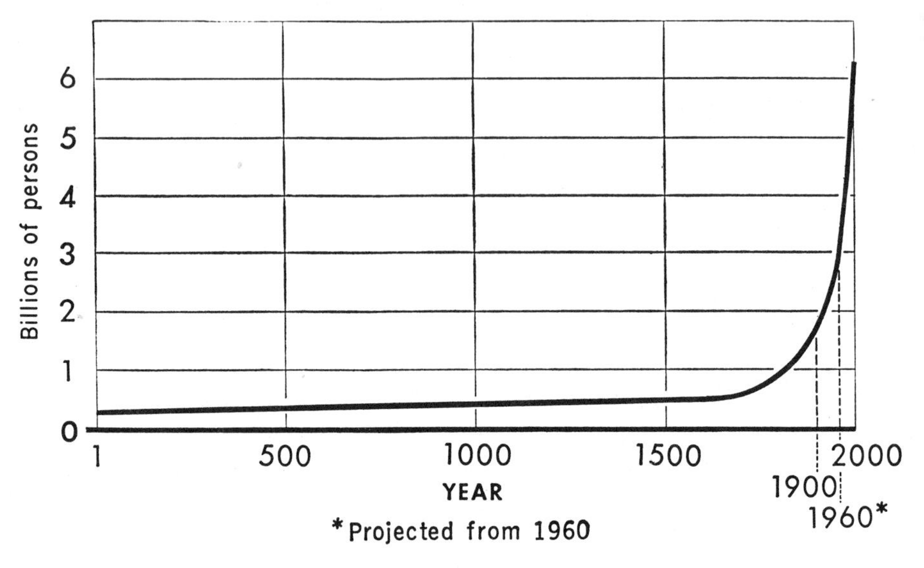


FIGURE 16–1. Twenty centuries of world population growth [U.S. Department of Agriculture].

by 3 percent per year double within a generation and multiply 18-fold in a century.

According to projections, world population, now just over 3 billion, will increase by another 3 billion over the remaining one-third of this century (Fig. 16–1). Even with the most optimistic assumptions concerning the effect of newly initiated family-planning programs in developing countries, we must still plan to feed an additional 1 billion people by 1980. The world has never before added 1 billion people in 15 years. More significantly, four-fifths of these will be added to the less-developed countries, where food is already in short supply.

Rising income levels throughout the world are generating additional demand on the world's food-producing resources. Virtually every country in the world today has plans for raising income levels among its people. In some of the more advanced countries the rise in incomes generates far more demand for food than the growth of population does.

Japan illustrates this well. There, population is increasing by only 1 percent per year but per capita incomes are rising by 7 percent per year. Most of the rapid increase in the demand for food now being experienced in Japan is due to rising incomes. The same may be true for several countries in western Europe, such as West Germany and Italy, where population growth is slow and economic growth is rapid.

Comparisons between population growth and increases in food production, seemingly in vogue today, often completely ignore the effect of rapidly rising incomes, in some instances an even more important demand-creating force than population growth.

The relationships between increases in per capita income and the consumption of grain are illustrated in Figure 16–2. The direct consumption of grain, as food, rises with income per person throughout the low-income brackets; at higher incomes it declines, eventually leveling off at about 150 pounds per year.

The more significant relationship, however, is that between total

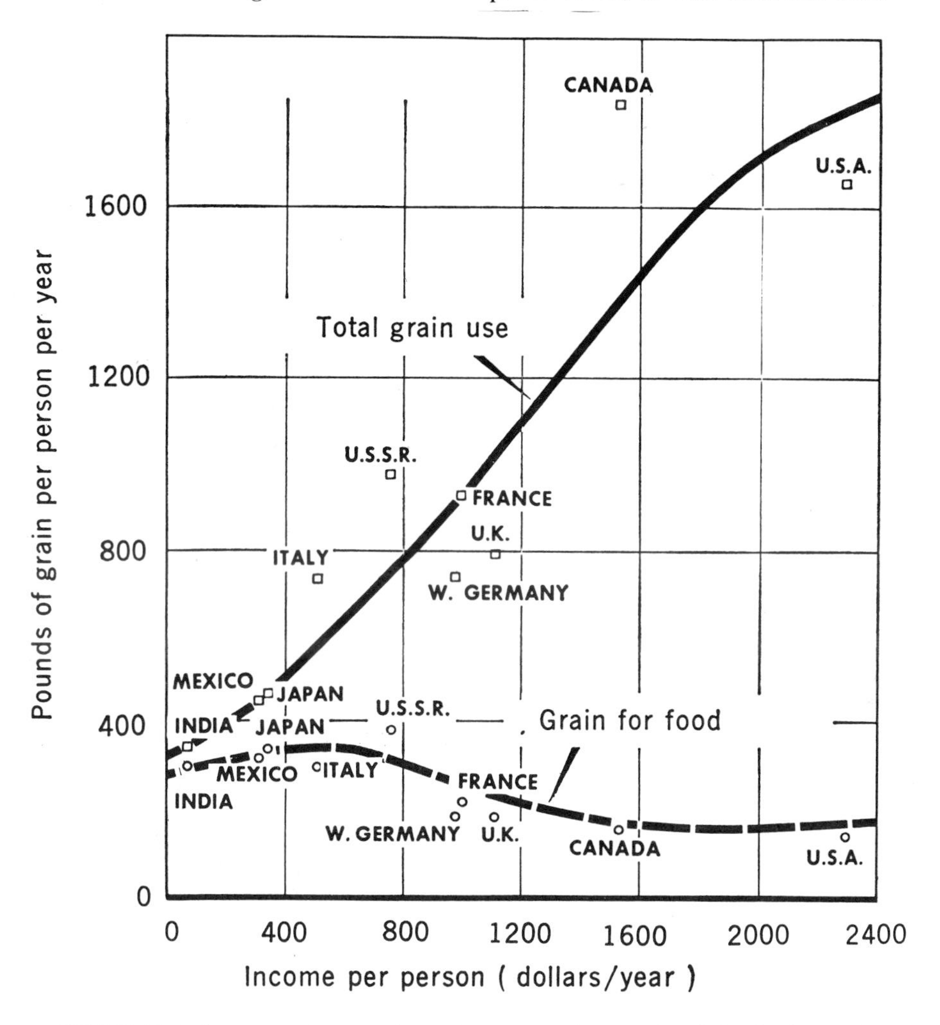


FIGURE 16–2. Income and per-capita grain consumption, total and for food (data for 1959–61) [U.S. Department of Agriculture]

grain use and income. Historically, as incomes have risen, the use of grain, both that consumed directly and that consumed indirectly in the form of meat, milk, and eggs, has risen also. The upper curve in Figure

16–2 indicates that every $2 gain in annual per capita income requires one pound of additional grain.

The rapid increases in both population and income are recent phenomena, in historical terms. Both have occurred since the war, and both are gaining momentum on a worldwide scale.

The effect of the resulting explosive increase in the demand for food is greater pressure on the world's food supplies. This rapid expansion of demand, together with the reduction of surplus grain stocks in North America, contributed to a rapid decline in world grain stocks during the 1960's (Fig. 16–3).

Between 1953 and 1961, world grain "carryover" stocks increased each year. The size of the annual buildup varied from a few million tons to nearly 20 million tons. After 1961, however, stocks began to decline, with the reduction or "drawdown" averaging 14 million tons per year.

A stock buildup, by definition, means that production is exceeding consumption; the converse is also true. The trend in grain stocks indicates clearly that 1961 marked a worldwide turning point; as population and income increases gained momentum, food consumption moved

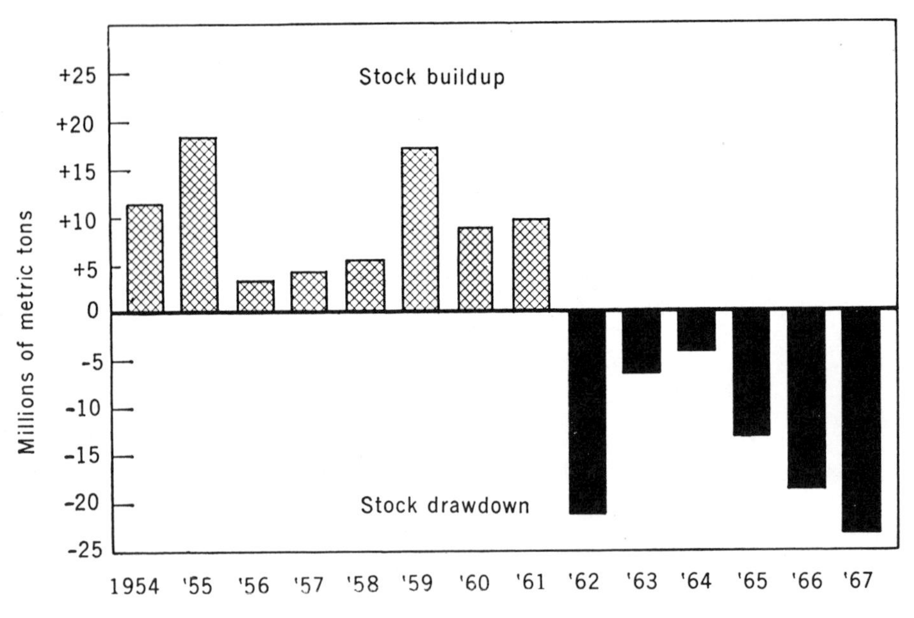


FIGURE 16–3. Changes in world grain stocks [U.S. Department of Agriculture].

ahead of production. Since 1961, the ever-widening excess of consumption over production has been compensated by "drawing down" stocks. But there is little opportunity for further reductions.

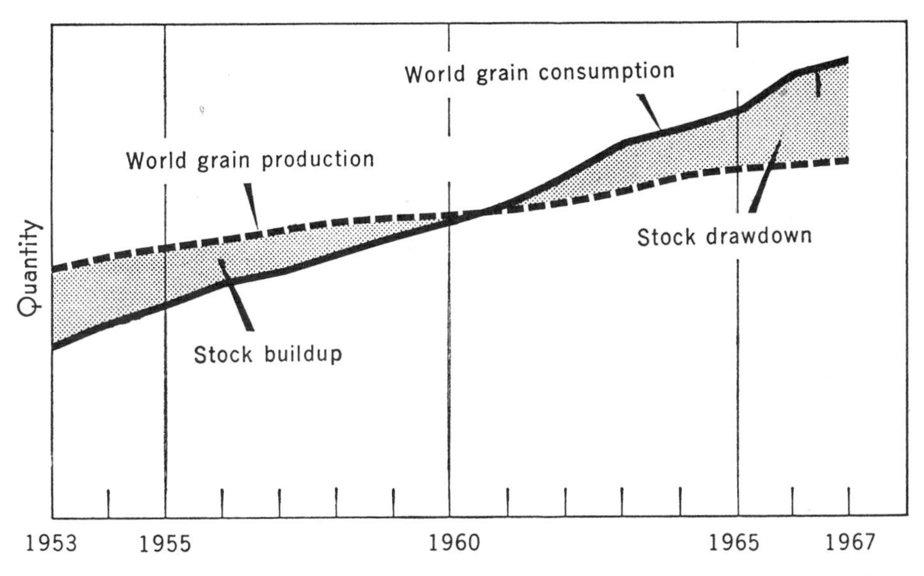


FIGURE 16–4. World grain production now lagging behind consumption (Schematic representation is not drawn to scale.) [U.S. Department of Agriculture].

This means that the two lines in Figure 16–4 cannot remain apart much longer. The question is: How will the lines be brought together? Will the production line go up, or will the consumption line come down? What are the implications of recent trends for world food price levels? Rising prices, a possible result, would act both to reduce consumption, particularly among the world's low-income peoples, and to stimulate production. At a time when hunger and, in some cases, severe malnutrition are commonplace in much of the world, reducing consumption is obviously not a desirable alternative. The effect would be to widen the food gap between the world's "haves" and "have-nots."

Meeting future food needs will require immense increases in output. The expected increase of 1 billion in world population over the next 15 years will require expansion of world grain production, now totaling about 1 billion tons, by about one-third, or 335 million tons. Additional demand generated by rising per capita incomes, even if only half as large as the population-generated component, could push the total needed increase toward 500 million tons.

What are the prospects of meeting these future increases in world food needs through conventional agriculture? There are two methods of increasing food production: expanding the cultivated area or raising the productivity (output per unit) of land already under cultivation. Throughout most of history, increases in food production have come largely from expanding the area under cultivation. Only quite recently, in historical terms, have some regions begun to rely on raising output per acre for most of the increases in their food supply (*1*).

Over the past 30 years, all of the increases in agricultural production in North America and western Europe have come from raising the productivity of land. Food output has about doubled in both regions, while the area cultivated has actually declined somewhat. Available technology has made it more profitable to raise output per acre than to increase the area under cultivation.

## Expanding the Cropland Area

The world's present cultivated land area totals some 3 billion acres (1.2 billion hectares). Estimates of the possibilities for expanding this area vary from a few hundred million acres to several billion. However, any such estimate of the area of new land likely to be brought under cultivation must, to be meaningful, specify at what cost this is to be accomplished.

Some land which was farmed a few decades ago has now been abandoned because it is no longer profitable. Much of the abandoned farmland in New England and Appalachia in the United States, or in other countries, such as portions of the Anatolian Plateau in Turkey, falls into this category.

In several countries of the world the area of cultivated land is actually declining. Japan, where the area of cultivated land reached a peak in 1920 and has declined substantially since, is a prominent example. Other countries in this category are Ireland, Sweden, and Switzerland.

Most of the world's larger countries are finding it difficult to further expand the area under cultivation. India plans to expand the cultivated land area by less than 2 percent over its Fourth Plan period, from 1966 to 1971; yet the demand for food is expected to expand by some 20 percent over this 5-year span. Mainland China, which has been suffering from severe population pressure for several decades, has plowed nearly all of its readily cultivable land.

Most of the countries in the Middle East and North Africa, which depend on irrigation or on dry-land farming, cannot significantly expand the area under cultivation without developing new sources of water for irrigation. The Soviet Union is reportedly abandoning some of the land brought under cultivation during the expansion into the "virgin-lands" area in the late 1950's.

The only two major regions where there are prospects for further significant expansion of the cultivated area in the near future are sub-Saharan Africa and the Amazon Basin of Brazil. Any substantial expansion in these two areas awaits further improvements in our ability to manage tropical soils—to maintain their fertility once the lush natural vegetation is removed.

Aside from this possibility, no further opportunities are likely to arise until the cost of desalinization is reduced to the point where it is profitable to use seawater for large-scale irrigation. This will probably not occur before the late 1970's or early 1980's at best.

The only country in the world which in recent years has had a ready reserve of idled cropland has been the United States. As recently as 1966, some 50 million acres were idled, as compared with a harvested acreage of 300 million acres. The growing need for imported food and feed in western Europe, the Communist countries, Japan, and particularly India is bringing much of this land back into production. Decisions made in 1966 and early 1967 to expand the acreage of wheat, feed grains, and soybeans brought some one-third of the idled U.S. cropland back into production in 1967.

Even while idled cropland is being returned to production in the United States and efforts are being made to expand the area of cultivated land in other parts of the world, farmland is being lost because of expanding urban areas, the construction of highways, and other developments. On balance, it appears that increases in world food production over the next 15 years or so will, because of technical and economic factors, depend heavily on our ability to raise the productivity of land already under cultivation.

## Increasing Land Productivity

Crop yield per acre in much of the world has changed little over the centuries. Rates of increase in output per acre have, in historical terms, been so low as to be scarcely perceptible within any given generation. Only quite recently—that is, during the 20th century—have certain countries succeeded in achieving rapid, continuing increases in output per acre—a yield "takeoff." Most of the economically advanced countries—particularly those in North America, western Europe, and Japan—have achieved this yield-per-acre takeoff (*2*).

The first yield-per-acre takeoff, at least the first documented by available data, occurred for rice in Japan during the early years of this century (Fig. 16–5). Yield takeoffs occurred at about the same time, or shortly thereafter, in several countries in northwestern Europe, such as Denmark, the Netherlands, and Sweden. Several other countries, such as the United Kingdom and the United States, achieved yield-per-acre takeoffs in the late 1930's and early 1940's.

Increasing food output per acre of land requires either a change in cultural practices or an increase in inputs, or both. Nearly all increases in inputs or improvements in cultural practices involve the use of more capital (*3*). Many (mechanization itself is an exception) require more labor as well (*4*).

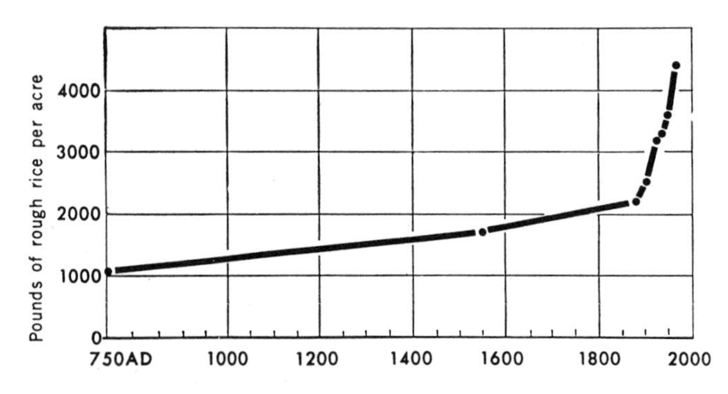


FIGURE 16–5. Rice yields in Japan from A.D. 750 to 1960. Historical estimates from Japanese ministry of agriculture [U.S. Department of Agriculture]

A review of the yield trends shown in Figures 16–5 and 16–6, or of any of several others for the agriculturally advanced countries, raises the obvious question of how long upward trends may be expected to continue. Will there come a time when the rate of increase will slow down or cease altogether? Hopefully, technological considerations, resulting from new research breakthroughs, will continue to postpone that date.

### *Differing Sources of Productivity*

One way of evaluating future prospects for continuing expansion in yields is to divide the known sources of increased productivity into two broad categories: "nonrecurring" and "recurring" sources of in-

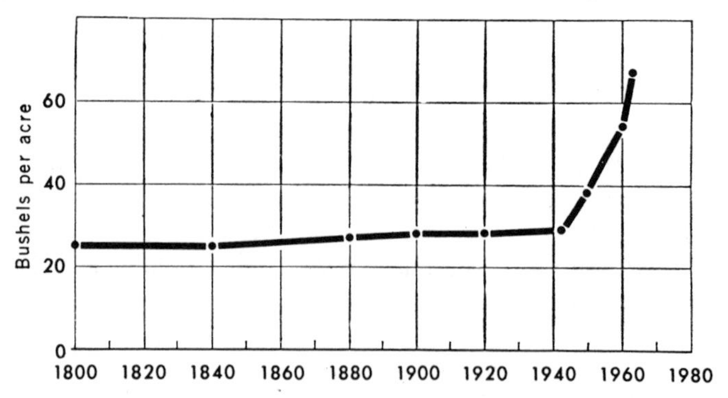


FIGURE 16–6. Corn yields in the United States [U.S. Department of Agriculture].

creased productivity (5). Nonrecurring inputs are essentially of a one-shot nature; once they are fully adopted, further increases in yields are limited. Recurring inputs, even when fully adopted, offer further annual increases in output through more intensive application.

Corn provides a good illustration. Yields have expanded sharply in the United States (Fig. 16–6). Total production now exceeds 100 million tons of grain annually, or about half the total U.S. grain crop. Much of the increase in corn yields, however, was due to two nonrecurring sources of productivity: the replacement of open-pollinated or traditional varieties with hybrids and, to a lesser extent, the use of herbicides.

Hybrid corn has now replaced open-pollinated varieties on more than 97 percent of the corn acreage in the United States (Fig. 16–7). Further improvements in hybrid varieties are to be expected. (Hybrids in use today are superior to hybrids developed in the mid-1930's). The big spurt in yields, however, is usually associated with the initial transition from open-pollinated or traditional varieties to hybrids. Consequently, the big thrust in corn yields in the United States resulting from the adoption of hybrids is probably a thing of the past. Likewise, once herbicides are widely used and virtually all weeds are controlled, there is little, if any, prospect of future gains in productivity from this source.

Some sources of increased yields are of a recurring nature. Among these, there is still ample opportunity for further yield increases as a result of the use of additional fertilizer. As plant populations increase, provided moisture is not a limiting factor, corn yields will rise further as more fertilizer is used.

Just how far the yield increase will go in the United States, however,

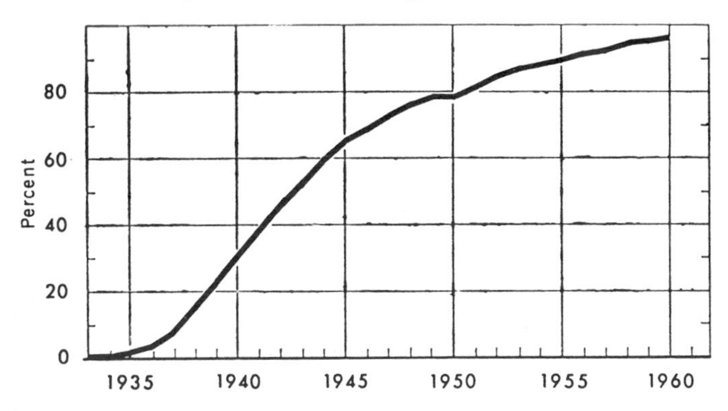


FIGURE 16–7. Share of U.S. corn acreage planted with hybrid seed [U.S. Department of Agriculture].

is not clear. Paul Mangelsdorf of Harvard University, speaking recently at the National Academy of Sciences, asked this vital question (*6*):

With more than 95 percent of the corn acreage already planted to hybrid corn, with the genetic potentials of the hybrids having reached a plateau, with 87 percent of the acreage in the Corn Belt and Lake States already using fertilizer, and with many farmers already employing herbicides, from where will come the future improvements that will allow us to continue our present rate of improvement?

The same question may be asked of other crops in some of the other agriculturally advanced countries.

### The S-shaped Yield Curve

As the nonrecurring sources of productivity are exhausted, the sources of increased productivity are reduced until eventually the rate of increase in yield per acre begins to slow. This might be depicted by that familiar biologic function the S-shaped growth curve (Fig. 16–8). John R. Platt of the University of Chicago recently explained the curve this way (7):

> Many of our important indices of technical achievement have been shooting up exponentially for many years, very much like the numbers in the biologists' colonies of bacteria, that double in every generation as each cell divides into two again. But such a curve of growth obviously cannot continue indefinitely in any field. The growth of the bacterial colony slows up as it begins to exhaust its nutrient. The exponential curve bends over and flattens out into the more general "S-curve" or "logistic curve" of growth.

We do not know with any certainty when the rate of yield increase for the major food crops on which man depends for sustenance will begin to slow, but we do know that ultimately it will.

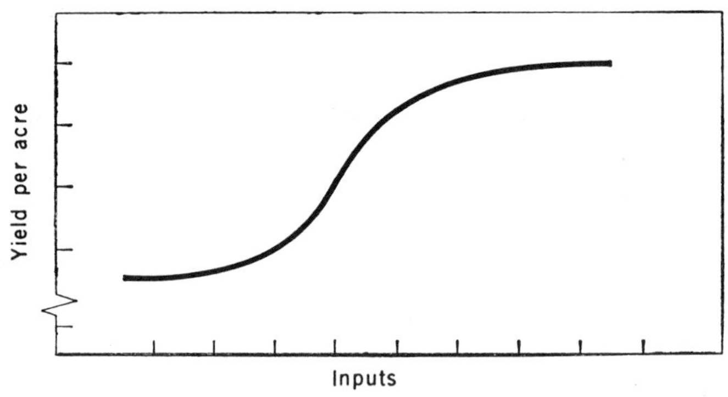


FIGURE 16–8. S-shaped yield curve (schematic representation) [U.S. Department of Agriculture].

The key questions are: Is the slowdown near for some of the major food crops in some of the agriculturally advanced countries? Will the slowdown come gradually, or will it occur abruptly and with little warning? Finally, to what extent can the level at which the final turn of the S-shaped yield curve occurs be influenced? Can the level be raised by increasing the prices received by farmers, by adopting technological innovations, and by stepping up investment in crop research?

Most of those countries which have achieved takeoffs in yield per acre are continuing to raise yields at a rapid rate. But there are indications that the rate of gain may be slowing for some crops in some of the more agriculturally advanced countries.

Projected per acre yield levels for the major grains in the United States show a substantial slowing of the rate of yield increase over the next 15 years as compared with the last 15. The rate of yield increase for wheat, averaging 3.5 percent yearly from 1950 to 1965, is projected to drop to less than 2 percent per year between 1965 and 1980 (Fig. 16–9). Sorghum yields, recently increasing at a rate of nearly 6 percent annually, are projected to increase at just over 2 percent per year between now and 1980 (Fig. 16–10). For corn, the projected slowdown is less dramatic, with yield increases dropping from about 4 percent to 3 percent. Per-acre yields of wheat and grain sorghum have apparently achieved their more rapid gains as the use of nonrecurring technologies becomes almost universal. In Platt's words (*7*), they may already be "past the middle of the S-curve."

The rate of increase could also be slowing down for certain crops elsewhere in the world. Rice yields in Japan may be a case in point. Yields were relatively static before 1900 but began to rise steadily shortly after the turn of the century. This rise continued until about 1959 (except for a brief period around World War II, and a period from 1949 to 1953, when production was disrupted by land reform). Since 1959, U.S. Department of Agriculture estimates (*8*) indicate, the rate of increase has slowed appreciably and, in fact, has recently nearly leveled off (Fig. 16–11). Whether or not this is a temporary plateau or a more permanent one remains to be seen. Interestingly, projections of per-acre rice yields made by the Japanese Institute of Agricultural Economic Research, using a 1958–1960 base period (*9*), did not anticipate the recent slowdown in the rate of increase in rice yields.

This recent leveling off of yields, however, may be caused by economic as well as technological factors. One key factor contributing to the very high yields obtained in Japan has been the intensive use of what was once low-cost labor. In recent years there has been a withdrawal of labor from rice production as rural workers have found more remunerative urban jobs. If economic development continues, it is unlikely that recent trends in labor costs will ever be reversed. Thus, it may well be that per-acre rice yields in Japan are approaching what is, in the immediately foreseeable future at least, a plateau.

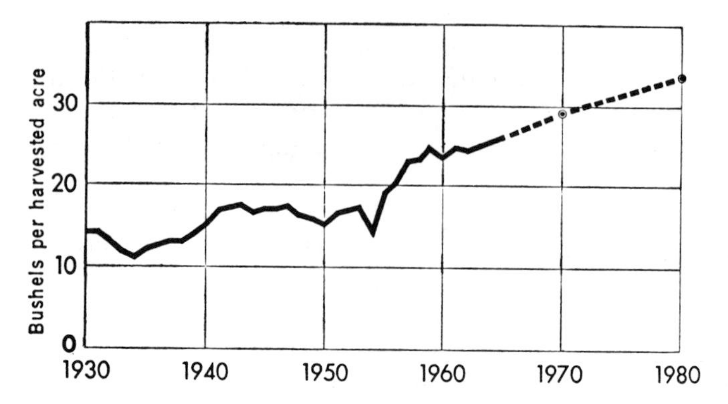


FIGURE 16–9. Wheat yields in the United States, with projections. Plotted as a 3-year sliding average [U.S. Department of Agriculture]

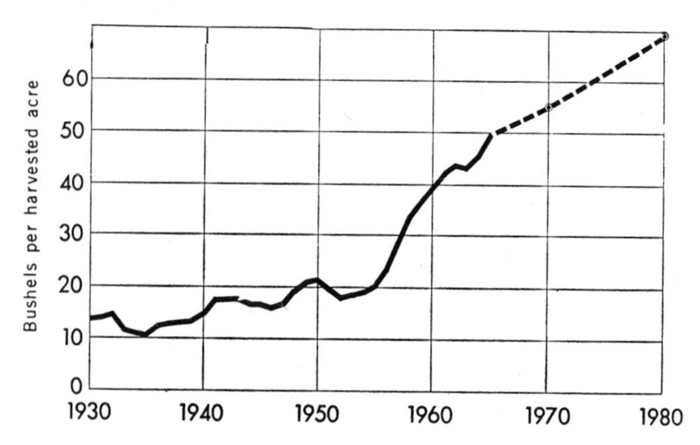


FIGURE 16–10. Grain sorghum yields in the United States, with projections. Plotted as a 3-year sliding average [U.S. Department of Agriculture].

A slowdown in the rate of yield increase seems also to be occurring for some of the grain crops in the Netherlands. This is not particularly surprising since yields there are already among the highest in the world. Further yield responses of some grains to the use of additional inputs, such as fertilizer, now seem limited by genetic constraints—the inherent ability of the plant to effectively use additional plant nutrients.

There are, on the other hand, some crops in the agriculturally developed countries which have not yet begun their upward advance on the growth curve. One of the major U.S. crops, the soybean, has thus far stubbornly resisted efforts to generate a yield-per-acre takeoff (*10*). The combination of near-static yields, on the one hand, and the very rapid growth in demand for soybeans, on the other, means that the necessary increases in the soybean supply are obtainable only through a rapid continuing expansion in the area planted to soybeans—an expansion which is steadily reducing the area available for other crops.

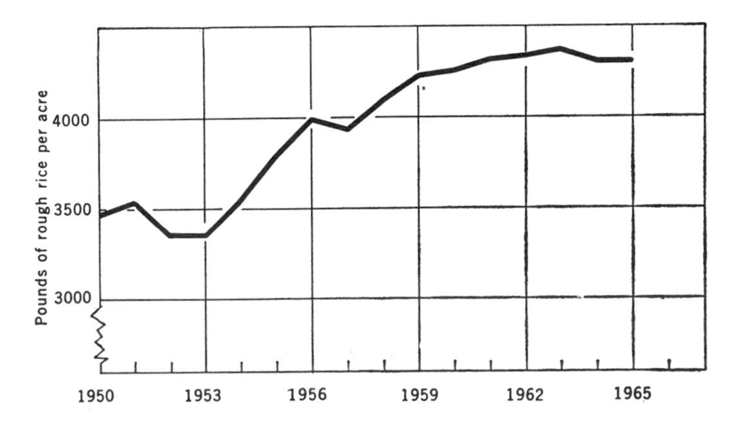


FIGURE 16–11. Rice yields in Japan, 1950–1965. Plotted as a 3-year sliding average [U.S. Department of Agriculture].

During the two decades since World War II, projections of increases in per acre yields in the United States have invariably underestimated the increases actually achieved. This may be due in part to the yield-raising effect of idling large areas of marginal cropland during this period. There is now a risk that our faith in technology will cause us to overestimate future increases in yields if, in fact, the rate of yield increase ultimately slows as the sources of further gains in productivity diminish.

It is significant that the major sources of increased agricultural productivity—the use of chemical fertilizer; the use of improved varieties, including hybrids; the use of pesticides and irrigation—have all been known for decades, if not longer. The key question now is: Are there any sources of increased productivity in existence or in the process of development comparable to the traditional ones listed above?

The concept of the S-shaped curve is not new, but its implications for future agricultural production have not been fully explored. Although the S-shaped yield curve for crops is, at this point, still an untested hypothesis, it is, in Platt's words (7), "at least as plausible as the uncritical assumption that changes like those of the twentieth-century will go on forever."

### *Photosynthetic Efficiency and Research*

The ultimate factor limiting crop output per acre is the crop's photosynthetic efficiency (*11*). Defined as the percentage of solar energy used relative to that which is available on a given area occupied by a particular crop, photosynthetic efficiency is always quite low, usually less than 3 percent. Density of plant population, actual position of the leaves on the plant, and temperature are key factors accounting for variations within this range.

In 1962, James Bonner of the California Institute of Technology stated (*11*):

> . . .the upper limit of crop yield, as determined by the factors that regulate photosynthetic efficiency, is already being approached today in those regions with the highest level of agricultural practice—in parts of Japan, of Western Europe, and of the United States.

Obviously, research into ways of increasing the upper limit of yield is needed. This increase could be achieved by developing plants which have greater photosynthetic efficiency or by improving present cultural practices so as to increase efficiency per acre, or by both means. The development of smaller and more efficient corn plants, along with reduction in the need for cultivation during the growing season, makes it possible to reduce the width between corn rows—a width that was initially determined by the width of a horse, in the age of the horse-drawn cultivator. The result is a dramatic gain in the number of corn plants per acre, and increased output.

More productive hybrid wheats have been developed, but they are still in the experimental stage and are not yet being grown commercially. Work on breeding new varieties with higher nutritive value—a potentially promising activity—is also under way. The adoption of a new technology takes time, even in an agriculturally advanced country. It took a quarter of a century for U.S. farmers to adopt hybrid corn (Fig. 16–7). Hybrid grain sorghum, introduced in the early 1950's, required about a decade to become widely disseminated.

Both corn and wheat have been the subject of many years of research in the United States and other developed nations. Much less work has been done in rice. To help rectify this situation, the Rockefeller and Ford foundations established the International Rice Institute in the Philippines several years ago. The Institute devotes its efforts not only to the development of new varieties but to the whole range of cultural practices as well.

The need for such research is further emphasized by a recent statement by Harvey Brooks, chairman of the Committee on Science and Public Policy of the National Academy of Sciences (*12*):

> Future food production, even for domestic purposes, will be strongly dependent on the quality and direction of both the basic and applied research undertaken within the next few years. Most of the potential of past basic research has already been realized, and new knowledge will be needed even to maintain present levels of productivity.

Clearly, much more research is essential if we are to (i) get the underdeveloped nations to the yield takeoff point, and (ii) maintain the upward thrust of yields in the developed countries by postponing the final turn on the S-shaped curve (*13*).

## Research and Reality

Two groups of factors should be kept in mind in evaluating the real potential of research results for significantly increasing food output on a worldwide basis. The first group centers about the pronounced variations in natural resources and managerial abilities, which can lead to wide differences between record yields and average national yields obtained by individual farmers under localized conditions. The second group concerns the matter of costs and returns, which spells the difference between technical potential and economic reality.

### *Record Yields versus Average Yields*

It is often assumed that record yields attained on experimental plots can be easily and quickly translated into national average yields. Such is not, however, the case. Maximum yields obtained on experimental plots under closely controlled conditions usually far exceed those generally obtained in practice. Average yields of wheat in this country, for example, are far below those attained on experimental plots during the latter part of the last century. The same is true for many other crops.

Equally common and equally unwarranted is the assumption that all countries will eventually attain the average yield prevailing in the nation which now has the highest yield. Potential yield levels attainable by individual countries vary widely with variations in rainfall, temperature, soil types and topography, production costs, managerial abilities of farmers, and other factors.

Wheat yields in the United Kingdom now average about 60 bushels per acre (52 hectoliters per hectare) as contrasted with only 18 bushels per acre in Australia. This does not mean that wheat-production technology is less advanced in Australia than in the United Kingdom. The yield differences do reflect the difference between growing conditions in Australia, where rainfall in the wheat-growing regions averages 12 to 15 inches (30 to 38 centimeters) annually, and those in the United Kingdom, where rainfall may average 40 to 50 inches. Although wheat yields in both the United Kingdom and Australia may continue to rise, there is no reason to assume that the differences in yields between the two countries will narrow appreciably in the foreseeable future.

The average national rice yield in Japan is nearly four times that in India. A large part of this difference is accounted for by a much greater volume of inputs, including labor as well as modern practices and management. Not to be overlooked, however, is the fact that virtually all of the rice crop produced in Japan is irrigated, whereas only part of India's rice crop is irrigated. A large share of India's rice fields are rainfed, thus the yield levels attained depend greatly on the vagaries of the monsoon.

There are also very wide variations in yield within individual countries. Variations in corn yields within various corn-producing states in the United States are almost as pronounced as variations in corn yields between the various corn-producing countries of the world. Average yields in principal U.S. corn-producing states in 1965, for instance, varied from more than 90 bushels per acre in some states in the Midwest to less than 40 bushels in some states in the southern Mississippi Valley.

It is significant that the leveling off of rice yields in Japan has occurred at a time when average rice yields in the more productive and the less productive prefectures vary widely. Some individual villages in Japan obtain rice yields at least double the national average.

Per acre yields obtained by individual farmers in the same area may vary even more than do those for various states or prefectures. It is often assumed that the performance of the best farmers can be emulated by all. There are and will continue to be some very basic differences in the innate capacities or motivation of farmers. There is no more reason for assuming that all farmers can or really want to attain a record yield of corn or wheat than to assume that all students can or want to become Harvard Phi Beta Kappas. The distribution of talent and motivation is probably at least as wide within the world's rural communities as in any other area.

### *Technical Potential versus Economic Reality*

The failure to distinguish between the technical potential for expanding food production and the economically profitable possibilities for doing so has resulted in confusing variations in estimates of future food production. The difference between estimates based on these two criteria is often very great. The earlier discussion of the experience in Japan—where rice yields seem to have leveled off in recent years—suggests the importance of economic relationships.

A recent reduction in milk production in the United States closely parallels the Japanese experience with rice yields. Through the early months of 1966, milk production in the United States was 3 to 5 percent below production in comparable months of the preceding year. At prevailing prices it was not profitable for dairy farmers to use some of the existing resources. During 1966, dairy farmers in New York State received scarcely 40 cents an hour for their labor (when allowance is made for interest on their investment), and farmers in Wisconsin received only 50 cents an hour. At a time when slaughter prices were high and there were many job opportunities to choose from—with a 5-day, 40-hour week in industry and a minimum wage of $1.25 per hour (*14*)—it comes as no surprise to learn that many dairy farmers liquidated their holdings and took other jobs. In order to help increase returns to farmers and expand milk production, the Department of Agriculture raised milk support prices twice during 1966, for a total increase of 23 percent.

Both prices received by farmers and costs of production must be taken into consideration in assessing potential increases in production. As farmers move up the per acre yield curve, the point of diminishing returns is eventually reached. Additional costs begin to exceed additional returns. Thus it is unrealistic to expect farmers to produce up to the full technical potential.

Therefore, while many farmers can produce much more under a given technology, it is sometimes uneconomic, at existing prices and costs, for them to do so. If society is willing to pay higher prices—and it may have to some day—much greater production may be expected.

## Conclusions

1) The worldwide demand for food will continue to be strong in the coming decades. Two forces—rapidly growing population and, in much of the world, rapidly rising incomes—are expected to result in increases in the demand for food even more rapid than those that have occurred during the past.

2) Conventional agriculture has assured an adequate food supply for the economically advanced one-third of the world. The challenge now is to assure an adequate food supply for the remaining two-thirds, where population is now increasing at the rate of 1 million people per week and where malnutrition is already widespread.

3) Economically feasible prospects for significantly expanding the world's area of cultivated land in the 1960's and 1970's are limited and largely confined to sub-Saharan Africa and the Amazon Basin. Even here, agronomic problems will limit the rate of expansion. When the cost of desalting sea water is substantially reduced—probably not before the late 1970's or early 1980's at best—it may become feasible to irrigate large areas of desert.

4) Given the limited possibilities for expanding the area of land under cultivation, most of the increases in world food needs must be met, for the foreseeable future, by raising the productivity of land already under cultivation. Food output per acre, rather static throughout most of history, has begun to increase rapidly in some of the more advanced countries in recent decades. All of the increases in food production over the past quarter century in North America, western Europe, and Japan have come from increasing the productivity of land already

under cultivation. The area under cultivation has actually declined.

5) Achieving dramatic gains in land productivity requires a massive investment of capital and the widespread adoption of new technology. A similar effort must now be made in the less developed nations if these nations are to feed their people. The most important single factor influencing this rate of investment is food prices, more particularly the relationship between the price farmers receive for their food products and the cost of modern inputs such as fertilizer.

6) In some of the more developed countries where per acre yields have been rising for a long time, there is now evidence that the rate of yield increase may be slowing. Nonrecurring inputs may have made their maximum contribution to output in the case of some crops, pushing yield levels past the middle of the S-shaped logistic curve. Although this cannot be determined with any certainty, the possibility that the middle of the curve has been passed in some instances should be taken into account in viewing the long-term future.

7) If the rate of increase in yield per acre does in fact begin to slow in some of the agriculturally advanced countries, additional pressure will be put on the less-developed countries—which have much of the world's unrealized food-production potential—to meet the continuing future increases in world food needs.

8) Man has not yet been able to bypass the process of photosynthesis in the production of food. This dependence on photosynthesis plays a significant role in determining the upper levels of the S-shaped yield curve. Additional research is urgently needed to increase the photosynthetic efficiency of crops and to raise the upper levels of economically feasible yields.

## References and Notes

1. I have previously examined these matters in some detail in "Man, Land and Food," *U.S. Dept. Agr. Foreign Agr. Econ. Rep. No. 11* (1963).
2. I have discussed this concept at length in "Increasing World Food Output," *U.S. Dept. Agr. Foreign Agr. Econ. Rep. No. 25* (1965).
3. For further discussion of this point, and the role that may be played by private industry, see L. R. Brown, *Columbia J. World Business* 2, No. 1, 15 (1967).
4. As one leading agricultural economist recently stated, there is considerable evidence that in most low-income countries "technological advance requires a complementary input of labor" [ J.Mellor, *The Economics of Agricultural Development* (Cornell Univ. Press, Ithaca, N.Y., 1966), p. 157].
5. The "nonrecurring" concept was introduced by Paul C. Mangelsdorf (see 6).
6. P. C. Mangelsdorf, *Proc. Nat. Acad. Sci. U.S.* 56, 370 (1966).
7. J. R. Platt, *The Road to Man* (Wiley, New York, 1966) [originally published in *Science* 149, 607 (1965)].

8. Estimates published by the Food and Agriculture Organization show a continued increase in rice yields up until 1963–64, followed by successive declines in each of the three following seasons; see annual issues of *Production Yearbook* (Rome) and *Monthly Bull. Agr. Economics Statistics* 15, No. 12, 26 (1966).
9. *Japanese Import Requirement: Projections of Agricultural Supply and Demand for 1965, 1970 and 1975* (Institute of Agricultural Economic Research, University of Tokyo, (1964), p. 84.
10. Soybeans cannot be commercially hybridized and show only limited response to nitrogen: see *The World Food Problem* (Government Printing Office, Washington, D.C., 1967), vol. 2, p. 197.
11. J. Bonner, *Science* 137, 11 (1962).
12. *The Plant Sciences Now and in the Coming Decade* National Academy of Sciences, Washington, D.C., 1966), *p. iv*.
13. A detailed discussion of the technical problems and issues faced in intensifying plant production in the developing nations is presented in *The World Food Problem* (Superintendent of Documents, Government Printing Office, Washington, D.C., 1967), pp. 215–233.
14. The minimum wage was recently raised to $1.40 per hour.
15. I am indebted to Dana G. Dalrymple of the U.S. Department of Agriculture for his suggestions and assistance.

# 17

# Photosynthesis and fish production in the sea

*John H. Ryther*

Numerous attempts have been made to estimate the production in the sea of fish and other organisms of existing or potential food value to man (1–4). These exercises, for the most part, are based on estimates of primary (photosynthetic) organic production rates in the ocean (5) and various assumed trophic-dynamic relationships between the photosynthetic producers and the organisms of interest to man. Included in the latter are the number of steps or links in the food chains and the efficiency of conversion of organic matter from each trophic level or link in the food chain to the next. Different estimates result from different choices in the number of trophic levels and in the efficiencies, as illustrated in Table 17–1 (2).

Implicit in the above approach is the concept of the ocean as a single ecosystem in which the same food chains involving the same number of links and efficiencies apply throughout. However, the rate of primary production is known to be highly variable, differing by at least two full orders of magnitude from the richest to the most impoverished regions. This in itself would be expected to result in a highly irregular pattern of food production. In addition, the ecological conditions which determine the trophic dynamics of marine food chains also vary widely and

Reprinted with permission from *Science*, 166:72–76 (3 October 1969).  The author is a member of the staff of the Woods Hole Oceanographic Institution, Woods Hole, Massachusetts.

TABLE 17–1

*Estimates of potential yields (per year) at various trophic levels, in metric tons* [After Schaeffer (2)]

| Trophic level | Ecological efficiency factor | | | | | |
|---|---|---|---|---|---|---|
| | 10 percent | | 15 percent | | 20 percent | |
| | Carbon (tons) | Total weight (tons) | Carbon (tons) | Total weight (tons) | Carbon (tons) | Total weight (tons) |
| 0. Phytoplankton (net particulate production) | $1.9 \times 10^{10}$ | | $1.9 \times 10^{10}$ | | $1.9 \times 10^{10}$ | |
| 1. Herbivores | $1.9 \times 10^{9}$ | $1.9 \times 10^{10}$ | $2.8 \times 10^{9}$ | $2.8 \times 10^{10}$ | $3.8 \times 10^{9}$ | $3.8 \times 10^{10}$ |
| 2. 1st stage carnivores | $1.9 \times 10^{8}$ | $1.9 \times 10^{9}$ | $4.2 \times 10^{8}$ | $4.2 \times 10^{9}$ | $7.6 \times 10^{8}$ | $7.6 \times 10^{9}$ |
| 3. 2nd stage carnivores | $1.9 \times 10^{7}$ | $1.9 \times 10^{8}$ | $6.4 \times 10^{7}$ | $6.4 \times 10^{8}$ | $15.2 \times 10^{7}$ | $15.2 \times 10^{8}$ |
| 4. 3rd stage carnivores | $1.9 \times 10^{6}$ | $1.9 \times 10^{7}$ | $9.6 \times 10^{6}$ | $9.6 \times 10^{7}$ | $30.4 \times 10^{6}$ | $30.4 \times 10^{7}$ |

in direct relationship to the absolute level of primary organic production. As is shown below, the two sets of variables—primary production and the associated food chain dynamics—may act additively to produce differences in fish production which are far more pronounced and dramatic than the observed variability of the individual causative factors.

## Primary Productivity

Our knowledge of the primary organic productivity of the ocean began with the development of the $C^{14}$-tracer technique for *in situ* measurement of photosynthesis by marine plankton algae (*6*) and the application of the method on the 1950–52 *Galathea* expedition around the world (*5*). Despite obvious deficiencies in the coverage of the ocean by *Galathea* (the expedition made 194 observations, or an average of about one every 2 million square kilometers, most of which were made in the tropics or semitropics), our concept of the total productivity of the world ocean has changed little in the intervening years.

While there have been no more expeditions comparable to the *Galathea*, there have been numerous local or regional studies of productivity in many parts of the world. Most of these have been brought together by a group of Soviet scientists to provide up-to-date world coverage consisting of over 7000 productivity observations (*7*). The result has been modification of the estimate of primary production in the world ocean from 1.2 to 1.5 $\times$ $10^{10}$ tons of carbon fixed per year (*5*) to a new figure, 1.5 to 1.8 $\times$ $10^{10}$ tons.

Attempts have also been made by Steemann Nielsen and Jensen (*5*), Ryther (*8*), and Koblentz-Mishke et al. (*7*) to assign specific levels or ranges of productivity to different parts of the ocean. Although the approach was somewhat different in each case, in general the agreement between the three was good and, with appropriate condensation and combination, permit the following conclusions.

1) Annual primary production in the open sea varies, for the most part, between 25 and 75 grams of carbon fixed per square meter and averages about 50 grams of carbon per square meter per year. This is true for roughly 90 percent of the ocean, an area of $326 \times 10^6$ square kilometers.

2) Higher levels of primary production occur in shallow coastal waters, defined here as the area within the 100-fathom (180-meter) depth contour. The mean value for this region may be considered to be 100 grams of carbon fixed per square meter per year, and the area, according to Menard and Smith (*9*), is 7.5 percent of the total world ocean. In addition, certain offshore waters are influenced by divergences, fronts, and other hydrographic features which bring nutrient-rich subsurface water into the euphotic zone. The equatorial divergences are examples of such regions. The productivity of these offshore areas is comparable to that of the coastal zone. Their total area is difficult to assess, but is considered here to be 2.5 percent of the total ocean. Thus, the coastal zone and the offshore regions of comparably high productivity together represent 10 percent of the total area of the oceans, or $36 \times 10^6$ square kilometers.

3) In a few restricted areas of the world, particularly along the west coasts of continents at subtropical latitudes where there are prevailing offshore winds and strong eastern boundary currents, surface waters are diverted offshore and are replaced by nutrient-rich deeper water. Such areas of coastal upwelling are biologically the richest parts of the ocean. They exist off Peru, California, northwest and southwest Africa, Somalia, and the Arabian coast, and in other more localized situations. Extensive coastal upwelling also is known to occur in various places around the continent of Antarctica, although its exact location and extent have not been well documented. During periods of active upwelling, primary production normally exceeds 1.0 and may exceed 10.0 grams of carbon per square meter per day. Some of the high values which have been reported from these locations are 3.9 grams for the southwest coast of Africa (*5*), 6.4 for the Arabian Sea (*10*), and 11.2 off Peru (*11*). However, the upwelling of subsurface water does not persist throughout the year in many of these places—for example, in the Arabian Sea, where the process is seasonal and related to the monsoon winds. In the Antarctic, high production is limited by solar radiation during half the year. For all these areas of coastal upwelling throughout the year, it is probably safe, if somewhat conservative, to assign an annual value of 300 grams of carbon per square meter. Their total area in the world is again difficult to assess.

On the assumption that their total cumulative area is no greater than 10 times the well-documented upwelling area off Peru, this would amount to some $3.3 \times 10^5$ square kilometers, or 0.1 percent of the world ocean. These conclusions are summarized in Table 17–2.

TABLE 17–2

*Division of the ocean into provinces according to their level of primary organic production*

| Province | Percentage of ocean | Area ($km^2$) | Mean productivity (grams of carbon/$m^2$/yr) | Total productivity ($10^9$ tons of carbon/yr) |
|---|---|---|---|---|
| Open ocean | 90 | $326 \times 10^6$ | 50 | 16.3 |
| Coastal zone* | 9.9 | $36 \times 10^6$ | 100 | 3.6 |
| Upwelling areas | 0.1 | $3.6 \times 10^5$ | 300 | 0.1 |
| Total | | | | 20.0 |

* Includes offshore areas of high productivity.

## Food Chains

Let us next examine the three provinces of the ocean which have been designated according to their differing levels of primary productivity from the standpoint of other possible major differences. These will include, in particular, differences which relate to the food chains and to trophic efficiences involved in the transfer of organic matter from the photosynthetic organisms to fish and invertebrate species large and abundant enough to be of importance to man.

The first factor to be considered in this context is the size of the photosynthetic or producer organisms. It is generally agreed that, as one moves from coastal to offshore oceanic waters, the character of these organisms changes from large "microplankton" (100 microns or more in diameter) to the much smaller "nannoplankton" cells 5 to 25 microns in their largest dimensions (*12*, *13*).

Since the size of an organism is an essential criterion of its potential usefulness to man, we have the following relationship: the larger the plant cells at the beginning of the food chain, the fewer the trophic levels that are required to convert the organic matter to a useful form. The oceanic nannoplankton cannot be effectively filtered from the water by most of the common zooplankton crustacea. For example, the euphausid *Euphausia pacifica*, which may function as a herbivore in the rich subarctic coastal waters of the Pacific, must turn to a carnivorous habit in the offshore waters where the phytoplankton become too small to be captured (*13*).

Intermediate between the nannoplankton and the carnivorous zooplankton are a group of herbivores, the microzooplankton, whose ecological significance is a subject of considerable current interest (*14, 15*). Representatives of this group include protozoans such as Radiolaria, Foraminifera, and Tintinnidae, and larval nuplii of microcrustaceans. These organisms, which may occur in concentrations of tens of thousands per cubic meter, are the primary herbivores of the open sea.

Feeding upon these tiny animals is a great host of carnivorous zooplankton, many of which have long been thought of as herbivores. Only by careful study of the mouthparts and feeding habits were Anraku and Omori (*16*) able to show that many common copepods are facultative if not obligate carnivores. Some of these predatory copepods may be no more than a millimeter or two in length.

Again, it is in the offshore environment that these small carnivorous zooplankton predominate. Grice and Hart (*17*) showed that the percentage of carnivorous species in the zooplankton increased from 16 to 39 percent in a transect from the coastal waters of the northeastern United States to the Sargasso Sea. Of very considerable importance in this group are the Chaetognatha. In terms of biomass, this group of animals, predominantly carnivorous, represents, on the average, 30 percent of the weight of copepods in the open sea (*17*). With such a distribution, it is clear that virtually all the copepods, many of which are themselves carnivores, must be preyed upon by chaetognaths.

The oceanic food chain thus far described involves three to four trophic levels from the photosynthetic nannoplankton to animals no more than 1 to 2 centimeters long. How many additional steps may be required to produce organisms of conceivable use to man is difficult to say, largely because there are so few known oceanic species large enough and (through schooling habits) abundant enough to fit this category. Familiar species such as the tunas, dolphins, and squid are all top carnivores which feed on fishes or invertebrates at least one, and probably two, trophic levels beyond such zooplankton as the chaetognaths. A food chain consisting of five trophic levels between photosynthetic organisms and man would therefore seem reasonable for the oceanic province.

As for the coastal zone, it has already been pointed out that the phytoplankton are quite commonly large enough to be filtered and consumed directly by the common crustacean zooplankton such as copepods and euphausids. However, the presence, in coastal waters, of protozoans and other microzooplankton in larger numbers and of greater biomass than those found in offshore waters (*15*) attests to the fact that much of the primary production here, too, passes through several steps of a microscopic food chain before reaching the macrozooplankton.

The larger animals of the coastal province (that is, those directly useful to man) are certainly the most diverse with respect to feeding type. Some (mollusks and some fishes) are herbivores. Many others, including most of the pelagic clupeoid fishes, feed on zooplankton. Another large group, the demersal fishes, feed on bottom fauna which may be anywhere from one to several steps removed from the phytoplankton.

If the herbivorous clupeoid fishes are excluded (since these occur predominantly in the upwelling provinces and are therefore considered separately), it is probably safe to assume that the average food organism from coastal waters represents the end of at least a three-step food chain between phytoplankton and man.

It is in the upwelling areas of the world that food chains are the shortest, or—to put it another way—that the organisms are large enough to be directly utilizable by man from trophic levels very near the primary producers. This, again, is due to the large size of the phytoplankton, but it is due also to the fact that many of these species are colonial in habit, forming large gelatinous masses or long filaments. The eight most abundant species of phytoplankton in the upwelling region off Peru, in the spring of 1966, were *Chaetoceros socialis*, *C. debilis*, *C. lorenzianus*, *Skeletonema costatum*, *Nitzschia seriata*, *N. delicatissima*, *Schroederella delicatula*, and *Asterionella japonica* (*11, 18*). The first in this list, *C. socialis*, forms large gelatinous masses. The others all form long filamentous chains. *Thalossiosira subtili*, another gelatinous colonial form like *Chaetoceros socialis*, occurs commonly off southwest Africa (*19*) and close to shore off the Azores (*20*). Hart (*21*) makes special mention of the colonial habit of all the most abundant species of phytoplankton in the Antarctic—*Fragiloriopsis antarctica*, *Encampia balaustrium*, *Rhizosalenia alata*, *R. antarctica*, *R. chunii*, *Thallosiothrix antarctica*, and *Phaeocystis brucei*.

Many of the above-mentioned species of phytoplankton form colonies several millimeters and, in some cases, several centimeters in diameter. Such aggregates of plant material can be readily eaten by large fishes without special feeding adaptation. In addition, however, many of the clupeoid fishes (sardines, anchovies, pilchards, menhaden, and so on) that are found most abundantly in upwelling areas and that make up the largest single component of the world's commercial fish landings, do have specially modified gill rakers for removing the larger species of phytoplankton from the water.

There seems little doubt that many of the fishes indigenous to upwelling regions are direct herbivores for at least most of their lives. There is some evidence that juveniles of the Peruvian anchovy (*Engraulis ringens*) may feed on zooplankton, but the adult is predominantly if not exclusively a herbivore (*22*). Small gobies (*Gobius bibarbatus*) found at mid-water in the coastal waters off southwest Africa had their stomachs filled with a large, chain-forming diatom of the genus *Fragilaria* (*23*). There is considerable interest at present in the possible commercial utilization of the large Antarctic krill, *Euphausia superba*, which feeds primarily on the colonial diatom *Fragilariopsis antarctica* (*24*).

In some of the upwelling regions of the world, such as the Arabian

Sea, the species of fish are not well known, so it is not surprising that knowledge of their feeding habits and food chains is fragmentary. From what is known, however, the evidence would appear to be overwhelming that a one- or two-step food chain between phytoplankton and man is the rule. As a working compromise, let us assign the upwelling province a 1½-step food chain.

## Efficiency

The growth (that is, the net organic production) of an organism is a function of the food assimilated less metabolic losses or respiration. This efficiency of growth or food utilization (the ratio of growth to assimilation) has been found, by a large number of investigators and with a great variety of organisms, to be about 30 percent in young, actively growing animals. The efficiency decreases as animals approach their full growth, and reaches zero in fully mature or senescent individuals (*25*). Thus a figure of 30 percent can be considered a biological potential which may be approached in nature, although the growth efficiency of a population of animals of mixed ages under steady-state conditions must be lower.

Since there must obviously be a "maintenance ration" which is just sufficient to accommodate an organism's basal metabolic requirement (*26*), it must also be true that growth efficiency is a function of the absolute rate of assimilation. The effects of this factor will be most pronounced at low feeding rates, near the "maintenance ration," and will tend to become negligible at high feeding rates. Food conversion (that is, growth efficiency) will therefore obviously be related to food availability, or to the concentration of prey organisms when the latter are sparsely distributed.

In addition, the more available the food and the greater the quantity consumed, the greater the amount of "internal work" the animal must perform to digest, assimilate, convert, and store the food. Conversely, the less available the food, the greater the amount of "external work" the animal must perform to hunt, locate, and capture its prey. These concepts are discussed in some detail by Ivlev (*27*) and reviewed by Ricker (*28*). The two metabolic costs thus work in opposite ways with respect to food availability, tending thereby toward a constant total effect. However, when food availability is low, the added costs of basal metabolism and external work relative to assimilation may have a pronounced effect on growth efficiency.

When one turns from consideration of the individual and its physiological growth efficiency to the "ecological efficiency" of food conversion from one trophic level to the next (*22, 29*), there are additional losses to be taken into account. Any of the food consumed but not assimilated would be included here, though it is possible that undigested organic matter may be reassimilated by members of the same trophic level (*2*). Any other nonassimilatory losses, such as losses due to natural death, sedimentation, and emigration, will, if not otherwise accounted for, appear as a loss in trophic efficiency. In addition, when one considers a specific or selected part of a trophic level, such as a population of fish of use to man, the consumption of food by any other hidden member of the same trophic level will appear as a loss in efficiency. For example, the role of such animals as salps, medusae, and ctenophores in marine food chains is not well understood and is seldom even considered. Yet these animals may occur sporadically or periodically in swarms so dense that they dominate the plankton completely. Whether they represent a dead end or side branch in the normal food chain of the sea is not known, but their effect can hardly be negligible when they occur in abundance.

Finally, a further loss which may occur at any trophic level but is, again, of unknown or unpredictable magnitude is that of dissolved organic matter lost through excretion or other physiological processes by plants and animals. This has received particular attention at the level of primary production, some investigators concluding that 50 percent or more of the photoassimilated carbon may be released by phytoplankton into the water as dissolved compounds (*30*). There appears to be general agreement that the loss of dissolved organic matter is indirectly proportional to the absolute rate of organic production and is therefore most serious in the oligotrophic regions of the open sea (*11, 31*).

All of the various factors discussed above will affect the efficiency or apparent efficiency of the transfer of organic matter between trophic levels. Since they cannot, in most cases, be quantitatively estimated individually, their total effect cannot be assessed. It is known only that the maximum potential growth efficiency is about 30 percent and that at least some of the factors which reduce this further are more pronounced in oligotrophic, low-productivity waters than in highly productive situations. Slobodkin (*29*) concludes that an ecological efficiency of about 10 percent is possible, and Schaeffer feels that the figure may be as high as 20 percent. Here, therefore, I assign efficiencies of 10, 15, and 20 percent, respectively, to the oceanic, the coastal, and the upwelling provinces, though it is quite possible that the actual values are considerably lower.

## Conclusions and Discussion

With values assigned to the three marine provinces for primary productivity (Table 17–2), number of trophic levels, and efficiencies, it is now possible to calculate fish production in the three regions. The results are summarized in Table 17–3.

These calculations reveal several interesting features. The open sea—

TABLE 17–3

*Estimated fish production in the three ocean provinces defined in Table 17–2*

| Province | Primary production [tons (organic carbon)] | Trophic levels | Efficiency (%) | Fish production [tons (fresh wt.)] |
|---|---|---|---|---|
| Oceanic | $16.3 \times 10^9$ | 5 | 10 | $16 \times 10^5$ |
| Coastal | $3.6 \times 10^9$ | 3 | 15 | $12 \times 10^7$ |
| Upwelling | $0.1 \times 10^9$ | 1½ | 20 | $12 \times 10^7$ |
| Total | | | | $24 \times 10^7$ |

90 percent of the ocean and nearly three-fourths of the earth's surface—is essentially a biological desert. It produces a negligible fraction of the world's fish catch at present and has little or no potential for yielding more in the future.

Upwelling regions, totaling no more than about one-tenth of 1 percent of the ocean surface (an area roughly the size of California) produce about half the world's fish supply. The other half is produced in coastal waters and the few offshore regions of comparably high fertility.

One of the major uncertainties and possible sources of error in the calculation is the estimation of the areas of high, intermediate, and low productivity. This is particularly true of the upwelling area off the continent of Antarctica, an area which has never been well described or defined.

A figure of 360,000 square kilometers has been used for the total area of upwelling regions in the world (Table 17–2). If the upwelling regions off California, northwest and southwest Africa, and the Arabian Sea are of roughly the same area as that off the coast of Peru, these semitropical regions would total some 200,000 square kilometers. The remaining 160,000 square kilometers would represent about one-fourth the circumference of Antarctica seaward for a distance of 30 kilometers. This seems a not unreasonable inference. Certainly, the entire ocean south of the Antarctic Convergence is not highly productive, contrary to the estimates of El-Sayed (*32*). Extensive observations in this region by Saijo and Kawashima (*33*) yielded primary productivity values of 0.01 to 0.15 gram of carbon per square meter per day—a value no higher than the values used here for the open sea. Presumably, the discrepancy is the result of highly irregular, discontinuous, or "patchy" distribution of biological activity. In other words, the occurrence of extremely high productivity associated with upwelling conditions appears to be confined, in the Antarctic, as elsewhere, to restricted areas close to shore.

An area of 160,000 square kilometers of upwelling conditions with an annual productivity of 300 grams of carbon per square meter would result in the production of about $50 \times 10^6$ tons of "fish," if we follow the ground rules established above in making the estimate. Presumably these "fish" would consist for the most part of the Antarctic krill, which feeds directly upon phytoplankton, as noted above, and which is known to be extremely abundant in Antarctic waters. There have been numerous attempts to estimate the annual production of krill in the Antarctic, from the known number of whales at their peak of abundance and from various assumptions concerning their daily ration of krill. The evidence upon which such estimates are based is so tenuous that they are hardly worth discussing. It is interesting to note, however, that the more conservative of these estimates are rather close to figures derived independently by the method discussed here. For example, Moiseev (*34*) calculated krill production for 1967 to be $60.5 \times 10^6$ tons, while Kasahara (*3*) considered a range of 24 to $36 \times 10^6$ tons to be a minimal figure. I consider the figure $50 \times 10^6$ tons to be on the high side, as the estimated area of upwelling is probably generous, the average productivity value of 300 grams of carbon per square meter per year is high for a region where photosynthesis can occur during only half the year, and much of the primary production is probably diverted into smaller crustacean herbivores (*35*). Clearly, the Antarctic must receive much more intensive study before its productive capacity can be assessed with any accuracy.

In all, I estimate that some 240 million tons (fresh weight) of fish are produced annually in the sea. As this figure is rough and subject to numerous sources of error, it should not be condidered significantly different from Schaeffer's (*2*) figure of 200 million tons.

Production, however, is not equivalent to potential harvest. In the first place, man must share the production with other top-level carnivores. It has been estimated, for example, that guano birds alone eat some 4 million tons of anchovies annually off the coast of Peru, while tunas, squid, sea lions, and other predators probably consume an equivalent amount (*22, 36*). This is nearly equal to the amount taken by man from this one highly productive fishery. In addition, man must take care to leave a large enough fraction of the annual production of fish to permit utilization of the resource at something close to its maximum sustainable yield, both to protect the fishery and to provide a sound economic basis for the industry.

When these various factors are taken into consideration, it seems unlikely that the potential sustained yield of fish to man is appreciably greater than 100 million tons. The total world fish landings for 1967 were just over 60 million tons (*37*), and this figure has been increasing at an average rate of about 8 percent per year for the past 25 years. It is clear that, while the yield can be still further increased, the resource is not vast. At the present rate, the industry can continue to expand for no more than a decade.

Most of the existing fisheries of the world are probably incapable of contributing significantly to this expansion. Many are already overexploited, and most of the rest are utilized at or near their maximum sustainable yield. Evidence of fishing pressure is usually determined directly from fishery statistics, but it is of some interest, in connection with the present discussion, to compare landings with fish production as estimated by the methods developed in this article. I will make this comparison for two quite dissimilar fisheries, that of the continental shelf of the northwest Atlantic and that of the Peruvian coastal region.

According to Edwards (*38*), the continental shelf between Hudson Canyon and the southern end of the Nova Scotian shelf includes an area of 110,000 square miles ($2.9 \times 10^{11}$ square meters). From the information in Tables 17–2 and 17–3, it may be calculated that approximately 1 million tons of fish are produced annually in this region. Commercial landings from the same area were slightly in excess of 1 million tons per year for the 3-year period 1963 to 1965 before going into a decline. The decline has become more serious each year, until it is now proposed to regulate the landings of at least the more valuable species such as cod and haddock, now clearly overexploited.

The coastal upwelling associated with the Peru Coastal Current gives rise to the world's most productive fishery, an annual harvest of some $10^7$ metric tons of anchovies. The maximum sustainable yield is estimated at, or slightly below, this figure (*39*), and the fishery is carefully regulated. As mentioned above, mortality from other causes (such as predation from guano birds, bonito, squid, and so on) probably accounts for an additional $10^7$ tons. This prodigious fishery is concentrated in an area no larger than about $800 \times 30$ miles (*36*), or $6 \times 10^{10}$ square meters. By the methods developed in this article, it is estimated that such an upwelling area can be expected to produce $2 \times 10^7$ tons of fish, almost precisely the commercial yield as now regulated plus the amount attributed to natural mortality.

These are but two of the many recognized examples of well-developed commercial fisheries now being utilized at or above their levels of maximum sustainable yield. Any appreciable continued increase in the world's fish landings must clearly come from unexploited species and, for the most part, from undeveloped new fishing areas. Much of the potential expansion must consist of new products from remote regions, such as the Antarctic krill, for which no harvesting technology and no market yet exist.

## References and Notes

1. H. W. Graham and R. L. Edwards, in *Fish and Nutrition* (Fishing News, London, 1962), pp. 3–8; W. K. Schmitt, *Ann. N.Y. Acad. Sci.* **118**; 645 (1965).

2. M. B. Schaeffer, *Trans. Amer. Fish. Soc.* **94**, 123 (1965).
3. H. Kasahara, in *Proceedings, 7th International Congress of Nutrition, Hamburg* (Pergamon, New York, 1966), vol. 4, p. 958.
4. W. M. Chapman, "Potential Resources of the Ocean" (Serial Publication 89–21, 89th Congress, first session, 1965) (Government Printing Office, Washington, D.C., 1965), pp. 132–156.
5. E. Steemann Nielsen and E. A. Jensen, *Galathea Report*, F. Bruun et al., Eds. (Allen & Unwin, London, 1957), vol. 1, p. 49.
6. E. Steemann Nielsen, *J. Cons. Cons. Perma. Int. Explor. Mer* **18**, 117 (1952).
7. O. I. Koblentz-Mishke, V. V. Volkovinsky, J. G. Kobanova, in *Scientific Exploration of the South Pacific*, W. Wooster, Ed. (National Academy of Sciences, Washington, D.C., in press).
8. J. H. Ryther, in *The Sea*, M. N. Hill, Ed. (Interscience, London, 1963), pp. 347–380.
9. H. W. Menard and S. M. Smith, *J. Geophys. Res.* **71**, 4305 (1966).
10. J. H. Ryther and D. W. Menzel, *Deep-Sea Res.* **12**, 199 (1965).
11. ———, E. M. Hulburt, C. J. Lorenzen, N. Corwin, "The Production and Utilization of Organic Matter in the Peru Coastal Current" (Texas A & M Univ. Press, College Station, in press).
12. C. D. McAllister, T. R. Parsons, J. D. H. Strickland, *J. Cons. Cons. Perma. Int. Explor. Mer* **25**, 240 (1960); G. C. Anderson, *Limnol. Oceanogr.* **10**, 477 (1965).
13. T. R. Parsons and R. J. Le Brasseur, in "Symposium Marine Food Chains, Aarhus (1968)."
14. E. Steemann Nielsen, *J. Cons. Cons. Perma. Int. Explor. Mer* **23**, 178 (1958).
15. J. R. Beers and G. L. Stewart, *J. Fish. Res. Board Can.* **24**, 2053 (1967).
16. M. Anraku and M. Omori, *Limnol. Oceanogr.* **8**, 116 (1963).
17. G. D. Grice and H. D. Hart, *Ecol. Monogr.* **32**, 287 (1962).
18. M. R. Reeve, in "Symposium Marine Food Chains, Aarhus (1968)."
19. Personal observation; T. J. Hart and R. I. Currie, *Discovery Rep.* **31**, 123 (1960).
20. K. R. Gaarder, *Report on the Scientific Results of the "Michael Sars" North Atlantic Deep-Sea Expedition 1910* (Univ. of Bergen, Bergen, Norway).
21. T. J. Hart, *Discovery Rep.* **21**, 261 (1942).
22. R. J. E. Sanchez, in *Proceedings of the 18th Annual Session, Gulf and Caribbean Fisheries Institute, University of Miami Institute of Marine Science, 1966*, J. B. Higman, Ed. (Univ. of Miami Press, Coral Gables, Fla., 1966), pp. 84–93.
23. R. T. Barber and R. L. Haedrich, *Deep-Sea Res.* **16**, 415 (1952).
24. J. W. S. Marr, *Discovery Rep.* **32**, 34 (1962).
25. S. D. Gerking, *Physiol. Zool.* **25**, 358 (1952).
26. B. Dawes, *J. Mar. Biol. Ass. U.K.* **17,** 102 (1930–31): ibid., p. 877.
27. V. S. Ivlev, *Zool. Zh.* **18**, 303 (1939).
28. W. E. Ricker, *Ecology* **16**, 373 (1946).
29. L. B. Slobodkin, *Growth and Regulation of Animal Populations* (Holt, Rinehart & Winston, New York, 1961), chap. 12.
30. G. E. Fogg, C. Nalewajko, W. D. Watt, *Proc. Roy. Soc. Ser B Biol. Sci.* **162**, 517 (1965).
31. G. E. Fogg and W. D. Watt, *Mem. Inst. Ital. Idrobiol. Dott. Marco de Marshi Pallanza Italy* **18**, suppl., 165 (1965).
32. S. Z. El-Sayed, in *Biology of the Antarctic Seas III*, G. Llano and W. Schmitt, Eds. (American Geophysical Union, Washington, D.C., 1968), p. 15–47.
33. Y. Saijo and T. Kawashima, *J. Oceanogr. Soc. Japan* **19**, 190 (1964).
34. P. A. Moiseev, paper presented at the 2nd Symposium on Antarctic Ecology, Cambridge, England, 1968.
35. T. L. Hopkins, unpublished manuscript.
36. W. S. Wooster and J. L. Reid, Jr., in *The Sea*, M. N. Hill, Ed. (Interscience, London, 1963), vol. 2, p. 253.

37. *FAO Yearb. Fish. Statistics* **25** (1967).
38. R. L. Edwards, *Univ. Wash. Publ. Fish.* **4**, 52 (1968).
39. R. J. E. Sanchez, in *Proceedings, 18th Annual Session, Gulf and Caribbean Fisheries Institute, University of Miami Institute of Marine Science* (Univ. of Miami Press, Coral Gables, 1966), p. 84.
40. The work discussed here was supported by the Atomic Energy Commission, contract No. AT(30-1)-3862, Ref. No. NYO-3862-20. This article is contribution No. 2327 from the Woods Hole Oceanographic Institution.

# 18

# The biology and psychology of crowding in man and animals[1]

*Charles H. Southwick*

## Introduction

Of the great myriad of problems which man and the world face today, there are three significant trends which stand above all others in importance: (1) the unprecedented population growth throughout the world—a net increase of 1,400,000 people per week—and all of its associations and consequences; (2) the increasing urbanization of these people, so that more and more of them are rushing into cities and urban areas of the world; and (3) the tremendous explosion of communication and social contact throughout the world, so that every part of the world is now aware of every other part. All of these trends are producing increased crowding and the perception of crowding.

It is important to emphasize at the outset that crowding and density are not necessarily the same. Density is the number of individuals per unit area or unit space. It is simple physical measurement. Crowding is a product of density, communication, contact, and activity. It implies a pressure, a force and a psychological reaction. It may occur at widely different densities. The frontiersman may have felt crowded when someone built a homestead a mile away. The suburbanite may feel relatively

[1]This paper is modified from a presentation on April 18, 1970 at the Western Reserve Academy, Hudson, Ohio, in a Symposium entitled, "Search for Survival." I am indebted to Russell Hansen and Henry P. Briggs of Western Reserve Academy and David Swetland of the Sears Family Foundation for their organization and support of the Symposium.

Reprinted with permission from *Ohio Journal of Science*, 71:65–72 (1971). The author is a professor of pathobiology at The Johns Hopkins University, Baltimore, Maryland 21205.

uncrowded in a small house on a half-acre lot if it is surrounded by trees, bushes, and a hedgerow, even though he lives under much higher physical density than did the frontiersman. I have seen laboratory populations of white-footed mice (*Peromyscus leucopus*) showing severe crowding symptoms at a density of 1 mouse per 50 square feet, whereas others were uncrowded, in the behavioral sense, at a density of 2 mice for each square foot. The important difference was the social history and behavioral response of the mice, not the density per se. Hence, crowding is very much a psychological and ecological phenomenon, and not just a physical condition.

A clear and satisfactory definition of crowding is not possible at the present time. It is perhaps easier to recognize crowding than it is to define it. For the time being we might attempt an operational definition of crowding in the following terms: Crowded populations are those in which there seem to be excessive numbers of individuals per unit space in relation to the behavior and activity of the individuals and the quality of their environment. Future research and understanding will certainly modify this tentative definition and sharpen it more satisfactorily.

In human terms, not only is man getting packed into higher density, but he is also getting psychologically crowded at an accelerating rate. Much of the world's population is now directly aware of all other parts of the world community. This awareness may be based on very limited knowledge, often inaccurate knowledge, but this does not necessarily diminish the force of that awareness. It may, in fact, exaggerate it, or make it dangerously inappropriate. We are all, in a sense, spectators at every major critical event. A crisis in any one of the world's cities, or in any of the world's countries, may be witnessed by tens of millions of people around the world. It has often been said that the Vietnamese war is the first war to come visually in full living color into the living room of American homes, and the armed conflicts of the Middle East are vividly familiar to people throughout the world. Similarly, the U.S. urban riots of 1968 and the campus riots of 1970 were viewed and experienced in millions of homes far away from their original location. This may be good or bad, but in any case it spreads the influence of these events around the world, and tremendously alters the impact they might otherwise have.

Thus, I think we could agree that the dramatic multimedia approach of our communications networks affects the sense of crowding and crisis that individuals and social groups perceive. I am suggesting, then, that crowding is becoming one of the predominant social and ecological forces of our times, and that it will most certainly increase in the future. It behooves us to learn as much about it as we can.

The scientific study of crowding has interested biologists, ecologists, and psychologists for many years. Professor W. C. Allee of the University of Chicago was one of the first to emphasize that most animals have optimal levels of crowding, above which and below which biological and behavioral functions are impaired (Allee, 1931). He showed that many aquatic animals, from crustacea to fishes, have enhanced survival and reproduction in social groups at certain densities. At low densities or high densities, survival is often reduced and reproduction impaired. This was found to be attributable to both chemical and behavioral aspects of optimal densities. A certain population level conditioned the environment through metabolites and pheromones for the most favorable physiological function, and a certain frequency of social contact provided the most favorable stimulation for optimal behavioral function. Above or below these optimal levels, both physical and behavioral impairments became appearent. The remainder of this paper will concentrate on consequences of exceeding these levels, that is, the behavioral and physiological consequences of excessive crowding.

### *Effects of Crowding in Animal Populations*

About 20 years ago, there were a number of studies by ecologists [such as John Calhoun (1962a, 1962b), Jack Christian (1961 and, with Davis, 1964), Robert Brown (1953), Dennis Chitty (1960), and others] on populations of rats and mice which demonstrated the occurrence of a variety of events as populations became crowded. Rodent populations, like those of most living organisms, have a great capacity for reproductive increase. Theoretically, one pair of Norway rats can produce 250 progeny in one year, and in the same period, one pair of house mice can produce 4,000 mice if all of their offspring survive and breed (Southwick, 1966). Thus populations of these animals can increase exponentially under favorable circumstances and can achieve conditions of crowding very quickly.

As rodent populations become crowded, a variety of changes occur, both behaviorally and physiologically. Crowding is often accompanied by a breakdown of normal territorial behavior and an upset of dominance hierarchies which were formerly stable. These changes in turn lead to increased social contact and irritation. Aggressive behavior becomes more frequent and more intense, and it often changes from threat display to injurious fighting and violence. Under these circumstances, females are no longer able to maintain good nests, nest construction breaks down, and young are poorly cared for, often trampled, and eventually deserted. Sometimes there is a flurry of unusual activity in which young are carried from nest to nest, some of whom are dropped and abandoned between nests. Occasionally, the young are viciously attacked, bitten, and partially cannibalized. Infant survival falls sharply, and few young reach the age of weaning. Figure 18–1 shows data from experimental populations of

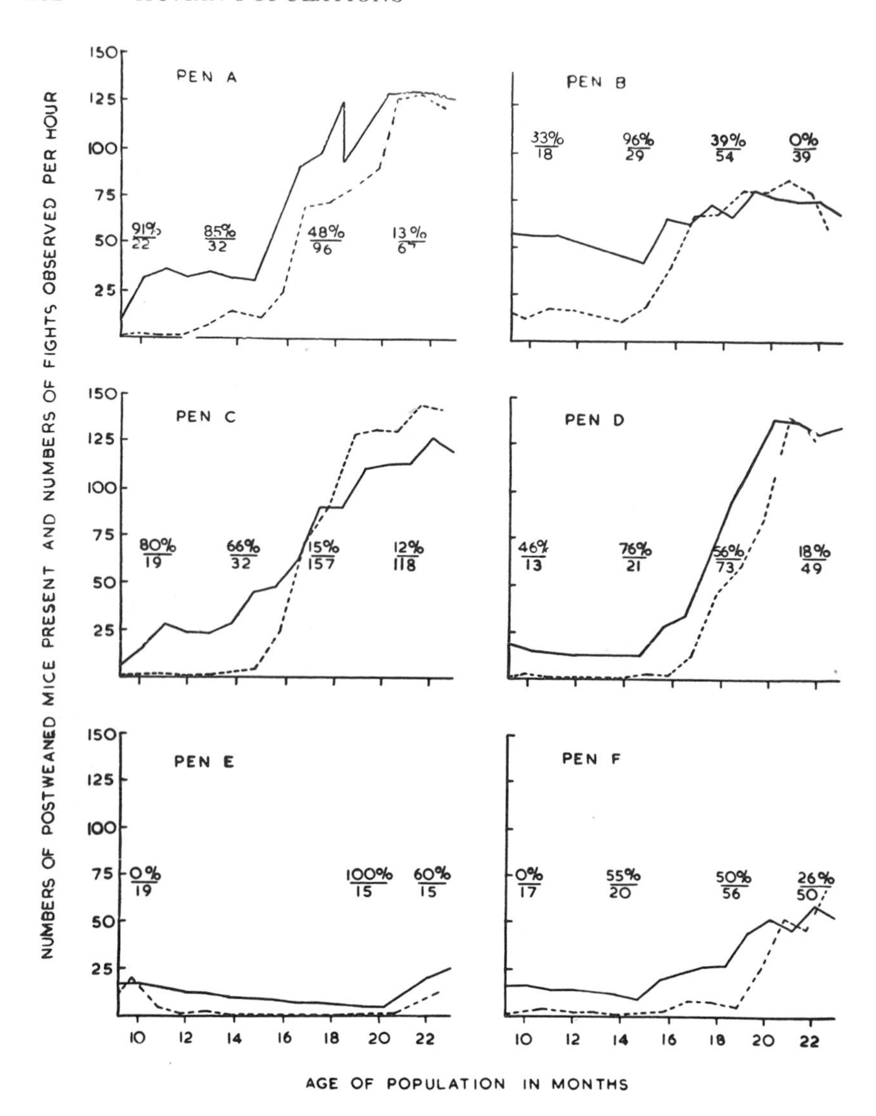


FIGURE 18–1. Population growth, aggressive behavior, and litter survival in seminatural populations of wild house mice, *Mus musculus* (from Southwick, 1955)
solid line—numbers of postweaned mice

dashed line—numbers of fights observed per hour

$\frac{91\%}{22} = \frac{\text{Percent survival to weaning of young born in a two-month period}}{\text{Number of young born in a two-month period}}$

wild house mice (*Mus musculus*) on population growth, the increase of aggression, and the decline of infant survival in crowded populations. Violence, if defined as destructive aggressive behavior, became more prevalent as the populations became crowded, and was expressed in terms of increased aggression between adults and increased cannibalism by adults on infants. As each population attained an aggression level of approximately one observed fight per mouse per hour, population growth ceased, primarily because of the high infant mortality. One population (Pen E) started out in this relatively violent manner and it failed to show population growth for 10 months after the level of aggression had subsided.

The precise density levels at which pathological behavior of this type occurs are highly variable for different populations. These density levels relate to the tolerance of individuals and of groups to crowding. This, in turn, is a function of the behavioral differences of individuals, the history of social groups, and the nature of environment.

It is important to note that crowding does not inevitably lead to increased aggression and violence in animal populations. Beyond a certain point of crowding, some animal populations show unusually passive and nonreactive behavior. In fact, in some very dense rodent populations studied at Columbia University by Kessler and in Washington, D.C. by Calhoun (Calhoun, 1971), there was remarkably little aggression. Highly crowded mice did not respond to social stimuli, and they failed to exhibit even normal irritability and aggressiveness. Not only aggression, but other social behaviors dropped out as well and the entire behavioral repertoire of the mice was reduced to very simple patterns of sleeping and feeding. An important area of future research is to study those circumstances in which crowding stimulates activity and aggression and those in which it inhibits them.

Other behavioral changes which occur in crowded rodent populations often involve abnormal clusterings of individuals. Though the individuals may already be crowded, they seem to seek and to exaggerate this crowding. Thus, they may group together in a few nest sites or at a few feeding spaces by the dozens, even though other sites and spaces are empty. There seems to be a loss of normal individual distances and social spacing characteristic of stable societies. Calhoun (1962 b) has called this "pathological togetherness," or a "behavioral sink" phenomenon. We have no satisfactory explanation for this phenomenon, but it is a conspicuous feature of crowded rodent populations. One theory states that crowded animals become conditioned or habituated to crowding in early life, and so they seek further crowding as adults. Many students of animal behavior have emphasized the importance of primary socialization and early experience in influencing adult behavior (Denenberg, 1962; Levine, 1961; Scott, 1962. 1967). Another theory suggests that, in crowded and disrupted populations, it is increasingly important for members to be able to predict crises and danger, which requires that each individual keep all other members of the population in view. Altmann (1967) has pointed out that the survival of social animals often depends upon instant recognition of, and correct response to, the social signals emanating from other individuals in the group. Thus, in critical social situations it may become necessary to crowd together to increase the probability of receiving important social signals as rapidly as possible. In other words, the individual who is not constantly in touch with the group may miss some essential social cue. It is at least reasonable to expect, therefore, that crowding may stimulate more crowding.

Still other types of deviant behavior found in crowded rodent populations include inappropriate sexual behavior, often misdirected in regard to sex and age, and social withdrawal, in which some individuals avoid contact and fail to display normal activity. Thus, Calhoun's study of wild Norway rat populations (1962 a) showed deviant sexual behavior by some individuals which he called "pansexuals." Their sexual behavior was indiscriminate in regard to the sex and age of the other individuals which they approached and mounted. He also observed the withdrawn individual or "social dropout"—one who entered a state of inactivity and depression and went into a spiral of deteriorating health.

A number of recent studies of nonhuman primates, primarily monkeys, have shown crowding to be a potent force in shaping social behavior. The common langur (*Presbytis entellus*) is a leaf-eating monkey of India which lives in peaceful social groups and shows a remarkably tranquil and well integrated social life when population density is low (in the order of 7 to 16 individuals per square mile). In contrast, this same species shows violent aggression and a high degree of social disruption under high population densities (in the order of 220 to 350 individuals per square mile) (Yoshiba, 1968). Factors other than density may have been operating to produce these differences, but certainly density was one of the most important. Rhesus monkeys and baboons provide other cases in which crowding aggravates aggression and social instability in both captive and natural populations (DeVore and Hall, 1965; Southwick, 1969).

As in rodents, however, primate population density must be related to environmental quality. Studies of vervet monkeys (*Cercopithecus aethiops*) in Africa have shown less aggression in dense vervet populations (230 per square mile) in an area of a rich, high quality environment, than in sparse populations (less than 50 per square mile) in a poor and deteriorated environment (Gartlan and Brain, 1968). Hence, the effects of density and crowding must always be considered in relation to environmental resources.

The physical consequences of crowding and of social stress are often profound and widespread. They clearly affect physical health and well being. Increased aggression often leads to wounding, and the wounds in rats and mice can become sites of infection of bacteria or ectoparasites.

Acarine dermatitis (a skin disease caused by a small mite) is a common occurrence in rodent populations which show high levels of fighting. Christian and Davis (1964) further believe that increased social contact and fighting cause endocrine changes in individuals through the mechanism of the Selye Stress Syndrome, often known as the General Adaptation Syndrome, or G.A.S.

This theory states that any general stress agent, physical or behavioral, alters the function of the hypothalamus, a portion of the midbrain, and thereby changes the secretory activity of the pituitary gland, the master gland of the body located at the base of the mid-brain. As a result, there is decreased production of the gonadotropic hormones, FSH, LH, and LTH, hormones which maintain proper functioning of the reproductive system, and there is increased production of ACTH (adrenocorticotropic hormone), which stimulates the adrenal glands to increased production of the adrenocorticosteroids. This enables the animal to maintain homeostatic conditions under chronic stress; blood-glucose levels and sodium and potassium balances are kept within normal limits, and other physiological demands are met. There is, however, a cost. The altered endocrine balances extract a cost in metabolic changes, some of which lead to what Selye (1950, 1955) calls the "Diseases of Adaptation." These involve adrenal enlargement or hypertrophy, lymphatic involution, declines in eosinophils and lymphocytes, hypertension or high blood pressure, increased susceptibility to infectious disease, and, in more severe forms, gastro-intestinal ulcers, sclerotic changes in the kidneys, and arteriosclerosis.

If the stress continues, it may lead to adrenal exhaustion, chronic shock, Addison's disease, myasthenia, and eventually coma and death. Thus, long-term stresses may be cumulatively fatal, though no single episode is lethal in itself.

All of these pathologic changes have been documented in experimental animals, and Christian and Davis (1964) believe that they can occur in natural populations of wild animals as well. This is a controversial point, and the evidence for this is mixed. In some natural populations, pathologies of the stress syndrome have been shown to occur; in other populations, often very crowded with intense aggression, however, such physical changes have not been found. This certainly does not refute the validity of the concepts, though it may refute their universal application. It is also not too surprising—anyone who works with biological systems understands that variability is an intrinsic part of these systems. If you do precisely the same thing to two different mice, you do not invariably get the same result. If two different populations are subjected to the same stresses, they do not necessarily respond in the same way. Evidence relating to this point and to other aspects of the social stress theory has been reviewed by Barnett (1964) and by Myers (1965).

### *Crowding in Human Populations*

Having seen this impressive array of behavioral and physiological upsets which can occur in crowded animal populations, the leading question is, of course, whether similar processes occur in man. There is no doubt about the fact that human populations around the world are drastically out of balance, and in a dangerously unstable condition. Not only is there excessive population growth in sheer numbers, but these populations are becoming dramatically concentrated in and around major cities. In the United States in 1800, only 5% of the population lived in towns containing more than 2,500 people. In 1960, over 65% lived in towns of that size, and now 73% live in cities of over 100,000 people. By the year 2000, it is predicted that 80 to 90% of our populace will be living in urban areas. In construction and urban development, we are devouring 4,000 acres per day in the United States alone.

The story is similar throughout the world. Calcutta developed from less than 850,000 people in 1901 to 6.5 million by 1965. Los Angeles grew from 102,000 in 1900 to over 3 million in 1965. Mexico City tripled in size in one generation, from 1.4 million in 1940 to 5.2 million in 1965. These rapid growth rates have involved qualitative changes as well; cities throughout the world are on the verge of financial collapse as their wealthy tax base moves outward and poverty-stricken components of the population pour inward.

Obviously, these patterns, in both quantitative and qualitative terms, cannot continue. They must reach more stable conditions at some point, or experience many of the behavioral and physical pathologies which have been found so frequently in animal populations.

Or, one might ask, are they already exhibiting some of these characteristics? There seems to be substantial evidence that optimal size in human terms has already been exceeded in many cities. There is little doubt that our inner cities have become concentrations of both physical disease and behavioral pathology. The realities of ghetto life, with inadequate housing, poor sanitation, severe crowding, ill health, high crime rates, and a breakdown of social services, force us to realize that our giant cities have already become a "behavioral sink" of human despair.

Yet people continue to flock toward the cities of the world by the millions as if irresistibly drawn. Obviously urban centers have an attraction and fascination for many people that rural areas cannot provide. It is doubtful, however, if the main driving force of urban expansion lies in the cultural excitement of cities. Rather, it is more likely to be found in the realities of modern economics: the prospect of a job or a welfare payment. As such, the urban movement may be very close to the struggle for survival, and thus far more of an ecologic phenomenon than we realize. Even in Calcutta, where the resident population density already exceeds 75,000 people per square mile and where there are hundreds of thousands of homeless people barely clinging to existence, people continue to pour into the city. They come in hopes of a job, a handout, or a better life. Although they may be victims of mistaken perception and misplaced hopes, they at least feel that they can find food and some form of collective security within the city. The disillusionment is often massive and profound.

A recent article by the psychologist Stanley Milgram (1970) pointed out that a resident of suburban Nassau county, outside of New York City, can meet a potential of 11,000 other people within a 10-minute radius of his office. In Newark, New Jersey, a moderate-sized city, he can meet 20,000 people within this radius, and in midtown Manhattan, he can meet fully 220,000. Thus the inner-city dweller has a potentially vast and oppressive number of social contacts with which to deal. City life, Dr. Milgram points out, constitutes a continuous set of encounters with with sensory overload. As a result, the sensory input becomes unmanageable, unless some adaptation arises to cope with this overload. City dwellers characteristically have several common ways of doing this. One adaptive response is to ignore and filter out many of the inputs. Thus the city dweller disregards many social opportunities. He obviously does not greet or acknowledge each person he passes on the street; he does not become involved in arguments; he may even ignore crises, or may stand idly by when some other person is in danger or distress. He increasingly uses unlisted telephone numbers; he limits his availability and contributions to social organizations. Such patterns of behavior protect the individual from undue demands and responsibilities, but they also estrange the individual from his social environment. It is no coincidence that most social surveys in cities have found urbanites to be more suspicious of and less helpful to strangers, to express a greater sense of vulnerability and insecurity, to become less readily involved in social-betterment projects, and to have less sense of social community.

These are all reasonable adaptations to the social and psychological overload conditions of crowded cities. They permit individuals to maintain themselves in socially stressful circumstances. But it is obvious that even these adaptations are not enough.

The records of inner cities in terms of physical and mental health are not good. Thirty years ago Faris and Dunham (1939) showed that, in Chicago, unusually high concentrations of schizophrenia occurred in the crowded populations around the loop of central Chicago. In the famous midtown Manhattan study of the 1950's, entitled "Mental Health in the Metropolis," Dr. Thomas Rennie and his colleagues showed that

80 percent of the people interviewed had detectable psychiatric disorders, and 25 percent of them had significant neuroses which made them indistinguishable from patients in mental hospitals (Srole, et al. 1962). More recently Myers and Bean (1968) showed a high prevalence of schizophrenia in the lower socioeconomic classes of several urban ghettos.

The extensive ecological studies on Philadelphia by Dr. Ian McHarg (1969) and his colleagues at the University of Pennsylvania have also shown the increased prevalence of behavioral pathologies in the inner-city environment. Not only mental illness, but various patterns of criminal assault, murder, rape, alcoholism, and drug abuse are also intensified in the inner-city ghettos. The prevalences of these are shown in a series of outline maps in McHarg's recent book on human ecology, "Design with Nature."

Many aspects of physical disease show similar concentrations in the inner city: tuberculosis, venereal disease, salmonellosis, dysentery, hypertension, and perinatal loss and infant mortality. Some of these are infectious diseases, clearly related to poor housing and unsanitary conditions; others, such as hypertension, are partially psychogenic in origin, and relate directly to the frequency and quality of social contacts.

It is, of course, not warranted to attribute all of these problems to crowding. There are so many variables in man which complicate inner-city disease patterns, such as socioeconomic status, vocation, malnutrition, poverty, race, education, family background, and the various environmental insults of pollution, that it is impossible to isolate crowding per se and to say that it alone is the cause of physical and behavioral pathology in man. The complexity of the problem has caused many scientists to despair of really understanding and sorting out the key variables. It is, in fact, a logical fallacy to seek a simple cause and effect in dealing with so complicated an issue. But this need not mask the essential significance of the syndrome—the reality of a package of related problems or a constellation of symptoms which seem to occur together in the recognizable form of what we might call the "inner-city syndrome."

The great challenge of our cities lies first in recognizing the interrelatedness of these problems, and secondly, in being able to mount some corrective attack on the predisposing factors. Thus, to the ecologist, planning for the future must direct itself to the basic issues of population and urbanization. The social, spatial, and behavioral needs of man must be more adequately understood and considered in urban planning. The quality of human life must be expressed in terms of all three components so often recognized by the ecologist, but readily forgotten in the pressure of modern economic growth; namely, that the physical, biological, and social environments of man compose an intensely interacting system structured in such a way that changes in any one component have major consequences for all.

## Bibliography


Allee, W. C. 1931. *Animal Aggregations, A Study in General Sociology*. University of Chicago Press. 431 p.

Altmann, S. A. 1967. *Social Communication Among Primates*. University of Chicago Press. 392 p.

Barnett, S. A. 1964. "Social stress." In Carthy, J. D., and C. L. Duddington, ed., *Viewpoints in Biology* (Chap. 5). Butterworths: Kingsway, London, pp. 170–218.

Brown, R. Z. 1953. Social behavior, reproduction, and population changes in the house mouse (*Mus musculus* L.). *Ecol. Mongr.* 23:217–240.

Calhoun, J. B. 1962 a. Population density and social pathology. *Sci. Amer.* 206:139–148.

———. 1962 b. "A behavioral sink." In Bliss, E. L., ed., *Roots of Behavior*. New York: Harper. p. 295–315.

———. 1971. Space and the strategy of life. In Esser, A. H., ed., *Behavior and Environment. The Use of Space by Animals and Men*. New York: Plenum Press, pp. 329–387.

Chitty, D. 1960. Population processes in the vole and their relevance to general theory. *Canad. J. Zool.* 28:99–113.

Christian, J. J. 1961. Phenomena associated with population density. *Proc. Nat. Acad. Sci.* 47:428–449.

Christian, J. J., and D. E. Davis. 1964. Endocrines, behavior and population. *Science* 146:1550–1560.

Denenberg, V. 1962. The effects of early experience. In Hafez, E. S. E., ed., *The Behavior of Domestic Animals* (Chap. 6). London: Balliere, Tindall and Cox. pp. 109–138.

Devore, I., and K. R. L. Hall. 1965. "Baboon ecology." In Devore, I., ed., *Primate Behavior: Field Studies of Monkeys and Apes*. New York: Holt, Rinehart and Winston, Inc., 20–52.

Faris, R. E. L., and W. H. Dunham. 1939. *Mental Disorders in Urban Areas: An Ecological Study of Schizophrenia and Other Psychoses*. University of Chicago Press. 270 p.

Gartlan, J. S., and C. K. Brain. 1968. Ecology and social variability in *Cercopithecus aethiops* and *C. mitis*. *In* Jay, P. ed., *Primates: Studies in Adaptation and Variability*. New York: Holt, Rinehart and Winston, pp. 253–292.

Levine, S. 1961. The psychophysiological effects of early stimulation. *In* Bliss, E., ed., *Roots of Behavior*. New York: Harper. pp. 246–253.

McHarg, Ian. 1969. *Design With Nature*. Garden City, New York: The Natural History Press. 197 p.

Milgram, S. 1970. The experience of living in cities. *Science* 167:1461–1468.

Myers, K. 1965. The effects of density on sociality and health in mammals. *Proc. Ecol. Soc. Australia*, 1:40–64.

Myers, J. K., and L. L. Bean. 1968. *A Decade Later: A Follow-up of Social Class and Mental Illness*. New York: John Wiley. 250 p.

Scott, J. P. 1962. Critical periods in behavioral development. *Science* 138:949–958.

———. 1967. "The development of social motivation." In *Nebraska Symposium on Motivation*, University of Nebraska. pp. 111–132.

Selye, H. 1950. *The physiology and pathology of exposure to Stress: a treatise based on the General Adaptation Syndrome and the diseases of adaptation*. Montreal: Acta Inc. 822 p.

———. 1955. Stress and disease. *Science* 122:625–631.

Southwick, C. H. 1955. Regulatory mechanisms of house mouse populations: social behavior affecting litter survival. *Ecology* 36:627–634.

———. 1966. Population characteristics of Murid Rodents. In Parrack, D. W., ed., *Indian Rodent Symposium*. Calcutta. pp. 152–176.

———. 1969. Population dynamics and social behavior of domestic rodents. In Sladen, B. K., and F. B. Bang, eds., *Biology of Populations* (Chap. 20). New York: American Elsevier. p. 284–298.

———. 1969. Aggressive Behaviour of rhesus monkeys in natural and captive groups. In Garattini, S., and E. B. Sigg, eds., *Aggressive Behaviour*. Amsterdam: Excerpta Medica, pp. 32–43.

Srole, L., T. S. Langner, S. T. Michael, M. K. Opler, and T. A. C. Rennie. 1962. *Mental Health in the Metropolis*. New York: McGraw Hill. Vol. 1. 428 p.

Yoshiba, K. 1968. Local and intertroop variability in ecology and social behavior of common Indian langurs. In Jay, P., ed., *Primates: Studies in Adaptation and Variability*. New York: Holt, Rinehart and Winston, pp. 217–252.

# 19

# Population policy: will current programs succeed?

*Kingsley Davis*

Throughout history the growth of population has been identified with prosperity and strength. If today an increasing number of nations are seeking to curb rapid population growth by reducing their birth rates, they must be driven to do so by an urgent crisis. My purpose here is not to discuss the crisis itself but rather to assess the present and prospective measures used to meet it. Most observers are surprised by the swiftness with which concern over the population problem has turned from intellectual analysis and debate to policy and action. Such action is a welcome relief from the long opposition, or timidity, which seemed to block forever any governmental attempt to restrain population growth: but relief that "at last something is being done" is no guarantee that what is being done is adequate. On the face of it, one could hardly expect such a fundamental reorientation to be quickly and successfully implemented. I therefore propose to review the nature and (as I see them) limitations of the present policies and to suggest lines of possible improvement.

---

Reprinted with permission from *Science*, 158:730–739 (10 November 1967). Copyright 1967 by the American Association for the Advancement of Science. The author is professor of sociology and director of International Population and Urban Research, University of California, Berkeley.

## The Nature of Current Policies

With more than 30 nations now trying or planning to reduce population growth and with numerous private and international organizations helping, the degree of unanimity as to the kind of measures needed is impressive. The consensus can be summed up in the phrase "family planning." President Johnson declared in 1965 that the United States will "assist family planning programs in nations which request such help." The Prime Minister of India said a year later, "We must press forward with family planning. This is a programme of the highest importance." The Republic of Singapore created in 1966 the Singapore Family Planning and Population Board "to initiate and undertake population control programmes" (*1*).

As is well known, "family planning" is a euphemism for contraception. The family-planning approach to population limitation, therefore, concentrates on providing new and efficient contraceptives on a national basis through mass programs under public health auspices. The nature of these programs is shown by the following enthusiastic report from the Population Council (*2*):

> No single year has seen so many forward steps in population control as 1965. Effective national programs have at last emerged, international organizations have decided to become engaged, a new contraceptive has proved its value in mass application, . . . and surveys have confirmed a popular desire for family limitation . . .
>
> An accounting of notable events must begin with Korea and Taiwan . . . Taiwan's program is not yet two years old, and already it has inserted one IUD [intrauterine device] for every 4–6 target women (those who are not pregnant, lactating, already sterile, already using contraceptives effectively, or desirous of more children). Korea has done almost as well . . . has put 2,200 full-time workers into the field, . . . has reached operational levels for a network of IUD quotas, supply lines, local manufacture of contraceptives, training of hundreds of M.D.'s and nurses, and mass propaganda . . .

Here one can see the implication that "population control" is being achieved through the dissemination of new contraceptives, and the fact that the "target women" exclude those who want more children. One can also note the technological emphasis and the medical orientation.

What is wrong with such programs? The answer is, "Nothing at all, if they work." Whether or not they work depends on what they are expected to do as well as on how they try to do it. Let us discuss the goal first, then the means.

## Goals

Curiously, it is hard to find in the population-policy movement any explicit discussion of long-range goals. By implication the policies seem to promise a great deal. This is shown by the use of expressions like *population control* and *population planning* (as in the passages quoted above). It is also shown by the characteristic style of reasoning. Expositions of current policy usually start off by lamenting the speed and the consequences of runaway population growth. This growth, it is then stated, must be curbed—by pursuing a vigorous family-planning program. That family planning can solve the problem of population growth seems to be taken as self-evident.

For instance, the much-heralded statement by 12 heads of state, issued by Secretary-General U Thant on 10 December 1966 (a statement initiated by John D. Rockefeller III, Chairman of the Board of the Population Council), devotes half its space to discussing the harmfulness of population growth and the other half to recommending family planning (*3*). A more succinct example of the typical reasoning is given in the Provisional Scheme for a Nationwide Family Planning Programme in Ceylon (*4*):

> The population of Ceylon is fast increasing.... [The] figures reveal that a serious situation will be created within a few years. In order to cope with it a Family Planning programme on a nationwide scale should be launched by the Government.

The promised goal—to limit population growth so as to solve population problems—is a large order. One would expect it to be carefully analyzed, but it is left imprecise and taken for granted, as is the way in which family planning will achieve it.

When the terms *population control* and *population planning* are used, as they frequently are, as synonyms for current family-planning programs, they are misleading. Technically, they would mean deliberate influence over all attributes of a population, including its age-sex structure, geographical distribution, racial composition, genetic quality, and total size. No government attempts such full control. By tacit understanding, current population policies are concerned with only the *growth* and *size* of populations. These attributes, however, result from the death rate and migration as well as from the birth rate; their control would require deliberate influence over the factors giving rise to all three determinants. Actually, current policies labeled population control do not deal with mortality and migration, but deal only with the birth input. This is why another term, *fertility control*, is frequently used to describe current policies. But, as I show below, family planning (and hence current policy) does not undertake to influence most of the determinants of human reproduction. Thus the programs should not be referred to as population control or planning, because they do not attempt to influence the factors responsible for the attributes of human populations, taken generally; nor should they be called fertility control, because they do not try to affect most of the determinants of reproductive performance.

The ambiguity does not stop here, however. When one speaks of controlling population size, any inquiring person naturally asks, What is "control"? Who is to control whom? Precisely what population size, or what rate of population growth, is to be achieved? Do the policies aim to produce a growth rate that is nil, one that is very slight, or one that is like that of the industrial nations? Unless such questions are dealt with and clarified, it is impossible to evaluate current population policies.

The actual programs seem to be aiming simply to achieve a reduction in the birth rate. Success is therefore interpreted as the accomplishment of such a reduction, on the assumption that the reduction will lessen population growth. In those rare cases where a specific demographic aim is stated, the goal is said to be a short-run decline within a given period. The Pakistan plan adopted in 1966 (*5*, p. 889) aims to reduce the birth rate from 50 to 40 per thousand by 1970; the Indian plan (*6*) aims to reduce the rate from 40 to 25 "as soon as possible"; and the Korean aim (*7*) is to cut population growth from 2.9 to 1.2 percent by 1980. A significant feature of such stated aims is the rapid population growth they would permit. Under conditions of modern mortality, a crude birth rate of 25 to 30 per thousand will represent such a multiplication of people as to make use of the term *population control* ironic. A rate of increase of 1.2 percent per year would allow South Korea's already dense population to double in less than 60 years.

One can of course defend the programs by saying that the present goals and measures are merely interim ones. A start must be made somewhere. But we do not find this answer in the population-policy literature. Such a defense, if convincing, would require a presentation of the *next* steps, and these are not considered. One suspects that the entire question of goals is instinctively left vague because thorough limitation of population growth would run counter to national and group aspirations. A consideration of hypothetical goals throws further light on the matter.

### *Industrialized Nations as the Model*

Since current policies are confined to family planning, their maximum demographic effect would be to give the underdeveloped countries the same level of reproductive performance that the industrial nations now have. The latter, long oriented toward family planning, provide a good yardstick for determining what the availability of contraceptives can do to population growth. Indeed, they provide more than a yardstick; they are actually the model which inspired the present population policies.

What does this goal mean in practice? Among the advanced nations there is considerable diversity in the level of fertility (*8*). At one extreme are countries such as New Zealand, with an average gross reproduction rate (GRR) of 1.91 during the period 1960–64; at the other extreme are countries such as Hungary, with a rate of 0.91 during the same period. To a considerable extent, however, such divergencies are matters of timing. The birth rates of most industrial nations have shown, since about 1940, a wavelike movement, with no secular trend. The average level of reproduction during this long period has been high enough to give these countries, with their low mortality, an extremely rapid population growth. If this level is maintained, their population will double in just over 50 years—a rate higher than that of world population growth at any time prior to 1950, at which time the growth in numbers of human beings was already considered fantastic. The advanced nations are suffering acutely from the effects of rapid population growth in combination with the production of ever more goods per person (*9*). A rising share of their supposedly high per capita income, which itself draws increasingly upon the resources of the underdeveloped countries (who fall farther behind in relative economic position), is spent simply to meet the costs, and alleviate the nuisances, of the unrelenting production of more and more goods by more people. Such facts indicate that the industrial nations provide neither a suitable demographic model for the nonindustrial peoples to follow nor the leadership to plan and organize effective population-control policies for them.

### Zero Population Growth as a Goal

Most discussions of the population crisis lead logically to zero population growth as the ultimate goal, because *any* growth rate, if continued, will eventually use up the earth. Yet hardly ever do arguments for population policy consider such a goal, and current policies do not dream of it. Why not? The answer is evidently that zero population growth is unacceptable to most nations and to most religious and ethnic communities. To argue for this goal would be to alienate possible support for action programs.

### Goal Peculiarities Inherent in Family Planning

Turning to the actual measures taken, we see that the very use of family planning as the means for implementing population policy poses serious but unacknowledged limits on the intended reduction in fertility. The family-planning movement, clearly devoted to the improvement and dissemination of contraceptive devices, states again and again that its purpose is that of enabling couples to have the number of children they want. "The opportunity to decide the number and spacing of children is a basic human right," say the 12 heads of state in the United Nations declaration. The 1965 Turkish Law Concerning Population Planning declares (*10*):

*Article 1.* Population Planning means that individuals can have as many children as they wish, whenever they want to. This can be ensured through preventive measures taken against pregnancy. . . .

Logically, it does not make sense to use *family* planning to provide *national* population control or planning. The "planning" in family planning is that of each separate couple. The only control they exercise is control over the size of *their* family. Obviously, couples do not plan the size of the nation's population, any more than they plan the growth of the national income or the form of the highway network. There is no reason to expect that the millions of decisions about family size made by couples in their own interest will automatically control population for the benefit of society. On the contrary, there are good reasons to think they will not do so. At most, family planning can reduce reproduction to the extent that unwanted births exceed wanted births. In industrial countries the balance is often negative—that is, people have fewer children as a rule than they would like to have. In underdeveloped countries the reverse is normally true, but the elimination of unwanted births would still leave an extremely high rate of multiplication.

Actually, the family-planning movement does not pursue even the limited goals it professes. It does not fully empower couples to have only the number of offspring they want because it either condemns or disregards certain tabooed but nevertheless effective means to this goal. One of its tenets is that "there shall be freedom of choice of method so that individuals can choose in accordance with the dictates of their consciences" (*11*), but in practice this amounts to limiting the individual's choice, because the "conscience" dictating the method is usually not his but that of religious and governmental officials. Moreover, not every individual may choose: even the so-called recommended methods are ordinarily not offered to single women, or not all offered to women professing a given religious faith.

Thus, despite its emphasis on technology, current policy does not utilize all available means of contraception, much less all birth-control measures. The Indian government wasted valuable years in the early stages of its population-control program by experimenting exclusively with the "rhythm" method, long after this technique had been demonstrated to be one of the least effective. A greater limitation on means is the exclusive emphasis on contraception itself. Induced abortion, for example, is one of the surest means of controlling reproduction, and one that has been proved capable of reducing birth rates rapidly. It seems peculiarly suited to the threshold stage of a population-control program—the stage when new conditions of life first make large families disadvantageous. It was the principal factor in the halving of the Japanese birth rate, a major factor in the declines in birth rate of East-European satellite countries after legalization of abortions in the early 1950's, and an important factor in the reduction of fertility in industrializing nations from 1870 to the 1930's (*12*). Today, according to *Studies in Family Planning* (*13*), "abortion is probably the foremost method of birth control throughout Latin America." Yet this method is rejected in nearly all national and international population-control programs. American foreign aid is used to help *stop* abortion (*14*). The United Nations excludes abortion from family planning, and in fact justifies the latter by presenting it as a means of combating abortion (*15*). Studies of abortion are being made in Latin America under the presumed auspices of population-control groups, not with the intention of legalizing it and thus making it safe, cheap, available, and hence more effective for population control, but with the avowed purpose of reducing it (*16*).

Although few would prefer abortion to efficient contraception (other things being equal), the fact is that both permit a woman to control the size of her family. The main drawbacks to abortion arise from its illegality. When performed, as a legal procedure, by a skilled physician, it is safer than childbirth. It does not compete with contraception but serves as a backstop when the latter fails or when contraceptive devices or information are not available. As contraception becomes customary, the incidence of abortion recedes even without its being banned. If, therefore, abortions enable women to have only the number of children they want, and if family planners do not advocate—in fact decry—legalization of abortion, they are to that extent denying the central tenet of their own movement. The irony of antiabortionism in family-planning circles is seen particularly in hair-splitting arguments over whether or not some contraceptive agent (for example, the IUD) is in reality an abortifacient. A Mexican leader in family planning writes (*17*):

> One of the chief objectives of our program in Mexico is to prevent abortions. If we could be sure that the mode of action [of the IUD] was not interference with nidation, we could easily use the method in Mexico.

The questions of sterilization and unnatural forms of sexual intercourse usually meet with similar silent treatment or disapproval, although nobody doubts the effectiveness of these measures in avoiding conception. Sterilization has proved popular in Puerto Rico and has had some vogue in India (where the new health minister hopes to make it compulsory for those with a certain number of children), but in both these areas it has been for the most part ignored or condemned by the family-planning movement.

On the side of goals, then, we see that a family-planning orientation limits the aims of current population policy. Despite reference to "population control" and "fertility control," which presumably mean determination of demographic results by and for the nation as a whole, the movement gives control only to couples, and does this only if they use "respectable" contraceptives.

## The Neglect of Motivation

By sanctifying the doctrine that each woman should have the number of children she wants, and by assuming that if she has only that number this will automatically curb population growth to the necessary degree, the leaders of current policies escape the necessity of asking why women desire so many children and how this desire can be influenced (*18*, p. 41; *19*). Instead, they claim that satisfactory motivation is shown by the popular desire (shown by opinion surveys in all countries) to have the means of family limitation, and that therefore the problem is one of inventing and distributing the best possible contraceptive devices. Overlooked is the fact that desire for availability of contraceptives is compatible with *high* fertility.

Given the best of means, there remain the questions of how many children couples want and of whether this is the requisite number from the standpoint of population size. That it is not is indicated by continued rapid population growth in industrial countries, and by the very surveys showing that people want contraception—for these show, too, that people also want numerous children.

The family planners do not ignore motivation. They are forever talking about "attitudes" and "needs." But they pose the issue in terms of the "acceptance" of birth control devices. At the most naive level, they assume that lack of acceptance is a function of the contraceptive device itself. This reduces the motive problem to a technological question. The task of population control then becomes simply the invention of a device that *will* be acceptable (*20*). The plastic IUD is acclaimed because, once in place, it does not depend on repeated *acceptance* by the woman, and thus it "solves" the problem of motivation (21).

But suppose a woman does not want to use *any* contraceptive until after she has had four children. This is the type of question that is seldom raised in the family-planning literature. In that literature, wanting a specific number of children is taken as complete motivation, for it implies a wish to control the size of one's family. The problem woman, from the standpoint of family planners, is the one who wants "as many as come," or "as many as God sends." Her attitude is construed as due to ignorance and "cultural values," and the policy deemed necessary to change it is "education." No compulsion can be used, because the movement is committed to free choice, but movie strips, posters, comic books, public lectures, interviews, and discussions are in order. These supply information and supposedly change values by discounting superstitions and showing that unrestrained procreation is harmful to both mother and children. The effort is considered successful when the woman decides she wants only a certain number of children and uses an effective contraceptive.

In viewing negative attitudes toward birth control as due to ignorance, apathy, and outworn tradition, and "mass-communication" as the solution to the motivation problem (*22*), family planners tend to ignore the power and complexity of social life. If it were admitted that the creation and care of new human beings is socially motivated, like other forms of behavior, by being a part of the system of rewards and punishments that is built into human relationships, and thus is bound up with the individual's economic and personal interests, it would be apparent that the social structure and economy must be changed before a deliberate reduction in the birth rate can be achieved. As it is, reliance on family planning allows people to feel that "something is being done about the population problem" without the need for painful social changes.

Designation of population control as a medical or public health task leads to a similar evasion. This categorization assures popular support because it puts population policy in the hands of respected medical personnel, but, by the same token, it gives responsibility for leadership to people who think in terms of clinics and patients, of pills and IUD's, and who bring to the handling of economic and social phenomena a self-confident naiveté. The study of social organization is a technical field; an action program based on intuition is no more apt to succeed in the control of human beings than it is in the area of bacterial or viral control. Moreover, to alter a social system, by deliberate policy, so as to regulate births in accord with the demands of the collective welfare would require political power, and this is not likely to inhere in public health officials, nurses, midwives, and social workers. To entrust population policy to them is "to take action," but not dangerous "effective action."

Similarly, the Janus-faced position on birth control technology represents an escape from the necessity, and onus, of grappling with the social and economic determinants of reproductive behavior. On the one side, the rejection or avoidance of religiously tabooed but otherwise effective means of birth prevention enables the family-planning movement to avoid official condemnation. On the other side, an intense preoccupation with contraceptive technology (apart from the tabooed means) also helps the family planners to avoid censure. By implying that the only need is the invention and distribution of effective contraceptive devices, they allay fears, on the part of religious and governmental officials, that fundamental changes in social organization are contemplated. Changes basic enough to affect motivation for having children would be changes in the structure of the family, in the position of women, and in the sexual mores. Far from proposing such radicalism, spokesmen for family planning frequently state their purpose as "protection" of the family—that is, closer observance of family norms. In addition, by concentrating on *new* and *scientific* contraceptives, the movement escapes taboos attached to old ones (the Pope will hardly authorize the condom, but may sanction the pill) and allows family planning to be regarded as a branch of medicine: overpopulation becomes a disease, to be treated by a pill or a coil.

We thus see that the inadequacy of current population policies with respect to motivation is inherent in their overwhelmingly family-planning character. Since family planning is by definition private planning, it eschews any societal control over motivation. It merely furnishes the means, and, among possible means, only the most respectable. Its leaders, in avoiding social complexities and seeking official favor, are obviously activated not solely by expediency but also by their own sentiments as members of society and by their background as persons attracted to the family-planning movement. Unacquainted for the most part with technical economics, sociology, and demography, they tend honestly and instinctively to believe that something they vaguely call population control can be achieved by making better contraceptives available.

## The Evidence of Ineffectiveness

If this characterization is accurate, we can conclude that current programs will not enable a government to control population size. In countries where couples have numerous offspring that they do not want, such programs may possibly accelerate a birth-rate decline that would occur anyway, but the conditions that cause births to be wanted or unwanted are beyond the control of family planning, hence beyond the control of any nation which relies on family planning alone as its population policy.

This conclusion is confirmed by demographic facts. As I have noted above, the widespread use of family planning in industrial countries has not given their governments control over the birth rate. In backward countries today, taken as a whole, birth rates are rising, not falling; in those with population policies, there is no indication that the government is controlling the rate of reproduction. The main "successes" cited in the well-publicized policy literature are cases where a large number of contraceptives have been distributed or where the program has been accompanied by some decline in the birth rate. Popular enthusiasm for family planning is found mainly in the cities, or in advanced countries such as Japan and Taiwan, where the people would adopt contraception in any case, program or no program. It is difficult to prove that present population policies have even speeded up a lowering of the birth rate (the least that could have been expected), much less that they have provided national "fertility control."

Let us next briefly review the facts concerning the level and trend of population in underdeveloped nations generally, in order to understand the magnitude of the task of genuine control.

## Rising Birth Rates in Underdeveloped Countries

In ten Latin-American countries, between 1940 and 1959 (*23*), the average birth rates (age-standardized), as estimated by our research office at the University of California, rose as follows: 1940–44, 43.4 annual births per 1000 population; 1945–49, 44.6; 1950–54, 46.4; 1955–59, 47.7.

In another study made in our office, in which estimating methods derived from the theory of quasi-stable populations were used, the recent trend was found to be upward in 27 underdeveloped countries, downward in six, and unchanged in one (*24*). Some of the rises have been substantial, and most have occurred where the birth rate was already extremely high. For instance, the gross reproduction rate rose in Jamaica from 1.8 per thousand in 1947 to 2.7 in 1960; among the natives of Fiji, from 2.0 in 1951 to 2.4 in 1964; and in Albania, from 3.0 in the period 1950–54 to 3.4 in 1960.

The general rise in fertility in backward regions is evidently not due to failure of population-control efforts, because most of the countries either have no such effort or have programs too new to show much effect. Instead, the rise is due, ironically, to the very circumstance that brought on the population crisis in the first place—to improved health and lowered mortality. Better health increases the probability that a woman will conceive and retain the fetus to term; lowered mortality raises the proportion of babies who survive to the age of reproduction and reduces the probability of widowhood during that age (*25*). The significance of the general rise in fertility, in the context of this discussion, is that it is giving would-be population planners a harder task than many of them realize. Some of the upward pressure on birth rates is independent of what couples do about family planning, for it arises from the fact that, with lowered mortality, there are simply more couples.

## Underdeveloped Countries with Population Policies

In discussions of population policy there is often confusion as to which cases are relevant. Japan, for instance, has been widely praised for the effectiveness of its measures, but it is a very advanced industrial nation and, besides, its government policy had little or nothing to do with the decline in the birth rate, except unintentionally. It therefore offers no test of population policy under peasant-agrarian conditions. Another case of questionable relevance is that of Taiwan, because Taiwan is sufficiently developed to be placed in the urban-industrial class of nations. However, since Taiwan is offered as the main showpiece by the sponsors of current policies in underdeveloped areas, and since the data are excellent, it merits examination.

Taiwan is acclaimed as a showpiece because it has responded favorably to a highly organized program for distributing up-to-date contraceptives and has also had a rapidly dropping birth rate. Some observers have carelessly attributed the decline in the birth rate—from 50.0 in 1951 to 32.7 in 1965—to the family-planning campaign (*26*), but the campaign began only in 1963 and could have affected only the end of the trend. Rather, the decline represents a response to modernization similar to that made by all countries that have become industrialized (*27*). By 1950 over half of Taiwan's population was urban, and by 1964 nearly two-thirds were urban, with 29 percent of the population living in cities of 100,000 or more. The pace of economic development has been extremely rapid. Between 1951 and 1963, per capita income increased by 4.05 percent per year. Yet the island is closely packed, having 870 persons per square mile (a population density higher than that of Belgium). The combination of fast economic growth and rapid population increase in limited space has put parents of large families at a relative disadvantage and has created a brisk demand for abortions and contraceptives. Thus the favorable response to the current campaign to encourage use of the IUD is not a good example of what birth-control technology can do for a genuinely backward country. In fact, when the program was started, one reason for expecting receptivity was that the island was already on its way to modernization and family planning (*28*).

At most, the recent family-planning campaign—which reached significant proportions only in 1964, when some 46,000 IUD's were inserted (in 1965 the number was 99,253, and in 1966, 111,242) (*29; 30*, p. 45)—could have caused the increase observable after 1963 in the rate of decline. Between 1951 and 1963 the average drop in the birth rate per 1000 women (see Table 19–1) was 1.73 percent per year; in the period 1964–66 it was 4.35 percent. But one hesitates to assign all of the acceleration in decline since 1963 to the family-planning campaign. The rapid economic development has been precisely of a type likely to accelerate a drop in reproduction. The rise in manufacturing has been much greater than the rise in either agriculture or construction. The agricultural labor force has thus been squeezed, and migration to the cities has skyrocketed (*31*). Since housing has not kept pace, urban families have had to restrict

TABLE 19–1

*Decline in Taiwan's fertility rate, 1951 through 1966*

| Year | Registered births per 1000 women aged 15–49 | Change in rate (percent)* |
|---|---|---|
| 1951 | 211 | |
| 1952 | 198 | −5.6 |
| 1953 | 194 | −2.2 |
| 1954 | 193 | −0.5 |
| 1955 | 197 | +2.1 |
| 1956 | 196 | −0.4 |
| 1957 | 182 | −7.1 |
| 1958 | 185 | +1.3 |
| 1959 | 184 | −0.1 |
| 1960 | 180 | −2.5 |
| 1961 | 177 | −1.5 |
| 1962 | 174 | −1.5 |
| 1963 | 170 | −2.6 |
| 1964 | 162 | −4.9 |
| 1965 | 152 | −6.0 |
| 1966 | 149 | −2.1 |

* The percentages were calculated on unrounded figures. Source of data through 1965, *Taiwan* Demographic Fact Book (1964, 1965); for 1966, *Monthly Bulletin of Population Registration Statistics of Taiwan* (1966, 1967).

reproduction in order to take advantage of career opportunities and avoid domestic inconvenience. Such conditions have historically tended to accelerate a decline in birth rate. The most rapid decline came late in the United States (1921–33) and in Japan (1947–55). A plot of the Japanese and Taiwanese birth rates (Fig. 19–1) shows marked similarity of the two curves, despite a difference in level. All told, one should not attribute all of the post-1963 acceleration in the decline of Taiwan's birth rate to the family-planning campaign.

The main evidence that *some* of this acceleration is due to the campaign comes from the fact that Taichung, the city in which the family-planning effort was first concentrated, showed subsequently a much fast-

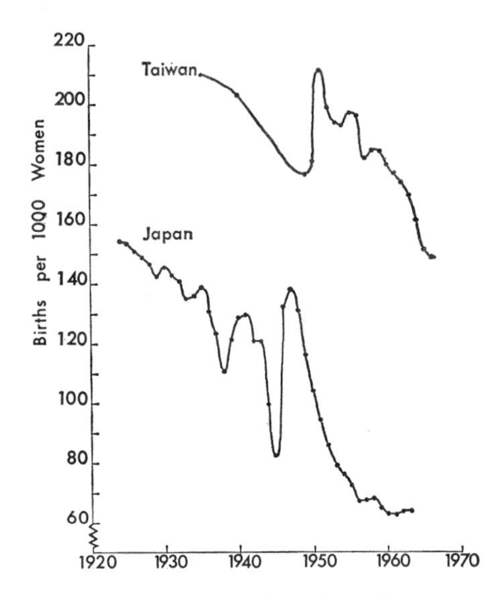


FIGURE 19–1. Births per 1000 women aged 15 through 49 in Japan and Taiwan.

er drop in fertility than other cities (*30*, p. 69; *32*). But the campaign has not reached throughout the island. By the end of 1966, only 260,745 women had been fitted with an IUD under auspices of the campaign, whereas the women of reproductive age on the island numbered 2.86 million. Most of the reduction in fertility has therefore been a matter of individual initiative. To some extent the campaign may be simply substituting sponsored (and cheaper) services for those that would otherwise come through private and commercial channels. An island-wide survey in 1964 showed that over 150,000 women were already using the traditional Ota ring (a metallic intrauterine device popular in Japan); almost as many had been sterilized; about 40,000 were using foam tablets; some 50,000 admitted to having had at least one abortion; and many were using other methods of birth control (*30*), pp. 18, 31).

The important question, however, is not whether the present campaign is somewhat hastening the downward trend in the birth rate but whether, even if it is, it will provide population control for the nation. Actually, the campaign is not designed to provide such control and shows no sign of doing so. It takes for granted existing reproductive goals. Its aim is "to integrate, through education and information, the idea of family limitation *within the existing attitudes, values, and goals* of the people" [*30*, p. 8 (italics mine)]. Its target is *married* women who do not want any more children; it ignores girls not yet married, and women married and wanting more children.

With such an approach, what is the maximum impact possible? It is the difference between the number of children women have been having and the number they want to have. A study in 1957 found a median figure of 3.75 for the number of children wanted by women aged 15 to 29 in Taipei, Taiwan's largest city; the corresponding figure for women from a satellite town was 3.93; for women from a fishing village, 4.90; and for women from a farming village, 5.03. Over 60 percent of the women in Taipei and over 90 percent of those in the farming village wanted 4 or more children (*33*). In a sample of wives aged 25 to 29 in Taichung, a city of over 300,000, Freedman and his co-workers found the average number of children wanted was 4; only 9 percent wanted less than 3, 20 percent wanted 5 or more (*34*). If, therefore, Taiwanese women used contraceptives that were 100-percent effective and had the number of children they desire, they would have about 4.5 each. The goal of the family-planning effort would be achieved. In the past the Taiwanese woman who married and lived through the reproductive period had, on the average, approximately 6.5 children; thus a figure of 4.5 would represent a substantial decline in fertility. Since mortality would continue to decline, the population growth rate would decline somewhat less than individual reproduction would. With 4.5 births per woman and a life expectancy of 70 years, the rate of natural increase would be close to 3 percent per year (*35*).

In the future, Taiwanese views concerning reproduction will doubtless change, in response to social change and economic modernization. But how far will they change? A good indication is the number of children desired by couples in an already modernized country long oriented toward family planning. In the United States in 1966, an average of 3.4 children was considered ideal by white women aged 21 or over (*36*). This average number of births would give Taiwan, with only a slight decrease in mortality, a long-run rate of natural increase of 1.7 percent per year and a doubling of population in 41 years.

Detailed data confirm the interpretation that Taiwanese women are in the process of shifting from a "peasant-agrarian" to an "industrial" level of reproduction. They are, in typical fashion, cutting off higher-order births at age 30 and beyond (*37*). Among young wives, fertility has risen, not fallen. In sum, the widely acclaimed family-planning program in Taiwan may, at most, have somewhat speeded the later phase of fertility decline which would have occurred anyway because of modernization.

Moving down the scale of modernization, to countries most in need of population control, one finds the family-planning approach even more inadequate. In South Korea, second only to Taiwan in the frequency with which it is cited as a model of current policy, a recent birth-rate decline of unknown extent is assumed by leaders to be due overwhelmingly to the government's family-planning program. However, it is just as plausible to say that the net effect of government involvement in population control has been, so far, to delay rather than hasten a decline in reproduction made inevitable by social and economic changes. Although the government is advocating vasectomies and providing IUD's and pills, it refuses to legalize abortions, despite the rapid rise in the rate of illegal abortions and despite the fact that, in a recent survey, 72 percent of the people who stated an opinion favored legalization. Also, the program is presented in the context of maternal and child health; it thus emphasizes motherhood and the family rather than alternative roles for women. Much is made of the fact that opinion surveys show an overwhelming majority of Koreans (89 percent in 1965) favoring contraception (*38*, p. 27), but this means only that Koreans are like other people in wishing to have the means to get what they want. Unfortunately, they want sizable families (*38*, p. 25):

> The records indicate that the program appeals mainly to women in the 30–39 year age bracket who have four or more children, including at least two sons. . .

In areas less developed than Korea the degree of acceptance of contraception tends to be disappointing, especially among the rural majority. Faced with this discouragement, the leaders of current policy, instead of reexamining their assumptions, tend to redouble their effort to find a contraceptive that will appeal to the most illiterate peasant, forgetting that he wants a good-sized family. In the rural Punjab, for example, "a disturbing feature . . . is that the females start to seek advice and adopt family planning techniques at the tag end of their reproductive period" (*39*). Among 5196 women coming to rural Punjabi family-planning centers, 38 percent were over 35 years old, 67 percent over 30. These women had married early, nearly a third of them before the age of 15 (*40*); some 14 percent had eight or more *living* children when they reached the clinic, 51 percent six or more.

A survey in Tunisia showed that 68 percent of the married couples were willing to use birth-control measures, but the average number of children they considered ideal was 4.3 (*41*). The corresponding averages for a village in eastern Java, a village near New Delhi, and a village in Mysore were 4.3, 4.0 , and 4.2, respectively (*42*, *43*). In the cities of these regions women are more ready to accept birth control and they want fewer children than village women do, but the number they consider desirable is still wholly unsatisfactory from the standpoint of population control. In an urban family-planning center in Tunisia, more than 600 of 900 women accepting contraceptives had four living children already (*44*). In Bangalore, a city of nearly a million at the time (1952), the number of offspring desired by married women was 3.7 on the average; by married men, 4.1 (*43*). In the metropolitan area of San Salvador (350,000 inhabitants) a 1964 survey (*45*) showed the number desired by women of reproductive age to be 3.9, and in seven other capital cities of Latin America the number ranged from 2.7 to 4.2. If women in the cities of underdeveloped countries used birth-control measures with 100-percent efficiency, they still would have enough babies to expand city populations senselessly, quite apart from the added contribution of rural-urban migration. In many of the cities the difference between actual and ideal number of children is not great; for instance, in the seven Latin American capitals mentioned above, the ideal was 3.4 whereas the actual births per women in the age range 35 to 39 was 3.7 (*46*). Bombay City has had birth-control clinics for many years, yet its birth rate (standardized for age, sex, and marital distribution) is still 34 per 1000 inhabitants and is tending to rise rather than fall. Although this rate is about 13 percent lower than that for India generally, it has been about that much lower since at least 1951 (*47*).

## Is Family Planning the "First Step" in Population Control?

To acknowledge that family planning does not achieve population control is not to impugn its value for other purposes. Freeing women from the need to have more children than they want is of great benefit to them and their children and to society at large. My argument is therefore directed not against family-planning programs as such but against the assumption that they are an effective means of controlling population growth.

But what difference does it make? Why not go along for awhile with family planning as an initial approach to the problem of population control? The answer is that any policy on which millions of dollars are being spent should be designed to achieve the goal it purports to achieve. If it is only a first step, it should be so labeled, and its connection with the next step (and the. nature of that next step) should be carefully examined. In the present case, since no "next step" seems ever to be mentioned, the question arises, Is reliance on family planning in fact a basis for dangerous postponement of effective steps? To continue to offer a remedy as a cure long after it has been shown merely to ameliorate the disease is either quackery or wishful thinking, and it thrives most where the need is greatest. Today the desire to solve the population problem is so intense that we are all ready to embrace any "action program" that promises relief. But postponement of effective measures allows the situation to worsen.

Unfortunately, the issue is confused by a matter of semantics. "Family *planning*" and "fertility *control*" suggest that reproduction is being regulated according to some rational plan. And so it is, but only from the standpoint of the individual couple, not from that of the community. What is rational in the light of a couple's situation may be totally irrational from the standpoint of society's welfare.

The need for societal regulation of individual behavior is readily recognized in other spheres—those of explosives, dangerous drugs, public property, natural resources. But in the sphere of reproduction, complete individual initiative is generally favored even by those liberal intellectuals who, in other spheres, most favor economic and social planning. Social reformers who would not hesitate to force all owners of rental property to rent to anyone who can pay, or to force all workers in an industry to join a union, balk at any suggestion that couples be permitted to have only a certain number of offspring. Invariably they interpret societal control of reproduction as meaning direct police supervision of individual behavior. Put the word *compulsory* in front of any term describing a means of limiting births—*compulsory sterilization*, *compulsory abortion*, *compulsory contraception*—and you guarantee violent opposition. Fortunately, such direct controls need not be invoked, but conservatives and radicals alike overlook this in their blind opposition to the idea of collective determination of a society's birth rate.

That the exclusive emphasis on family planning in current population policies is not a "first step" but an escape from the real issues is suggested by two facts. (i) No country has taken the "next step." The industrialized countries have had family planning for half a century without acquiring control over either the birth rate or population increase. (ii) Support and encouragement of research on population policy other than family planning is negligible. It is precisely this blocking of alternative thinking and experimentation that makes the emphasis on family planning a major obstacle to population control. The need is not to abandon family-planning programs but to put equal or greater resources into other approaches.

## New Directions in Population Policy

In thinking about other approaches, one can start with known facts. In the past; all surviving societies had institutional incentives for marriage, procreation, and child care which were powerful enough to keep the birth rate equal to or in excess of a high death rate. Despite the drop in death rates during the last century and a half, the incentives tended to remain intact because the social structure (especially in regard to the family) changed little. At most, particularly in industrial societies, children became less productive and more expensive (*48*). In present-day agrarian societies, where the drop in death rate has been more recent, precipitate, and independent of social change (*49*), motivation for having children has changed little. Here, even more than in industrialized nations, the family has kept on producing abundant offspring, even though only a fraction of these children are now needed.

If excessive population growth is to be prevented, the obvious requirement is somehow to impose restraints on the family. However, because family roles are reinforced by society's system of rewards, punishments, sentiments, and norms, any proposal to demote the family is viewed as a threat by conservatives and liberals alike, and certainly by people with enough social responsibility to work for population control. One is charged with trying to "abolish" the family, but what is required is selective restructuring of the family in relation to the rest of society.

The lines of such restructuring are suggested by two existing limitations on fertility. (i) Nearly all societies succeed in drastically discouraging reproduction among unmarried women. (ii) Advanced societies unintentionally reduce reproduction among married women when conditions worsen in such a way as to penalize childbearing more severely than it was penalized before. In both cases the causes are motivational and economic rather than technological.

It follows that population-control policy can de-emphasize the family in two ways: (i) by keeping present controls over illegitimate childbirth yet making the most of factors that lead people to postpone or avoid marriage, and (ii) by instituting conditions that motivate those who do marry to keep their families small.

## Postponement of Marriage

Since the female reproductive span is short and generally more fecund in its first than in its second half, postponement of marriage to ages beyond 20 tends biologically to reduce births. Sociologically, it gives women time to get a better education, acquire interests unrelated to the family, and develop a cautious attitude toward pregnancy (*50*). Individuals who have not married by the time they are in their late twenties often do not marry at all. For these reasons, for the world as a whole, the average age at marriage for women is negatively associated with the birth rate: a rising age at marriage is a frequent cause of declining fertility during the middle phase of the demographic transition; and, in the late phase, the "baby boom" is usually associated with a return to younger marriages.

Any suggestion that age at marriage be raised as a part of population policy is usually met with the argument that "even if a law were passed, it would not be obeyed." Interestingly, this objection implies that the only way to control the age at marriage is by direct legislation, but other factors govern the actual age. Roman Catholic countries generally follow canon law in stipulating 12 years as the minimum *legal* age at which girls may marry, but the actual average age at marriage in these countries (at least in Europe) is characteristically more like 25 to 28 years. The actual age is determined, not by law, but by social and economic conditions. In agrarian societies, postponement of marriage (when postponement occurs) is apparently caused by difficulties in meeting the economic prerequisites for matrimony, as stipulated by custom and opinion. In industrial societies it is caused by housing shortages, unemployment, the requirement for overseas military service, high costs of education, and inadequacy of consumer services. Since almost no research has been devoted to the subject, it is difficult to assess the relative weight of the factors that govern the age at marriage.

## Encouraging Limitation of Births within Marriage

As a means of encouraging the limitation of reproduction within marriage, as well as postponement of marriage, a greater rewarding of nonfamilial than of familial roles would probably help. A simple way of accomplishing this would be to allow economic advantages to accrue to the single as opposed to the married individual, and to the small as opposed to the large family. For instance, the government could pay people to permit themselves to be sterilized (*51*); all costs of abortion could be paid by the government; a substantial fee could be charged for a marriage license; a "child-tax" (*52*) could be levied; and there could be a requirement that illegitimate pregnancies be aborted. Less sensationally, governments could simply reverse some existing policies that encourage childbearing. They could, for example, cease taxing single persons more than married ones; stop giving parents special tax exemptions; abandon income-tax policy that discriminates against couples when the wife works; reduce paid maternity leaves; reduce family allowances (*53*); stop awarding public housing on the basis of family size; stop granting fellowships and other educational aids (including special allowances for wives and children) to married students; cease outlawing abortions and sterilizations; and relax rules that allow use of harmless contraceptives only with medical permission. Some of these policy reversals would be beneficial in other than demographic respects and some would be harmful unless special precautions were taken. The aim would be to reduce the number, not the quality, of the next generation.

A closely related method of de-emphasizing the family would be modification of the complementarity of the roles of men and women. Men are now able to participate in the wider world yet enjoy the satisfaction of having several children because the housework and childcare fall mainly on their wives. Women are impelled to seek this role by their idealized view of marriage and motherhood and by either the scarcity of alternative roles or the difficulty of combining them with family roles. To change this situation women could be required to work outside the home, or compelled by circumstances to do so. If, at the same time, women were paid as well as men and given equal educational and occupational opportunities, and if social life were organized around the place of work rather than around the home or neighborhood, many women would develop interests that would compete with family interests. Approximately this policy is now followed in several Communist countries, and even the less developed of these currently have extremely low birth rates (*54*).

That inclusion of women in the labor force has a negative effect on reproduction is indicated by regional comparisons (*18*, p. 1195; *55*). But in most countries the wife's employment is subordinate, economically and emotionally, to her family role, and is readily sacrificed for the latter. No society has restructured both the occupational system and the domestic establishment to the point of permanently modifying the old division of labor by sex.

In any deliberate effort to control the birth rate along these lines, a government has two powerful instruments—its command over economic planning and its authority (real or potential) over education. The first determines (as far as policy can) the economic conditions and circumstances affecting the lives of all citizens; the second provides the knowledge and attitudes necessary to implement the plans. The economic system largely determines who shall work, what can be bought, what rearing children will cost, how much individuals can spend. The schools define family roles and develop vocational and recreational interests; they could, if it were desired, redefine the sex roles, develop interests that transcend the home, and transmit realistic (as opposed to moralistic) knowledge concerning marriage, sexual behavior, and population problems. When the problem is viewed in this light, it is clear that the ministries of economics and education, not the ministry of health, should be the source of population policy.

## The Dilemma of Population Policy

It should now be apparent why, despite strong anxiety over runaway population growth, the actual programs purporting to control it are limited to family planning and are therefore ineffective. (i) The goal of zero, or even slight, population growth is one that nations and groups find difficult to accept. (ii) The measures that would be required to implement such a goal, though not so revolutionary as a Brave New World or a Communist Utopia, nevertheless tend to offend most people reared in existing societies. As a consequence, the goal of so-called population control is implicit and vague; the method is only family planning. This method, far from de-emphasizing the family, is familistic. One of its stated goals is that of helping sterile couples to *have* children. It stresses parental aspirations and responsibilities. It goes along with most aspects of conventional morality, such as condemnation of abortion, disapproval of premarital intercourse, respect for religious teachings and cultural taboos, and obeisance to medical and clerical authority. It deflects hostility by refusing to recommend any change other than the one it stands for: availability of contraceptives.

The things that make family planning acceptable are the very things that make it ineffective for population control. By stressing the right of parents to have the number of children they want, it evades the basic question of population policy, which is how to give societies the number of children they need. By offering only the means for *couples* to control fertility, it neglects the means for societies to do so.

Because of the predominantly pro-family character of existing societies, individual interest ordinarily leads to the production of enough offspring to constitute rapid population growth under conditions of low mortality. Childless or single-child homes are considered indicative of personal failure, whereas having three to five living children gives a family a sense of continuity and substantiality (*56*).

Given the existing desire to have moderate-sized rather than small families, the only countries in which fertility has been reduced to match reduction in mortality are advanced ones temporarily experiencing worsened economic conditions. In Sweden, for instance, the net reproduction rate (NRR) has been below replacement for 34 years (1930–63), if the period is taken as a whole, but this is because of the economic depression. The average replacement rate was below unity (NRR = 0.81) for the period 1930–42, but from 1942 through 1963 it was above unity (NRR = 1.08). Hardships that seem particularly conducive to deliberate lowering of the birth rate are (in managed economies) scarcity of housing and other consumer goods despite full employment, and required high participation of women in the labor force, or (in freer economics) a great deal of unemployment and economic insecurity. When conditions are good, any nation tends to have a growing population.

It follows that, in countries where contraception is used, a realistic proposal for a government policy of lowering the birth rate reads like a catalogue of horrors: squeeze consumers through taxation and inflation; make housing very scarce by limiting construction; force wives and mothers to work outside the home to offset the inadequacy of male wages, yet provide few childcare facilities; encourage migration to the city by paying low wages in the country and providing few rural jobs; increase congestion in cities by starving the transit system; increase personal insecurity by encouraging conditions that produce unemployment and by haphazard political arrests. No government will institute such hardships simply for the purpose of controlling population growth. Clearly, therefore, the task of contemporary population policy is to develop attractive substitutes for family interests, so as to avoid having to turn to hardship as a corrective. The specific measures required for developing such substitutes are not easy to determine in the absence of research on the question.

In short, the world's population problem cannot be solved by pretense and wishful thinking. The unthinking identification of family planning with population control is an ostrich-like approach in that it permits people to hide from themselves the enormity and unconventionality of the task. There is no reason to abandon family-planning programs; contraception is a valuable technological instrument. But such programs must be supplemented with equal or greater investments in research and experimentation to determine the required socioeconomic measures.

## References and Notes


1. *Studies in Family Planning, No. 16* (1967).
2. Ibid., *No. 9* (1966), p. 1.
3. The statement is given in *Studies in Family Planning* (*1*, p. 1), and in *Population Bull.* **23**, 6 (1967).
4. The statement is quoted in *Studies in Family Planning* (*1*, p. 2).
5. *Hearings on S. 1676, U.S. Senate, Subcommittee on Foreign Aid Expenditures, 89th Congress, Second Session, April 7, 8, 11* (1966), pt. 4.
6. B. L. Raina, in *Family Planning and Population Programs*, B. Berelson, R. K. Anderson, O. Harkavy, G. Maier, W. P. Mauldin, S. G. Segal, Eds. (Univ. of Chicago Press, Chicago, 1966).
7. D. Kirk, *Ann. Amer. Acad. Polit. Soc. Sci.* **369**, 53 (1967).
8. As used by English-speaking demographers, the word *fertility* designates actual reproductive performance, not a theoretical capacity.
9. K. Davis, *Rotarian* **94**, 10 (1959); *Health Educ. Monographs* **9**, 2 (1960); L. Day and A. Day, *Too Many Americans* (Houghton Mifflin, Boston, 1964); R. A. Piddington, *Limits of Mankind* (Wright, Bristol, England, 1956).
10. *Official Gazette* (15 Apr. 1965); quoted in *Studies in Family Planning* (*1*, p. 7).
11. J. W. Gardner, Secretary of Health, Education, and Welfare, "Memorandum to Heads of Operating Agencies" (Jan. 1966), reproduced in *Hearings on S. 1676* (*5*), p. 783.
12. C. Tietze, *Demography* **1**, 119 (1964); *J. Chronic Diseases* **18**, 1161 (1964); M. Muramatsu, *Milbank Mem. Fund Quart.* **38**, 153 (1960); K. Davis, *Population Index* **29**, 345 (1963); R. Armijo and T. Monreal, *J. Sex Res.* **1964**, 143 (1964); Proceedings World Population Conference, Belgrade, 1965; Proceedings International Planned Parenthood Federation.

13. *Studies in Family Planning, No. 4* (1964), p. 3.
14. D. Bell (then administrator for Agency for International Development), in *Hearings on S. 1676 (5)*, p. 862.
15. *Asian Population Conference* (United Nations, New York, 1964), p. 30.
16. R. Armijo and T. Monreal, in *Components of Population Change in Latin America* (Milbank Fund, New York, 1965), p. 272; E. Rice-Wray, *Amer. J. Public Health* **54**, 313 (1964).
17. E. Rice-Wray, in "Intra-Uterine Contraceptive Devices," *Excerpta Med. Intern. Congr. Ser. No. 54* (1962), p. 135.
18. J. Blake, in *Public Health and Population Change*, M. C. Sheps and J. C. Ridley, Eds. (Univ. of Pittsburgh Press, Pittsburgh, 1965).
19. J. Blake and K. Davis, *Amer. Behavioral Scientist* **5**, 24 (1963).
20. See "Panel discussion on comparative acceptability of different methods of contraception," in *Research in Family Planning*, C. V. Kiser, Ed. (Princeton Univ. Press, Princeton, 1962), pp. 373–86.
21. "From the point of view of the woman concerned, the whole problem of continuing motivation disappears, . . . " [D. Kirk, in *Population Dynamics*, M. Muramatsu and P. A. Harper, Eds. (Johns Hopkins Press, Baltimore, 1965)].
22. "For influencing family size norms, certainly the examples and statements of public figures are of great significance . . . also . . . use of mass-communication methods which help to legitimize the small-family style, to provoke conversation, and to establish a vocabulary for discussion of family planning." [M. W. Freymann, in *Population Dynamics*, M. Muramatsu and P. A. Harper, Eds. (Johns Hopkins Press, Baltimore, 1965)].
23. O. A. Collver, *Birth Rates in Latin America* (International Population and Urban Research, Berkeley, Calif., 1965), pp. 27–28; the ten countries were Colombia, Costa Rica, El Salvador, Ecuador, Guatemala, Honduras, Mexico, Panamá, Peru, and Venezuela.
24. J. R. Rele, *Fertility Analysis through Extension of Stable Population Concepts.* (International Population and Urban Research, Berkeley, Calif., 1967).
25. J. C. Ridley, M. C. Sheps, J. W. Lingner, J. A. Menken, *Milbank Mem. Fund Quart.* **45**, 77 (1967); E. Arriaga, unpublished paper.
26. "South Korea and Taiwan appear successfully to have checked population growth by the use of intrauterine contraceptive devices" [U. Borell, *Hearings on S. 1676* (5), p. 556].
27. K. Davis, *Population Index* **29**, 345 (1963).
28. R. Freedman, ibid. **31**, 421 (1965).
29. Before 1964 the Family Planning Association had given advice to fewer than 60,000 wives in 10 years and a Pre-Pregnancy Health Program had reached some 10,000, and, in the current campaign, 3650 IUD's were inserted in 1965, in a total population of 2½ million women of reproductive age. See *Studies in Family Planning, No. 19* (1967), p. 4, and R. Freedman *et al.*, *Population Studies* **16**, 231 (1963).
30. R. W. Gillespie, *Family Planning on Taiwan* (Population Council, Taichung, 1965).
31. During the period 1950–60 the ratio of growth of the city to growth of the noncity population was 5:3; during the period 1960–64 the ratio was 5:2; these ratios are based on data of Shaohsing Chen, *J. Sociol. Taiwan* **1**, 74 (1963) and data in the United Nations *Demographic Yearbooks.*
32. R. Freedman, *Population Index* **31**, 434 (1965). Taichung's rate of decline in 1963–64 was roughly double the average in four other cities, whereas just prior to the campaign its rate of decline had been much less than theirs.
33. S. H. Chen, *J. Soc. Sci. Taipei* **13**, 72 (1963).
34. R. Freedman et al., *Population Studies* **16**, 227 (1963); ibid, p. 232.

35. In 1964 the life expectancy at birth was already 66 years in Taiwan, as compared to 70 for the United States.
36. J. Blake, *Eugenics Quart.* **14**, 68 (1967).
37. Women accepting IUD's in the family-planning program are typically 30 to 34 years old and have already had four children. [*Studies in Family Planning No. 19* (1967), p. 5].
38. Y. K. Cha, in *Family Planning and Population Programs*, B. Berelson et al., Eds. (Univ. of Chicago Press, Chicago, 1966).
39. H. S. Ayalvi and S. S. Johl, *J. Family Welfare* **12**, 60 (1965).
40. Sixty percent of the women had borne their first child before age 19. Early marriage is strongly supported by public opinion. Of couples polled in the Punjab, 48 percent said that girls *should* marry before age 16, and 94 percent said they should marry before age 20 (H. S. Ayalvi and S. S. Johl, ibid., p. 57). A study of 2380 couples in 60 villages of Uttar Pradesh found that the women had consummated their marriage at an average age of 14.6 years [J. R. Rele, *Population Studies* **15**, 268 (1962)].
41. J. Morsa, in *Family Planning and Population Programs*, B. Berelson et al., Eds. (Univ. of Chicago Press, Chicago, 1966).
42. H. Gille and R. J. Pardoko, ibid., p. 515; S. N. Agarwala, *Med. Dig. Bombay* **4**, 653 (1961).
43. *Mysore Population Study* (United Nations, New York, 1961), p. 140.
44. A. Daly, in *Family Planning and Population Programs*, B. Berelson et al., Eds. (Univ. of Chicago Press, Chicago, 1966).
45. C. J. Goméz, paper presented at the World Population Conference, Belgrade, 1965.
46. C. Miro, in *Family Planning and Population Programs*, B. Berelson *et al.*, Eds. (Univ. of Chicago Press, Chicago, 1966).
47. *Demographic Training and Research Centre (India) Newsletter* **20**, 4 (Aug. 1966).
48. K. Davis, *Population Index* **29**, 345 (1963). For economic and sociological theory of motivation for having children, see J. Blake [Univ. of California (Berkeley)], in preparation.
49. K. Davis, *Amer. Economic Rev.* **46**, 305 (1956); *Sci. Amer.* **209**, 68 (1963).
50. J. Blake, *World Population Conference* [*Belgrade, 1965*] (United Nations, New York, 1967), vol. 2, pp. 132–36.
51. S. Enke, *Rev. Economics Statistics* **42**, 175 (1960); ——, *Econ. Develop. Cult. Change* **8**, 339 (1960); ——, ibid. **10**, 427 (1962); A. O. Krueger and L. A. Sjaastad, ibid., p. 423.
52. T. J. Samuel, *J. Family Welfare India* **13**, 12 (1966).
53. Sixty-two countries, including 27 in Europe, give cash payments to people for having children [U.S. Social Security Administration, *Social Security Programs Throughout the World, 1967* (Government Printing Office, Washington, D.C., 1967), pp. xxvii-xxviii].
54. Average gross reproduction rates in the early 1960's were as follows: Hungary, 0.91; Bulgaria, 1.09; Romania, 1.15; Yugoslavia, 1.32.
55. O. A. Collver and E. Langlois, *Econ. Develop. Cult. Change* **10**, 367 (1962); J. Weeks, [Univ. of California (Berkeley)], unpublished paper.
56. Roman Catholic textbooks condemn the "small" family (one with fewer than four children) as being abnormal [J. Blake, *Population Studies* **20**, 27 (1966)].
57. Judith Blake's critical readings and discussions have greatly helped in the preparation of this article.

# VII

# Pollutants and Ecosystems

# PESTICIDES AND RELATED MATERIALS

# 20

# Effects of insecticides

*Charles F. Wurster*

## Introduction

World opinion has been awakening to a variety of man-made environmental problems during the past few years, and few of these have received more public debate than 'the pesticide problem.' Strong emotions are evoked in people when they get the idea that they or other animals are being poisoned, regardless of whether the threat is real or imagined. 'The pesticide problem' is by no means a single problem, but instead consists of a host of individual and highly diverse problems depending on the pesticide and circumstances involved. In considering the effects of pesticides on nontarget organisms, I propose to organize my remarks into several categories, touching only on those aspects of pesticides that seem most important, and omitting much that appears less critical in my view.

The word 'pesticide' is generally applied to any chemical that is used to kill pests, but I shall restrict myself to insecticides. The various herbicides, fungicides, rodenticides, and nematocides certainly introduce problems of their own, but it is my impression that they are generally less serious and widespread than are the insecticide problems. In any event, I cannot discuss these other pesticides with any real competence, and will therefore leave their consideration to others (Moore, 1967).

---

Reprinted from Nicholas Polunin: *The Environmental Future*, published by Macmillan, London, and Barnes & Noble, New York, 1972 of which it comprised pp. 293–310 (followed by Discussion). 

## Non-Persistent Insecticides

The insecticides can roughly be divided into two large groups—those that are stable (persistent) and those that are not. The nonpersistent insecticides currently in use are mainly organophosphates and carbamates (O'Brien, 1967). Being chemically unstable, they do not retain their original identity and associated biological activity sufficiently long to permit them to be transported to distant regions. Most of them break down rapidly into non-toxic products. The effects of these nonpersistent insecticides are therefore primarily restricted to the treated areas, and their residues do not accumulate extensively in the biosphere.

This is not to suggest that organophosphates and carbamates do not pose problems, however. Some of the organophosphates, such as Parathion, Systox, and TEPP are extremely toxic, making them potentially hazardous to farm personnel and other nontarget organisms that may be present in the vicinity of the application (U. S. Department of Health, Education & Welfare, 1969). Other organophosphates, including Malathion, Chlorthion, Dibrom, Ronnel, and Dipterex have lower acute toxicities to vertebrates than has DDT, and DDT is not an especially toxic material when compared with most pesticides.

Another major problem with some nonpersistent insecticides, in common with many other insecticides, is their tendency to be disruptive within insect communities, often aggravating, rather than alleviating insect control problems (Bosch, 1970; Huffaker, 1971). By destroying beneficial insects that are natural enemies of the pests, and by eliminating the susceptible individuals, use of insecticides may generate outbreaks of insecticide-resistant, injurious insects that are far worse than those which existed prior to the treatment. The solution to most of the problems with nonpersistent insecticides involves careful regulation and wise usage when necessary, rather than a complete prohibition of their use.

However severe the effects of nonpersistent insecticides may be, they are principally restricted to the vicinity and time of application. They cannot become a world problem by contaminating or affecting nontarget organisms in areas that are remote in distance or in time from the treated areas. In this regard they contrast dramatically with the persistent insecticides, and I will therefore accord primary attention to the latter in this paper.

## Persistent Chlorinated Hydrocarbon Insecticides

Stability or persistence confers an entirely new dimension on an insecticide because its unintended effects can, and sometimes do extend thousands of miles and many years beyond the area and time of application, thereby involving a wide variety of nontarget organisms. The persistent chlorinated hydrocarbon (CH) or organochlorine insecticides have thus become one of the world's most serious pollution problems, involving many nonagricultural interests and values. This family of insecticides includes DDT, Aldrin, Dieldrin, Isodrin, Endrin, Chlordane, Telordrin, Heptachlor, Strobane, Toxaphene, Mirex, and a few others.

Of this group DDT has been by far the most widely manufactured and applied; nearly 3 billion ($10^9$) pounds ($1.36 \times 10^9$ kg) of DDT have been produced in the United States alone since 1944 (Fig. 20–1; U.S. Department of Agriculture, 1970). Production was 123 million pounds ($55.8 \times 10^6$ kg) in 1969 (the latest year for which data are available), which is just under the average rate of 145 million pounds ($65.8 \times 10^6$ kg) produced annually during the 1960s. The world's largest DDT manufacturing plant, and now the only one in the United States, is that of the Montrose Chemical Corporation in Los Angeles. The data in Fig. 20–1 further show that, except in one year (1967), more DDT has been

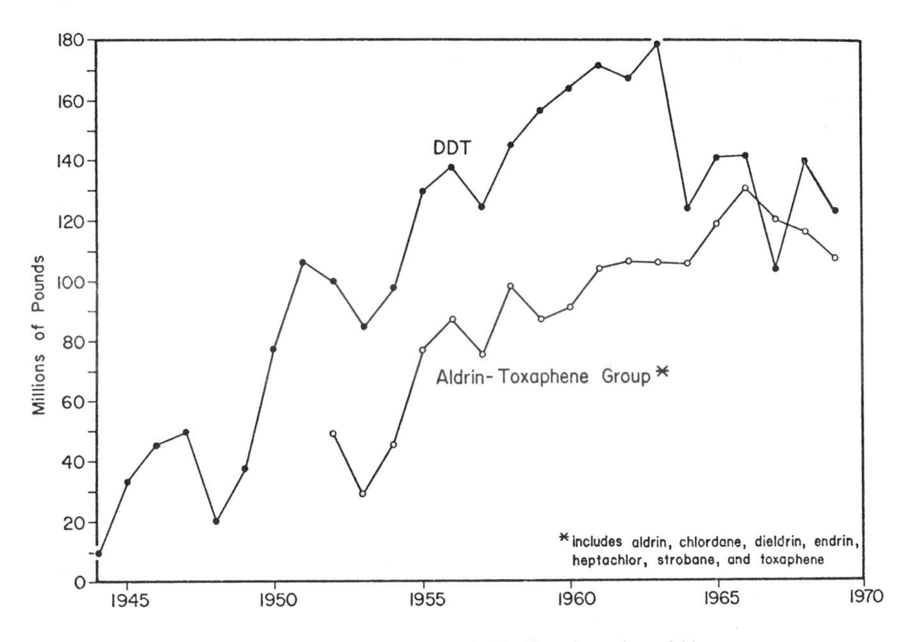


FIGURE 20–1. Production of DDT and Aldrin-Toxaphene insecticides in the United States (U.S. Department of Agriculture, 1970).

and continues to be produced annually in the United States than the total for Aldrin, Chlordane, Dieldrin, Endrin, Heptachlor, Strobane, and Toxaphene combined. Although we read that DDT is being 'restricted' and 'phased out' in many countries, it continues to be used in greater quantities than any other insecticide.

As we continue to add to the 3 billion ($10^9$) pounds of DDT already released into the environment, and with some unknown fraction of it still circulating in the biosphere, it should hardly be surprising that DDT poses by far the most serious of our persistent pesticide problems. I believe that DDT has had a greater impact on nontarget organisms and ecosystems than has any other pesticide, possibly greater than all other pesticides combined. For these reasons, and to bring the issue up to date and correct misinformation, I shall discuss the CH insecticide problem in some detail, with primary emphasis on DDT (Wurster, 1969*a*). The polychlorinated biphenyls (PCBs), a mixture of CH compounds of industrial origin (used as plasticizers, flame retardants, insulating fluids, adhesives, etc.) that are widespread in nature and exhibit environmental behaviour which is similar to that of the persistent insecticides, may also represent a serious environmental hazard (Peakall & Lincer, 1970). The PCBs are another example of the large scale release of a material into the environment before adequate research on their potentiial environmental impact had been performed.

### Properties of the Chlorinated Hydrocarbons

To appreciate the environmental behaviour of the CHs, along with the associated hazards which they pose to various nontarget organisms, it is necessary to understand their properties. Their behaviour results from the combination of the following four properties:

1. *Mobility*. CHs unfortunately do not remain where they are applied. By various mechanisms they enter the air and surface waters to be dispersed over great distances within world circulation patterns (Wurster, 1969*a*).

Although these materials have very low vapour pressures, volatilization is an important mechanism whereby they pass into the air (Edwards, 1966). Codistillation with water facilitates their passage into air from wet surfaces (Bowman et al., 1964), and CHs adsorbed to particulates, especially soil particles, and existing as suspensions, are also dispersed to remote regions by the winds (Risebrough et al., 1968).

Exceptionally low water solubilities do not prevent these compounds from being transported in very dilute solution by flowing water, but the transport capacity of water is greatly increased by the tendency of CHs to form suspensions in water and to adsorb to particulates, all of which pass downstream within watersheds (Bowman et al., 1964; Edwards, 1966).

These dispersal mechanisms transport CHs to most parts of the world after they have been released into the environment, thus explaining their presence in remote, untreated regions such as Antarctica (Peterle, 1969; Wurster, 1969 *a*).

2. *Persistence*. The persistence of the CHs varies with the compound and the conditions. DDT is slowly converted to DDD, then to other metabolites, and eventually to the non-toxic DDA by various organisms and conditions (O'Brien, 1967). DDT is also converted to DDE, which is very stable though not very toxic, but which shows a variety of environmentally important enzyme effects. DDE is the most widespread of all CHs in the environment. DDT, DDD, and DDE are commonly called 'DDT residues', and these biologically active materials remain in the biosphere for many years and probably decades after their use (Edwards, 1966; Nash & Woolson, 1967).

In the environment, Aldrin and Isodrin are gradually converted into Dieldrin and Endrin, respectively, both of which are very persistent (Nash & Woolson, 1967). Mirex appears to be unusually stable (Valin et al., 1968).

3. *Solubility Characteristics*. Being typical non-polar organic compounds, CHs have extremely low solubilities in water, but high solubilities in lipids or fatty tissues. Water is saturated with DDT at only 1.2 parts per billion ($10^9$) (ppb), and Mirex is evidently still less soluble; Aldrin and Dieldrin have somewhat higher (but still very low) solubilities of 27 and 186 ppb, respectively (Park & Bruce, 1968; Wurster, 1969 *a*). Since all living organisms contain lipids, CHs are much more soluble in living tissues than in water, and the partition coefficient strongly favours accumulation and retention of CHs by the tissues of living organisms.

The solubility properties explain why CHs are not 'lost' by dilution in the inorganic components of the environment—the water, air, and soil. Instead, CH residues, especially DDT, have become nearly universal in animal tissues in most regions of the world. Birds, fish, and mammals, including man, are almost invariably contaminated.

4. *Broad Biological Activity*. The toxicity and biological activity of the CHs are by no means limited to the target species of insect, but instead can affect a great variety of animals, including all classes of vertebrates; furthermore, they can operate by several different mechanisms. They are nerve poisons—an effect that can be lethal (O'Brien, 1967). In addition, most are inducers of hepatic enzyme systems and inhibitors of certain other enzymes (Conney, 1967; Kupfer, 1967; Peakall, 1970 *b*). DDT and Aldrin have oestrogenic activity (Welch et al., 1969), and several CHs inhibit photosynthesis, possibly by inhibiting electron transport (Wurster, 1968; Menzel et al., 1970). DDT, Aldrin, Dieldrin, Heptachlor, Strobane,

and Mirex are carcinogenic in rodents (U. S. Department of Health, Education & Welfare, 1969), and DDT has shown mutagenic activity as well (Legator, in preparation). Although effects are by no means predictable, contamination of non-target organisms with compounds that have such a broad spectrum of biological activity clearly carries the potential for affecting those organisms. The realization of that potential is the major topic of this paper.

### *Biological Concentration*

When CHs are absorbed into food chains, they tend to remain in them because of their solubility characteristics. Each organism feeds on many organisms from the next lower trophic level; food organisms are metabolized and excreted, but the CHs are retained, thus leading to a higher concentration of CHs in the higher organism than was present in the food organism. Concentrations of CHs thereby increase with each step in the food chain, reaching the highest values in carnivores at the ends of long food chains. This biological or trophic concentration causes CHs to reach levels that are many thousands or even millions of times higher in organisms than are found in the surrounding inorganic environment (Woodwell et al., 1967; Korschgen, 1970). Analyses of soil, water, or air, showing only minute quantities of CHs to be present, can therefore be misleading indicators of environmental quality because they ignore the importance of these highly efficient biological concentrating mechanisms. Analyses of predatory organisms that are high in the food chain are a more relevant measure of environmental contamination with CHs.

### *Significance of these Properties*

In summary, the CHs can travel great distances from application sites, are sufficiently stable to retain their identity for years, are accumulated by non-target organisms because of their solubility characteristics, reach the highest levels of contamination in carnivores, and are hazardous to these organisms because they have a broad spectrum of biological activity. CHs are therefore inherently uncontrollable materials after they have been released into the environment. Few major, widespread environmental pollutants combine these properties.

### *Effects on Birds*

It has long been known that CHs can cause extensive mortality among birds. Bird mortality following attempts to control Dutch elm disease with DDT has been documented on numerous occasions (D. H.

Wurster et al., 1965), and other CHs, especially Dieldrin, have frequently killed birds following their use (Rudd, 1964; Stickel et al., 1969). Rather than being restricted only to treated areas, mortality of certain raptors has sometimes resulted from general environmental contamination with CHs. Deaths of Bald Eagles (*Haliaeetus leucocephalus*) from Dieldrin poisoning in the United States is an example (Mulhern et al., 1970). It is hard to know just how extensive direct mortality among carnivorous birds from general environmental contamination might be; it may be far more important than is immediately obvious. The available evidence suggests, however, that the sublethal effects of CHs on avian reproduction have a greater overall impact on bird populations than does acutely lethal direct mortality, even though a bird kill may seem more spectacular.

The effects of DDT residues on avian reproduction have only recently become well understood, and the many years of research that developed this knowledge make a fascinating and alarming 'science detective story' (Wurster, 1969*b*; Peakall, 1970*a*). By 1967 it had become clear that extensive and widespread declines in populations of many species of carnivorous birds and DDT contamination of these birds were correlated, but only in the past four years have cause-effect relationships been confirmed and mechanisms of action partially clarified.

DDT contamination inhibits avian reproduction by causing the birds to lay abnormally thin-shelled eggs, which break prematurely in the nest and therefore do not produce chicks (Peakall, 1970*a*). Additional symptoms include late ovulation and nesting, abnormal behaviour, hatching failure, and failure to lay eggs. The thinning of the eggshells may be caused by inhibition of carbonic anhydrase, an enzyme that is essential to the formation of calcium carbonate eggshells in the shell gland of the oviduct (Peakall, 1970 *b*). DDE, the most widespread CH pollutant that was considered to be innocuous for many years, is an inhibitor of this enzyme. This explanation has recently been challenged, however (Dvorchik et al., 1971). Jefferies and French (1971) recently suggested that the thin eggshells result from a hypothyroidal condition caused in the birds by DDT.

DDT residues, as well as most other CHs, are inducers of hepatic hydroxylating enzymes that metabolize steroids, including sex hormones (Conney, 1967; Kupfer, 1967; Peakall, 1970*b*). Enzyme induction by DDT thus reduces the level of circulating endogenous estradiol, which is partially responsible for various secondary sex characteristics and breeding behaviour in female birds (Peakall, 1970 *a*; 1970 *b*); this apparently explains the additional symptoms mentioned above. Evidently by simultaneously affecting different organs in different ways, contamination with DDT residues can reduce the reproductive success in wild bird populations to only a small fraction of what is normal.

Abnormally thin-shelled eggs have become commonplace among

populations of many species of carnivorous birds in recent years (Blus, 1970; Ratcliffe, 1970; Cade et al., 1971). Controlled experiments have shown that the levels of contamination with DDT residues regularly found in wild populations cause the thin-shelled eggs and other symptoms of reproductive failure that have now become typical of those populations (Wiemeyer & Porter, 1970). As a consequence of these phenomena, DDT has suppressed or even extirpated some populations of many species of birds of prey and sea birds in the United States, including the Bald Eagle, Peregrine Falcon, Prairie Falcon, Sharp-shinned Hawk, Cooper's Hawk, Marsh Hawk, Black-crowned Night Heron, Double-crested Cormorant, Osprey, Common Murre, Brown Pelican, Ashy Petrel, and Common Egret.

The role of CHs other than DDT in avian reproduction is less clear, partly because they are much less widespread in the environment and partly because they have been studied less. Various evidence does not indicate that complacence about them is justified, however. Dieldrin also causes birds to lay thin-shelled eggs (Lehner & Egbert, 1969); it inhibits carbonic anhydrase (Verrett & Desmond, 1959), is a powerful hepatic enzyme inducer (Peakall, 1970 *b*), and was evidently the major factor in the low reproductive success of the Golden Eagle in Scotland (Ratcliffe, 1970). Since Dieldrin is very widespread in nature, it is presumably a contributor to the above problems.

### Effects on Fish

CHs have long been known to be highly toxic to fish. Widespread mortality of salmon (*Salmo salar*) resulted from spraying of the coniferous forests of New Brunswick with DDT in the 1950s (Wurster, 1969 *a*), and large, spectacular fish kills resulted from the application of Dieldrin to a Florida salt marsh and the discharge of Endrin into the Mississippi River (Harrington & Bidlingmayer, 1958). As with birds, however, it is probable that acutely lethal effects of CHs on fish are less important to fish populations than are the less obvious sublethal effects.

CHs can inhibit reproduction in fish, but the mechanism is quite different from that which inhibits avian reproduction. Macek (1968 *a*) showed, by controlled experiments, that concentrations of DDT in the diet that are sublethal to adult fish may be lethal to fry after they hatch from contaminated eggs. The DDT is passed into the egg yolk; the embryo develops and hatches and, after hatching, at the stage of final yolk sac absorption, the fry will die if the concentration of DDT in the yolk is sufficiently high. This form of reproductive failure has occurred several times in nature, has sometimes been severe, and is probably more widespread than published accounts indicate. Reproduction of Lake Trout (*Salvelinus namaycush*) in Lake George and other New York lakes has failed completely for the past dozen years, with 100 percent mortality of the fry occurring annually (Burdick et al., 1964). Similar, though less severe fry mortality has involved the Coho Salmon (*Oneorhynchus kisutch*) in Lake Michigan (Johnson & Pecor, 1969), trout (several salmonids) in Alberta and New Zealand, and Spotted Seatrout (*Cynoscion nebulosus*) in the Gulf oof Mexico (Butler et al., 1970). Since the concentrations of DDT at which fry mortality has been shown to occur both under contrrolled and field conditions are now being approached or equalled in some freshwater and marine fisheries, it is hard to escape the conclusion that these fisheries are threatened by contamination with DDT (Risebrough, 1969).

Unfortunately, few data are available on the possible effects of CHs on the reproduction of marine fish, so it is hard to know what might be happening to marine fishery resources.

CHs have other sublethal effects on fish. A few ppb of DDT in the water upset the temperature regulating mechanism of young salmon (Anderson & Peterson, 1969), and controlled experiments by Macek (1968 *b*) showed that the stresses of falling temperature and starvation killed most Brook Trout (*Salvelinus fontinalis*) that were subjected to sublethal amounts of DDT, whereas very few of the control fish died. DDT-induced susceptibility to stress presumably explains the delayed salmon mortality in New Brunswick that coincided with colder weather several months after the DDT application and initial fish kill (Wurster, 1969 *a*). Sublethal exposure to Toxaphene and DDT also affects the behaviour of fish (Anderson & Peterson, 1969).

### *Chlorinated Hydrocarbons and Photosynthesis*

Recent studies have shown that a few ppb of DDT, Dieldrin, Endrin, or PCBs in the water decreases photosynthesis, as measured by $^{14}C$ uptake, in certain species of marine phytoplankton, an effect that could result from inhibition of electron transport by these materials (Wurster, 1968; Menzel et al., 1970). I do not, however, subscribe to the theory frequently advanced in the public media that these findings indicate that DDT will ultimately eliminate oxygen production in the oceans, thereby greatly diminishing the oxygen supply in the atmosphere.

This effect of CHs on algae appears to be highly selective, affecting certain susceptible species and not others. Selective poisoning of some algal species in areas near sites of CH application could lead to an undesirable imbalance within the flora, and a bloom of the resistant species might then occur. An alteration in species composition of phytoplankton communities could have profound ecological consequences, but too little research has yet been done on these phenomena to draw any conclusions.

### Effects on Ecosystems

Many of the activities of man tend to simplify ecosystems by reducing species diversity. An agricultural ecosystem generally contains fewer species of organisms than the ecosystem it replaced, and the same is usually true of cities, towns, and the human environment generally. Fire, ionizing radiation, a variety of pollutants including pesticides, and other forms of human disturbance all tend to degrade and simplify the structure of ecosystems in a somewhat similar manner. They tend to favor small, rapidly reproducing organisms low in the food chain at the the expense of larger but more slowly reproducing carnivores higher in the food pyramid. These processes have been articulated with the usual clarity by Woodwell (1970).

The effects of persistent CHs on ecosystems are better understood than are those of most other pollutants. For a variety of reasons, their maximum impact tends to involve organisms high in the food chain (Moore, 1967; Harrison et al., 1970). Biological concentration subjects predators to the greatest dosages of CHs, and by virtue of their fewer numbers and slow rates of reproduction the predators are less capable of sustaining increased mortality than are the smaller, more rapidly reproducing organisms further down in the food pyramid. The latter can quickly develop populations that are resistant to CHs, whereas the predators cannot afford the mortality that is always the price of resistance.

The effects of CHs on predators is apparent in a number of instances (Moore, 1967). The large predatory fish are diminished by decreases in reproductive success, and declines in populations of carnivorous birds for the same reason are even more obvious. Perturbations among other components of the ecosystem inevitably follow decreases of predators.

Losses of predators and parasites following the use of insecticides are especially marked among insect communities, a phenomenon that has been well documented (Ripper, 1956). Phytophagous (plant-eating) insects are well equipped for chemical warfare, but entomophagous (insect-eating) insects are not so endowed (Krieger et al., 1971). Losses of predatory and parasitic insects frequently cause enormous outbreaks of phytophagous insects, an effect that may be highly detrimental to agriculture (Huffaker, 1971). The strategy of attempting to control insects by inflicting mortality with broad spectrum, persistent poisons would appear to be counterproductive over the long term because it is ultimately beneficial to populations of phytophagous insects and other herbivores, and deleterious to many higher, non-target organisms, thus degrading the structure, diversity, and stability of ecosystems.

## Human Health Implications of Chlorinated Hydrocarbon Insecticides

Apparently all human beings carry residues of DDT in their tissues, and most of them contain Dieldrin and other CH residues as well (U.S. Department of Health, Education & Welfare, 1969). What is the significance of these residues to human health? After several decades of using these materials there is still no adequate answer to this question.

Humans are very poor experimental animals. They tend not to volunteer for experiments that terminate with their sacrifice and dissection; yet such experiments must be done to evaluate the biological effects of a chemical or drug. Laboratory animals therefore 'volunteer' their services in these experiments, and most of our knowledge of these subjects depends on their dedication and self-sacrifice. Mice and rats are not men, but their similarities to man are greater than their dissimilarities. In actual practise, the correlation between findings in laboratory animals and those in man is quite high. Results in animals therefore indicate the *probability* that man would react similarly. Often we have no choice but to accept results on experimental animals, and prudence dictates that we base our actions on these experiments where experimentation with human subjects is inadequate or lacking. Studies on laboratory animals suggest at least four areas in which human populations may be affected by current levels of exposure to CHs, especially to DDT residues.

### *Behavioural Effects*

CHs are nerve toxins (O'Brien, 1967). At concentrations well below those producing obvious toxic symptoms, a variety of behavioural changes are known to occur in experimental animals (Desi et al., 1966; Anderson & Peterson, 1969; Wurster, 1969a). At the concentrations of CHs which are present in the tissues of the general human population, such effects would be extremely difficult to detect, especially in the absence of controls, in the presence of a host of other variables, and with inadequate testing procedures. We do not know whether or not a threshold exists—a concentration of CH above which effects occur but below which they do not. Nevertheless, until better evidence is available we should assume that behavioural changes in laboratory animals are indicative of comparable effects within the human population. Such changes could be of great importance to human society, but they have not been studied. The absence of knowledge, however, is not evidence of safety.

### *Hepatic Enzyme Induction*

Most CHs are enzyme inducers, i.e., they induce a substantial increase in the synthesis of relatively nonspecific hepatic enzymes that hydroxylate steroid hormones and various other substrates (Conney, 1967; Kupfer, 1967; Peakall, 1970*b*; U. S. Department of Health, Education & Welfare, 1969). The effect has been noted in experimental animals at dietary levels of a few parts per million, concentrations that are legally permissible in some human foods in the United States. Enzymes induced by DDT have reduced the concentration of endogenous estrogen in the blood of birds, an effect that causes a variety of behavioural and reproductive changes (Peakall, 1970*a*; 1970*b*). Induction of these enzymes has been shown to occur in men who had been occupationally exposed to DDT and Endrin (Jager, 1970; Poland et al., 1970). In man it is possible that these enzymes can reduce the level of circulating steroid hormones sufficiently to affect behaviour and other parameters that are influenced by these steroids, and to affect various drug interactions, including the function of oral contraceptives containing steroids, and other biochemical processes. Furthermore, it is not known whether there is a threshold concentration for CHs below which these effects would not occur. But, once again, the absence of knowledge is no indication of safety, and animal experimentation suggests that hepatic enzyme induction by CHs, especially by DDT residues, may be a human health hazard.

## Cancer

The evaluation of chemicals, drugs, and pesticides for carcinogenic activity in humans presents numerous inherent difficulties. These include (1) the inability to use human subjects, necessitating the employment of laboratory animals as substitutes, (2) the presence of cancer from numerous other, usually unidentified causes within the human population, and (3) the difficulty of testing at environmental exposure levels and attaining statistical significance when those levels might produce only a single tumour in thousands of individuals. Detection of carcinogenesis at environmental concentrations would require enormous and unmanageable numbers of animals to demonstrate a statistically significant increase in the frequency of tumour induction (Epstein, 1970). Nevertheless, induction of cancer in the general population at a rate of 1 in 10,000 individuals would involve about 20,000 persons in the United States alone—a number that would clearly be of major public health significance.

Some technique must be employed to increase the sensitivity of the testing procedure so that smaller, more practical numbers of test animals can be used. One way of doing so is to increase dosage levels, thereby also increasing the incidence or frequency of tumour formation. There is no evidence that an increase in dosage converts noncarcinogenic materials into carcinogens. Carcinogenesis is a specific, relatively rare biological event and the ability to induce it is possessed by few chemicals. The standard procedure for evaluating carcinogenesis involves high dosage levels with both positive and negative controls in laboratory animals, usually rodents (Epstein, 1970). This procedure shows 'a remarkable degree of concurrence . . . between chemical carcinogenesis in animals and that in man where it has been studied closely' (U. S. Department of Health, Education & Welfare, 1969). In the absence of better techniques, prudent public policy should be based on results obtained in this way.

Four different laboratory experiments on mice, rats, and trout have shown an elevated rate of carcinogenesis by DDT, while Aldrin, Dieldrin, Mirex, Strobane, and Heptachlor proved carcinogenic to mice (U. S. Department of Health, Education & Welfare, 1969). Carcinogenicity thus appears to be rather common among CH insecticides. Cancer induction in these animals indicates a high probability, but not a certainty, that these chemicals are also carcinogenic to humans. Evidence is lacking that there is a threshold or safe tolerance level for carcinogens, above which they cause cancer and below which they do not. The frequency or incidence of cancer induction by such chemicals may fall to zero only at a zero concentration of the chemical concerned. In the United States the Food, Drug and Cosmetic Act prohibits the presence of carcinogenic additives in human foods, but the federal government has not enforced this law adequately.

A further suggestion that DDT and Dieldrin are human carcinogens is found in two studies showing that victims of terminal cancer contain substantially elevated concentrations of DDT and Dieldrin residues in their adipose tissues, as compared with the general population (Casarett et al., 1968; Radomski et al., 1968). These elevated CH levels did not correlate with, and are therefore not explainable by, the loss of weight in these people prior to death. The presence of these elevated CH residues does not prove that they caused the cancers, but the findings are certainly consistent with the hypothesis that they did, and also with the results obtained with test animals.

One must conclude, based on the standard procedures for evaluating carcinogenesis, that these CHs represent a significant cancer hazard in the human environment.

## Mutagenesis

Genetic toxicity of a chemical can manifest itself as carcinogenesis, mutagenesis, or teratogenesis, depending on the location and maturity of the damaged cell, and often two or three of these phenomena may be caused by the same chemical (Legator, 1970). In addition to being a carcinogen, DDT has been shown to be a mutagen as well, as indicated by recent studies by the U. S. Food and Drug Administration using the dominant lethal test in rats (M. S. Legator, in preparation). The argument regarding mutagenesis is similar to that for carcinogenesis: mutagenesis in rats indicates a high probability, though not a certainty, that DDT is a human mutagen. The present level of contamination of human tissues with DDT could mean that future generations of human beings will be burdened by an increased incidence of genetic defects.

The ubiquity of CH residues in the human environment, together with the above-cited evidence of genetic toxicity, at least of the most prevalent of these chemicals, rather strongly suggests that man has, during the past quarter-century, increased the burden of genetically toxic agents throughout the entire human population of the earth. Although thousands or millions of people may be affected, however, a cause-effect relationship may remain unproven for many years, if, in fact, it can ever be proven. The logistic difficulties of such experimentation with humans may be insurmountable.

Several studies of the physiological effects of DDT, Aldrin, Dieldrin, and Endrin have involved human subjects (Jager, 1970; Hayes et al., 1971). These studies are deficient in experimental design, failed to study the most relevant parameters, and were more concerned with levels of CH storage than with physiological or biochemical effects. They establish only that under current environmental conditions, excluding accidents and suicides, members of the general population are not dying of acute CH insecticide poisoning, nor are they suffering overt, toxic symptoms. Long term, chronic effects were inadequately studied.

To be more specific, the investigations by Hayes et al. (1971) and those conducted in the Shell laboratories (Jager, 1970) had only men in their samples; women, children, and infants were not studied. The small numbers of men involved were completely inadequate to evaluate biological events (such as carcinogenesis or mutagenesis) that may occur once in many thousands of individuals. Periods of exposure were too short to detect biological effects involving induction periods that may be many years or decades. Emphasis was given to reviewing the men's attendance records at work, and many of the other simple blood and other routine tests performed were largely irrelevant. When 2 of 22 men who were being fed high dosages of DDT became severely ill after months on this diet, they were dropped from the experiment and excluded from the data with the conclusion that 'at no time was there any objective finding to indicate a relationship between illness and DDT storage' (Hayes et al., 1971).

It is unlikely that these tests on men could have detected behavioural changes, hepatic enzyme induction, carcinogenesis, mutagenesis, or other effects that might be anticipated in man because they occurred in experiments with laboratory animals. The authors concluded, nevertheless, that exposure to these CH insecticides involved no ill effects on human health—a conclusion that has been widely quoted by the pesticide industry. It seems remarkable that, although hundreds of millions of people have been exposed to these substances for more than two decades, their effects have been so inadequately tested by such primitive studies on such a small number of men!

## Insect Control with Chlorinated Hydrocarbons

The introduction of the CH insecticides during and shortly after World War II was accompanied by the optimistic belief by some people that these 'miracle' substances would eliminate our insect pest problems. These dreams proved naïve. Although the control of insect populations is a fundamentally ecological problem, these materials were developed and introduced by chemists and medical authorities with almost no ecological sophistication (Smith & Bosch, 1967). The CH insecticides are ecologically crude, powerful, and highly disruptive poisons within insect communities, so it is hardly surprising that problems soon appeared among populations of such rapidly reproducing and adaptive organisms.

The following are among the more serious problems that have occurred following the use of these materials in an attempt to control insect populations.

1. *Resistance*. When a high rate of mortality is inflicted by a poison on an insect population, a few insects survive because they have certain traits (detoxifying enzymes, behavioural mechanisms, less permeable cuticles than the others, etc.) that protect them from the poison (Brown, 1961). These survivors repopulate the region, and so the protective traits become more prevalent. Repeated insecticide applications further the process, resulting in heavy selection for those traits with survival value. The population soon consists of insects that can no longer be killed by the original insecticide at the original dosage. Insect resistance to CHs is now widespread, rendering these insecticides far less effective than they once were.

2. *Pest Resurgence*. Even in the 'monocultures' of modern agriculture, insects live in complex communities containing hundreds of different species (Smith & Bosch, 1967). Most of these species are maintained under biological control, so that only a very few achieve pest status, do economic damage, and require human intervention. The potential pest species are phytophagous (plant-eating) ones, and primary among their control agents are the entomophagous insects—the insect parasites and predators of other insects.

Most CH insecticides destroy phytophagous and entomophagous insects alike because they are broad spectrum, highly toxic poisons to all arthropods (Ripper, 1956). The entomophagous insects cannot recover, for lack of food or hosts, until after recovery of the phytophagous insects, which may vigorously rebound with an ample food supply (the crop), less intraspecific competition, and the absence of biotic pressure from their natural enemies. The pest insects may thus resurge to much greater proportions and numbers than were present before the insecticide was applied, thereby making the pest problem worse and creating the apparent need for more insecticide (Ripper, 1956; Smith & Bosch, 1967; Huffaker, 1971).

3. *Creation of New Pests.* A phytophagous insect species that was not previously present at pest densities may be elevated to pest status by the resurgence in numbers that follows destruction of its natural enemies by broad spectrum toxins such as the CH insecticides (Smith & Bosch, 1967).

In many instances, then, the use of certain insecticides aggravates the insect pest problems that they are intended to solve, or creates new ones. Without realizing what has happened to him, the farmer may find himself with nightmarish insect problems such as he has never known before. Farmers sometimes get 'addicted' to this insecticide treadmill, just as a person becomes addicted to drugs. The farmer, the consumer, and the environment all suffer while the insecticide manufacturer benefits. Since insect control is an ecological problem, it requires the employment of ecological principles and techniques to achieve a long range, satisfactory solution for the agricultural community and the environment.

## The Alternative: Modern Integrated Control

In contrast with the purely chemical approach to insect pest problems, modern integrated control employs an ecological approach to pest management by combining and integrating biological, chemical, and other effective measures into a single, unified pest management system. Insecticides are used only when and where necessary, and in a manner that is least disruptive to beneficial regulating agents in the environment (Smith & Bosch, 1967). Crop yields and farmers' profits are thereby generally increased, and environmental damage is minimized or eliminated. Integrated control is not a dream for the distant future, but is already available in many instances and can readily be developed in most other cases (Bosch, 1970; Huffaker, 1971). Most CH insecticides are incompatible with integrated control and, in fact, they destroy its operation.

Modern agriculture must adopt effective, economical and ecologically sound integrated insect pest management systems to avoid the numerous shortcomings, hazards, and high costs of complete reliance on insecticides. An increasingly hungry and polluted world can ill afford to continue on its present course, for the adaptable insects will be the ultimate winners.

## Choices for the Future

The current excessive dependence on insecticides, especially persistent CHs, seems filled with troubles for man. CH residues, particularly those of DDT, are diminishing the richness and diversity of the human environment by causing widespread declines in populations of many species of carnivorous birds, in some cases to very low levels or extinction. These residues threaten freshwater and marine fisheries by inhibiting the reproductive success of the larger predatory fish, thereby threatening an important protein source. They exhibit genetic toxicity to experimental animals, and are therefore probably adding to the existing burden of cancer and mutations within human populations. And finally, these materials are frequently counterproductive in achieving their intended objective of controlling insect populations.

Man can choose to continue these trends by maintaining current pesticide practises, with further deterioration indicated. Or he can choose to reverse them by adopting ecologically sound insect pest management using modern integrated control systems. These choices may seem simple, but they are less simple than they appear. It will not be easy to reverse the momentum of current policies, to adopt existing knowledge, to seek new knowledge via imaginative and unbiased research, and to re-educate a whole generation of people who believe that insecticides represent the only approach to insect control. The adoption of new approaches will be resisted by many and actively opposed by powerful economic interests, and the faint-hearted will not prevail over these forces. Ecological pest control without environmentally dangerous materials is essential, however, if we are to preserve the integrity of the biosphere as we know it.

## References *


Anderson, J. & Peterson, M. (1969). DDT: sublethal effects on Brook Trout nervous system. *Science*, 164, pp. 440–1.

Blus, L. J. (1970). Measurement of Brown Pelican eggshells from Florida and South Carolina. *BioScience*, 20, pp. 867–9.

Bosch, R. Van Den (1970). Pesticides: prescribing for the ecosystem. *Environment*, 12 (3), pp. 20–5.

Bowman, M. C., Acree, F., Lofgren, C. S. & Beroza. (1964). Chlorinated insecticides: fate in aqueous suspensions containing mosquito larvae. *Science*, 146, pp. 1480–1.

Brown, A. W. A. (1961). The challenge of insecticide resistance. *Bull. Entomol. Soc. Am.*, 7, pp. 6–19.

Burdick, G. E., Harris, E. J., Dean, H. J., Walker, T. M., Skea, J. & Colby, D. (1964). The accumulation of DDT in Lake Trout and the effect on reproduction. *Trans. Amer. Fish. Soc.*, 93, pp. 127–36.


*Documentation is not exhaustive, but is intended to guide the reader to important papers, reviews, or recent sources in which additional references will be found.

Butler, P.A., Childress, R. & Wilson, A. J. (1970). The association of DDT residues with losses in marine productivity. Food and Agriculture Organization (U. N.) Technical Conference on Marine Pollution and its Effects on Living Resources and Fishing, 9–18 December 1970, Rome, Italy. FIR: MP/70/E-76, 11 Nov. 1970, 13 pages.

Cade, T. J., Lincer, J. L., White, Roseneau, D. G. & Swartz, L. G. (1971). DDE residues and eggshell changes in Alaskan falcons and hawks. *Science*, 172, pp. 955–7.

Casarett, L. J., Fryer, G. C., Yauger, W. L. & Klemmer, H. W. (1968). Organochlorine pesticide residues in human tissue—Hawaii. *Arch. Environ. Health*, 17, pp. 306–11.

Conney, A. H. (1967). Pharmacological implications of microsomal enzyme induction. *Pharmacol. Rev.*, 19, pp. 317–66.

Desi, I., Farkas, I. & Kemeny, T. (1966). Changes of central nervous function in response to DDT administration. *Acta Physiol. Acad. Scient. Hungaricae*, 30, pp. 275–82.

Dvorchik, B. H., Istin, M. & Maren, T. H. (1971). Does DDT inhibit carbonic anhydrase? *Science*, 172, pp. 728–9.

Edwards, C. A. (1966). Insecticide residues in soils. *Residue Rev.*, 13, pp. 83–132.

Epstein, S. S. (1970). Control of chemical pollutants. *Nature (London)*, 228, pp. 816–9.

Harrington, R. W. & Bidlingmayer, W. L. (1958). Effects of Dieldrin on fishes and invertebrates of a salt marsh. *J. Wildl. Mgmt.*, 22, pp. 76–82.

Harrison, H. L., Loucks, O. L., Mitchell, J. W., Parkhurst, D. F., Tracy, C. R., Watts, D. G. & Yannacone, V. J. (1970). Systems studies of DDT transport. *Science*, 170, pp. 503–8.

Hayes, W. J., Dale, W. E. & Pirkle, C. I. (1971). Evidence of safety of long-term, high, oral doses of DDT for man. *Arch. Environ. Health*, 22, pp. 119–35.

Huffaker, C. B. (Ed.) (1971). *Biological Control*. Plenum Press, New York: 468 pp.

Jager, K. W. (1970). *Aldrin, Dieldrin, Endrin and Telodrin: an Epidemiological and Toxicological Study of Long-term Occupational Exposure*. Elsevier, Amsterdam: 234 pp.

Jefferies, D. J. & French, M. C. (1971). Hyper- and hypothyroidism in pigeons fed DDT: an explanation for the thin eggshell phenomenon.' *Environ. Pollut.*, 1, pp. 235–42.

Johnson, H. E. & Pecor, C. (1969). Coho Salmon mortality and DDT in Lake Michigan. *Trans. 34th North Amer. Wildl. Nat. Res. Conf.*, March 3–5, Washington,. C.

Korschgen, L. J. (1970). Soil-food-chain-pesticide wildlife relationships in Aldrin-treated fields. *J. Wildl. Mgmt.*, 34, pp. 186–99.

Krieger, R. I., Feeny, P. P. & Wilkinson, C. F. (1971). Detoxication enzymes in the guts of caterpillars: an evolutionary answer to plant defenses? *Science*, 172, pp. 579–81.

Kupfer, D. (1967). Effects of some pesticides and related compounds on steroid function and metabolism. *Residue Rev.*, 19, pp. 11–30.

Legator, M. S. (1970). Mutagenic effects of environmental intrusions. *Assoc. Food & Drug Offic. U. S.*, 34, pp. 3–5.

Lehner, P. N. & Egbert, A. (1969). Dieldrin and eggshell thickness in ducks. *Nature (London))*, 224, pp. 1218–9.

Macek, K. J. (1968 *a*). Reproduction in Brook Trout (*Salvelinus fontinalis*) fed sublethal concentrations of DDT. *J. Fish. Res. Bd. Canada*, 25, pp. 1787–96.

Macek, K. J. (1968 *b*). Growth and resistance to stress in Brook Trout fed sublethal levels of DDT. *J. Fish. Res. Bd. Canada*, 25, pp. 2443–51.

Menzel, D. W., Anderson, J. & Randtke, A. (1970). Marine phytoplankton vary in their response to chlorinated hydrocarbons. *Science*, 167, pp. 1724–6.

Moore, N. W. (1967). A synopsis of the pesticide problem. *Adv. Ecol. Res.*, 4, pp. 75–129.

Mulhern, B. M., Reichel, W. L., Locke, L. N., Lamont, T. G., Belisle, A., Cromartie, E., Bagley, G. E. & Prouty, R. M. (1970). Organochlorine residues and autopsy data from Bald Eagles, 1966–68. *Pesticides Monitoring J.*, 4, pp. 141–4.

Nash, R. G. & Woolson, E. A. (1967). Persistence of chlorinated hydrocarbon insecticides in soils. *Science*, 157, pp. 924–7.

O'Brien, R. D. (1967). *Insecticides, Action and Metabolism*. Academic Press, New York: 332 pp.

Park, K. S. & Bruce, W. N. (1968). The determination of the water solubility of Aldrin, Dieldrin, Heptachlor, and Heptachlor epoxide. *J. Econ. Entomol.*, 61, pp. 770–4.

Peakall, D. B. (1970 *a*). Pesticides and the reproduction of birds. *Scientific Amer.*, 222 (4), pp. 72–8.

Peakall, D. B. (1970 *b*). p,p'-DDT: effect on calcium metabolism and concentration of estradiol in the blood. *Science*, 168, pp. 592–4.

Peakall, D. B. & Lincer, J. L. (1970). Polychlorinated biphenyls: another longlife widespread chemical in the environment. *BioScience*, 20, pp. 958–64.

Peterle, T. J. (1969). DDT in Antarctic snow. *Nature (London)*, 224, p. 620.

Poland, A., Smith, D., Kuntzman, R., Jacobson, M. & Conney, A. H. (1970). Effect of intensive occupational exposure to DDT on phenylbutazone and cortisol metabolism in human subjects. *Clin. Pharmacol. Therap.*, 11, pp. 724–32.

Radomski, J. L., Deichmann, W. B. & Clizer, E. E. (1968). Pesticide concentrations in the liver, brain and adipose tissue of terminal hospital patients. *Fd. Cosmet. Toxicol.*, 6, pp. 209–20.

Ratcliffe, D. A. (1970). Changes attributable to pesticides in egg breakage frequency and eggshell thickness in some British birds. *J. Appl. Ecol.*, 7, pp. 67–115.

Ripper, W. E. (1956). Effect of pesticides on balance of arthropod populations. *Ann. Rev. Entomol.*, 1, pp. 403–38.

Risebrough, R. W., Huggett, R. J., Griffin, J. J. & Goldberg, E. D. (1968). Pesticides: Transatlantic movements in the Northeast trades. *Science*, 159, pp. 1233–6.

Risebrough, R. W. (1969). Chlorinated hydrocarbons in marine ecosystems. Pp. 5–23 in Miller, M. W. & Berg, G. G. (Eds.), *Chemical Fallout: Current Research of Persistent Pesticides*. C. C. Thomas, Springfield, Ill.: 531 pp.

Rudd, R. L. (1964). *Pesticides and the Living Landscape*. University of Wisconsin Press, Madison: 320 pp.

Smith, R. F. & Bosch, R. Van Den (1967). Integrated control. Pp. 295–340 in *Pest Control: Biological, Physical and Selected Chemical Methods*, Ed. W. W. Kilgore & R. L. Doutt. Academic Press, New York: 477 pp.

Stickel, W. H., Stickel, L. F. & Spann, J. W. (1969). Tissue residues of Dieldrin in relation to mortality in birds and mammals. Pp. 174–204 in *Chemical Fallout: Current Research on Persistent Pesticides*, Eds. M. W. Miller & G. G. Berg. C. C. Thomas, Springfield, Ill.: 531 pp.

U. S. Department of Agriculture (1970). *The Pesticide Review 1970*. Agric. Stabiliz. Conserv. Serv., Washington, D. C.: 46 pp.

U. S. Department of Health, Education & Welfare (1969). *Report of the Secretary's Commission on Pesticides and Their Relationship to Environmental Health*. E. M. Mrak, Chairman, Parts I and II, Washington, D. C.: 677 pp.

Valin, C. C. Van, Andrews, A. K. & Eller, L. L. (1968). Some effects of Mirex on two warm-water fishes. *Trans. Amer. Fish. Soc.*, 97, pp. 185–96.

Verrett, M. J. & Desmond, A. H. (1959). Inhibition of carbonic anhydrase by chlorinated hydrocarbons. *Pharmacologist*, 1, p. 72.

Welch, R. M., Levin, W. & Conney, A. H. (1969). Estrogenic action of DDT and its analogs. *Toxicol. Appl. Pharmacol.*, 14, pp. 358–67.

Wiemeyer, S. N. & Porter, R. D. (1970). DDE thins eggshells of captive American Kestrels. *Nature (London)*, 227, pp. 737–8.

Woodwell, G. M. (1970). Effects of pollution on the structure and physiology of ecosystems. *Science*, 168, pp. 429–33.

Woodwell, G. M., Wurster, C. F. & Isaacson, P. A. (1967). DDT residues in an East Coast estuary: a case of biological concentration of a persistent insecticide. *Science*, 156, pp. 821–4.

Wurster, C. F. (1968). DDT reduces photosynthesis by marine phytoplankton. *Science*, 159, pp. 1474–5.

Wurster, C. F. (1969 *a*). Chlorinated hydrocarbon insecticides and the world ecosystem. *Biological Conservation*, 1 (2), pp. 123–9.

Wurster, C. F. (1969 *b*). Chlorinated hydrocarbon insecticides and avian reproduction: how are they related? Pp. 368–89 in *Chemical Fallout: Current Research on Persistent Pesticides*, Eds. M. W. Miller & G. G. Berg. C. C. Thomas, Springfield, Ill.: 531 pp.

Wurster, D. H., Wurster, C. F. & Strickland, W. N. (1965). Bird mortality following DDT spray for Dutch elm disease. *Ecology*, 46, pp. 488–99.

# 21

# Polychlorinated biphenyls: another long-life widespread chemical in the environment

*David B. Peakall*
*Jeffrey L. Lincer*

The recent finding that pelagic birds dying on the coasts of Great Britain had polychlorinated biphenyls (PCBs) in their livers in concentrations of several hundred parts per million (Bourne and Mead, 1969) shows that these compounds are present in the ecosystem in large amounts. Thus, it seems worthwhile to summarize and evaluate our current knowledge of these compounds.

## Structural and Physical Properties

The picture is complicated by the fact that we are not dealing with a single compound. The basic structure of PCBs is shown in Figure 21–1. Any of the positions marked with an $x$ can be substituted by chlorine. Widmark (1968) has calculated that of the 210 possible combinations 102 are probable. His criteria for these limitations are those compounds containing five to eight chlorine atoms per molecule and the number of chlorine atoms per ring differing by not more than one.

The commercially available Aroclors (Monsanto Company Trade-

Reprinted with permission from *BioScience*, 20:958–964 (1970). The authors are currently at the Langmuir Laboratory, Cornell University, Ithaca, N.Y.

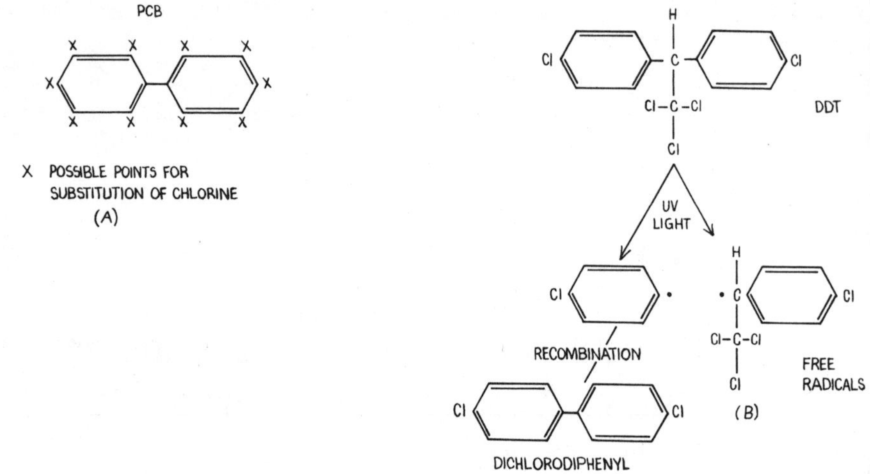

FIGURE 21–1. Basic structure of PCBs (A) and possible reactions of PCBs (B).

mark) are designated by numbers (Monsanto Technical Bulletins). The first two digits represent the molecular type: 12–chlorinated biphenyls; 25 and 44—blends of chlorinated biphenyls and chlorinated terphenyls (75% biphenyl and 60% biphenyl, respectively); 54—chlorinated terphenyls. The last two digits give the weight per cent of chlorine. Thus, Aroclor 1242 is a chlorinated biphenyl containing 42% chlorine. The biphenyls commercially available from Monsanto range from 21% to 68% chlorine. Mass spectrographic studies of Aroclor 1260 (Koeman et al., 1969a) show the presence of 11 isomers: five containing six chlorine atoms, five containing seven chlorine atoms, and one containing eight. Bagley et al. (1970), studying Aroclor 1254, found 18 compounds: one containing three chlorine atoms, four containing four chlorines, four containing five chlorines, five containing six chlorines, and four containing seven chlorines. Thus, the number of compounds present is, fortunately, much smaller than is theoretically possible.

PCBs are chemically inert, are not hydrolyzed by water, and resist alkalies, acids, and corrosive chemicals. They have low volatility, their boiling points ranging from 278 C for Aroclor 1221 to 415 C for Aroclor 1268 (Penning, 1930). All are stable to prolonged heating at 150 C, and the lower Aroclors can be distilled at atmospheric pressure without appreciable decomposition. PCBs are described as insoluble in water and very soluble in hydrocarbon solvents, although no exact figures appear to be available. Nothing is known about the biological decomposition of PCBs, but it is likely that they are more stable than DDT and its metabolites since they lack the ethane component between the aromatic rings, which is the site of action of most of the transformations of DDT. Thus, PCBs have the necessary physical and chemical characteristics for persistence and accumulation up the food chain.

## Use of PCBs

Polychlorinated biphenyls were first described in the literature in 1881 (Schmidt and Shultz, 1881), and the successful commercial production was achieved by the Swann Company in 1930. In that year the physical characteristics and commercial possibilities of these materials were described by Penning (1930). The uses lidted in the table are suggested applications taken from Technical Bulletin 306 of the Monsanto Chemical Company. PCBs are now manugactured by Monsanto in the United States (Trade name Aroclor), Prodelèe in France (Phenochlor), and Bayer in Germany (Colphen). Other Manufacturers are located in Japan and the Soviet Union. U.S. production for the decade 1961–1970 was 310,000 tons, with maximum annual values of 44,530 tons (1970) and 41,430 tons (1968).

TABLE 21–1

*Some uses of polychlorinated biphenyls taken from Monsanto Technical Bulletin O/PL-306*

| Material with which Aroclor is combined | Aroclor used with ut % | Use |
|---|---|---|
| Polyvinyl chloride | Aroclor 1248, 1254 & 1260 (7-8%) | Secondary plasticizer to improve flame retardance and chemical resistance. |
| Nitrocellulose lacquers | Aroclor 1262 (7%) | Co-plasticizer to enhance resistance. |
| Polyvinyl acetate | Aroclor 1221, 1232, 1242 (11%) | Improve quick-track and fiber-tear properties |
| Ethylene vinyl acetate | Aroclor 1254 (41%) | Pressure-sensitive adhesive. |
| Epoxy resins | Aroclor 1221 & 1248 (20%) | Increase chemical and oxidation resistance and adhesive qualities. |
| Polyester resins | Aroclor 1260 (10-15%)<br>Aroclor 1260 (10-20%) | Effective and economical fire retardant.<br>Increases strength of fiber-glass reinforced polyester resins. |
| Polystyrene | Aroclor 1221 (2%) | Plasticizer. |
| Chlorinated rubber | Aroclor 1254 (5-10%) | Enhances resistance, flame retardance, and improves electrical insulating properties. |
| Styrene-butadiene co-polymer | Aroclor 1254 (8%) | Improves chemical resistance. |
| Neoprene | Aroclor 1268 (40%)<br>Aroclor 1268 (1.5%) | Fire retardant.<br>Injection moldings. |
| Crepe Rubber | Aroclor 1262 (5-50%) | Plasticizer in paint compositions. |
| Varnish | Aroclor 1260 (25% of oil) | Improves water and alkali resistance. |
| Wax | Aroclor 1242 (5%) | Moisture and flame resistance. |

The use of highly chlorinated Aroclors for extending the kill-life of formulations containing chlordane, aldrin, and dieldrin is mentioned under miscellaneous applications of Aroclors. This possibility has been considered in a number of papers. Sullivan and Hornstein (1953), Hornstein and Sullivan (1953), and Tsao et al. (1953) found that the chlorinated terphenyl Aroclor 5460 increased the residual persistence of lindane. Tsao et al. (1953) considered that there was some indication that chlorinated polyphenyl may have a synergistic effect with lindane. Duda (1957) sprayed elm saplings with a lindane Aroclor 5460 mixture (1:4) and then tested the residual effect on the leaves to infestation with elm leaf beetle (*Galerucella xanthomelaena*). He found that leaves treated with lindane plus Aroclor were protected longer and that the insecticidal effect was more rapid than with lindane alone. Peaks corresponding to Aroclor 5460 have not been detected in the environment. However, under conditions of temperature and flow-rate commonly used for gas chromatographic analysis these compounds would not be detected (Reynolds, 1970).

Nearly 25 years ago it was noted that Aroclor 1242 was one of 175 compounds out of 6000 tested that was effective against mosquito (*Aegyps*) larvae (Deonier et al., 1946). Lichtenstein et al. (1969) tested the effect of adding a variety of biphenyls and terphenyls to DDT and dieldrin with respect to their toxicity to house and fruit flies. It was found that PCBs had very low toxicity to house flies when given alone but PCBs increased the toxicity of dieldrin and DDT, especially the latter. The effectiveness of these compounds decreased as the chlorinated level increaed. For example, Aroclor 1221 increased the mortality of fruit flies from 59% to 93%, whereas Aroclor 1268 increased it to only 77%.

## Analytical Methods

### 1) Identification

Since various compounds with electron-capturing properties have been tentatively identified in atmospheric samples in the past (Abbott et al., 1966), and some pesticides are capable of hybridizing in the soil to form a new compound (Bartha, 1969), it is important that the presence of PCBs in field samples be proven beyond doubt. Identification by means of a combination of high resolution gas chromatography and mass spectrometry has been carried out by three independent laboratories in Sweden (Widmark, 1967), Holland (Koeman et al., 1969 a), and the United States (Bagley et al., 1970). Widmark's report states that all peaks were identified by mass spectrometry (although no data were given), but Koeman and co-workers gave full experimental details, and Bagley et al. (1970) demonstrated that most chemicals in the eagle samples examined were components of Aroclor 1254. Thus, despite a recent statement by the Monsanto Chemical Company (in Risebrough, 1970) that the case for labeling the peaks in question as PCBs was not proven, there is enough evidence to convince an unbiased scientific jury.

### 2) *Separation*

The chemical techniques preliminary to quantitation of residues in samples containing both PCBs and chlorinated hydrocarbon pesticides fall into two groups—those necessitating the destruction or alteration of one or more of the compounds, and those which do not.

Included in the first group is nitration. Treatment with a 1:1 mixture of sulfuric acid-nitric acid at 0° C for 5 min destroys or alters aldrin, p.p′-DDE, p,p′-DDD (TDE), p,p′-DDT, and dieldrin such that they can no longer be detected at the original position on the chromatogram. This procedure leaves unaffected PCB, lindane, and BHC (Jensen and Widmark, 1967). A more rigorous nitration with a 1:1 mixture of sulfuric acid-fuming nitric acid for 15 min at room temperature removes, in addition to DDT and its related products, aldrin, heptachlor, Kelthane, Perthane, Tedion, Telodrin, and Trithion while lindane, heptachlor epoxide, toxaphene, and Strobane are not removed (Erro et al., 1967). Risebrough et al. (1969) reported that this nitration also removed the chromatographic peaks of PCBs. Reynolds (1969) reported that his attempts to nitrate samples were not fully successful in that there appeared to be loss of some of the more volatile PCBs while peaks with longer retention times appeared. Armour and Burke (1969) reported that complex chromatograms resulted after nitration which could not be related to the unreacted DDT-PCB mixture, and nitration was not pursued as a practical means of separating DDT and PCB for further tests.

Saponification with alcoholic NaOH or KOH will dehydrochlorinate Perthane, Toxaphene, DDD, and DDT to their respective olefins (Archer and Crosby, 1966; Klein and Watts, 1964). Risebrough et al. (1969) reported that PCB peaks are not removed or displaced but gave no data.

The second, and in some instances more desirable, group of analytical techniques allows the special separation of many chlorinated hydrocarbon pesticides from PCBs. Reynolds (1969, 1970) reported on an activated Florisil column technique which separated heptachlor, aldrin, DDE, and PCB with the first elution (60 ml n-hexane) from lindane; heptachlor epoxide, DDD, and DDT with the second elution (40 ml 50% ethyl ether in hexane). Armour and Burke (1970) developed a method utilizing a silicic acid-Celite column eluting aldrin and PCB with the first fraction (250 ml petroleum ether), and lindane, heptachlor, heptachlor epoxide, dieldrin, endrin, p,p′-DDE, o,p′-DDT, p.p′-DDT, and p,p′-DDD with the second fraction (200 ml acetonitrile hexane methylene chloride—1:19:80). Koeman et al. (1969a), using an activated Florisil column, eluted the apolar compounds including DDE and PCB with hexane, and then dieldrin and endrin with 10% diethyl ether in hexane. Snyder and Reinert (1971) using a silica gel column and eluting with pentane (first fraction) and benzene (second fraction) were able to clearly separate Aroclors 1254 and 1260 from DDT, TDE, and DDE. Recovery values generally exceeded 92%. Mulhern (1968) reported on a method which utilized silica gel-coated thin-layer plates and a hexane/ethyl ether (98.2) solvent system. The plates were developed, sprayed with a silver nitrate solution, and exposed to UV light. The plates were then divided into five horizontal sections. Dieldrin, endrin, γ-BHC, heptachlor epoxide, p,p'-DDD, p,p'-DDD, o,p'-DDT, and p,p'-DDE (in that order) were found in the first four fractions, while most of the interfering compounds found in wildlife samples were found in the fifth zone. Bagley et al. (1970) used this method but mentioned that zones three and four contained practically all unknown components as well as p,p'-DDT (zone three) and p,p'-DDE (zone four). Armour and Burke (1969) used precoated (aluminum oxide) sheets and n-heptane and 2% acetone/n-heptane for solvent systems. PCBs (Aroclors 1254 and 1260) and DDE were not separated, but p,p'-DDT and p,p'-DDD were completely separated from PCB by both solvent systems. Another possible means of separating interfering substances is to use a series of differing polarity columns in the gas chromatograph at the time of determination. This is less time consuming and may be useful when operating conditions can be selected such that PCB peaks are absent in the region where sought pesticides emerge (Simmons and Tatton, 1967).

### 3) *Quantitation*

Koeman et al. (1969 a) semiquantitatively measured the residues in Japanese quail fed phenochlor DP6 by using one of the peaks in a phenochlor DP6 mixture as a standard. Risebrough (1969) quantitated relative levels of PCBs by assuming that each PCB compound produced the same peak height with the electron capture detector as the same amount by weight of p,p'-DDE. After summing the heights of the individual peaks, the total was multiplied by a factor derived from measurements of standard solutions with electron capture and microcoulometric detectors. Jensen et al. (1969) reported PCB amounts as the sum of all PCB components and based the estimate on a combination of mass spectrometry and microcoulometric and electron capture detection. Even with this elaborate approach, these investigators suggest that the method is still rough and may be correct only within a factor of 2. Anderson et al. (1969) devised a method for obtaining a crude estimate of PCB residues as Aroclor 1254 from chromatograms where separation had not been attempted. On an empirical basis, it was found that Aroclor 1254 could be quantitated by considering peak 10 as p,p'-DDT and multiplying that value by 10. Since this peak had originally been quantitated as p,p'-DDT, and in fact many of the original samples contained little or no p,p'DDT, it was possible to estimate relative PCB values. Anderson and his co-workers also saponified samples to remove interfering p,p'-DDT and p,p'-TDE and then quantitated PCBs as Aroclor 1254 by relating sample peaks 9 and 10 to the corresponding peaks of an Aroclor 1254 standard. Reynolds (1970) employed a method similar to Koeman et al. (1969 a) but based his quantitation on an average of two or more peaks. In addition, his results were reported as Aroclor 1254 or 1260 depending on the overall pattern of the chromatographic peak profile. It is clear that we are still relatively unsophisticated in our PCB quantitation methodology and will continue to estimate only relative amounts of PCBs in field samples until we synthesize the individual PCB components commonly found in the ecosystem and are able to speak in terms of these individual peaks as we do for most pesticide residues. However, it has been kindly pointed out (Risebrough, pers. comm.) that, with reference to biological significance, the correct order of magnitude and accurate relative amounts of PCB give the essential information. This is because biological effects, such as enzyme induction, are related to degree of chlorination, and the existing methods give some indication of this activity. Therefore, that information would be lost if stress were placed only upon quantitating individual peaks.

### 4) *Magnitude of Error*

Ever since PCB peaks were recognized for what they are, residue chemists and other researchers interested in pesticide residues have asked what magnitude of error is likely to result from ignoring the presence of PCBs. Only recently have studies either directly or indirectly resulted in an estimate of this error. Anderson et al. (1969) quantitated p,p'-DDE, p,p' - DDD, and p,p' - DDT in five egg samples before and after saponification. There was no appreciable change in the p,p'-DDE, but apparent p,p'-DDD was reduced by approximately 58% and p,p' - DDT by 90%. Reynolds (1970) looked at a large number of samples before and after his PCB-Florisil separation and found that the actual p,p'-DDD residue (relative to the apparent residue) represented from 0 to 7% in California gull (*Larus occidentalis*) fat, 0% in cormorant (*Phalacrocorax auritus*) eggs, 80% in 10 pooled mallard (*Anas platyrhynchos*) duck eggs, and from 0 to 104% in great blue heron (*Ardea cinerea*) eggs. Respective values for p,p'-DDT were 0 to 53%, 13 to 41%, 100%, and 18 to 102%. Actual residue levels of heptachlor epoxide generally represented a majority of the apparent values. Before and after values for p,p'-DDE and dieldrin were not significantly different. Ob-

TABLE 21–2

*Pathologic changes induced by PCBs*

| Treatment | Animal | Liver | Kidney | Pericarduim and Peritoneum | Other Observable Changes | References |
|---|---|---|---|---|---|---|
| Single oral dose of 69 mg (42% cl) | Guinea Pig<br>Rat<br>Rabbit | Small fat droplets through lobules, slight to moderate central atrophy, focal necrosis noted in a few animals. | Essentially normal | No noteworthy changes | Adrenals, spleen, and pancreas showed no noteworthy changes. | Miller (1944) |
| 300 mg daily for 6 days (65% Cl) | Rat | Cells swollen, hyaline granules present, most died within few days. | | | | Bennett et al. (1938) |
| 50 mg daily for up to 6 months 65% Cl) | Rat | Enlarged (33% weight increase), large number of hyaline globules in cytoplasm. Several died during experiment. | | | | Bennett |
| 25, 50, & 100 ppm in diet for 15 days (21-68% Cl Aroclors) | Rat | Increase in weight, effect increasing with increasing chlorine content. Aroclor 1232-10%, 1242-12%, 1254-14%, 1268-24% at 50 ppm. | | | | Street et al. (In press) |
| 100 ppm in diet<br>200 ppm in diet<br>400 ppm in diet<br>800 ppm in diet<br>(Aroclor 1242) | Chicken | No effect<br>No effect<br>Enlarged and mottled<br>Damaged | <br><br><br>Damaged | Slight<br>Hydropericardium<br>Hydropericardium<br>Hydropericardium, hydroperitoneum, Enlarged. | | McCune et al. (1962) |
| 200 & 400 ppm in diet for 3 weeks (42%, Aroclor) | Chicken | No changes noted | Paleness at 200 ppm, extensive hemorrhage, and enlargement at 400 ppm. | Increased fluid in pericardial sac at the higher concentration. | Paleness of pancreas, enlargement of adrenal and small spleen at low concentrations. At higher concentrations pale cream-colored pancreas, adrenals hemorrhagic. | Flick et al. (1965) |
| Various doses (54% Cl, Aroclor) | Bengalese Finch | No weight changes | Weight was 32.4% of brain weight for controls and 53.5% for those dying from PCB poisoning. | Slight weight increase, a few showed liquid in pericardial sac. | | Presst et al. (In press) |
| 400 ppm in diet for 60 days (60% Cl[a]) | Chicken | Centrolobalar necrosis (compd 1 and 2). Liver weight increased from 2.76 g/100 g to 4.31 g/100 g (compd. 3). Fatty degeneration. | Tubular dilatation, (compd. 1 and 2) Rare with compd. 3. | Hydropericardium common with compds. 1 and 2. Rare with compd. 3. | Increased porphyria, spleen small with reduction of red pulp and atrophy of white pulp (compd. 1 and 2). Spleen decreased from 0.146 g/100 g to 0.136 g/100g, compd. 3). | Vos and Koeman (In press) |

[a]Phenoclor DP 6 (compd. 1), Clophen A60 (compd. 2) and Aroclor 1260 (compd. 3) were used. Differential effects noted under compd. numbers. All chickens died on compd. 1 and 2 within 60 days., only 15% mortality on compd. 3.

viously, the magnitude of error may vary depending on the trophic level sampled and certainly with the area from which a sample is collected (Risebrough et al., 1968, Jensen et al., 1969). Since the publishing of adequate separation techniques, there is no reason why there should be any error in pesticide quantitation contributed by PCBs.

## Toxicology

Despite some recent studies, the toxicology of PCBs remains rather poorly known as compared to that of the chlorinated hydrocarbon pesticides. For example, no definite work has been done to establish LD50 values for the various formulations of PCBs.

### 1) *Acute, single dose experiments*

Tucker and Crabtree (1970) found that a single dose of 100 mg/kg Aroclor 1254 (stomach tubed in oil) was fatal to two out of three rats, while 500 mg/kg was not fatal to three rats. Aroclor 1268 killed one rat out of three at 500 mg/kg, 1000 mg/kg, 2000 mg/kg, and 4000 mg/kg. Smyth (1931) found that a 4 g/kg (degree of chlorination not stated) was nontoxic to guinea pigs and rabbits, but this appears to be due to the fact that material which was given as a paste passed through the intestine unabsorbed. Tucker and Crabtree (1970) found that Aroclor 1242, 1254, 1260, and 1268 at a dose of 2000 mg/kg was not fatal to mallard ducks (four groups of three birds each).

### 2) *Acute, feeding experiments*

Monsanto Bulletin 306 states that 100 ppm diet had no effect on rats, although no details of the studies were given. The experiments of Bennett et al. (1938) in which 0.05 g/rat of 65% chlorine biphenyl was given orally every other day led to 50% mortality. If one assumes a body weight of 200 g, then the alternate day dose is roughly 250 mg/kg, which can be compared to oral LD50 for DDT of 113 mg/kg (Frear, 1968). Miller (1944) found that two oral doses of 69 mg of 42% chlorine biphenyl a week apart were fatal to guinea pigs. At an estimated body weight of 400 g, this gives a dose of 170 mg/kg. Tucker and Crabtree (1970) fed rats diets containing 10 and 1000 ppm Aroclor 1254. One rat out of six on the low dose died; this mortality was considered to be due to other causes. Four out of four of the high group died within 53 days and the calculated intake was 1330–1520 mg/kg. The food intake of the 1000 ppm group was only 79% of the control group.

Presst et al. (1970) have examined the toxicity of Aroclor 1254 to Bengalese finches (*Lonchura striata*). This is a difficult species for which to calculate the dietary intake. Due to their dependence upon unshelled food, it is only possible to present the PCB-laden food for a few hours a day (Jefferies, 1967). Loss by evaporation and loss by spillage must be allowed for, and increase of weight by defecation must be kept to a minimum. Thus, the calculated dietary intake is subject to more error than is usually the case. Presst et al. (1970) also measured the concentration of PCBs in the liver; they found that the range was large (i.e., 70–697 ppm in birds that died compared to 3–634 ppm in those that survived). These authors conclude that Aroclor 1254 has only 1/13 the toxicity of DDT, although the different shape of the mortality curves—steep with DDT, gradual with PCB—makes comparison difficult. Jefferies and Walker (1966) found good correlation between calculated dietary intake and live concentrations for p,p′-DDT in the Bengalese finch. However, other workers (Dale et al., 1963; Stickel et al., 1966; Stickel and Stickel, 1969) have considered that levels in the brain are a more reliable index of toxic levels than those in the liver or whole carcass.

De Vos and Koeman (1970) fed Phenoclor DP6, Clophen A60, and Aroclor 1260 to chickens at a dosage of 400 ppm. Mortality was complete (20/20) for those birds on Phenoclor in 12–58 days and in 13–29 days for Clophen. The mortality for Aroclor was only 3/20 for a 60-day period. This differential effect is unexplained and is surprising in view of the fact that all three formulations contain 60% chlorine. These workers measured the residue levels in the liver, and, for some birds, in the brain, for chickens dying during the experiment. Although there was considerable variation, most of the brain levels were between 210 and 420 ppm, which can be compared to 50–80 ppm for DDT (Stickel et al., 1966). On the basis of this work, PCB is 1/4–1/5 as toxic as DDT. Little difference was noted between the three different PCB formulations, although the number of determinations involved was rather small. If this finding is borne out by subsequent work, it would suggest that differential absorption from the gut or differential penetration of the blood brain barrier is involved. The liver values were more variable. There was a considerable number in the 200–400 ppm range, but there were several values in excess of 2000 ppm. This finding is in agreement with the findings of Dale, Stickel, and co-workers that the brain levels are the best indication of acute toxic levels.

McCune et al. (1962) found no mortality with chickens fed 100 or 200 ppm Aroclor 1242 in their diet for a 4-week period. On diets of 400 ppm and 800 ppm, the mortalities over a 4-week period were, respectively, 50% and 90%. During the first 3 weeks, the mortality figures were 10% and 50% respectively. Flick et al. (1965), using the same material at 400 ppm in the diet, had three birds out of 24 die in a 3-week period. Koeman et al. (1969a) found that a diet containing 2000 ppm Phenoclor DP6 caused complete mortality with Japanese quail (5/5) in 5–13 days and rats (8/8) in 1–56 days.

Schoettger (unpublished) found that the 96-hr TLm for Aroclor 1221 using cutthroat trout was 1.2 mg/1 and for Aroclor 1260 was 60.9 mg/1. In general, they found that the toxicity of Aroclors was inversely proportional to their percentage chlorination and directly proportional to their solubilities. These figures suggest that PCBs are two to three orders of magnitude less toxic to fish than DDT. Work by Lichtenstein et al. (1969) on house and fruit flies found that PCBs were 40–300 times less toxic than DDT, the toxicity decreasing as the chlorine content increases. Thus, it appears that the lower vertebrates and invertebrates are much less susceptible than mammals to direct toxicity from PCBs.

### 3) *Sublethal effects*

As with the chlorinated hydrocarbon pesticides, the most important effects are long-range sublethal effects. The pathologic changes in various organs are summarized in Table 21–2. The table shows some interesting differences between mammals and birds. The most striking finds in mammals are alterations to the liver, whereas fluid in the pericardial sac, kidney damage, and reduced spleen were found in birds.

McLaughlin et al. (1963) found that 25 mg Aroclor 1242 injected into the yolk sac of chicken eggs caused complete mortality, whereas 10 mg caused 95% failure and teratogenetic effects were noted among the young that hatched (beak deformity, edema, and growth retardant).

Induction of hepatic hydroxylating enzymes has been demonstrated in the pigeon (Risebrough et al., 1968), rat (Street et al., 1969), and American kestrel (*Falco sparverius*) (Lincer and Peakall, 1970). Street and co-workers studied the effects of a diet of 50 ppm and 100 ppm on sleeping time induced by a standard dose of hexobarbital, in vitro rates of aniline hydroxylation and demethylation of p-nitroanisole, and the rate of excretion of dieldrin. These workers studied 10 compounds ranging in chlorine content from 21% to 68% and found that all the effects increased with increasing chlorine content. For example, 50 ppm of Aroclor 1221 reduced hexobarbital sleeping time by 11%, whereas for Aroclor 1248 and 1268 the figures were, respectively, 35% and 48%. Thus, the sublethal effects have direct correlation with chlorine content, while the lethal effects appear to be inversely correlated (Tucker and Crabtree, in press). Lincer and Peakall (1970) noted an increase of the in vitro rate of metabolism of estradiol in kestrels fed 0.5 and 5 ppm Aroclor 1254 in their diet and also demonstrated increased levels of cytoplasmic RNA with the higher dietary level using a cytophotometric technique.

Tucker (unpublished) found that a single oral dose of 500 mg/kg Aroclor 1254 caused regular egg laying of Japanese quail (*Coturnix coturnix*) to turn first to scattered egg production and then to stop completely for a week. The scattered eggs had shells 9% thinner than normal, but thickness returned to control values when regular laying was resumed. Mallard ducks dosed with 1000 mg/kg of Aroclor laid one or no eggs before stopping for 1–2 weeks. The few eggs laid after the single large dose of Aroclor had shells 18% thinner, and again eggshell thickness was normal after resumption of regular laying. The period before egg laying has been noted to be increased in the ring dove (Peakall, unpublished). It is possible that the mechanism involved is increased rate of metabolism of circulating estradiol in the liver (Peakall, 1970).

## Levels of PCBs Found in Nature

Roburn (1965), comparing total chlorine (by concentration cell techniques) with results calculated from gas chromatography, found that some unknown chlorine compounds were present in several tissue samples and eggs of wild birds in Great Britain. The first identification of these materials was by Jensen and it is stated (Anonymous, 1966) that residues in feathers collected in Sweden go back to 1944. PCB residues have been reported in wildlife from Canada (Holden and Marsden, 1967; Anderson et al., 1969; Reynolds, 1970), Germany (Fiuczynski and Wendland, 1968; Koeman et al., 1967), Great Britain, (Holmes et al., 1967; Presst and Jefferies, 1969; Presst et al., 1970; Holden and Marsden, 1967), Netherlands (Koeman et al., 1967; 1969), Sweden (Anonymous, 1966; Jensen et al., 1969), and the United States (Anderson et al., 1969; Risebrough et al., 1968; Risebrough, 1969).

## Biological Magnification

No detailed studies, such as those for DDD at Clear Lake (Hunt and Bischoff, 1960) and DDT and its metabolites in Lake Michigan (Hickey et al., 1966), have yet been made for PCBs.

The most detailed studies currently available are those of Jensen et al. (1969). The figures (mean, range of values, and sample size) given in the table below are taken from their paper.

| | PCBs (ppm in extractable fat) | | | |
|---|---|---|---|---|
| | Baltic | | Stockholm Archipelago | |
| Mussel | 4.3 (1.9-8.6) | (40) | 5.2 (3.4-7.0) | (15) |
| Herring | 6.8 (0.5-23) | (18) | 5.1 (3.3-8.5) | ( 4) |
| Seal | 34 (16-44) | ( 3) | 30 (16-56) | ( 3) |
| Guillemot eggs | 250 (140-360) | ( 9) | ---- | |
| White-tailed Eagle | | | | |
| Pectoral muscle | ---- | | 14,000 (8400-17,000) | ( 4) |
| Brain | ---- | | 910 (490-1500) | ( 3) |
| Eggs | ---- | | 540 (250-800) | ( 5) |
| Heron | ---- | | 9,400 | ( 1) |

The levels in three species of fish in Clear Lake in 1968 were 0.03–0.005 ppm (wet weight) compared to 0.098 ppm in the breast muscle of a western grebe (Risebrough et al., 1969). Anderson et al. (1969) found that in most fish extracts the levels of PCBs were less than 0.1 ppm, whereas the levels in the eggs of cormorants (*Phalacrocorax auritus*) were 5–9 ppm. Presst, Jefferies, and Moore (unpublished) found that the livers of fish-eating birds in the British Isles ranged up to a maximum of 900 ppm (wet weight), bird feeders up to 70 ppm, mammal eaters to 50 ppm, and insectivores to 1 ppm. Unfortunately, no average values or prey items were included.

The physical properties of PCB and the available residue data clearly indicate that these materials are capable of biological magnification up the food chain.

## Ratio of DDT to PCB

Risebrough et al. (1968) and Risebrough (1970) have examined the ratio of total DDT, i.e., DDT and its metabolites to PCB. He has found that the ratio DDT/PCB was 1–2 in San Francisco Bay and 5–10 for seabirds in the Pacific; in the Gulf of California, a region relatively remote from contamination, the ratio was 9–10. Vermeer (quoted in Reynolds, 1970) in western Canada found a DDE/PCB ratio of 7 for California gull tissues and 13 for great blue heron eggs. In the Baltic the ratio was 1–2 (Jensen et al., 1969), although along the west coast of Sweden the ratio was as low as 0.15. In grebes in the British Isles the ratio was 0.4–0.8 (Presst and Jefferies, 1969). For sea-bird eggs, Presst et al. (unpublished) found ratios of 0.06 to 0.5.

The overall impression is that the amount of PCB in tissues tends to parallel that of DDE, at least in local ecosystems, and the DDE/PCB ratio is lowest near industrial areas suggesting that PCB is not carried quite so readily to remote areas. Nevertheless, the variation of the ratio is small enough to suggest that the routes of dispersal are similar. Since the evidence points to aerial fallout as the route of dispersal of the chlorinated hydrocarbon pesticides (Risebrough et al., 1968; Frost, 1969), it is likely that this is also the main route for PCBs. The pathways by which PCBs escape into the ecosystem are poorly known, although the possibilities have been recently discussed at some length (Risebrough, 1970; Reynolds, 1970). Since a large number of plastics and resins may contain PCBs (Table 21–1), the most likely route is combustion of these materials. This supposition remains to be tested. The possibility that PCBs could be derived from DDT should also be considered. The possibility that this conversion occurs in tissue is most unlikely for two reasons. First, it has never been detected despite the detailed work on the metabolism of DDT. Second, the only mechanism likely to give rise to a biphenyl is via free radicals and this is unlikely to occur in tissue. However, in the atmosphere under the influence of UV light, such a breakdown is more probable. A possible reaction is shown in Figure 21–1. Tautomeric shift could lead to a variety of isomers of dichlorophenyl, but it is difficult to envision the formation of more highly chlorinated biphenyls by this route. Since PCBs extracted from biological material match well with higher Aroclors (i.e., 1254), it seems unlikely that PCBs found in nature could be derived from other materials.

## Significance of Current Levels

In view of the similarity of PCBs to DDT and its metabolites, the addition of PCB residues to the environment is roughly equivalent to an increase of DDE residues. However, since a synergistic effect of PCBs on DDT has been demonstrated in insects (Tsao et al., 1953), this possibility should not be overlooked in higher organisms. The enzyme induction effects of PCBs have been well documented. The effect of PCBs on photosynthesis is a critical experiment that has not been done. Wurster's (1968) experiments with phytoplankton should be repeated with PCBs.

The most critical area for research on PCBs is to discover the major source(s) of escape into the environment. Legislative control of a material escaping as a side effect of its use may have a different set of problems than the control of pesticides which are broadcast as a function of their normal use. The evidence suggests that it is important to find the leak and stop it.

## Acknowledgments

We are grateful to Dr. Koeman (University of Utrecht), Drs. Presst and Jefferies (Nature Conservancy, Great Britain), Dr. Reynolds (Ontario Research Foundation), Dr. Street (Utah State University), and Dr. Tucker (U.S. Fish and Wildlife Service) for advance copies of important material. Some of the work presented here was carried out under NIH Grant ES00306, Dr. T. J. Cade, Principal Investigator. The review was written while one of us (D.B.P.) was an Established Investigator, American Heart Association. Thanks go to Dr. T. J. Cade for reviewing the manuscript.

## References

Abbott, D. C., R. B. Harrison, J. O'G. Tatton, and J. Thomson. 1966. Organochlorine pesticides in the atmosphere. *Nature*, **211:** 259–261.

Anderson, D. W., J. J. Hickey, R. W. Risebrough, D. F. Hughes, and R. E. Christensen. 1969. Significance of chlorinated hydrocarbon residues to breeding pelicans and cormorants. *Can. Field Natur.*, **83:** 89–112.

Anonymous, 1966. Report of a new chemical hazard. *New Sci.*, **32:** 612.

Archer, T. E., and G. G. Crosby. 1966. Gas chromatographic measurement of toxaphene in milk, fat, blood, and alfalfa hay. *Bull. Environ. Contam. Toxicol.*, **1:** 70–75.

Armour, J., and J. Burke. 1969. Polychlorinated biphenyls as potential interference in pesticide residue analysis. F.D.A. Laboratory Information Bull. No. 918, 11 p.

———. 1970. A method for separating polychlorinated biphenyls from DDT and its analogs. *J. Assoc. Offic. Anal. Chem.* (in press).

Bagley, G. E., W. L. Reichel, and E. Cromartie. 1970. Identification of polychlorinated biphenyls in two bald eagles by combined gas-liquid chromatography-mass spectrometry. *J. Assoc. Offic. Anal. Chem.*, **53**: 251–261.

Bartha, R. 1969. Pesticide residue interaction creates hybrid residue. *Science*, **166:** 1299–1300.

Bennett, G. A., C. K. Drinker, and M. F. Warren. 1938. Morphological changes in the livers of rats resulting from exposure to certain chlorinated hydrocarbons. *J. Ind. Hyg. Toxicol.*, **20:** 97–123.

Bourne, W., and C. Mead. 1969. Seabird slaughter. *B.T.O. News*, No. 36, p. 1–2.

Dale, W. E., T. B. Gaines, W. J. Hayes, Jr., and G. W. Pearce. 1963. Poisoning by DDT: Relation between clinical signs and concentration in rat brain. *Science*, **142:** 1474–1476.

Deonier, C. C., H. A. Jones, and H. H. Incha. 1946. Organic compounds effective against *Anopheles quadrimaculatus*. Laboratory test. *J. Econ. Entomol.*, **39:** 459–462.

Duda, E. J. 1957. The use of chlorinated polyphenyls to increase the effective insecticidal life of lindane. *J. Econ. Entomol.*, **50:** 218–219.

Erro, F., A. Bevenue, and H. Beckman. 1967. A method for the determination of toxaphene in the presence of DDT. *Bull. Environ. Contam. Toxicol.*, **2:** 372–380.

Fiuczynski, V. D., and V. Wendland. 1968. The population of *Milyus migrans* in Berlin 1952–1967. *J. Ornithol.*, **109:** 462–471.

Flick, D. F., R. G. O'Dell, and V. A. Childs. 1965. Studies of the chick edema disease. 3. Similarity of symptoms produced by feeding chlorinated biphenyl. *Poultry Sci.*, **44:** 1460–1465.

Frear, D. E. H. 1968. *Pesticide Handbook-Entoma.* College Science Publishers.

Frost, J. 1969. Earth, air, and water. *Environment*, **11:** 15–33.

Hickey, J. J., J. A. Keith, and F. B. Coon. 1966. An exploration of pesticides in a Lake Michigan ecosystem. *J. Appl. Ecol.*, **3** (Suppl): 141–154.

Holden, A. V., and K. Marsden. 1967. Organochlorine residues in seals and porpoises. *Nature*, **216:** 1274–1276.

Holmes, D. C., H. J. Simmons, and J. O.'G. Tatton. 1967. Chlorinated hydrocarbons in British wildlife. *Nature*, **216:** 227–229.

Hornstein, I., and W. N. Sullivan. 1953. The role of chlorinated polyphenyls in improving lindane residues. *J. Econ. Entomol.*, **46:** 937–940.

Hunt, E. G., and A. I. Bischoff. 1960. Inimical effects on wildlife of periodic DDD applications to Clear Lake. *Calif. Fish Game*, **46:** 91–106.

Jefferies, D. J. 1967. The delay in ovulation produced by pp′-DDT and its possible significance in the field. *Ibis*, **109:** 266–272.

Jefferies, D. J., and C. H. Walker. 1966. Uptake of pp′-DDT and its post-mortem breakdown in the avian liver. *Nature*, **212:** 533–534.

Jensen, S., and G. Widmark. 1967. Organochlorine residues, OECD preliminary study 1966–67. Report given at the OECD Pesticide Conference.

Jensen, S., A. G. Johnels, S. Olsson, and G. Otterlind. 1969. DDT and PCB in marine animals from Swedish waters. *Nature*, **224:** 247–250.

Klein, A. K., and J. O. Watts. 1964. Separation and measurement of Perthane, DDD (TDE), and DDT in leafy vegetables by electron capture gas chromatography. *J. Assoc. Offic. Anal. Chem.*, **47:** 311–316.

Koeman, J. H., A. A. G. Oskamp, J. Veen, E. Brouwer, J. Rooth, P. Zwart, E. V. D. Broek, and H. Van Genderen. 1967. Insecticides as a factor in the mortality of the Sandwich tern (*Sterna sandvicensis*). A preliminary communication. *Meded. Riijksfac. Landbouwwetensch. Gent.*, **32:** 841–853.

Koeman, J. H., M. C. Ten Noever de Brauw, and R. H. de Vos. 1969 a. Chlorinated biphenyls in fish, mussels and birds from the River Rhine and the Netherlands coastal area. *Nature*, **221:** 1126–1128.

Koeman, J. H., J. A. J. Vink, and J. J. M. de Goeij. 1969 b. Causes of mortality in birds of prey and owls in the Netherlands in the winter of 1968–1969. *Ardea*, **57:** 67–76.

Lichtenstein, E. P., K. R. Schulz, T. W. Fuhremann, and T. T. Liang. 1969. Biological interaction between plasticizers and insecticides. *J. Econ. Entomol.*, **62:** 761–765.

Lincer, J. L., and D. B. Peakall. 1970. Induced hepatic steroid metabolism and increased cytoplasmic RNA by polychlorinated biphenyls (PCB) in the American kestrel (*Falco sparverius*). *Nature* (in press).

McCune, E. L., J. E. Savage, and B. L. O'Dell. 1962. Hydropericardium and ascites in chicks fed a chlorinated hydrocarbon. *Poultry Sci.*, **41:** 295–299.

McLaughlin, J., Jr., G. P. Marliac, M. J. Verrett, M. K. Mutchler, and O. G. Fritzlaugh. 1963. The injection of chemicals into the yolk sac of fertile eggs prior to incubation as toxicity test. *Toxicol. Appl. Pharmacol.* **5:** 760–771.

Miller, J. W. 1944. Pathologic changes in animals exposed to a commercial chlorinated diphenyl. *U.S. Pub. Health Rec.*, **59:** 1085–1093.

Monsanto Technical Bulletins O/PL-311A, O/PL-306, and O-FF/1.

Mulhern, B. 1968. An improved method for the separation and removal of organochlorine insecticides from thin-layer plates. *J. Chromatogr.*, **34:** 556–558.

Peakall, D. B. 1970. Pesticides and the reproduction of birds. *Sci. Amer.*, **222:** 73–78.

Penning, C. H. 1930. Physical characteristics and commercial possibilities of chlorinated diphenyl. *Ind. Eng. Chem.*, **22:** 1180–1182.

Presst, I., and D. J. Jefferies. 1969. Winter numbers, breeding success, and organochlorine residues in the great crested grebe in Britain. *Bird Study*, **16:** 168–185.

Presst, I., D. J. Jefferies, and N. W. Moore. 1970. Polychlorinated biphenyls in wild birds in Britain and their avian toxicity. *Environ. Pollut.* (in press).

Reynolds, L. M. 1969. Polychlorobiphenyls (PCBs) and their interference with pesticide residue analysis. *Bull. Environ. Contam. Toxicol.*, **4:** 128–143.

———. 1970. Pesticide residue analysis in the presence of polychlorobiphenyls (PCBs). *Residue Rev.* (in press).

Risebrough, R. W. 1969. Chlorinated hydrocarbons in the global ecosystem. In: *Chemical Fallout*, G. G. Berg and M. W. Miller (Eds.), Charles C Thomas, Springfield, Ill., p. 5–23.

———. 1970. More letters in the wind. *Environment*, **12:** 16—27.

Risebrough, R. W., P. Reiche, D. B. Peakall, S. G. Herman, and M. N. Kirven. 1968. Polychlorinated biphenyls in the global ecosystem. *Nature*, **220:** 1098–1102.

Risebrough, R. W., R. Reiche, and H. S. Olcott. 1969. Current progress in the determination of polychlorinated biphenyls. *Bull. Environ. Contam. Toxicol.*, **4:** 192–201.

Roburn, J. 1965. A simple concentration-cell technique for determining small amounts of halide ions and its use in the determination of residues of organochlorine pesticides. *Analyst*, **90:** 467–475.

Schmidt, H., and G. Shultz. 1881. Einwirkung von Fünffach-Chlorphosphor auf das γ-diphenol. *Ann. Chem.*, **207:** 338–344.

Simmons, J. H., and J. O'G Tatton. 1967. Improved gas chromatographic systems for determining organochlorine pesticide residues in wildlife. *J. Chromatogr.*, **27:** 253–255.

Smyth, H. F. 1931. The toxicity of certain benzene derivatives and related compounds. *J. Ind. Hyg. Toxicol.*, **13:** 87–96.

Stickel, L. F., W. H. Stickel, and R. Christensen, 1966. Residues of DDT in brains and bodies of birds that died on dosage and in survivors. *Science*, **151:** 1549–1551.

Stickel, L. and W. Stickel. 1969. Distribution of DDT residues in tissues of birds in relation to mortality, body conditions, and time. *Ind. Med. Surg.*, **38:** 44–53.

Street, J. C., F. M. Urry, D. J. Wagstaff, and A. D. Blau. 1969. Comparative effects of polychlorinated biphenyls and organochlorine pesticides in induction of hepatic microsomal enzyme. Presented at ACS meeting September 8–12, New York.

Sullivan, U.N., and I. Hornstein. 1953. Chlorinated polyphenyls to improve lindane residues. *J. Econ. Entomol.*, **46:** 158–159.

Tsao, Ching-Hsi, W. N. Sullivan, and I. Hornstein. 1953. A comparison of evaporation rates and toxicity to houseflies of lindane and lindane-chlorinated polyphenyl deposits. *J. Econ. Entomol.*, **46:** 882–884.

Tucker, R. K., and D. G. Crabtree. 1970. *Handbook of Toxicity of Pesticides to Wildlife*. U.S. Dept. of Interior, Bureau of Sport Fisheries and Wildlife, Resources Publication No. 84.

de Vos, J. G., and J. H. Koeman. 1970. Comparative toxicological study with polychlorinated biphenyls in chickens with special reference to porphyria, edema formation, liver necrosis and tissue residues. *Toxicol. Appl. Pharmacol.* (in press).

Widmark, G. 1967. Possible interference by chlorinated biphenyls. *J. Assoc. Offic. Anal. Chem.*, **50:** 1069.

———. 1968. Determination of number of compounds which can result from the chlorination of biphenyl, and development of a simple system by which these may be codified. OECD report Sweden.

Wurster, C. F. 1968. DDT reduces photosynthesis by marine phytoplankton. *Science*, **159:** 1474–1475.

# RADIOACTIVE MATERIALS

# 22

# Radiation and the patterns of nature

*George M. Woodwell*

The partial answers we have to the question of what radiation does *in* and *to* nature are revealing not only of the effects of radiation on living systems, but also of the architecture of the systems themselves. My object is to show the patterns of the effects of radiation on natural communities, and how the patterns parallel and help to explain the normal patterns of structure, function, and development of these communities. It is important in this discussion to remember that most life as we know it has evolved in environments in which total exposures to ionizing radiation have amounted to less than a few tenths of 1 roentgen per annum, and that ionizing radiation is generally thought to have played a very minor role among the selective processes of evolution. It is somewhat surprising therefore that the effects of radiation on natural communities follow predictable patterns apparently related to the evolution of life.

The significance of natural communities to biology and to man is not immediately apparent. For my purposes it is important to recognize that all organisms have evolved as functional units in communities of organisms, and that the structure and function of these communities have determined in some measure the structure and function of the organisms themselves. So we can think of Darwin's struggle for existence as operative in the evolution of not only species but also groups of species

---

Reprinted with permission from *Science*, 156(3774):461–470 (28 April 1967).  Work carried out at Brookhaven National Laboratory under the auspices of the USAEC.

and whole communities. This is not a new concept; it was set down by Darwin in his *Origin of Species*, published in 1859.

The evolutionary implications of Darwin's struggle for existence at the community level are shown most clearly by a simple example, which is based rather freely on Darwin's own studies in the Galápagos Islands. Let us assume a small group of islands in the tropics, volcanic, and therefore young in a geologic sense, but supporting the limited flora and fauna that have arrived from the mainland some 1000 kilometers away. The climate is diversified, ranging from desert to moist forest. The islands have trees, grasses, and shrubs, but no mammals and few birds.

Over the years, probably hundreds of years, chance, possibly in the form of westering storms, brought small flocks of birds. From among these flocks at various times some finches survived and found a favorable habitat, rich in a diversity of foods and free of both mammalian and avian predators; reproducing rapidly, each new immigrant population became a plague, much as the Japanese beetle, the sparrow, the starling, the gypsy moth, and a host of other introductions have become plagues in our own experience. Food, although at first abundant, quickly became limited, and the struggle for existence intensified. Competition for food was fierce and a premium attached to any ability to exploit new food supplies—foods different from those exploited by competitors. Small differences in behavior or in size or shape of beak resulted in small differences in survival and in ability to rear young. These differences, when hereditary and useful, were passed on and amplified in the population, and on each island there developed a population of finches peculiarly adapted to that environment and different from populations on other islands.

There was one additional complication. Exchanges of individuals or small groups of individuals occurred occasionally among the islands, continually testing the degree of genetic isolation achieved by the evolution of different races. Frequently these transported populations failed on the new island or were absorbed into the now indigenous population; occasionally, however, a small one found itself partially isolated ecologically, by behavior, food supply, or local preference of habitat, from the indigenous population and survived as a distinct population, competition and evolution tending to accentuate the isolating mechanisms. Thus the islands gradually acquired a diverse bird fauna consisting largely of races of finches: ground finches, tree finches, a warbler finch, a woodpecker finch—each race using a set of resources used elsewhere in the world by a totally different species. Ecologists call the resources used by any one species a niche; where niches overlap and resources are shared, they say that competition occurs.

We see from this example, which is a grossly simplified version of *Darwin's Finches* (1), that the evolution of life proceeds toward reduction of competition, toward utilization of space and other resources, toward diversity in form and function, toward the filling of niches. We see, moreover, that the evolution of a race is affected not only by its physical environment, but also by the evolution of other races whose evolutions are in turn affected. Thus the chain of cause and effect here becomes entangled in bewildering ways. The product is a complex and, in some degree, mutable array of plants and animals which, itself, has clear and predictable patterns of structure, function, and development; these are "natural communities." Thus physical environments that are similar tend to support organisms that are similar in form and function, if not in species. So certain climates support forests the world over; others, grasslands; others, desert; and these words—forest, grassland, desert, and tundra—have meaning for us in terms of climate and flora and fauna.

Thus, where environments are similar, we find organisms that may have little or no common genetic past performing parallel functions. In Australia the marsupials, for instance, fill the grazing niches filled by placental mammals elsewhere; and the genus *Eucalyptus* has filled the tree niches occupied elsewhere by a score of other genera. The communities in which these organisms participate are one answer, tested through millions of years of evolution, to the very fundamental question: How can the resources of environment be used to perpetuate life? This is, of course, a fundamental objective of man: the use of environment to best advantage.

The evolutionary answer is a magnificently durable one and, in terrestrial communities, usually a surprisingly stable one, free of plagues or rapid changes in sizes of population. By this I mean that controls of population size have evolved, building stability into these complex biological systems—putting the "balance" into nature.

Now let us consider for a moment certain other characteristics of natural communities. It is clear that the communities have developed over long periods and are very much a product of the evolution of life; and that they vary in a spatial sense with geography, climate, topography, and a host of other environmental factors. They also vary with time.

To show the variation with time, let us assume for a moment that after we harvest our corn crop in the eastern United States we simply abandon the land. The weeds of the garden take over; crabgrass, at first; later, grasses; then, pine forest; and finally, after 100 years or so, an oak forest. The general pattern is familiar; environmental circumstance may modify details. The change from herbaceous weed field to forest involves not only changes in the species forming the communities, but also changes in the total weight of living matter on a unit of land, in the total amount of essential nutrients available, in the total amount of water used, in the total amount of niches available, and probably in the rates of biologic evolution itself. This process—succession—becomes one of the great central principles of biology.

We can examine one succession, from abandoned field to forest, most easily by considering stored energy in plants over time. By plotting such data (my own, and those produced by workers elsewhere in eastern North America) we obtain an S-shaped curve similar to the growth curve of a single organism (Fig. 22–1). It rises slowly during the early herba-

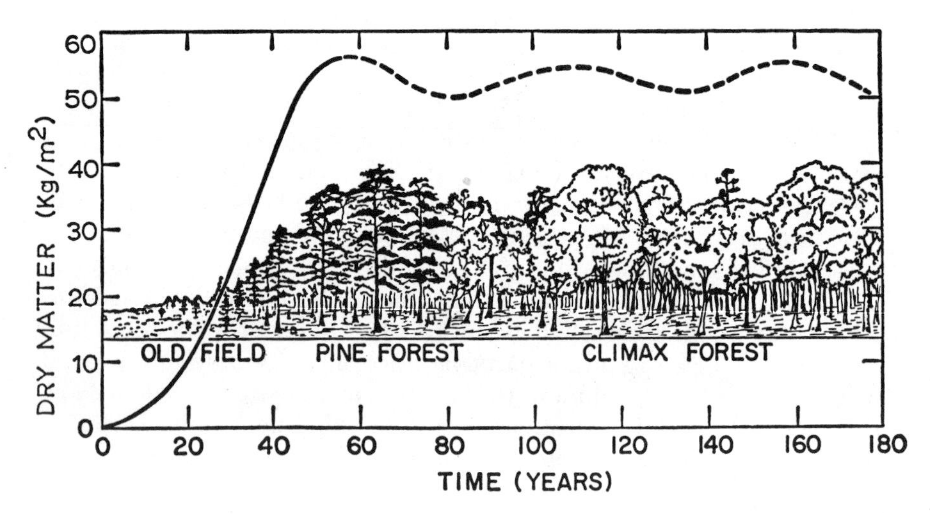


FIGURE 22–1. Field-to-forest succession in the eastern United States. The oscillations of climax are assumed.

ceous stages, rises much more rapidly during the pine-forest stage, and levels during the deciduous-forest stage as the degree of stability increases. Along this curve several very fundamental changes occur in community structure and function. There is, of course, a shift in species from herbaceous plants to trees. But there is also thought to be increase in diversity—total numbers of species present—from the few of the abandoned field to the many of the ultimate forest. There is change in degree of stability from the field, with its patches of ragweed and crabgrass which may be rapidly replaced by any of several species, to the forest with its spatial uniformity and slow replacement. There is increase in the total pool of minerals within the system: small amounts held within the herbaceous communities, large amounts in the forest. Total respiration and total photosynthesis increase, but at different rates, producing a regular change with time in the ratio of photosynthesis to respiration during the course of succession. We assume in addition that the total amount of water used increases along this succession (2).

If, in the course of such a succession, one or more factors essential to the system become exhausted or available only in short supply, the rate of succession is slowed and the climax is diverted, possibly by as much as from forest to grassland. Thus, in areas of low rainfall, succession ends in a stable grassland or woodland; where little mineral nutrient is available, whatever the reason, the succession is slowed and the S-shaped curve levels.

If, on the other hand, the environment is changed drastically by erosion or by sudden change in climate, or catastrophically by fire or windstorm or even by fallout from a bomb, then the changes that occur in these arrays tend to be just the reverse of those occurring during a normal succession: the communities are simplified, niches are opened, the nutrient inventory accumulated during succession is lost at least partially, the community becomes less stable, and a new succession begins, possibly marked by large fluctuations in populations that reproduce rapidly (such as insects) and can exploit the open niches.

Succession, then, is such a fundamental part of biology that it forms the logical core for appraisal of the effects of any change in environment, most especially a change that has such far-reaching and basic implications for life as ionizing radiation.

At first glance the problems in appraisal of the effects of ionizing radiation on the communities along a successional gradient seem so complex as to be impossible. But we can borrow a trick from the mathematicians and examine the effects on the extremes: we can use a gradient of exposures from very high to very low and examine the early stages of succession, which, in eastern North America, are abandoned fields, and the later stages, which are forests. The question we ask is, in each of these stages: What are the effects of irradiation on the community? In the forest, for instance, we need to know what exposure to radiation changes the composition of the plant community. When the composition does change, how does it change? Do species behave individually, or are there groups of species having similar characteristics? After what esposures do we expect insect populations to change? Do we affect metabolism, use of water? How do we affect them? Are there any patterns of radiosensitivity that may be useful for prediction of effects of radiation or for interpretation of the structure and function of unirradiated communities? The overriding question is: What are the patterns of radiation effects on the structure, function, and development of natural communities? This was the question posed in 1961 when the work at Brookhaven, which I shall discuss, was started.

We had then considerable information on radiation effects on many species of plants (3). It was known that the amount of damage caused by any exposure was related to the size and number of chromosomes in the cell nucleus (4). Sparrow had observed that certain species of pine trees are killed by exposures in the same general range as those killing man. Other data had shown a very great range, more than 1000-fold, in the sensitivity of plants to damage by radiation (5). The sensitivity of pines had been confirmed (6), and it had been shown that forests are generally more sensitive than had been known (7). Field observations, however, were most limited, and there was good reason to explore the problems experimentally and in detail.

Our approach entailed the establishment of two experiments, in each of which we used a single large source of $\gamma$-radiation (equivalent to about 9500 curies of $Cs^{137}$), arranged in such a way that it could be lowered into a shield (for safe approach) or suspended several meters above ground to provide radiation over a large area. The sources were large enough to administer several thousand roentgens per day within a few meters, the dose approaching background levels beyond 300 meters. The two experiments were conducted in an irradiated old field in the now well-known $\gamma$-radiation field established in 1949 (8), and in an irradiated forest—a completely new installation (9, 10); thus they gave us a sample from each end of the successional curve that I have discussed.

A section of the $\gamma$-radiation field was abandoned in the fall, after harvest, and the herbaceous communities common to abandoned gardens were allowed to develop. On Long Island about 40 herbaceous species participate in colonizing land prepared in this way; one of the most conspicuous is pigweed (*Chenopodium album*) because of its height (up to 1 meter) and abundance. During the 2nd year, horseweed (*Erigeron canadensis*) is the most conspicuous and one of the most abundant. In subsequent years, grasses such as broom sedge (*Andropogon* spp.) and asters (*Aster ericoides*) become dominants, to be followed by pine, and oak-hickory forest (11, 12).

Irradiation produced striking changes in the communities of the early stages of the succession. Although we have studied several of these communities over five summers at Brookhaven, I shall discuss here only the 1st-year communities. The most conspicuous change was drastic simplification at high exposures. We can measure simplification as a reduction in numbers of species per unit area, or in "diversity." Figure 22–2 is a plot of diversity along the radiation gradient. Irradiation at 1000 roentgens per day reduced diversity to about 50 percent of that of the unirradiated community, another field 2 kilometers distant. This decrease was continuous along the radiation gradient and was not marked by any abrupt decline indicating exclusion of several species in a narrow range of rates of exposure. Certain species survived daily exposures that exceeded 2000 roentgens.

The pattern of distribution of standing crop, or total weight of plants, at the end of the growing season, was strikingly different (Fig.

22–3). Total standing crop along the radiation gradient ranged between about 400 grams per square meter in the control community and 800 grams at 1000 roentgens per day, with a consistent increase with increase in exposure between these extremes. While the significance of this increase is not entirely clear, it is plain that, at exposures exceeding 1000 roentgens per day, total standing crop dropped abruptly to a few grams per square meter and, although some species survived even higher expo-

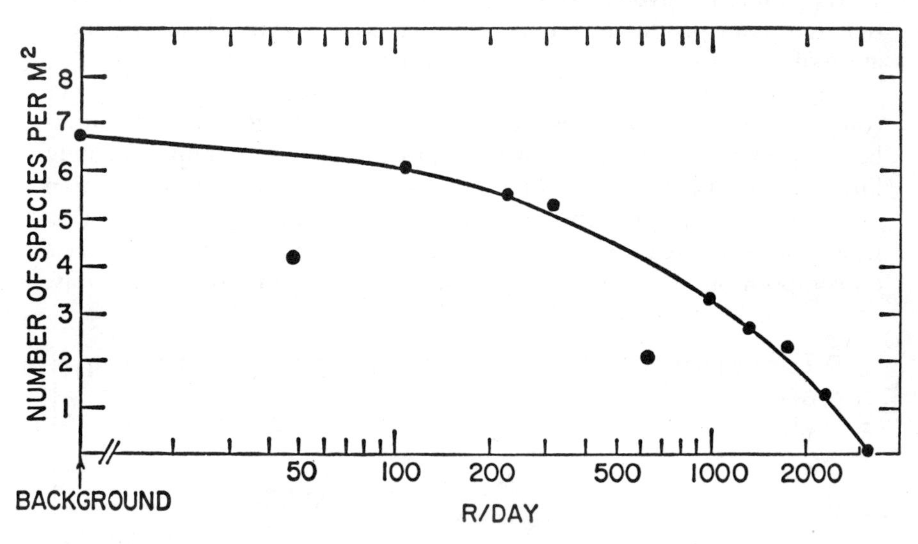


FIGURE 22–2. Diversity in the 1st-year-old field.

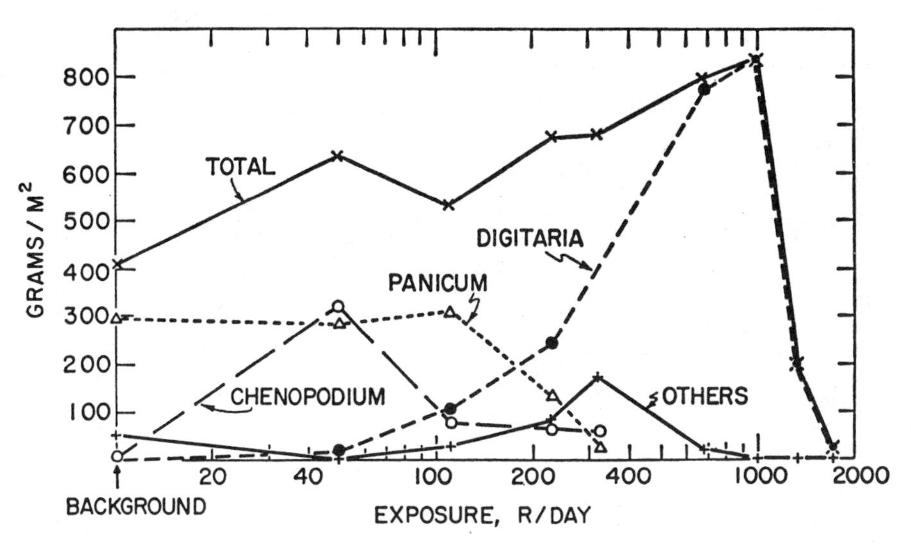


FIGURE 22–3. Total dry weight of plants, by species, in the irradiated old field. Dry weights were measured at the end of the season and do not represent total production.

sures, production of plant mass was very low indeed. There is clear evidence that at intermediate exposures exclusion of one species freed resources for others, crabgrass being by far the most benefited; at exposures exceeding 200 roentgens per day it was the major contributor to the total standing crop. Thus these old-field communities appear to be plastic, maintaining and possibly even increasing the total amount of energy fixed, despite a reduction in diversity of up to 50 percent. It also appears that diversity of species is more sensitive to radiation effects than is organic production. This relation is borne out by a brief consideration of coefficient of community, and percentage similarity.

The coefficient of community is simply the total number of species common to two communities, expressed as a percentage of the total number of species in both communities; Figure 22–4 shows an approximately linear relation between the coefficients of community along the radiation gradient, calculated for the control community, and the logarithm of radiation-exposure rate (11). There appears to be no threshold for effects on composition by species at exposures as low as 50 roentgens per day.

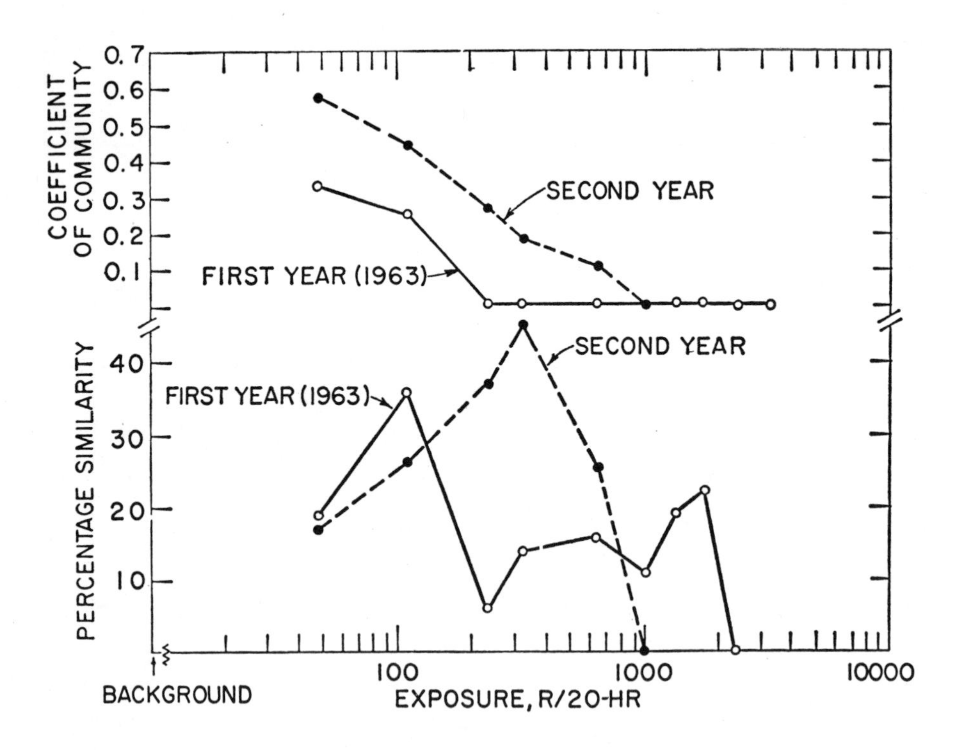


FIGURE 22–4. Coefficient of community, and percentage similarity for irradiated communities compared with the control community (2 kilometers distant). The linear relation between coefficient of community and the logarithm of exposure rate shows that species composition, alone, independent of density, is a useful criterion of the severity of disturbance by radiation.

If we weight the coefficient of community with a measure of abundance of each species, we can calculate what is called percentage similarity, and Figure 22–4. shows that there is no simple relation between these figures and radiation exposure, an observation that seems to confirm the earlier observation that the relative abundance of any species, however measured, is primarily controlled by competition with other species. Thus diversity and coefficient of community (and probably any other index of species diversity) emerge as relatively sensitive measures of radiation damage—and probably of any type of environmental change; abundance, density, and standing crop are insensitive.

Let us examine somewhat more closely the characteristics of plants that survive high rates of exposure. Two characteristics seem particularly significant: first, at high exposures the incidence of species that normally grow close to the ground [prostrate, decumbent, or depressed (13)] is substantially higher than in unirradiated communities (Fig. 22–5); second, there appears to be sorting on the basis of chromosome size, plants with large chromosomes being excluded from the areas receiving high exposures (Fig. 22–6). While it is difficult to venture a reason for apparent correlation between small size of chromosomes and a prostrate or decumbent growth habit among plants of old-field communities, these observations suggest that such a pattern may exist.

Thus the first year of succession is characterized by a loose array of herbaceous plants, most of them annuals or biennials, of varying life-forms and physiologies. Diversity in form and function allows rapid colonization of a wide variety of disturbed areas, and contributes toward making the community as a whole resilient in the face of disaster—such as a gardener's hoe or a gradient of ionizing radiation. The primary effect of stresses, including irradiation, is reduction of diversity. In the case of radiation, the reduction is continuous along the radiation gradient and not characterized by simultaneous exclusion of two or more closely associated species, an observation testifying to the looseness of the community organization. Although it is true that the plasticity of the community as a whole makes it resistant to radiation damage, it is certainly not true that all species in the community are equally resistant. Daily irradiation at 50 or more roentgens produced continuous sorting of species according to life-form and according to the average volumes of their chromosomes.

Irradiation of the forest commenced in November 1961 after a detailed series of preirradiation studies. The approach was to make a case-history study of one relatively complex ecological system by examining as many aspects as possible of its structure and function, both normal and pathological (10). Six months after installation of the source the forest appeared as in Figure 22–7; my data, with few exceptions, apply to the forest as it was in the summer of 1962, after approximately the same period of exposure as the old field.

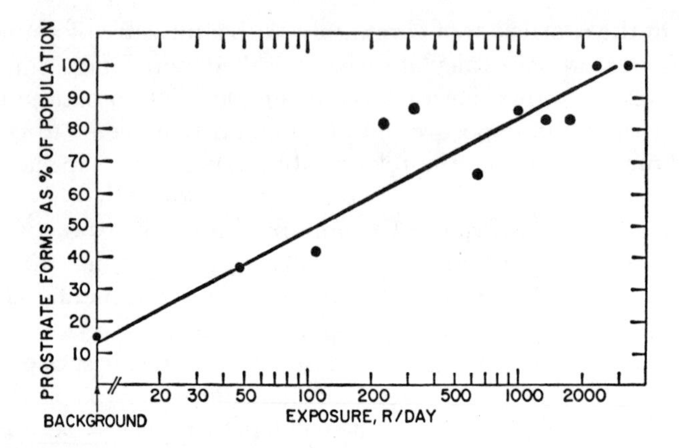


FIGURE 22–5. Life-forms in an irradiated field (1963, 1st year); "prostrate" forms include forms labeled normally prostrate, decumbent, or geniculate by Fernald (13).

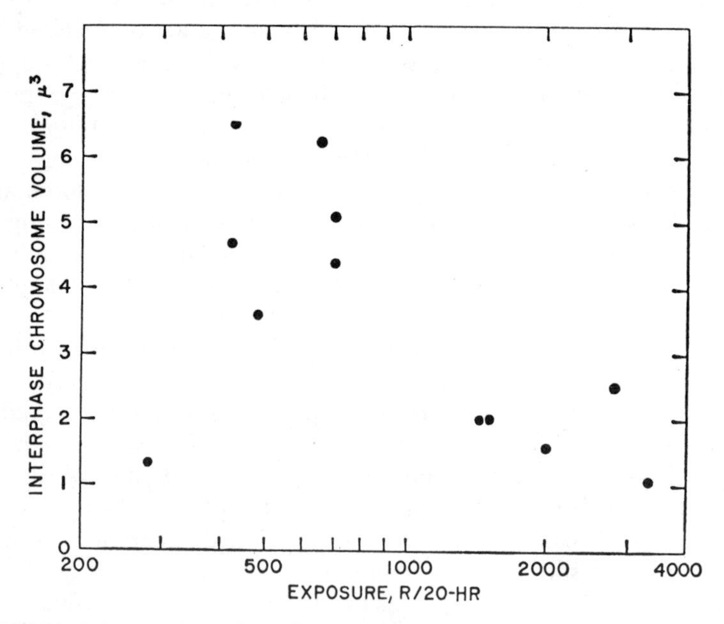


FIGURE 22–6. Average interphase chromosome volumes of 12 species of plants, showing maximum exposure at which any individual lived [from Wagner (*25*) ].

Five zones were apparent along the radiation gradient: a central zone in which no higher plants survived; a sedge zone containing *Carex pensylvanica* and a few sprouts of the heathshrub layer; a shrub zone where the two blueberries and huckleberry survived; an oak-forest zone at daily exposures less than about 40 roentgens; and the oak-pine forest in which radiation effects on growth were apparent, without change in species composition (14).

FIGURE 22–7. Effects of 6-month exposure to gamma radiation ranging in intensity from several thousand roentgens per day near the center of the circle to about 60 roentgens at the perimeter of the defoliated area. The experiment is part of a study at Brookhaven National Laboratory of the effects of chronic exposure to ionizing radiation.

The zoning of vegetation reflected the decline in diversity along the gradient (Fig. 22–8). If the normal "plot" in this forest be accepted as having 5.5 species, then 50-percent diversity occurred at 160 roentgens per day, or less than one-fifth the exposure to reduce diversity by 50 percent in the herbaceous community. Shielding by the stems of large trees in the forest allowed survival by species at average exposures substantially greater than the normally lethal exposures. Therefore the differential is probably even greater, and the forest may have its diversity depressed by 50 percent at exposures as low as one-tenth of those required in the herb field (15).

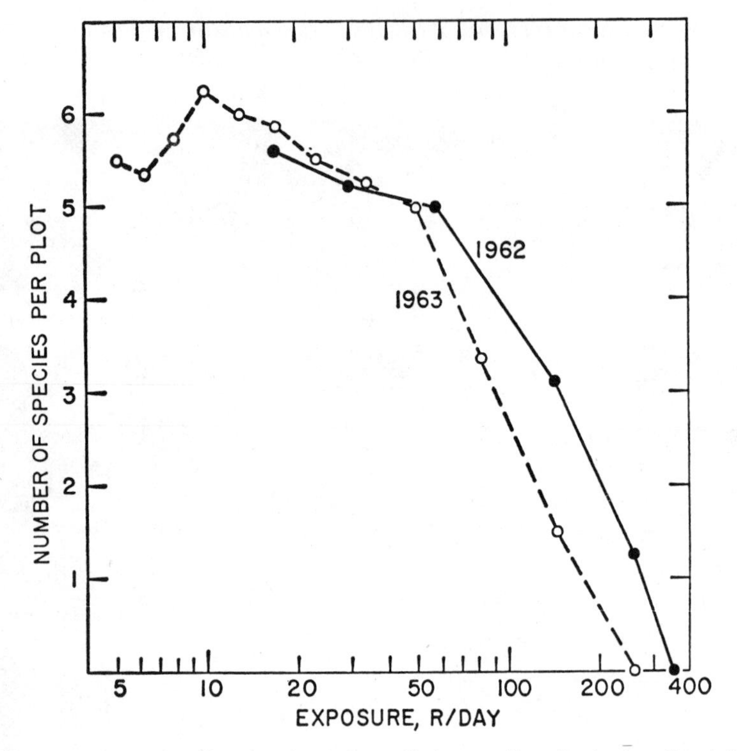


FIGURE 22–8. Species diversity along the radiation gradient in the irradiated forest in 1962 and 1963. Measurement of diversity in a forest requires differently sized samples for differently sized plants; thus the unit of diversity here is "species per plot" [from Woodwell and Rebuck (*15*) ].

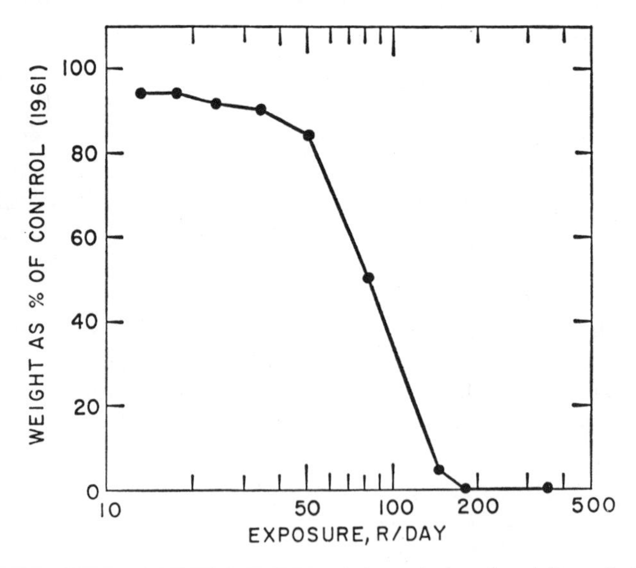


FIGURE 22–9. Total weight of above-ground shoots in irradiated forest in 1962.

Unlike the old field, standing crops in the forest declined at approximately the same rate as diversity (Fig. 22–9). This relation between diversity and abundance measured by standing crop is, of course, to be expected, since there is no possibility of a population of oak trees, 9 meters in height, expanding within a year to fill a niche vacated by pine. Nor was there invasion by any of the herbaceous species more resistant to radiation. There was, however, expansion of the population of *Carex*, a plant that normally occurs as a ubiquitous but very sparse herb, to cover as much as 20 percent of the total ground surface. This expansion was in response to the demise of the tree and shrub cover; it points to the potential importance of rare, or at least inconspicuous, species, capable of rapid regeneration, in maintaining certain aspects of function in disturbed communities.

Other examples of rapid response to the changed resources in the damaged community abound, especially among insect populations. In general these populations have followed quite closely change in food supply (16). Populations that utilize dead organic matter and decay organisms increased in the central zone of high mortality to the vegetation; bark lice are a good example (Fig. 22–10). While this type of change

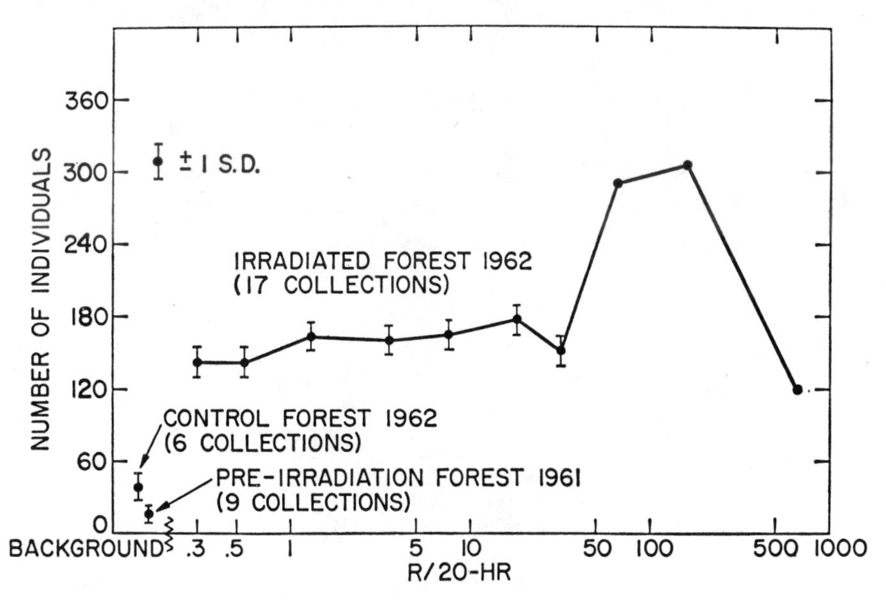


FIGURE 22–10. Abundance of bark lice (*Psocoptera*), which feed on decay organisms and dead organic matter, along the radiation gradient [from Brower (*16*) ].

seems quite straightforward and predictable, all changes in insect populations were not: during the 2nd year of the experiment, for instance, there was an unexpected and still-unexplained population explosion of aphids on white oaks exposed to 5 to 10 roentgens per day (Fig. 22–11). Aphids share with certain fungi, such as wheat rust, ability to reproduce asexually very rapidly to exploit any available resource. Although mobile, they are not strong fliers and do not migrate far; it is unlikely that the high populations resulted from migration from neighboring forests. It seems much more probable that leaves of trees exposed to 5 to 10 roentgens per day differed qualitatively from leaves of unirradiated trees sufficiently to support large populations of aphids: the difference appears to be not in either total sugars or total proteins, but in some more subtle factor detectable by aphids but not yet by man (17).

FIGURE 22–11. Abundance of aphids (*Myzocallis* sp.) on oak leaves in 1963 along the radiation gradient. At 9.5 roentgens per day, populations were more than 200 times normal.

profile (Fig. 22–12) showing the five vegetation zones and their approximate distribution along the radiation gradient in 1962. The most striking observation is the relative sensitivity of all the higher plants. No higher plant indigenous to the forest survived the 1st year off exposure exceeding 350 roentgens per day; in the old field, certain species survived more than 3000 roentgens per day. The 50-percent diversity point occurred in the forest at less than 160 roentgens per day; in the field, at 1000 roentgens per day. It seems abundantly clear that the forest as a unit is substantially more sensitive than the herb field. A second important relation is that there is sorting by size along the radiation gradient, smaller forms of life being generally more resistant than trees; this relation also extends to mosses and lichens.

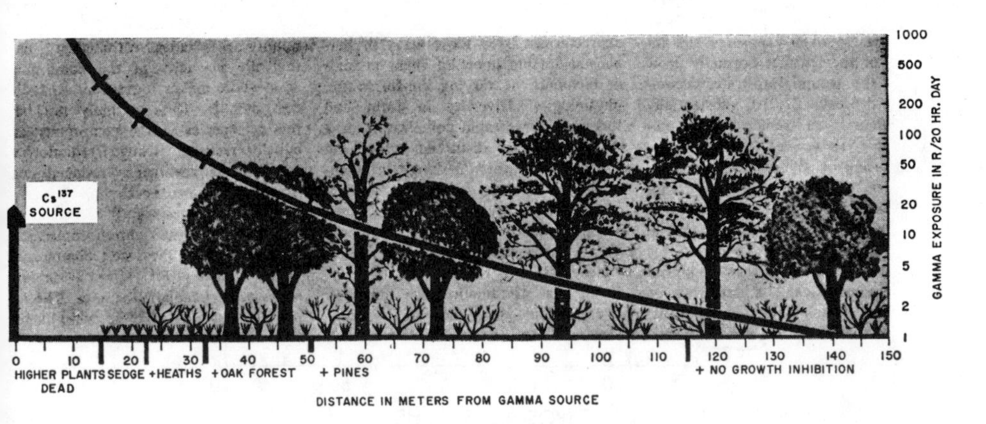


FIGURE 22–12. Pattern of radiation damage to oak-pine forest in 1962 after about 6 months' exposure.

This sorting by size, which now seems to be a well-established characteristic of radiation damage, has interesting parallels elsewhere in nature. It occurs along gradients of increasing climatic severity, such as the transition from forest to tundra in the north, and on mountain slopes. At such transitions, forest is replaced by low-growing shrubs, frequently blueberries and other members of the heath-plant family. In more extreme environments the heath shrubs are replaced by a sedge mat formed by a species of *Carex;* in the most extreme, the *Carex* is restricted to protected spots, and mosses and lichen are the only vegetation. The parallel with the irradiated forest is quite remarkable, holding even to genera and species, in certain instances. The conclusion to be drawn from this relation is merely that characteristics that confer resistance to certain types of environmental extremes also, curiously enough, confer resistance to damage by radiation.

We can test the hypothesis a little more rigorously by examining in detail the shrub layer of the forest, which is itself a small community containing two species of blueberries, the huckleberry, and the sedge. Changes in this community after burning have been studied intensively (18); their general pattern appears in Figure 22–13: with increased frequency of fire, the huckleberry populations decline, the blueberries increase, and the sedge increases. Under irradiation the pattern is strikingly similar until the point at which radiation kills the blueberry. The parallelism between the effects of fire and of radiation should not be expected to be universal, for many factors are implicated. Nonetheless

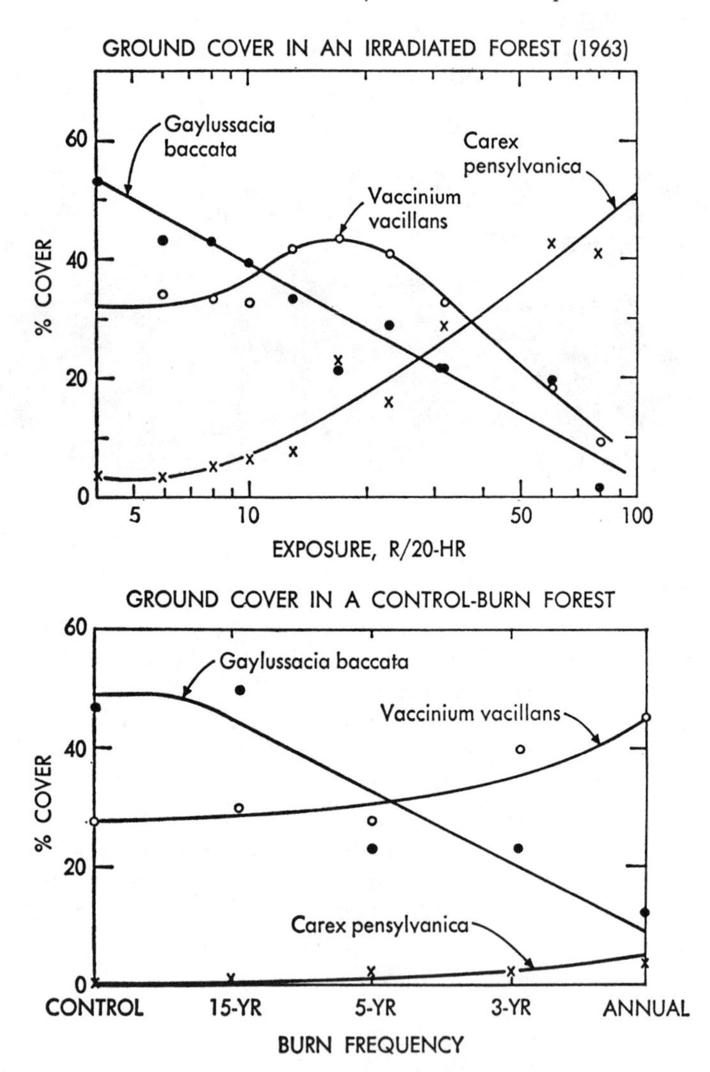


FIGURE 22–13. Comparison of effects of ionizing radiation and different frequencies of burning on the shrub-and-sedge community [from Brayton and Woodwell (*19*)].

there seems to be a strong parallel between the effects of radiation and the effects of another extreme; and in both instances, as well as in the herb field, the correlation between durability and small stature applies (19).

While there is no completely satisfactory explanation of the parallels, one important contributory factor may be simply the size of the plant. Perennialism, height and complexity of structure all represent investments of energy in nonphotosynthesizing tissue, tissue that requires energy for maintenance. We might think of this tissue as a mortgage that must be paid off with income from photosynthesis. As the size of a plant increases, both mortgage and total income increase, but at different rates. In Figure 22–14 are plotted the total weights of trees against $h \times d^2$, a measure of size (20). Since total weight of the tree is not a proper measure of total living tissue (there being considerable nonliving tissue

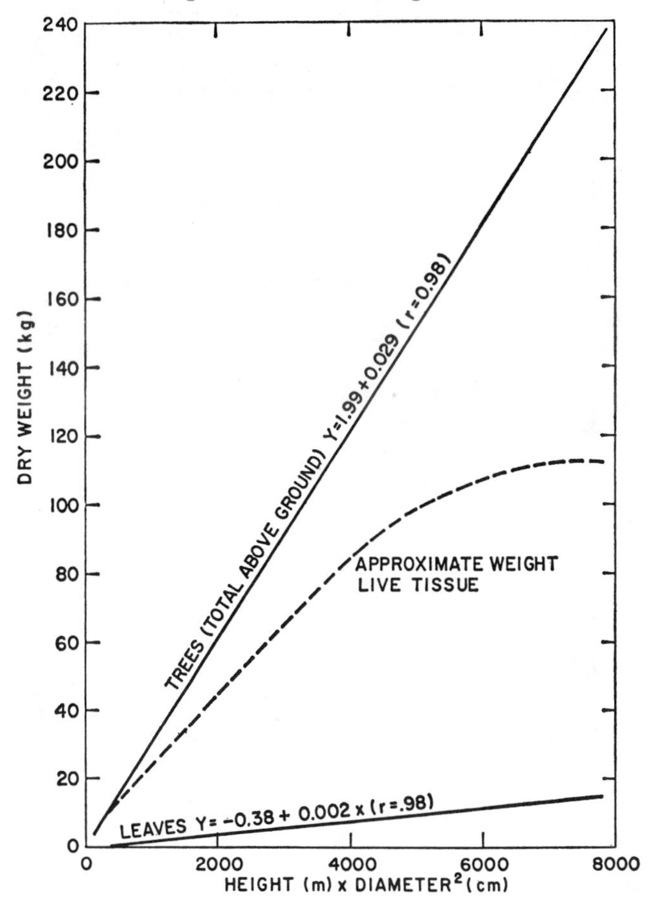


FIGURE 22–14. Relation between weights of leaves, weights of trees, and tree sizes for an oak-pine forest on Long Island.

in a tree), we have also plotted an estimate of the weight of tissue that may normally be considered living. It seems clear that in small trees leaves represent a substantially larger fraction of the total weight of live tissue than in large trees. An increase in size thus puts greater demands on the photosynthetic mechanism simply for maintenance, leaving less for growth and repair.

A similar relation applies along the successional gradient that we have discussed (Fig. 22–15). In the early stages of succession most of the tissue produced is green, and the mortgage payments to support respiration are small. As succession progresses, the complexity of structure increases, but total living tissue increases more rapidly than the weight of leaves, which supply the energy for respiration; the mortgage

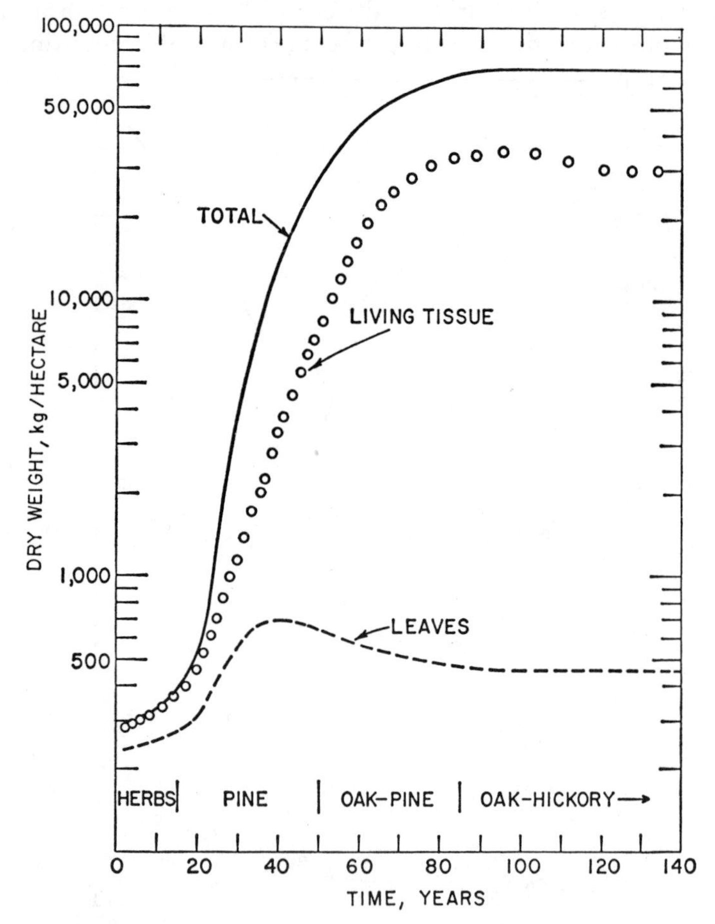


FIGURE 22–15. Approximate relations between total above-ground standing crop of plants, weight of living tissue, and weight of leaves in a normal field-forest succession of eastern North America.

increases, but income does not increase proportionally. It is true that the existence of the structure allows greater use of space, greater total photosynthesis up to a point, and greater diversity of species. But it is also true that the maintenance of the structure and diversity hinges on the annual interest paid from photosynthesis into the structural mortgage, and, if for any reason the interest is not paid, the structure begins to decay.

And here lies the crux of the matter: the mechanisms related to energy fixation—bud set, bud burst, leaf production, photosynthesis—are at much greater hazard than mechanisms related to energy use. Almost any disturbance of a forest may reduce its capacity for fixing energy; it either increases respiration or reduces it relatively slightly. If the disturbance is chronic, the vegetation comes to a new equilibrium, supporting a less complex structure. For this reason we might expect a forest to be more sensitive to disturbance than is an herb field because the forest is less plastic in species composition and because its capacity for fixing energy must remain substantially intact or it will burn up more than it fixes and deteriorate. And that is exactly what happens, but it is far from the whole explanation.

We have shown for the herbaceous field that there was sorting along the radiation gradient, dependent on chromosome volume: plants with large average chromosome volumes are sensitive; those with small volumes, resistant. The pattern in the forest was similar. If we plot the average chromosome volumes (8, 21) against the daily exposure required to inhibit growth to 10 percent of growth of unirradiated plants (Fig. 22–16), it is abundantly clear that radiosensitivity correlates with size of the chromosomes, and that this correlation applies to populations in nature as well as to cultivated populations. Also, the larger plants tend to have larger chromosome volumes; the smaller plants, smaller. Clearly, chromosome volume has played a role in the persistence of plants along the radiation gradient in the forest as well as in the field.

Now let me recapitulate briefly: the successional gradient we have used to explore effects of radiation on natural communities is characterized in the early stages by a loosely structured community or series of communities shifting in species composition, diversity, dominance, density, total mass of living matter, and probably in every other measurable parameter, within relatively broad limits, in response to disturbance. It is also true that the species of the early successional communities, including mosses and lichens, tend generally to be resistant to radiation. The forest does not share the plasticity of communities having simpler structure; in this respect the forest is more sensitive to any disturbance. In plants, large size alone, because of its effects on the ratio of photosynthesis to respiration, contributes to this type of sensitivity; but, more

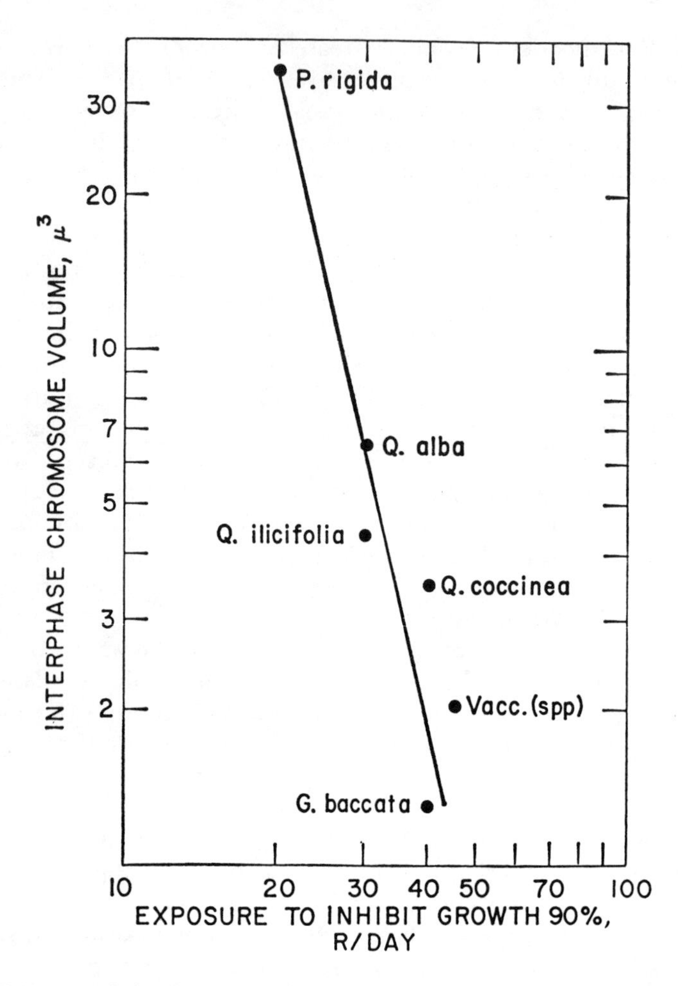


FIGURE 22–16. Relation between radiosensitivity, measured as inhibition of growth (90 percent), and interphase chromosome volume in irradiated forest (*21*).

importantly, plants of the forest are inherently sensitive to radiation damage because they have large chromosomes and because woody species in general are more sensitive than herbaceous plants having the same-sized chromosomes (22).

Thus there seems to be a shift toward greater sensitivity to radiation as succession progresses. The shift is due to at least three factors: (i) what I term the relative plasticity of the communities; (ii) increase in the amount of structure in the communities, with its implications for the photosynthesis:respiration ratio; and (iii) changes in the intrinsic characteristics of the plants participating in these communities, including changes in size of chromosomes. All these factors work in the same direction, contributing toward greater sensitivity to radiation and probably to other types of disturbance later in succession.

What do the patterns of radiosensitivity mean? Could they be sheer coincidence, on the one hand, or a useful new clue to the mechanisms of evolution on the other?

It is difficult to discard them as mere coincidenc; true, they are imperfect; there are radiation-resistant plants in the forest and radiosensitive plants in the field; furthermore, the pines, the most sensitive of all, are a minor part of the mature forest, and the pattern of increasing radiosensitivity along succession is imperfect in detail. Yet the difference in sensitivity between field and forest spans a factor of nearly 10; if we include lichen communities, which sometimes precede herbaceous communities in succession, there is a factor of 10 of additional resistance beyond that of the herbs (23). And the correlations between life-form and size and radiosensitivity, and the parallels between radiation effects and known effects of environmental gradients are too strong to be set aside lightly. There is no evidence at all that the enormous range of radiosensitivities among higher plants correlates in any way with the distribution of radioactivity in nature; nor is there reason to believe that radiation levels have changed appreciably during the quarter-billion years or so of existence of the higher plants. Rather, it seems that we must look further for other environmental factors or combinations of factors that have affected the evolution of that constellation of characteristics we measure when we measure radiosensitivity, including especially chromosome size.

It is an intriguing if somewhat over-simplified hypothesis that sensitivity to radiation damage is a measure of sensitivity to environmentally induced mutation (mutation is used in its broadest sense). It seems reasonable to accept the concept that rates of mutation tend toward some optimum, which is under hereditary control (24). If the rate were too high, there would be reduction in reproductive success; if too low, there would be insufficient variability to meet the evolutionary demands of constantly changing environments.

Certainly it is conceivable that environments vary in capacity to produce mutations. If, on the basis of current evidence about mutations, one were to seek a mutagenic natural environment (independent of radiation intensity), he would probably seek one characterized by extremes: extremes of temperature, moisture availability, and solar radiation. One thinks immediately of surfaces exposed to the sun: soil, rock, bark. The evidence that I report suggests strongly that plants that normally inhabit such surfaces—algae, lichens, mosses, and prostrate-growing vascular plants—are more resistant to ionizing radiation (and doubtless to many other stresses) than plants of more ameliorated environments such as forests.

Whether this suggestion will prove to be true when examined in a larger context than has yet been possible remains to be seen. Nonetheless, we now infer that ability to survive such rigorous environmental conditions also confers in some degree, at least, resistance to ionizing radiation. The factors that confer resistance involve growth form, length of life cycle, regenerative capacity, and cytological characteristics, especially average interphase chromosome volume. Experimental examination of this question is a current challenge to radiation research. Only by willingness to look at such really tough questions will we gain further insight into both radiation and the patterns of nature.

## References and Notes


1. D. Lack, *Darwin's Finches* (Cambridge Univ. Press, London, 1947).
2. This series of general statements is a synthesis of concepts current in ecology; for further discussion see R. Margalef, *Amer. Naturalist* 97, 357 (1963); F. E. Clements, *Plant Succession and Indicators* (Wilson, New York 1928).
3. Largely because of the efforts of A. H. Sparrow and colleagues at Brookhaven National Laboratory.
4. This work has been recently summarized: A. H. Sparrow, R. C. Sparrow, K. H. Thompson, and L. A. Schairer, in *Proc. Use of Induced Mutations in Plant Breeding* (Pergamon, Oxford, 1965), pp. 101–32.
5. A. H. Sparrow and G. M. Woodwell, *Radiat. Bot.* 2, 9 (1962).
6. By a 3-year study (Emory Univ., Atlanta; directed by R. B. Platt) of the effects of radiation on vegetation surrounding an unshielded reactor in Georgia.
7. R. B. Platt, in *Ecological Effects of Nuclear War, BNL 917 (C-43)*, G. M. Woodwell, Ed. (Brookhaven National Laboratory, Upton, N.Y., 1965), pp. 39–60.
8. By A. H. Sparrow and colleagues, Brookhaven National Laboratory.
9. A. H. Sparrow, in *Large Radiation Sources in Industry* (Intern. Atomic Energy Agency, Vienna, 1960), Vol. 3, pp. 195–219; ——— and W. R. Singleton, *Amer. Naturalist* 87, 29 (1953).
10. G. M. Woodwell, *Radiat. Bot.* 3, 125 (1963).
11. ——— and J. K. Oosting, ibid. 5, 205 (1965).
12. G. E. Bard, *Ecol. Monographs* 22, 195 (1952); H. J. Oosting, *Amer. Midland Naturalist* 28, 1 (1942).
13. M. L. Fernald, *Gray's Manual of Botany* (American Book, New York, 1950).
14. G. M. Woodwell, *Science* 138, 572 (1962).
15. ——— and A. L. Rebuck, *Ecol. Monographs*, 37, 53 (1967).
16. J. H. Brower, dissertation, Univ. of Massachusetts, 1964.
17. G. M. Woodwell and J. H. Brower, *Ecology*, in press.
18. M. F. Buell and J. E. Cantlon, ibid. 34, 520 (1953).
19. R. D. Brayton and G. M. Woodwell, *Amer. J. Bot.* 53, 816 (1966).
20. G. M. Woodwell and P. Bourdeau, in *Proc. Symp. Methodol. Plant Ecophysiol.* (UNESCO, 1965), pp. 519–27.

21. G. M. Woodwell and A. H. Sparrow, *Radiat. Bot.* 3, 231 (1963).
22. R. C. Sparrow and A. H. Sparrow, *Science* 147, 1449 (1965).
23. G. M. Woodwell and T. P. Gannutz, *Amer. J. Bot.*, in press.
24. J. F. Crow, *Sci. Amer.* 201, 138 (1959); M. Kimura, *J. Genet.* 52, 21 (1960).
25. R. H. Wagner (Brookhaven National Laboratory), unpublished data.
26. Research carried out at Brookhaven National Laboratory under the auspices of the AEC. Many have contributed for years to this work. I thank especially F. H. Bormann of Yale University, R. H. whittaker of the University of California, Irvine, and A. H. Sparrow of Brookhaven National laboratory for many long discussions.

# 23

# Cesium-137 in Alaskan lichens, caribou and eskimos

*W. C. HANSON*

Cesium-137 is generally recognized as a most important constituent of worldwide radioactive fallout resulting from nuclear weapons tests. Its importance is nowhere greater than in the arctic and subarctic regions of the northern hemisphere, where the unique features of the lichen-caribou (reindeer)-man food chain have created a situation of increasing interest to workers in the fields of radiation ecology and radiological health.

Our radiation ecology studies in Alaska began in 1959 and have progressed from a description of the general spectra of fallout radionuclides in arctic biota to a definition of seasonal cycles, rates and routes of radionuclide movement within the ecosystems, with particular attention to $^{90}Sr$ and $^{137}Cs$. Cesium-137 has been of greatest importance in these studies, not only because of its public health aspects, but also because its convenient gamma emission and biological concentration permits effective study of its transfer within the natural systems. The results of these studies and those of other northern regions have been remarkably consistent. The $^{137}Cs$ situation in the arctic regions may be briefly described as depending upon (1) the extremely effective retention of $^{137}Cs$ by lichens, (2) the importance of lichens as a winter food for

Reproduced from *Health Physics*, 13:383–389 (1967) by permission of the Health Physics Society. This paper is based on work performed under United States Atomic Energy Commission Contract AT(45–1)–1830.

caribou and reindeer, and (3) the dependence upon caribou and reindeer for food by many northern peoples.

The transfer of $^{137}Cs$ up the food chain in this series of steps or stages, known as trophic levels, illustrates the simple and thus delicately balanced structure of arctic ecosystems.

## Cesium-137 in Lichens

It is apparent that lichens represent a most important reservoir of $^{137}Cs$ and other fallout radionuclides because of their longevity (decades, even one century), persistence of aerial parts and their dependence upon nutrients dissolved in precipitation (1). These properties result in a tenacious holding of $^{137}Cs$. Recent field experiments in Alaska

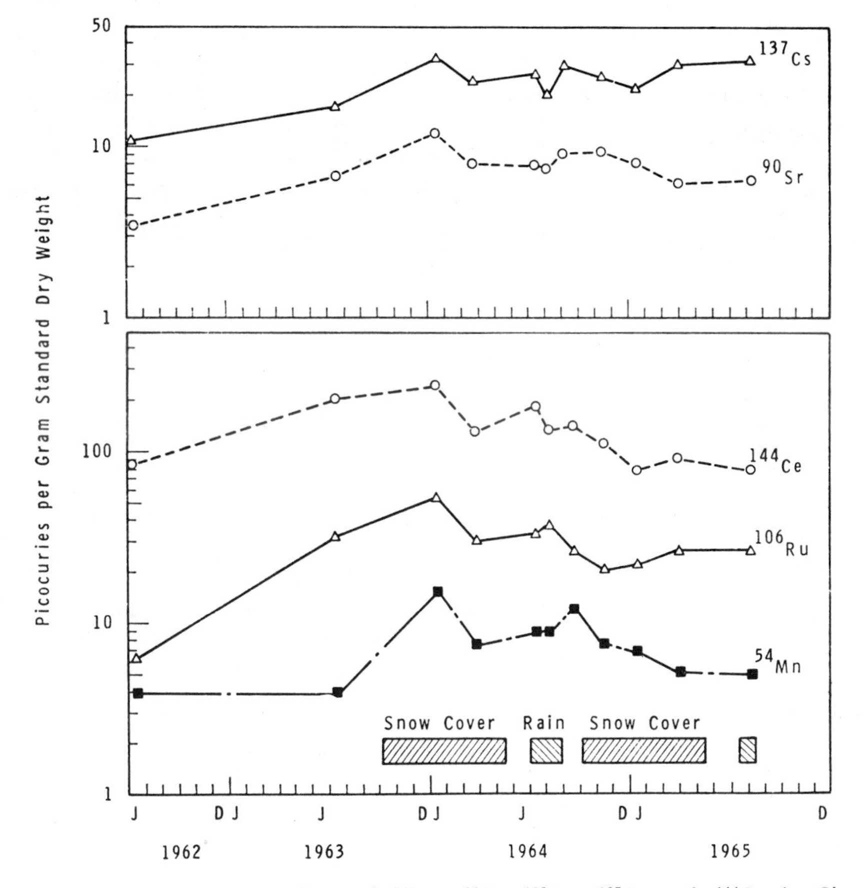


FIGURE 23–1. Concentrations of $^{54}Mn$, $^{90}Sr$, $^{106}Ru$, $^{137}Cs$ and $^{144}Ce$ in *Cladonia-Cetraria* lichens at Anaktuvuk Pass, Alaska during 1962–1965.

utilizing $^{134}Cs$ as a tracer suggest a biological half-time of about 13 years for cesium in lichens (2). This value is median to half-time estimates of 17 years (3) and 6–10 years (4) obtained by comparison of changes in $^{137}Cs$ levels in lichens and $^{137}Cs$ deposition in northern Scandinavia.

Further evidence recently reported (5) showed that $^{137}Cs$ applied to the green living top part of lichens remained there rather than being translocated throughout the plant, as was $^{90}Sr$. The continuing long-term accumulation of $^{137}Cs$ in Alaskan lichens is shown in Figure 23–1, which illustrates that concentrations of other fallout radionuclides $^{54}Mn$, $^{90}Sr$, $^{106}Ru$ and $^{144}Ce$ have decreased since early 1964, particularly during periods of snow cover when the lichens were shielded from direct fallout deposition. A comparison of $^{137}Cs$ concentrations in eight lichen species suggests that $^{137}Cs$ levels tend to be ranked according to a species' ecological "niche," structural diversity and growth pattern. This is illustrated here by three examples (Fig. 23–2). Highest concentrations occur in *Cornicularia divergens*, which grows on windswept ridgetops that are nearly free of snow during winter and are first snow-free in the spring; intermediate levels were found in *Cetraria cuculata* collected from the sides of the ridges and which were covered by an

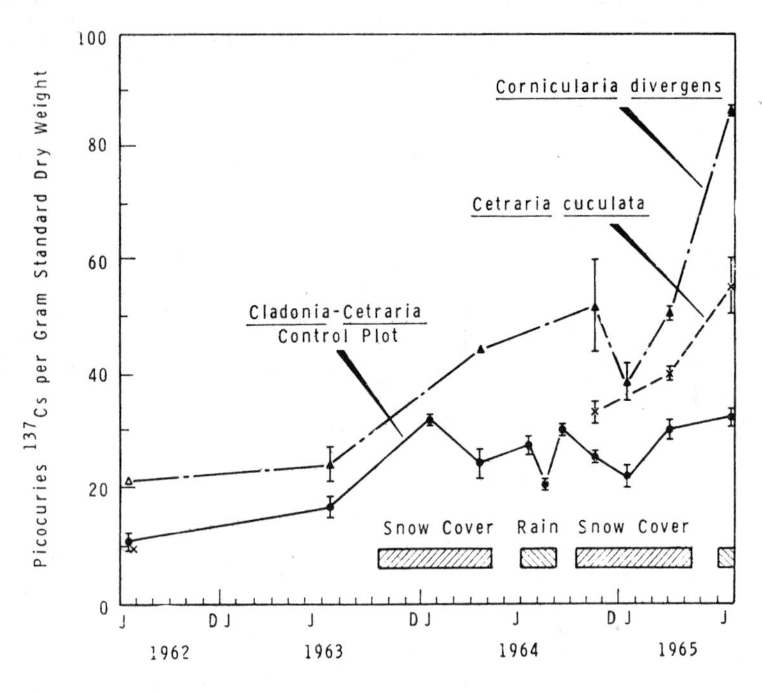


FIGURE 23–2. Cesium-137 concentrations in three lichen sample types at Anaktuvuk Pass, Alaska during 1962–1965. Values shown are mean ± one standard error.

average of 5 cm of snow for about 6 months each year, and least amounts were in mixed *Cladonia-Cetraria* samples which were snow-covered to a depth of 2–3 dm during winter.

The quantities of fallout $^{137}Cs$ per square meter of *Cladonia-Cetraria* lichens at a site in Anaktuvuk Pass, Alaska during the month of July of the years 1962–1965 were 18, 28, 41, and 48 nCi, respectively. These values are nearly identical to those reported from northern Sweden (3) at comparable times and are slightly greater than cumulative $^{137}Cs$ fallout estimates obtained from HASL soil and precipitation samples from other Alaskan locations (6, 7). These data reinforce other observations that lichen mats are efficient accumulators of fallout and may well serve as natural fallout meters (8).

## Cesium-137 in Caribou and Reindeer

The heavy utilization of lichens by caribou and reindeer as a basic winter food results in the rapid accumulation of $^{137}Cs$ levels in flesh to levels that exceed those of other arctic herbivores.

Within any one herd of Alaskan caribou there has usually been a coefficient of variation of about 30 percent in concentrations of $^{137}Cs$ (9). Finnish investigators have recently reported greater variability among $^{137}Cs$ concentrations in male reindeer flesh compared to females due to physiological parameters (4). Controlled metabolic studies of orally administered $^{137}Cs$ in reindeer suggest that muscle physiological activity may influence accumulation and distribution (10).

Caribou and reindeer samples collected from several locations in northern Alaska suggest that animals from the Noatak and Kobuk River drainages may contain higher $^{137}Cs$ concentration than do animals from other locations, although it is difficult to characterize $^{137}Cs$ concentrations among the several caribou and reindeer herds in Alaska because of the extensive migrations of the animals and large area involved. There is, however, a definite seasonal cycle of $^{137}Cs$ concentrations in caribou and reindeer, with maximum values during the winter period when lichens are their main food and minima during the summer pasturing period (3, 5, 8, 11, 12).

The rapid decrease of flesh $^{137}Cs$ concentrations with the shift to summer diets, as well as other experimental data, indicates a biological half-time of about 3–5 weeks in caribou and reindeer. Recent experiments in the USSR have been cited (5) in which 50 percent of the $^{137}Cs$ body burden had a biological half-time of 1.5 days and 50 percent had a half-time of 20 days. Thus, $^{137}Cs$ concentrations in Alaskan caribou flesh often have varied over a threefold range from winter high to summer low, while $^{137}Cs$ levels in lichens were essentially unchanged. Maximum $^{137}Cs$ values in caribou flesh have been consistently associated with animal samples obtained during April or early May (the spring kill of animals migrating from the winter range). These values usually have been slightly greater than those in caribou collected throughout the winter period near Anaktuvuk Pass, and many represent animals moving northward from ranges containing greater $^{137}Cs$ concentrations than those of the Anatuvuk Pass area. The ratio of summer range:winter range caribou flesh values generally has been about 1:3 to 1:10, and is near that reported in Swedish and Russian reindeer herds individually identified and related to their respective ranges (3, 5).

A sharp increase of $^{137}Cs$ concentrations in caribou flesh sampled during April and May was apparent in 1964 and 1965, and emphasized the possible importance of caribou migration patterns in regulating the $^{137}Cs$ body burdens in Anaktuvuk Pass Eskimos. These spring-killed animals are the main food source during the summer months when few other game animals are available. Although we have no conclusive proof the increases were due to sampling of animals from distant ranges, we assume they were associated with the caribou migration patterns.

## Cesium-137 in Eskimos

The dependence upon wildlife and other natural resources for substantial amounts of food results in major differences in seasonal and geographic utilization of animals in various villages (13). These differences appear to control the $^{137}Cs$ levels in the natives, as shown by our initial surveys of radionuclides in northern Alaskan natives and their foods (14, 15, 16). In general there was a linear relationship between reported caribou or reindeer consumption and the $^{137}Cs$ body burdens. The higher levels usually have been found at the inland village at Anaktuvuk Pass. These people have a very limited economic base and are thus more dependent upon their own resources than other native Alaskan villages. Caribou meat forms about one-half of the total diet at Anaktuvuk Pass, because of necessity and preference to other foods. The Anaktuvuk Pass people represent the last major remnant of nunamiut Eskimos that were formerly nomadic and now remain at their mountain village site. Their culture has always centered on the caribou and the welfare of the people is still highly dependent upon the local environment. Average adults estimate they eat about 5–6 kg of caribou meat per week. Body burdens are about 50–100 times those of persons on an average temperate zone diet.

The seasonal pattern of $^{137}Cs$ concentrations in Anaktuvuk Pass Eskimos is inversely related to the concentration in the caribou because of the timing of the kill and the stockpiling practices. Animals killed during the fall season are returning from summer ranges and contain low levels of $^{137}$Cs. This meat supply forms the base of the winter diet of the Eskimos.

The spring caribou kill is made during April and May as the animals migrate northward from their winter range, at which time they contain the highest $^{137}$Cs concentrations of the cycle. This meat is stored in underground caches for use throughout the summer. Maximum $^{137}$Cs levels in the Eskimos usually have occurred during July and August and minimums during January.

The relationship of the $^{137}$Cs seasonal cycles in caribou, Eskimos and lichens is shown in Figure 23–3. The graph of Eskimo body burdens prior to January 1964 and after August 1965 is drawn point-to-point and does not show seasonal variations defined by the bimonthly measurements made during the period January 1964–August 1965. Comparison of $^{137}$Cs body burdens from one summer to the next shows a pronounced increase of about 50 percent from 1962 to 1963 and a doubling between 1963 and 1964, followed by a decrease of 30 percent from 1964 to 1965. This is consistent with the pattern of $^{137}$Cs concentrations in caribou flesh forming the food base of people. But, $^{137}$Cs concentrations in lichens, the main determinants of the level in caribou flesh, have shown a steady increase with time. Thus, the 30 percent decrease from 1964 to 1965 was as inconsistent with the $^{137}$Cs trend in lichens as was the abrupt increase during the spring of 1964 and emphasizes the probable importance of caribou migration patterns in the determination of $^{137}$Cs levels in the Eskimos.

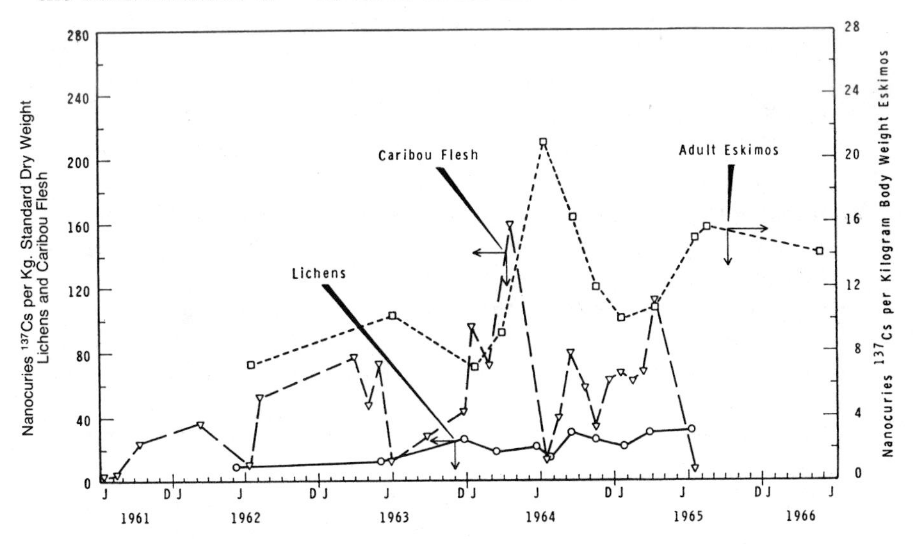


FIGURE 23–3. Cesium-137 concentrations in lichens, caribou flesh and Eskimos at Anaktuvuk Pass, Alaska during 1961–1963.

Adult males between the ages of 20 and 50 ordinarily have the highest body burdens of $^{137}Cs$, apparently due to their greater proportion of muscle and greater consumption of caribou. Children (3–14 years old) usually contained about half as much $^{137}Cs$ per kilogram as adults (>21 years old). Minors (15–20 years old) contained about 70 percent as much as adults. Comparison of males and females during the 1964 and 1965 seasonal maxima indicated no important differences in $^{137}Cs$ body burdens until about the age of 20, males then consistently contained more $^{137}Cs$ than did females, maximum differences occurred at about the age of 40, and the difference then decreased during advanced age (17).

The biological half-time of $^{137}Cs$ in adult Eskimos at Anaktuvuk Pass during the summer of 1965 was about 65 days, comparable to that reported for other arctic peoples (18, 19, 20, 21). Nevstrueva *et al.* (5) have reported that measurements of 60 reindeer breeders yielded biological half-times of 97 ± 50 days in summer and 57 ± 35 days in winter; but details of the study are not yet available to explain the differences.

Concentrations of $^{137}Cs$ in various arctic peoples in Finland, Norway, Sweden, the Soviet Union and Alaska are presented in Figure 23–4. These data show that in the three countries where measurements have been performed regularly since 1962, the Finnish Lapps have consistently exhibited the highest values and Alaskan Eskimos have shown the lowest values until recent months. The Swedish investigators are presently defining the seasonal cycle of $^{137}Cs$ in their northern peoples by frequent counting begun during mid-1965, and initial results appear to follow the same general pattern observed in Alaska during 1964–1965. Values from the three countries were quite comparable during 1962 and 1963, and then diverged during 1964 and 1965. Now a decreasing trend has apparently begun in the Finnish and Swedish Lapps, while the Alaskan Eskimos continue to increase. This contrast may be explained by changes in reindeer herding practices in Scandinavian countries, that may result in the animals regrazing ranges where the top parts of the lichens containing the greater concentrations of $^{137}Cs$ have previously been removed. The Alaskan trend is consistent with the pattern of steadily increasing $^{137}Cs$ concentrations in lichens and caribou, which do not occupy such specific herd areas as do reindeer. The single points for Norwegian and Russian reindeer breeders have been included to provide reference to other northern populations.

Wolf flesh obtained from animals of the Anaktuvuk Pass region, presumably dependent upon caribou for their main diet during the period September–March, contained about twice the $^{137}Cs$ concentration in caribou flesh. Estimates of $^{137}Cs$ concentrations in Eskimo muscle were about the same as those for the caribou flesh used for food. Because

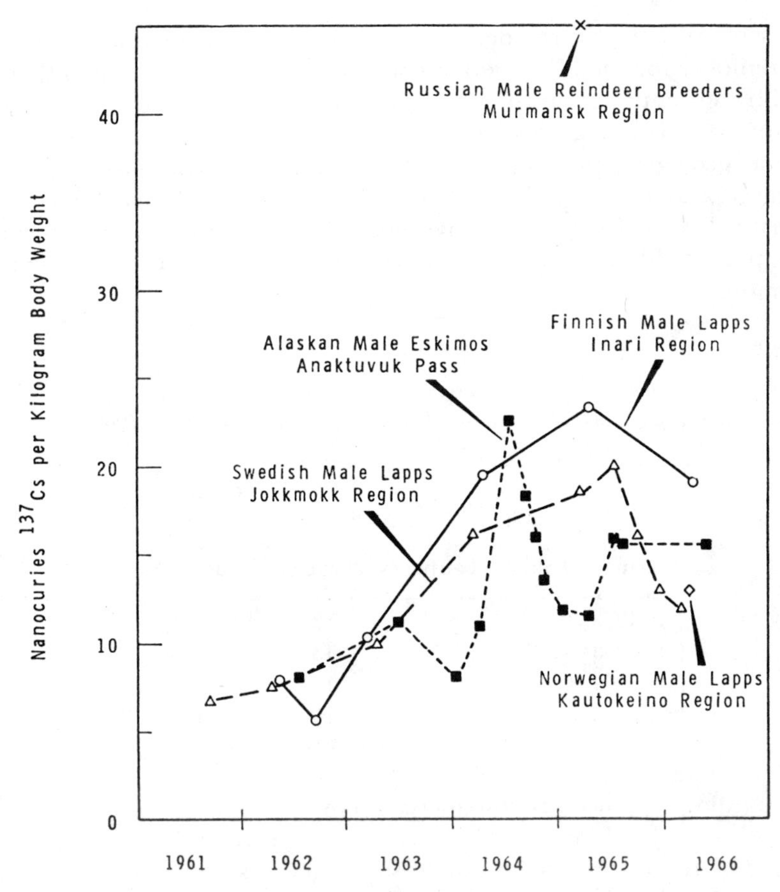


FIGURE 23–4. Cesium-137 body burdens in northern human populations of Sweden,[3] Finland,[4] USSR,[5] Norway[22] and Alaska (U.S.A.) during 1961–1966.

about half the Eskimo's food was caribou meat, it is reasonable to assign a concentration factor of two to the caribou-carnivore (including man) link of the Alaskan food chain.

## Summary and Conclusions

Studies of $^{137}Cs$ in arctic ecosystems have been conducted to provide information about transfer routes and rates of this important fallout radionuclide in a region where wildlife resources are necessary to human welfare. The lichen-caribou (reindeer)-man food chain has been observed to constitute an important $^{137}Cs$ concentration process. This results from (1) the unusual capacity of lichens to absorb and retain fallout materials, especially $^{137}Cs$; (2) the utilization of lichens for food in winter by caribou and reindeer; and (3) the dependence upon caribou and reindeer for food by several northern populations. Although there are differences in radionuclide concentrations in plants and animals, including man, from various locations within the arctic region, the same general ecological processes have been found to efficiently transfer appreciable amounts of $^{137}Cs$ through the food chain. Climatic factors, animal behavior, food habits, physiological parameters and human cultural practices have important effects upon $^{137}Cs$ accumulation and retention.

TABLE 23-1

*Cumulative $^{137}Cs$ fallout in Alaska ($nCi/m^2$) derived from soil and precipitation samples at Barrow, Fairbanks and Palmer* (6,7) *and from lichens at Anaktuvuk Pass during 1962–65*

| Date | Soil and Precipitation | | | Lichens |
|---|---|---|---|---|
| | Palmer | Fairbanks | Barrow | Anaktuvuk Pass |
| July 1962 | 23 | 20 | 8 | 18 |
| July 1963 | 32 | 29 | 13 | 28 |
| July 1964 | 38 | 33 | 16 | 41 |
| July 1965 | 45 | 38 | 16 | 48 |

Cesium-137 concentrations increased by a factor of about two at each successive trophic level of the Alaskan lichen-caribou-man(wolf) food chain. The trend with time is one of steady increase, and contrasts with a recent decrease of $^{137}Cs$ body burdens in Finland and Sweden. Recent measurements in Alaska suggest that $^{137}Cs$ whole body burdens will continue to increase during the summer of 1966.

## References

1. D. C. Smith, *Biol. Rev.* 37, 537 (1962).
2. W. C. Hanson, D. G. Watson, and R. W. Perkins, *Proceedings of the International Symposium on Radioecological Concentration Processes*, Stockholm, 1966. Pergamon Press, oxford (1967).
3. K. Lidén and M. Gustafsson, *Proceedings of the International Symposium on Radioecological Concentration Processes*, Stockholm, 1966. Pergamon Press, Oxford (1967).
4. J. K. Miettinen and E. Häsänen, *Proceedings of the International Symposium on Radioecological Concentration Processes*, Stockholm, 1966. Pergamon Press, Oxford (1967).
5. M. A. Nevstrueva, P. V. Ramsaev, A. A. Moiseev, M. S. Ibatullin, and L. A. Teplykh, *Proceedings of the International Symposium on Radioecological Concentration Processes*, Stockholm, 1966. Pergamon Press, Oxford (1967).

6. E. P. Hardy, Jr. and J. Rivera, HASL Fallout Program—Quarterly Summary Report for June 1, 1964 through September 1, 1964, p. 29 HASL-149. Health and Safety Laboratory, U.S. AEC New York Operations Office (1964).
7. Health and Safety Laboratory, Fallout Program—Quarterly Summary Report, p. 5, HASL-165. U.S. AEC New York Operations Office (1966).
8. G. K. Svensson and K. Lidén, *Health Phys.* 11, 1393 (1965).
9. L. E. Eberhardt, *Nature, Lond.* 204, 238 (1964).
10. L. Ekman and U. Griest, *Proceedings of the International Symposium on Radioecological Concentration Processes*, Stockholm, 1966. Pergamon Press, Oxford (1967).
11. W. C. Hanson and H. E. Palmer, *Health Phys.* 11, 1401 (1965).
12. J. K. Miettinen, *Proceedings of the Third International Conference on the Peaceful Uses of Atomic Energy*, Geneva, 1964, Vol. 14, p. 122. United Nations, New York (1965).
13. C. A. Heller, *J. Am. Dietetic Assoc.* 45, 425 (1964).
14. W. C. Hanson, H. E. Palmer, and B. I. Griffin, *Health Phys.* 10, 421 (1964).
15. W. C. Hanson and H. E. Palmer, *Trans. N. Am. Wildlife and Nat. Resources Conf.* 29, 215 (1964).
16. H. E. Palmer, W. C. Hanson, B. I. Griffin, and L. A. Braby, *Science* 147, 620 (1965).
17. W. C. Hanson, *Proceedings of a Conference on the Pediatric Significance of Peacetime Radioactive Fallout*, San Diego, 1966. American Academy of Pediatrics, Chicago. In press.
18. J. K. Miettinen, A. Jokelainen, P. Roine, K. Lidén, Y. Naversten, G. Bengtsson, E. Häsänen, and R. C. McCall, *Ann. Acad. Sci. Fennicae AII. Chemica* 120, 35 (1963).
19. L. G. Bengtsson, Y. Naversten, and K. G. Svensson, *Assessment of Radioactivity in Man*, Vol. II, p. 21. I.A.E.A., Vienna (1964).
20. K. Lidén, *Assessment of Radioactivity in Man*, Vol. II, p. 33. I.A.E.A., Vienna, (1964).
21. Y. Naversten and K. Lidén, *Assessment of Radioactivity in Man*, Vol. II, p. 79, I.A.E.A., Vienna (1964).
22. K. Madshus, T. Berthelsen, and E. Westerlund, *Proceedings of the International Symposium on Radioecological Concentration Processes*, Stockholm, 1966. Pergamon Press, Oxford (1967).

# HEAVY METALS

# 24

# Lead in the environment

*Stephen K. Hall*

Lead has been mined and worked by man for many years. The widespread early use of lead is readily understood in light of its many desirable properties. It is relatively easily refined from natural ores. Its ductility, high resistance to corrosion and other properties make it one of the most useful metals.

Lead found extensive application in both Greek and Roman culture. It was used for the manufacture of the cooking utensils of the wealthy and in the lead pipes for the extensive plumbing systems of their homes. It has been *suggested* that Roman civilization deteriorated very largely as a result of *extensive* lead poisoning in the ruling class. A high incidence of infant mortality, mental retardation and sterility was found among the wealthy. This theory is supported by data showing a high concentration of lead in the bones of their remains (*1*).

Of the nonferrous metals, lead is one of the most widely used in industry and everyday life. The annual consumption in the United States alone is well above one million tons. The storage battery industry consumes the largest amount, about 40%. The second largest consumer is the petroleum industry, which uses about 20% in producing lead alkyls as gasoline additives. These 300,000 tons of lead are, of course, added directly into the ambient air (*2*). By contrast, only about 45%

---

Reprinted in modified form from *Environmental Science and Technology*, 6(1):31–35 (1972).  The author is a member of the Chemistry Department at Southern Illinois University, Edwardsville, Illinois 62025.

of the total lead consumed is recovered from metal products and batteries.

Today lead is a ubiquitous element present in the food we eat, the water we drink and the air we breathe. Lead aerosol is a common air contaminant. A recent study of annual snow strata in northern Greenland has revealed some interesting data (*3*). Dated samples of snow indicate that up to 1750 there were about 20 μg of lead per ton of ice. A marked increase occurred with the Industrial Revolution. By 1860, lead had increased to 50 μg per ton of ice. The proliferation of the automobile since World War II caused a very sharp increase: in 1940 to 80 μg; in 1950 to 120 μg; and in 1965 to 210 μg. It is thus evident that in the past two decades, man's continuing use of lead has resulted in an environmental level far above that which would exist naturally, and that this could have grave consequences in human health.

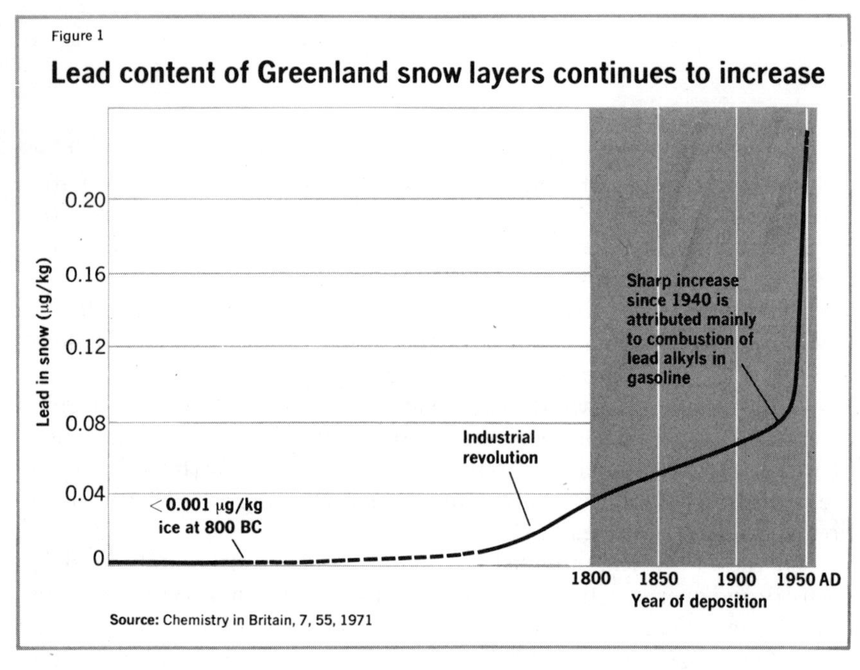


FIGURE 24–1. Lead content of Greenland snow layers continues to increase. Source: Chemistry in Britain, 7, 55, 1971.

## Sources of Lead

Lead is a natural constituent of soil, water, vegetation, animal life, and air, although the levels of the natural concentrations are not certain. Significant sources of naturally occurring lead include dust from soils and particles from volcanoes. In contrast to certain other metals such as mercury, lead in its elemental form is not believed to be a major source of poisoning.

In medieval times, the practice of "sweetening" wine with lead or lead acetate became such a serious toxicological problem that the death penalty was imposed on occasion for adulteration of wine with lead. Even today, alcoholic beverages of illicit origin are occasionally incriminated as the source of poisoning. Old automobile radiators are often used as the condenser component in illicit stills. These contain enough solder to cause dangerous contamination of the moonshine. Lead encephalopathy (disease of the brain), nephritis (disease of the kidneys), together with gout and other lead-related conditions have been reported in moonshine consumers. The problem of diagnosis is complicated by the fact that the symptoms of acute lead poisoning and acute alcoholism are similar in many ways.

Until the 1950s most interior paints contained lead pigments and this was a major source of childhood lead poisoning. Since that time lead oxide has been replaced by titanium dioxide. However, some old houses were painted countless times with leaded paint. As a result, thick chips of leaded paint fall off as the walls and ceilings peel. Putty also contains lead and is even more likely to be found in substandard dwellings. Children between the ages of one and four or five ingest nonfood particles of all types, partly out of curiosity and partly because of a behavior syndrome called *pica*. As a result, about 8% of children in ghetto areas suffer from some degree of lead poisoning.

Another major source of lead is earthenware improperly glazed with lead. Large amounts of lead are leached out of the glaze when certain food is stored. The danger of poisoning from such earthenware has been well documented. Today, lead water pipes and soldered joints are no longer in use. However, man continues to pollute his environment with lead from other sources such as manufacturing, use of pesticides, incineration of refuse and combustion of coal and leaded gasoline. The available data indicate that combustion of leaded gasoline is the major source of atmospheric lead in urban areas (*4*).

## Atmospheric Lead

Anti-knock agents of lead alkyls, in the form of tetraethyllead (TEL) or tetramethyllead (TML), have been added to most gasoline since 1923. Their use very rapidly became a cause for concern. Lead intoxication and even death occurred among those occupationally exposed. Apprehension also developed concerning possible toxic effects in the general public. In accordance with the recommendation from a conference held in Washington on May 20, 1925, the US Surgeon General appointed a committee to investigate the possibility of a public health hazard in the manufacture, distribution or use of leaded gasoline. A set of regulations was proposed by the committee in the following year. They were concerned only with precautions to be taken in the manufacture and distribution. The proposed regulations were adopted voluntarily by the petroleum industry.

About 1930, under the leadership of Kehoe of the University of Cincinnati, an attack on the significance of lead as a hazard to the general public began. Unfortunately, for the following thirty years, hardly anyone else was much interested in the potential environmental pollution problem of lead. Then in 1958, the Ethyl Corporation sought and was granted an increase in the concentration of lead in gasoline by the Surgeon General. Today, the amounts of TEL range from 2 to 4 grams per gallon of gasoline.

Since the addition of lead to gasoline, a number of articles on health problems have appeared in old literature. During the early years there were some fatalities in manufacturing plants but in the last three decades the record has been quite good. A survey in 1964 indicated there have been only 88 cases of TEL poisoning reported in the United States and Canada, subsequent to adoption of the regulations in 1926, and only 16 of these have been fatal (*5*). This is not bad, considering that several million tons of TEL has been manufactured and distributed since 1926.

The lead exhausted by automobiles originates in the anti-knock fluid in the gasoline. The fluid contains lead alkyls and organic scavengers, ethylene dichloride and dibromide, whose function is to combine with the lead to form inorganic lead salts, chiefly the bromochloride, that enters into the atmosphere as part of the exhaust gases. Since only negligible amounts of the lead alkyls in the gasoline are exhausted directly into the atmosphere (*6*), it is unlikely that the health of the general public would be affected.

The amount of lead discharged to the atmosphere in the exhaust gases varies, depending on driving conditions, from about 25% to above 75% of the lead intake in the fuel. At low speeds the lead tends to be retained in the exhaust system and is discharged at some later period when the engine is run at greater speeds.

The level of atmospheric lead varies directly with the volume of traffic and the size of the community (*7*). Los Angeles, with a population of more than 2.5 million, has the highest concentration of the communities studied, with a mean of approximately 5 $\mu g/m^3$ of air. Other urban communities with a population greater than 2 million have values of about 2.5 $\mu g$. Communities that are smaller than a million have a mean atmospheric concentration of about 2 $\mu g$, and communities with a population of less than 100,000 have a mean value of about 1.7 $\mu g$ (*8*). The mean is today on the rise by as much as 5% per year and week-long averages of 8 $\mu g$ now occur in San Diego (*9*).

From the atmospheric precipitation samples collected by a nationwide network of 32 stations throughout the United States, the concentration of lead was found to be correlated with the amount of gasoline consumed in the area, and the amount of lead in surface water supplies (*10*). Thus, any further rise in the dissemination of lead wastes into the environment can cause adverse effects on human health.

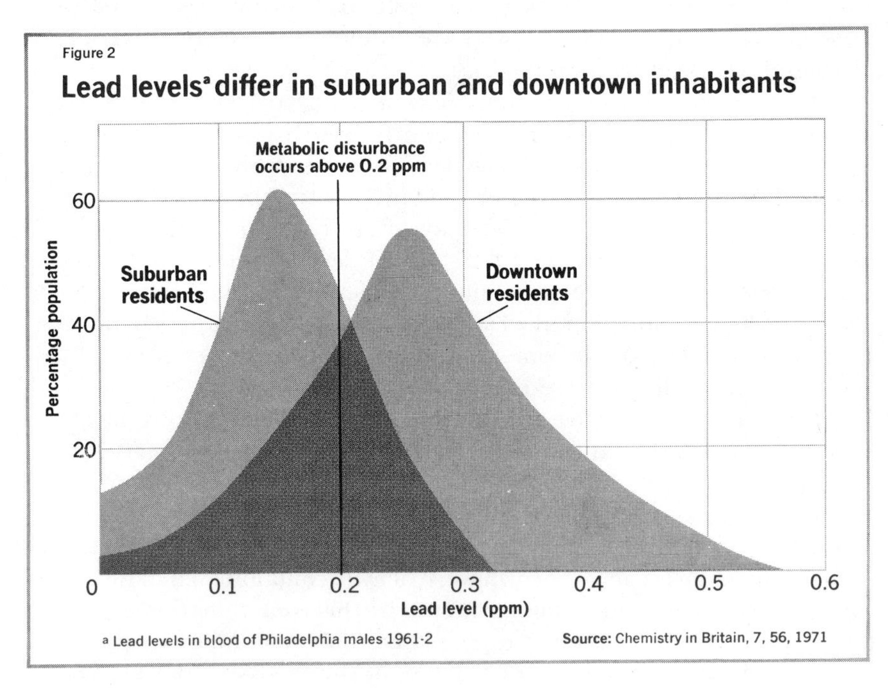


FIGURE 24–2. Lead levels[a] differ in suburban and downtown inhabitants
aLead levels in blood of Philadelphia males 1961–
Source: Chemistry in Britain, 7, 56, 1971

## Absorption of Lead

The two major routes that lead enters into the human body are the alimentary and respiratory tracts. Absorption of inorganic lead through the skin is insignificant.

Trace amounts of lead are found in normal daily diets. The mean dietary intake of lead for adult Americans is approximately 300 $\mu g$ per day. The majority of this intake passes straight through the intestinal tract. The amount absorbed depends on such factors as the concentration in the diet and the presence of calcium and phosphate. It has been estimated that approximately 6–7% of the metal ingested is actually absorbed and that the percent absorption does not appear to be materially influenced by the amount of intake (*11*).

In addition to the dietary intake, absorption of lead from the respiratory tract has long been under study. The amount of lead absorbed from a given air concentration varies according to the particle size. Particles of approximately 1 $\mu$ or smaller will probably be retained by the lung. Of the lead with such particle size that is inhaled, up to 50% may be retained in the body, the remaining portion being carried out by the expired air. Particles larger than 2 $\mu$ impinge on the mucous lining of the airways. These are swallowed and can be handled as though ingested by mouth (*12*).

As a community air pollutant, lead has a particle size that is likely to be retained in the lung. It has been found that about 75% of particulate lead in vehicle exhaust is less than 0.9 $\mu$ in mean diameter, a size easily retained by the lung (*13*). Typically, an urban adult breathing air containing, e.g., 3 $\mu g$ of lead and respiring about 15 $m^3$ of air per day, will absorb about 20–25 $\mu g$ of lead from the respiratory tract. This is roughly the same amount absorbed from a normal daily diet. Thus, the contribution of atmospheric lead to the body burden approaches closely, and in some communities, even surpasses the contribution made by dietary intake.

Cigarette smokers have a slight but consistently higher concentration of lead in their serum and blood than the nonsmokers (*14*). Lead is concentrated in tobacco leaves by lead arsenate residues in the soil from previous spraying or atmospheric fallout. Although these sprays are no longer used for tobacco and have not been used for many years, the soils in which the tobacco is grown were heavily contaminated many years ago and the tobacco still shows evidence of this contamination and continues to transmit some of it to smokers. This phenomenon provides a most-instructive prototype of the long-term effects of the pesticides used in agriculture.

## Metabolism of Lead

Lead absorbed from the digestive tract and lungs moves initially into tissues more or less in proportion to their vascularity and affinity. The manner in which lead circulates in the blood has been studied extensively. More than 95% of the lead circulating in the blood is associated with the red blood corpuscles and the remainder is in the plasma. It is then distributed throughout the viscera, especially in the liver and kidneys, until it is taken up and stored in the skeleton as an insoluble, biologically inert, tertiary lead phosphate. Under normal conditions, more than 90% of the lead retained in the body is in the skeleton. The amount of lead which is not retained in the body is excreted back into the gut in the bile. Lead is also excreted from the body in urine, sweat, hair and nails. The most commonly used measure on the total environmental exposure and the resulting body burden of lead are the concentrations in blood and urine, though the relative contribution of the various sources of lead can never be clear cut.

Long-term exposure at some daily level of lead is potentially hazardous because lead is a highly cumulative poison. Under conditions of abnormally high calcium metabolism such as feverish illness or during cortisone therapy, lead may be mobilized and a toxic amount released from the skeleton. It should be emphasized that only a small fraction of the total bone lead need be mobilized to add an appreciable increment to the small soft tissue pool.

Early investigations showed that the amount of lead gained is proportional to calcium lost. The similarity of bone lead metabolism to bone calcium metabolism has been demonstrated in numerous ways. Like calcium, lead can be mobilized from the skeleton in metabolic states such as acidosis, or conversely, can be deposited from the circulation during alkalosis. It has also been found that, shortly after absorption, lead concentrates in areas of active bone formation. With the passage of time, absorbed lead becomes progressively more deeply buried in the bone matrix. This develops a potentially dangerous pool of exchangeable lead which can persist for months or even years. It is not uncommon to encounter acute lead poisoning long after exposure to abnormal amounts of lead has ceased.

## Effects of Lead Poisoning

The use of lead has resulted in outbreaks of lead poisoning in humans from time to time since antiquity. The toxic properties of lead were recognized as far back as the second century B.C. when Nicander, a Greek poet-physician, described classical symptoms of plumbism (from Latin for lead) among men mining and refining lead. Lead poisoning is the result of very high levels of lead in the human body, particularly in the soft tissues. While in the soluble form, lead is both mobile and toxic, and most often affects the blood, kidneys and nervous system. At one time, lead poisoning was common. As a result, there is a wealth of clinical information on the toxicological effects of lead from direct observation of man rather than from extrapolation from animal experimentation.

The effect of excessive amounts of lead in the blood is anemia. Lead inhibits the activity of certain enzymes that are dependent on the presence of free –SH groups for their activity. Lead interacts with these groups in such a way that they are not available to the enzymes. The consequent decrease in heme synthesis leads at first to a decrease in the 120-day average life-span of red cells and later to a decrease in the number of red cells and in the amount of hemoglobin per cell. As a result, immature red cells and stippled cells appear in the circulation. The presence of stippled cells is the most characteristic finding in the blood of a patient with lead poisoning.

Chronic nephritis, which is sometimes accompanied by gout, is a disease characterized by a scarring and shrinking of kidney tissues. It is a result of long-term severe industrial exposure and most of the patients have a history of repeated acute lead poisoning. In severe cases, it may lead to ultimate kidney failure.

The effects of lead on the central nervous system are seen in behavior problems characterized by convulsions or swelling of the brain. A single attack of acute encephalopathy can cause brain hemorrhage, mental retardation and other forms of permanent neurological injury. Similarly, repeated bouts of lead poisoning can also cause permanent brain damage ranging from subtle learning deficits to profound mental incompetence and epilepsy.

Chronic lead poisoning is due to either the result of a slow buildup of lead over a period of years or the aftereffects of acute lead poisoning. Acute lead poisoning has easily recognizable symptoms and if treatment is administered in time, permanent damage can be avoided. Chronic lead poisoning, however, is either without symptoms or has vague symptoms that could be ascribed to any of a dozen causes. It is difficult to be diagnosed and it may be recognized only after irreversible damage has been done.

## Childhood Lead Poisoning

Childhood lead poisoning is primarily a metropolitan disease seen almost exclusively in children of preschool age who live in deteriorated housing. Many of these houses were originally well-built, well-painted, and well-maintained with thick layers of leaded paint. As they were converted into slum tenements, they began to deteriorate and much of the thick leaded paint began to peel and crumble and was readily available in the immediate environment of the young child. A chip of paint about the size of an adult nail can contain between 50 and 100 mg of lead. A child swallowing a few small chips a day easily ingests beyond the tolerable adult intake of the metal.

While lead poisoning ordinarily suggests a slowly developing occupational disease of adults, it is in children that it is most dangerous. During the first few weeks of abnormal ingestion of lead, there are generally no symptoms. After about six weeks, minor symptoms such as decreased appetite, constipation, clumsiness, fatigue, irritability, lethargy, headache, abdominal pain and sometimes vomiting begin to occur. These are, of course, quite nonspecific symptoms of many childhood diseases and are frequently dismissed as behavior problems in children. Repeated ingestion of lead can lead to intermittent drowsiness and stupor, more persistent and forceful vomiting, and convulsions. With more prolonged or intense ingestion, the course of the disease can culminate in intractable convulsions and coma—the symptoms of an encephalopathy—and sometimes sudden death.

Acute encephalopathy is most common in children between 15 and 30 months of age. Older children tend to suffer recurrent but less severe episodes. The symptoms of even acute encephalopathy are nonspecific. Diagnosis depends on a high level of suspicion. To make a positive diagnosis it is necessary to show high blood level as well as the adverse effects of lead.

It has been suggested that the minimum blood concentration of lead below which it is most unlikely that poisoning will occur is 80 μg of lead per 100 ml of blood for adults (*15*). In children, however, the threshold should be much lower. The blood lead concentrations for normal children have been reported from 12.3 μg (*16*) to 25 μg per 100 ml of blood (*17*). In a group of 122 mentally retarded children 55 of them had blood lead levels above 36 μg (*17*). Mild symptoms of lead poisoning have been found in children at 60–80 μg. As the blood lead concentration rises above 80 μg, the risk of severe symptoms increases sharply. Even in the absence of symptoms, in children blood lead concentrations exceeding 60 μg call for immediate treatment and separation of the child from the source of lead. It is the asymptomatic aspect of chronic lead poisoning that can be most dangerous in children. It is estimated that as many as 50,000 children in the United States today suffer from the asymptomatic form of lead poisoning. Even though physical recovery is possible, brain damage and mental retardation often ensue, for the ages of one to four or five are critical years in the growth and development of the brain.

## Treatment of Lead Poisoning

The most frequently used treatment is to bind the soluble lead to chemical compounds known as chelating agents. Chelating agents remove lead ions from soft tissues and prevent deposition of lead on the surface of the red cells. With chelating agents very high levels of lead in the tissues can be rapidly reduced to levels approaching normal and the adverse metabolic effects can be promptly suppressed. However, chelating agents must be used with great care. If there is a large residue of lead in the stomach, chelates may encourage such rapid absorption of lead as to cause death. Chelates may also release lead already stored in the bones, causing further complications. The three frequently used chelating agents in the treatment of lead poisoning are calcium disodium ethylenediamine tetraacetate (EDTA), 2, 3-dimercaptopropanol or British Anti-Lewisite (BAL), and *d*-penicillamine.

In the treatment of patients with acute lead poisoning, the simultaneous use of EDTA and BAL has been found to be far more effective than in the injection of EDTA or BAL alone. After the blood lead level has been reduced, *d*-penicillamine may be administered orally as a follow-up therapy. Treatment by EDTA and BAL can eliminate lead from the soft tissues but not from the skeleton. Oral *d*-penicillamine as a follow-up therapy will continuously remove significant amounts of lead from the bones as well as from the soft tissues (*18*). In the chronic cases, oral *d*-penicillamine treatment has been successful in reducing lead concentrations to the normal range.

The major routes of lead excretion are through the kidneys and through the liver. The excretion of lead into the urine is quite slow and not as great as in excretion in the bile. Other routes of lead excretion probably play only an extremely minor role.

Before chelating agents were available for treatment of lead poisoning, about two-thirds of all children with lead encephalopathy died. Now the mortality rate is less than 5%. Unfortunately the improvement in treatment has not substantially reduced the incidence of brain damage and mental retardation in the survivors. Many children saved from one episode of lead poisoning are returned to the same dangerous environment only to succumb to the same syndrome again. Therefore, after treatment, it is essential for the patient not to return to the same environment until all hazardous lead sources have been removed.

## Control of Lead Pollution

Among the various substances that man concentrates in his immediate environment, lead wastes have been accumulating rapidly during the last century. As far as is known, lead is not a trace metal essential to nutrition. There is need for man to know whether or not his uncontrolled dissemination of lead into the environment presents some subtle risk to health in the not-too-distant future. Furthermore, there is a more urgent need to control this environmental hazard in certain groups of people who are being poisoned and suffering from many tragic consequences.

Concerning the health effect of lead in man, one must remember that the lead absorbed in seemingly harmless trace quantities over a long period of time can accumulate to exceed the threshold level for potential poisoning and therefore produce long delayed toxic effects. It has been reported that the concentrations of lead in various soft tissues as well as in bone continues to rise from birth through the fourth or fifth decade of life and that the concentrations are higher in the United States than in other parts of the world (*19*).

Normal concentrations of lead in adults are below 40 μg per 100 ml of blood. The range between 40–80 μg is still acceptable and it represents increased absorption resulting from occupational or abnormal exposure. The range between 80–120 μg is unacceptable and represents excessive absorption from occupational or other exposure. Recognizable mild lead poisoning symptoms usually result. Above 120 μg is dangerous and severe symptoms as well as long-term sequelae are probable (*20*). The average concentration of lead in Americans is 27 μg. The question is whether this average is normal or whether it already represents a serious increase beginning with the addition of TEL to gasoline and thereby reflecting the documented increase in environmental lead. If this is so, then any further increase might well be dangerous. Unfortunately, we have no baseline with which to compare contemporary populations. Whereas we know that 80 μg can cause lead poisoning, we cannot be sure that 70, 60, or even 50 μg are innocuous. Low level effects are yet to be recognized.

The increase in airborne lead is particularly important because up to 50% of this lead can be absorbed by inhalation, in contrast with the 6–7% absorbed from diet. Hopefully, the amount of airborne lead will remain constant instead of increasing steadily since the automobile and oil companies have started to eliminate leaded gasoline this year. Ironically, this has not occurred because of any recognition of the health effects from airborne lead but rather because leaded gasoline fouls antipollution devices.

Childhood lead poisoning is essentially an environmental problem. The environment of these children contains a high concentration of a toxic agent with which these children come in contact unavoidably as long as they live in that environment. Ghetto children subject to possible lead poisoning must be identified before they become poisoned and be treated as promptly as possible. Lead poisoning is probably a major source of brain damage, mental deficiency and serious behavior problems among ghetto children.

Some states have laws prohibiting the use of leaded paints for interior painting. However, the application of a coat of lead-free paint over old leaded paint does not solve the problem. When paint peels, it all comes off the bare woodwork and the leaded paint again becomes available to the child. Old leaded paint must be removed from substandard housing. Where rehabilitation is impossible, urban redevelopment projects should replace substandard housing. We know enough to act with regard to childhood lead poisoning. It is therefore impermissible for a humane society, such as ours, to fail to do what is necessary to eliminate a wholly preventable environmental pollution.

The standard dithizone method of determining blood lead level requires between 5 and 10 ml of blood. This is all right for adults but may be a traumatic experience for very small children. What is needed today is a portable machine that requires only a finger-prick to get 2 or 3 drops of blood for the test. Although a couple of manufacturers have such machines in the market, the efficiency of such machines has yet to be proven.

Present American industrial standards consider 200 $\mu g/m^3$ to be an acceptable level of occupational exposure. A known effect of chronic exposure to lead among industrial workers is peripheral nerve disease, affecting primarily the motor nerves of the extremities, while symptoms of acute lead poisoning are strikingly absent. Federal and state governments have much to do to institute effective control measures in industry.

The manifestation in regard to lead pollution is quite similar to the one resulting from the widespread use of chlorinated hydrocarbon biocides. Both substances are acknowledgedly poisonous and both are being widely disseminated in the environment. The similarity does not end there. Both lead and the chlorinated hydrocarbon biocides persist for long periods in man's body and his environment. Whether or not the concentrations of lead measured are of themselves hazardous to human health, any further increase of lead in the environment will result in further concentration in some food chains, leading ultimately to toxic doses for man or for some other important organisms. Lead pollution, therefore, must be controlled.

## Additional Reading

1. Gilfillan, S. C., Lead Poisoning and the Fall of Rome, *J. Occupational Medicine* 7, 53 (1965).
2. Morse, R. S., *The Automobile and Air Pollution, A Program for Progress, Part II*, U. S. Department of Commerce, Washington, D. C., 1967.
3. Murozumi, M., Chow, T. J., and Patterson, C. C., Chemical Concentrations of Pollutant Lead Aerosols, Terrestrial Dusts and Sea Salts in Greenland and Arctic Snow Strata, *Geochim. Cosmochim.* Acta 33, 1247 (1969).
4. Sterba, M. J., in Symposium on Environmental Lead Contamination, *U. S. Public Health Service Publication* No. 1440 (1966).
5. Sanders, L. W., Tetraethyl Lead Intoxication, *Arch. Environ. Health*, 8, 270 (1964).
6. Hirschler, D. A. and Gilbert, L. F., Nature of Lead in Automobile Exhaust Gas, *Arch. Environ. Health* 8, 297 (1964).
7. Daines, R. H., Motto, H. and Chilko, D. M., Atmospheric Lead: Its Relationship to Traffic Volume and Proximity to Highways, *Environ. Sci. Technol.* 4, 318 (1970).
8. Cholak, J. Further Investigations of Atmospheric Concentration of Lead, *Arch. Environ. Health* 8, 314 (1964).
9. Chow, T. J. and Earl, J. L., Lead Aerosols in the Atmosphere: Increasing Concentrations, *Science* 169, 577 (1970).

10. Lazrus, A. L., Lorange, E. and Lodge, J. P. Jr., Lead and Other Metal Ions in United States Precipitation, *Environ. Sci. Technol.* 4, 55 (1970).
11. Kehoe, R. A., The Metabolism of lead in Man in Health and Disease, *J. Roy. Inst. Public Health Hyg.* 24, 101 (1961).
12. Kehoe, R. A., Metabolism of Lead under Abnormal Conditions, *Arch. Environ. Health* 8, 235 (1964).
13. Habibi, K. Characterization of Particulate Lead in Vehicle Exhaust Experimental Techniques, *Environ. Sci. Techol.* 4, 239 (1970).
14. Butt, E. M., Nusbaum, R. E., Gilmour, T. C., idio, S. L. and Sister Mariano, Trace Metal Levels in Human Serum and Blood, *Arch. Environ. Health* 8, 52 (1964).
15. Kehoe, R. A., Toxicological Appraisal of Lead in Relation to the Tolerable Concentration in the Ambient Air, *J. Air Pollution Control Assn.* 19, 690 (1969).
16. Millar, J. A., Battistini, V., Cumming, R. L. C., Carswell, F. and Goldberg, A., Lead and $\pi$-Aminolevulinic Acid Dehydrase Levels in Mentally Retarded Children and in Lead-Poisoned Suckling Rats. *Lancet* II, 695 (1970).
17. Moncrieff, A. A., Koumides, O. P., Clayton, B. E., Patrick, A. D., Renwick, A. G. C., and Roberts, G. E., Lead Poisoning in Children, *Arch. Disease Childhood* 39 1 (1964).
18. Harrison, H. E., Lead Poisoning. In *Drugs and Poisons in Relation to the Developing Nervous System*, U. S. Department of Health, Education and Welfare p. 245 (1968).
19. Schroeder, H. A. and Balassa, J. J., Abnormal Trace Metals in Man, *J. Chronic Diseases* 14, 408 (1961).
20. Anonymous, Diagnosis of Inorganic Lead Poisoning: A Statement, *British Medical Journal* p. 501 (1968).

# 25

# Ecological implications of mercury pollution in aquatic systems

*Robert C. Harriss*

## Introduction

During the past quarter-century, in the United States alone, about 30 million kilograms of mercury have been discharged into the natural environment, yet negligible effort has been made to study the pathways or concentration of this toxic element in the world ecosystem. Mercury pollution now poses serious economic, legal, and public health problems in many parts of the world. Hundreds of humans have died as a result of eating fish from mercury-contaminated coastal areas in Japan (Kurland *et al.*, 1960), rapid declines in bird populations in Sweden were traced to the use of mercurial compounds as fungicides (Johnels & Westermark, 1969), and at least 20 states in the United States have banned the sale of commercial fish or warned against public consumption of fish from waters contaminated with mercury.

The global extent of the problem became apparent with the discovery of mercury concentrations exceeding the United States Food and Drug Administration and World Health Organization safe level for food of 0·5 parts per million (ppm) in seals from the Pribilof Islands, deep-sea tuna, and swordfish. These aquatic organisms all live far removed

---

Reproduced from *Biological Conservation*, *3*:279–283 (1971) with permission of the Author and Publishers. The author is Director of the Marine Laboratory, Department of Oceanography, Florida State University, Tallahassee, Florida 32306.

from the activities of man, yet they are sufficiently contaminated with mercury to prohibit their sale or that of associated products for human consumption.

## Sources and Distribution

Some of the principal users of mercury in the United States are listed in Table 25–1. The proportionate consumption of mercury by different countries depends also on their degree of industrialization. The United States uses about 27 per cent of the world output of mercury, while producing only 10 per cent from mining. The uncontrolled discharge of 4000–5000 metric tons of mercury per year by human society throughout the world is approximately equal to the 5000 metric tons per year that are calculated to be released from natural rocks by chemical weathering (Goldberg, 1970).

TABLE 25–1

*Mercury Consumption in the United States in 1969 by Some Major Users* (Source: Bureau of Mines)

| *User* | *Quality (kilograms)* |
|---|---|
| Electrolytic chlorine | 713,000 |
| Electrical apparatus | 626,000 |
| Paints | 335,000 |
| Agriculture | 95,000 |
| Pharmaceuticals | 23,600 |
| Paper industries | 19,100 |

A summary of available data on estimated natural concentrations, and on concentrations in samples from areas receiving mercury discharge, is presented in Table 25–2. The limited data available on air suggest a direct relationship between general air pollution and mercury concentrations in air (Williston, 1968). Organic-rich materials such as coal and petroleum contain relatively high natural levels of mercury, which are released into the atmosphere during burning.

The wide range in mercury concentrations in open-sea water has been observed by Hosohara (1961) and by Robertson (1970). Mercury concentrations are generally higher in deep than in shallow water, perhaps due to uptake by plankton in the euphotic zone and to bacterial degradation of sinking organic matter at depth which releases elements back into solution. Levels of dissolved mercury in sea water exceeding 0·0005 ppm have only been noted in coastal waters in the immediate vicinity of a discharge, as in the case of Minamata Bay, Japan.

The geochemical factors determining the distribution of mercury in soils and sediments are not well known. Studies by Andersson (1967) have shown that pH and organic content are important in determining the distribution of mercury in Swedish soils. At low pH values, most of the soil mercury is adsorbed on humus; but with increasing pH, adsorption on soil minerals becomes a dominant factor. The mercury content of cultivated soils in Sweden averages about twice that of non-cultivated fields, reflecting the use of persistent organomercurial fungicides up to 1966 when the license for all alkylmercury compounds used

TABLE 25–2

*Mercury Concentrations Reported in Environmental Samples*

| *Sample* | *Estimated natural levels* | *Concentrations measured in contaminated samples* |
|---|---|---|
| Air | <2 ng/m$^3$ | 2–20 ng/m$^3$ |
| Water: | | |
| Sea water | 0·00006–0·0003 ppm | 0·0005–0·030 ppm |
| Fresh water | <0·00006 ppm | 0·0001–0·040 ppm |
| Soils* | 0·04 ppm | 0·08–40 ppm |
| Lake sediments* | <0·06 ppm | 0·08–1800 ppm |
| Biological materials: | | |
| Fish | <0·02 ppm | 0·5–17 ppm |
| Human blood | <0·0008 ppm | 0·001–0·013 ppm |

* Note: Mercury concentration is dependent on organic content.

in agriculture was revoked. The mercury in lake, estuarine, and marine deposits is generally derived from the erosion of soils, input from the atmosphere, or discharge of industrial or domestic wastes (Fimreite, 1970). Klein & Goldberg (1970) found approximately from two to eight times as high concentrations of mercury in marine sediments near an ocean sewer outfall as compared with similar sediments farther away.

## Biological Properties

Inorganic and phenyl mercury compounds discharged into the aquatic environment are rapidly adsorbed or absorbed by particulate inorganic and organic material. Within a few hours after incorporation of inorganic or phenyl mercury into sediment, a conversion of these forms to methyl ($CH_3Hg+$) or dimethyl ($(CH_3)_2Hg$) compounds is observed in the presence of Bacteria (Jensen & Jernelov, 1969). The exact rates and mechanisms of the biological methylation of mercury in sediments have not been well studied, but it is known that the organic content and pH of the sediment are important variables.

Experimental and field investigations on the accumulation of mercury in aquatic organisms have indicated that all the plants and animals tested could concentrate mercury, in biological tissues, above the concentrations existing in the surrounding water (Johnels *et al.*, 1967; Hannerz, 1968; Glooschenko, 1969). A summary of some representative concentration factors is given in Table 25–3. The uptake of mercury by aquatic animals can occur both directly from water and indirectly from food. Mercury compounds—in particular the organomercurials—exhibit a higher solubility in biological tissues than in water, and ingested mercurials are primarily stored in such organs as the liver, brain, and kidneys, resulting in biological concentration.

TABLE 25–3

*Some Representative Concentration Factors (Hg in water/Hg in organism) for Aquatic Organisms*

| *Organism* | *Environment* | *Concentration Factors* | *Ref.** |
|---|---|---|---|
| Algae | experimental ponds | 200–1200 | 1 |
| Large plants | experimental ponds | 4–2400 | 1 |
| Invertebrates | experimental ponds | 400–8400 | 1 |
| Fish (Pike) | natural lakes | 3000 | 2 |

* References: 1. Hannerz (1968).
2. Johnels *et al.* (1967).

A phenomenon termed biological magnification results when each organism consumes, from a lower trophic level, many organisms containing mercury. The food organisms are digested and partially excreted, but most of the mercury which they contained is retained in the tissues of the predator. For example, the smaller plankton-feeding species of fish in polluted areas of the Great Lakes show typical values of mercury ranging from 0·2–1·0 ppm; the top predators such as the Walleye, Pike, and White Bass, show typical values ranging from 1·0–2·0 ppm of total mercury. Waterfowl and other fish-eating animals in the vicinity of Lake St Clair have been discovered to contain up to 20 ppm of mercury. Mercury in aquatic ecosystems shows many similarities to the behaviour of other global contaminants such as DDT and $^{90}Sr$, which have been discussed in several excellent articles (Woodwell, 1967; Woodwell et al., 1967; Wurster, 1969).

## Effects on the Ecosystem

The effects of acute mercury pollution, such as massive fish-kills and poisoning of humans, generally result in legal and political action directed towards the protection of man. The long-term effects of sublethal quantities of mercury are subtle, and are not noticed until serious and often irreversible damage is done to the natural environment. The following are some examples of possible long-term effects and their ecological implications.

### *Changes in Community Structure and Food-webs*

The amount of mercury required to cause death in organisms is species-dependent. Certain species of phytoplankton are extremely sensitive to organomercurials, growth and photosynthesis being completely inhibited at concentrations below 0·0001 ppm) Harriss et al., 1970). The acute toxicity of mercury to phytoplankton depends on the chemical nature of the mercury compound and on the abundance of particulate material in the water. Figure 25–1 illustrates the effects of four different mercury compounds on growth and photosynthesis by *Nitzschia delicatissima* (Cleve), a common species of marine diatom, after 24 hours of exposure to these mercurials in laboratory cultures. Inorganic mercury was also tested but proved to be approximately only one hundredth as toxic as any of the organomercurials, and so is not included in the illus-

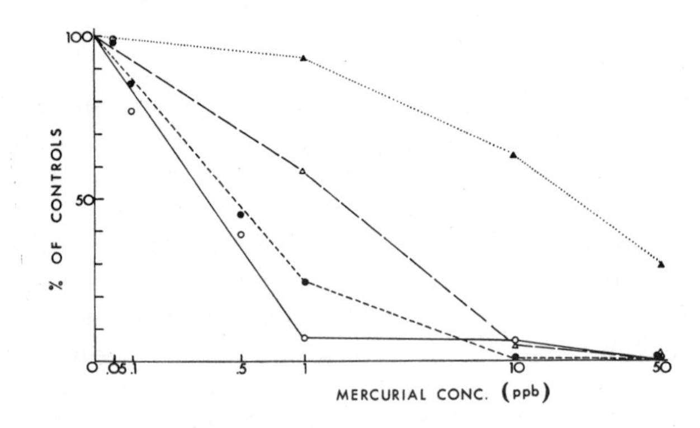


FIGURE 25–1. Effect of mercurials on photosynthesis by the marine diatom *Nitzschia delicatissima* after 24 hours' exposure to the following compounds: filled triangles, diphenylmercury; open triangles, phenylmercuric acetate; filled circles, methylmercury dicyandiamide; open circles, MEMMI (N-methylmercuric-1,2,3,6-tetrahydro-3, 6-methano-3,4,5,6,7,7-hexachlorophthalimide)

tration. Similar toxicity has been measured in natural populations of phytoplankton (Harriss *et al.*, 1970). Increasing the amount of suspended matter in the water will decrease the mercury uptake per unit of algal cell surface and decrease the toxicity of a given concentration of a mercurial.

The exact mechanisms of photosynthetic inhibition in phytoplankton by mercurials are not known; however, studies by Thimann & Bonner (1949) demonstrated the action of several organomercurials as enzyme inhibitors in terrestrial plant cells, and similar reactions can be expected in aquatic plants. In contrast to phytoplankton, certain species of fish, such as the Coho, can tolerate up to 1 ppm of mercury. Low levels of mercury pollution may not result in an obvious kill of macro-fauna, but rather in the removal of the plankton on which they feed and, consequently, in a migration of plankton feeders out of the polluted area. Selective poisoning of certain species of phytoplankton, and not others, could lead to an undesirable imbalance in the plankton populations. Reduced competition may result in a 'population explosion' of the species which are less susceptible to mercury poisoning, with subsequent changes in the rates of mineral cycling and in the type of food that is available to higher trophic levels.

### *Behavioural, Physiological, and Genetic Effects*

The biological magnification of chlorinated hydrocarbons in food-webs results in the highest concentrations being developed in top carnivores. It has been conclusively demonstrated that sub-lethal amounts of DDT and Dieldrin cause reproductive failure in many birds, owing to disturbances in hormonal and enzyme systems (Peakall, 1970). Similar effects can be expected from heavy-metal compounds, while mercury and other metals inhibit the action of liver enzymes in the Killifish Jackim et al., 1970). Studies on the genetic and metabolic effects of mercurials on plant cells, insects, and mice, have been reviewed by Löfroth (1969).

The most subtle and least-studied aspects of the biological effects of chemical pollutants are the effects on behaviour. Weir & Hine (1970) found that 0·003 ppm of mercury in water could impair the conditioned performance of Goldfish (*Carassius auratus*).

### *Resistant Populations*

While there is no known literature on the development of resistance to mercury poisoning in aquatic organisms, studies on pesticide-resistant populations of fishes indicate the type of problems which can be expected. Small fish having a relatively high reproductive rate, which are subjected to a continuous exposure to low concentrations of chlorinated hydrocarbon pesticides, develop a tolerance for withstanding concentrations in their tissues that are many times the normal lethal levels (Ferguson, 1966; Ferguson *et al.*, 1966). The resistant fishes are lethal to unresistant predators, and in some areas in the southeastern United States top predators such as the Large-mouth Bass have been completely eliminated through devouring contaminated specimens from lower trophic levels.

### *Synergism*

Most watersheds in populated areas receive a wide variety of contaminants including pesticides, industrial effluents, and excess nutrients. In many cases the combined effects of two or more pollutants will be more detrimental to living organisms than the effects of any single pollutant. Corner & Sparrow (1956) found that sub-lethal doses of copper increased the toxicity of inorganic mercury and ethylmercuric chloride to *Artemia*. Copper and mercury also showed additive toxic effects on copepods (Barnes & Stanbury, 1948).

## The Need for Environmental Standards

The present water-quality standards for mercury are inadequate for the protection of aquatic ecosystems. The Bureau of Water Hygiene, United States Public Health Service, and the Soviet Union, have tentatively adopted a standard of 0·005 ppm for mercury in drinking water. In Japan a maximum allowable concentration of 0·01 ppm for methylmercury in industrial waste water has been adopted for the protection of humans. While these standards are probably sufficient to prevent acute poisoning of humans, the studies by Harriss et al. (1970) and Weir & Hine (1970) have clearly demonstrated that concentrations of mercurial compounds well below the proposed water-quality standards can have detrimental effects on phytoplankton growth and fish behaviour. With the limited data available on the biological effects of long-term exposure to sub-lethal concentrations of mercury, it is impossible to establish whether a 'threshold dose' exists above natural levels which is safe for aquatic organisms. The most promising techniques for evaluating the possible environmental impact of mercury pollution, and for establishing adequate standards, is the bioassay of chronic effects on behavioural patterns and on phytoplankton photosynthesis.

The present procedure of water analysis for the detection of mercury pollution is also inadequate for the protection of the natural system. Mercury, like chlorinated hydrocarbon insecticides, accumulates in solid phases, and low concentrations in water are not an indication that the environment is free from contamination. Mercury that is bound to sediment can be extracted in the digestive system of bottom feeders, or can be released from the sediment back into the water as a result of a change in chemical conditions. Even after a source of mercury pollution is cut off, the contaminated sediment can act as a reservoir, slowly releasing mercury into the biosphere. At present there are no satisfactory techniques for restoring the quality of contaminated sediments. Proper monitoring procedures must be based on an understanding of the pathways and sinks for mercurial compounds in various ecosystems.

Federal action in the United States against known industrial sources of mercury pollution resul ted in an almost immediate 86 percent reduction in their mercury discharge. These industries effected this reduction in mercury discharge with existing technology and without insurmountable economic problems, but did not do so until questions of public health were raised and legal action was initiated against gross pollution. Yet even the reduced levels of mercury discharge still present a serious threat to the aquatic ecosystem. Until the chronic effects of mercurial compounds are understood, and adequate standards and monitoring procedures have been established which are based on the concept of protection of nature, we can expect to see many more changes in the structure and behavioural pattern of various aquatic ecosystems due to mercury pollution.

## References

Andersson, A. (1967). Kvicksilvret i marken. *Grundforbattring*, **20**, pp. 95–105.

Barnes, H. & Stanbury, F. (1948). The toxic action of copper and mercury salts both separately and when mixed on the harpactacid copepod, *Nitocra spinipes*. *J. Exp. Biol.*, **25**, pp. 270–5.

Corner, E. D. & Sparrow, B. W. (1956). The modes of action of toxic agents. 1. Observations on the poisoning of certain crustaceans by copper and mercury. *J. Mar. Biol. Assoc. UK*, **35**, pp. 531–48.

Ferguson, D. E. (1966). The effects of combinations of insecticides on susceptible and resistant mosquito fish. *Bull. Environ. Contamination and Toxicology*, **1**, pp. 97–103.

Ferguson, D. E., Ludke, J. L. & Murphy, G. G. (1966). Dynamics of endrin uptake and release by resistant and susceptible strains of mosquito fish. *Trans. Amer. Fish. Soc.*, **95**, pp. 335–44.

Fimreite, N. (1970). Mercury uses in Canada and their possible hazards as sources of mercury contamination. *Environ. Pollution*, **1**(2), pp. 119–31.

Glooschenko, W. A. (1969). Accumulation of $^{203}Hg$ by the marine diatom *Chaetoceros costatum*. *J. Phycol.*, **5**, pp. 224–7.

Goldberg, E. D. (1970). The chemical invasion of the ocean by Man. Pp. 63–73 in *McGraw-Hill Yearbook of Science and Technology*.

Hannerz, L. (1968). Experimental investigations on the accumulation of mercury in water organisms. *Inst. of Freshwater Res., Drottningholm, Sweden*, Rept N 48, pp. 120–76.

Harriss, R. C., White, D. & Macfarlane, R. (1970). Mercury compounds reduce photosynthesis by plankton. *Science*, **170**, pp. 736–8.

Hosohara, K. (1961). Mercury in seawater. *Nippon Kagaku Zasshi*, **82**, pp. 1107–8.
Jackim, E., Hamlin, J. M. & Sonis, S. (1970). Effects of metal poisoning on five liver enzymes in the Killifish (*Fundulus heteroclitus*). *J. Fish. Res. Bd. Can.*, **27**, pp. 383–90.
Jensen, S. & Jernelov, A. (1969). Biological methylation of mercury in aquatic organisms. *Nature.* (London) **223**, pp. 753–4.
Johnels, A. G., Westermark, T., Berg, W., Persson, P. & Sjostrand, B. (1967). Pike (*Esox lucius* L.) and some other aquatic organisms in Sweden as indicators of mercury contamination in the environment. *Oikos,* **18**, pp. 323–33.
Johnels, A. G. & Westermark, T. (1969). Mercury contamination of the environment in Sweden. Pp. 221–44, in M. W. Miller and G. Berg (Eds.) *Chemical Fallout.* Charles C Thomas Publishers, Springfield, Illinois.
Klein, D. H. & Goldberg, E. D. (1970). Mercury in the marine environment. *Environ. Sci. and Tech.*, **4**, pp. 765–8.
Kurland, L., Faro, S. & Siedler, H. (1960). Minamata disease. *World Neurol.*, **1**, pp. 320–5.
Löfroth, G. (1969) Methylmercury. *Swedish Natural Science Research Council, Ecological Res. Comm. Bull.*, N 4, 29 pp.
Peakall, D. B. (1970). Pesticides and the reproduction of birds. *Sci. Amer.*, **223**, pp. 73–6.
Robertson, D. (1970). Personal Communication.
Thimann, K. V. & Bonner, W. D. (1949). Experiments on the growth and inhibition of isolated plant parts. II The action of several enzyme inhibitors on the growth of the avena coleoptile and on pisum internodes, *Amer. J. Botany,* **36**, pp. 214–21.
Weir, P. A. & Hine, C. H. (1970). Effects of various metals on behavior of conditioned goldfish. *Arch. Environ. Health*, **20**, pp. 45–51.
Williston, S. H. (1968). Mercury in the atmosphere. *J. Geophys. Res.*, **73**, pp. 7051–5.
Woodwell, G. M. (1967). Toxic substances and ecological cycles. *Sci. Amer.*, **216**, pp. 24–31.
Woodwell, G. M., Wurster, C. F. & Isaacson, P. A. (1967). DDT residues in an East Coast estuary: a case of biological concentration of a persistent insecticide. *Science*, **156**, pp. 821–4.
Wurster, C. F. (1969*b*). Chlorinated hydrocarbon insecticides and avian reproduction: how *Conserv.*, **1**, (2) pp. 123–9.

# AIR POLLUTANTS

# 26

# Decline and mortality of smog-injured ponderosa pine

*Fields W. Cobb, Jr.*
*R. W. Stark*

Pollutants resulting primarily from photochemical reduction of auto exhaust gases in the atmosphere can cause a serious disease of plants. The reduction process generates ozone and other oxidants which cause rapid loss of chlorophyll (*4*). When excessive, the photosynthetic capacity is reduced below that necessary to sustain plant growth (*5*). Photochemical air pollution injury to ponderosa pine results in premature chlorosis, senescence, stunting, and casting of foliage (*6*).

In 1965, observations indicated that smog-injured ponderosa pines in the San Bernardino Mountains were frequently attacked by the western pine beetle, *Dendroctonus brevicomis* (LeConte), and the mountain pine beetle, *D. ponderosae* (Hopkins). Studies were initiated on the San Bernardino National Forest in 1966 to determine the relationships between air pollution injury and bark beetle attacks.

## Smog-Injured Pines More Susceptible

These studies (*1*, *2*, *3*, *6*) showed that ponderosa pines exhibiting advanced smog-injury symptoms were more frequently attacked by bark beetles than those with less severe symptoms. Only 3.5 percent of the

---

Reprinted by permission from the *Journal of Forestry*, 68:147–149 (March, 1970).

The authors are assistant professor, Dept. of Plant Pathol. and professor, Dept. of Entomol. and Parasitol., Univ. of Calif. Berkeley, and assistant plant pathologist and entomologist, respectively, in the Calif. Agr. Exp. Sta. These studies were supported by Nat. Sci. Found. Grant No. GB-7363.

most healthy trees (all trees were affected to some degree) were infested, compared to 46.0 percent of the severely affected trees. The mountain pine beetle was not found infesting any of the healthy trees (*6*).

These studies also showed that smog injury reduced tree growth and caused significant reductions in the live crown ratio (length of live crown/total height). In addition, smog injury reduced oleoresin exudation pressure, resin yield, and rate of resin flow. Crystallization rate, on the other hand, was increased (*1*). These effects are believed to facilitate invasion of the trees by bark beetles (*2*).

Further effects of smog injury included reductions in sapwood and phloem moisture, phloem thickness, and phloem carbohydrates. Phloem pH and oleoresin quality (based on major monoterpene constituents) were not affected (*3*). Reduction in moisture content was also believed to enhance the probable success of beetle attack, but it was speculated that the reduction of phloem thickness and amount of carbohydrates would be unfavorable for bark beetle brood development (*2*).

## Methods

During the 1966 studies a plot of 150 ponderosa pines was established. Fifty of the trees were relatively healthy, 50 had light to moderate injury symptoms (intermediate), and 50 had severe symptoms (advanced). The classification was based on a scale of 0 to 9 using branch mortality, needle retention, color, length, and complement as indicators (*6*). These trees were re-examined approximately 3 years later in late June 1969. Trees that had died were examined for evidence of bark beetle infestation and the condition of living trees was updated using the same rating scale.

## Results

Thirty-six of the 150 trees were killed during the three-year period (Table 26–1), all of which were infested by bark beetles prior to death.

TABLE 26–1

*Morality and Change in Disease Rating of Smog-Injured Ponderosa Pine Between 1966 and 1969*

| Status and Change in disease rating | Disease Rating | | |
|---|---|---|---|
| | Healthy | Intermediate | Advanced |
| | | *Number of trees* | |
| Status July, 1966 | 50 | 50 | 50 |
| Change to intermediate | 21 | — | — |
| Change to advanced | 1 | 34 | — |
| Mortality | 0 | 3 | 33 |
| Status June, 1969 | 28 | 34 | 52 |

Thirty-three of these trees were in the advanced disease category in 1966, and three were in the intermediate category. All trees with the most advanced symptoms were killed (Fig. 26–1). Mortality became progressively less as severity of symptoms decreased; none of the healthy trees or least diseased of the intermediate group were killed.

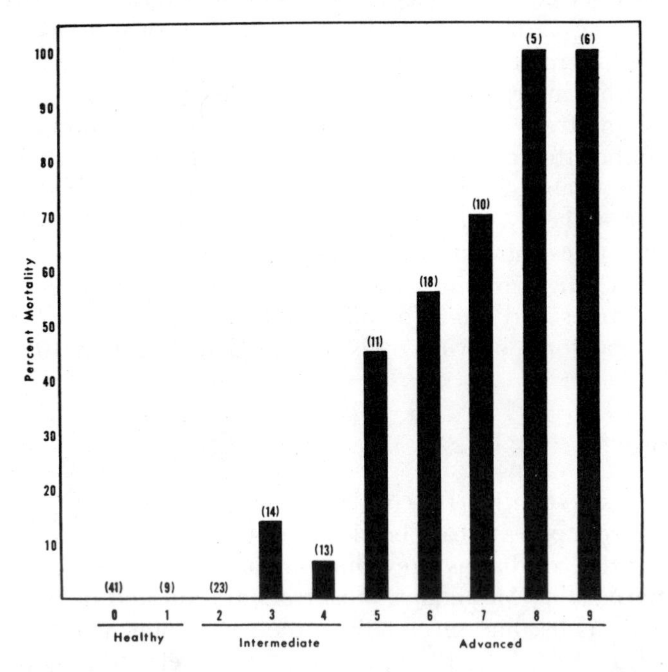


FIGURE 26–1. Mortality of smog-injured ponderosa pines in relation to 1966 numerical rating of symptoms. (Total number of trees per rating in parentheses.)

Tree diameters ranged from 9 inches to more than 32 inches; 60 trees were less than 17 inches and 90 were 17 inches or greater (Table 26–2). More of the lower diameter trees (35 percent) were killed than those in higher diameter classes (16.7 percent).

TABLE 26–2

*Mortality in Relation to Diameter Class of Smog Injured Ponderosa Pine*

| Diameter Class (inches) | No. Trees | No. Trees Killed | Percent Mortality |
|---|---|---|---|
| 9–12 | 11 | 7 | 63.6 |
| 13–16 | 49 | 14 | 28.5 |
| 17–20 | 47 | 8 | 17.0 |
| 21–24 | 25 | 4 | 16.0 |
| 25> | 18 | 3 | 16.7 |

Sixteen of the killed trees had been removed during sanitation-salvage operations. Of the remaining 20 trees, 11 were infested by the western pine beetle, five by the mountain pine beetle, and four by both species. These proportions of attack by beetle species are roughly equivalent to those found in 1966 (*6*).

Of the 114 trees still living in 1969, only 29 had the same numerical symptom rating as in 1966; 15 of these were healthy trees. Another 29 trees had increased symptoms but stayed in the same disease categories (healthy, intermediate, or advanced) assigned to them in 1966. The remaining trees exhibited enough increase in symptom severity to move from healthy to intermediate or advanced, or from intermediate to advanced (Table 26–1). Thus, despite the death of 33 advanced diseased trees, there were 52 trees in this category in 1969. The 1966 intermediate trees showed the greatest increase in disease severity with an average change in numerical rating of +2.5, compared to +1.1 for healthy trees and +0.5 for advanced diseased trees. The overall average rating for the living trees increased from 2.1 in 1966 to 3.7 in 1969.

## Discussion

These results confirm the conclusion that air pollution injury predisposes ponderosa pines to bark beetle infestations (*6*). They further show that, under the conditions prevailing in the Lake Arrowhead area of the San Bernardino Mountains, most trees are infested by beetles only after injury has become severe. Even the three intermediate diseased trees which had died (in 1969), had time to develop the more severe symptoms of advanced diseased trees before being infested by beetles, probably in fall 1967 or spring 1968.

The results also indicate that a rating system based on needle length, color, retention, and overall complement and on branch mortality can be used to predict those trees that will become infested by bark beetles. Such a system could be used in choosing trees for cutting, thus removing poor risk trees before beetle infestation.

The oleoresin exudation pressures (OEP) of all and the moisture contents of sapwood and phloem tissues of many of the plot trees were determined in 1966 (*1*). When these data from trees with a disease rating of 3 or more are related to the results of the current study, we find that 63 percent of the trees with zero OEP were killed, compared to only 31 percent of the high (125 psi and greater) OEP trees. Also, the moisture contents of both sapwood (108 vs. 118 percent dry wt.) and phloem (54.8 vs. 60.0 percent saturation) were less in those trees subsequently killed.

## Conclusions

The mortality and increase in severity of smog injury in ponderosa pine that have occurred during the period 1966–1969 indicate the threat of air pollution to forest stands. Mortality in certain areas of the San Bernardino Mountains has been high for at least 10 years, and many of the more susceptible ponderosa pines were killed prior to 1966. Thus, the trees being killed at the present time are probably those more resistant to air pollution. Yet, only about one-half of the trees classified as healthy in 1966 can still be considered as such; and only 11 of 41 trees with a disease rating of zero in 1966 have maintained that rating.

If the present level of mortality continues, it can be expected that only a small number of highly resistant ponderosa pines will be alive within 10–15 years. Ponderosa pine is the predominant species in most stands in the area, but there are other conifers such as sugar pine, incense cedar, white fir, and Douglas-fir. These appear to be less susceptible to pollution injury although they are also beginning to show symptoms of injury. If air pollution from the Los Angeles basin continues unabated, there will be a conversion from the originally well stocked ponderosa pine stands to poorly stocked stands of the less susceptible species. Only rigorous control of pollution or reforestation with highly resistant species can retain the area as a productive forest region, or even as a favored recreation area. As air pollution increases in other areas of the country, similar results can be expected, especially if the forest species are as susceptible to smog injury as is ponderosa pine.

No attempt was made to estimate the actual and potential loss of timber or the cost of removal of beetle-killed trees in the highly developed recreational area near Lake Arrowhead. However, the economic impact has been substantial.

## Literature Cited

1. Cobb, F. W. Jr., D. L. Wood, R. W. Stark and P. R. Miller, 1968. Photochemical oxidant injury and bark beetle (*Coleoptera: Scolytidae*) infestation of ponderosa pine. II. Effect of injury upon physical properties of oleoresin, moisture content, and phloem thickness. Hilgardia 39(6):127–134.
2. ———, D. L. Wood, R. W. Stark and J. R. Parameter Jr. 1968. Photochemical oxidant injury and bark beetle (*Coleoptera: Scolytidae*) infestation of ponderosa pine. IV. Theory on the relationships between oxidant injury and bark beetle infestation. Hilgardia 39(6):141–152.
3. Miller, P. R., F. W. Cobb Jr. and É. Zavarin. 1968. Photochemical oxidant injury and bark beetle (*Coleoptera: Scolytidae*) infestation of ponderosa pine. III. Effect of injury upon oleoresin composition, phloem carbohydrates and phloem pH. Hilgardia 39(6):135–140.

4. ———, J. R. Parameter, Jr., B. H. Flick, and C. W. Martinez. 1969. Ozone dosage response of ponderosa pine seedlings. J. Air Pollution Control Ass. 19(6):435–438.
5. Parameter, J. R., Jr., and P. R. Miller. 1968. Studies relating to the cause of decline and death of ponderosa pine in southern California. Plant Disease Reporter. 52(9):707–711.
6. Stark, R. W., P. R. Miller, F. W. Cobb Jr., D. L. Wood and J. R. Parameter Jr. 1968. Photochemical oxidant injury and bark beetle (*Coleoptera: Scolytidae*) infestation of ponderosa pine. I. Incidence of bark beetle infestation in injured trees. Hilgardia 39(6):121–126.

# 27

# Effects of air pollutants on L. A. basin citrus

*C. R. Thompson*
*O. C. Taylor*
*B. L. Richards*

During the late 1950's a unique cooperative organization, the Agricultural Air Research Program, was formed with headquarters at the then Citrus Experiment Station of the University of California, Riverside. Sponsors of this group, industry, agricultural organizations and state and county government, had decided that a concerted effort was needed to find out what was happening to the citrus production in the Los Angeles Basin (1). Yield records from Los Angeles, San Bernardino, Riverside, and Orange Counties indicated a drop in average production per acre in mature orchards during the previous 15 years. This decline had not been recognized as a problem in other citrus-growing areas of the state.

They wrote: "The basic, overall objective is to measure under field conditions the effects of certain atmospheric phytotoxicants on agricultural crops growing in the upper Santa Ana Drainage Basin. Such measurements include plant growth, crop yield, longevity, and the quality of fruits. The concept of atmospheric phytotoxicants in this research program includes fluorides, ozone, and various other oxidants." A further objective was to find out not only *if* but, if so, *how much* injury was being caused to crops.

---

Reprinted with permission from the *Citrograph*, 55:165–66, 190–192 (1970).

The authors are with the Statewide Air Pollution Research Center, University of California, Riverside, California 92502.

FIGURE 27–1. Plastic-covered greenhouses at one experimental installation.

A primary problem was to design an inexpensive greenhouse which could be installed over bearing citrus trees—one that would alter the environment least. An aluminum-framed, plastic-covered structure equipped with a blower which changed the inside atmosphere twice per minute was devised (2). The greenhouses (Figure 27–1) were covered with a plastic film of polyvinyl fluoride, Tedlar (E. I. Dupont de Nemours). A bank of three Filterfold (Barnaby Cheney Co.) cannisters filled with activated coconut charcoal and/or limestone, depending upon the treatment, was installed in the blower cabinet.

Airborne gaseous fluoride, total fluoride and total oxidants were monitored continuously at all three experimental installations. Accumulation of fluoride in citrus leaves was also measured during each three-month period.

Three plots of uniform, vigorously growing citrus trees, two of lemons and one of navel oranges, were leased. Individual greenhouses were installed over 24 of the trees in each of the three groves. The trees were then divided at random into six lots of four replicates each. The treatments, Table 27–1, were designed to test the effects of the total air pollutants in ambient air (fluoride, ozone, peroxyacyl nitrates, and oxides of nitrogen) and each of these constituents by either elimination or addition to treated ambient air.

Tensiometers were installed in all houses about 1½ meters from the tree at a depth of ½ meter to monitor soil moisture. Tabulations of soil suction, a measure of moisture content, were made daily.

The first experimental installation was located in Upland, California, over Eureka lemons. This was followed by a second lemon study at Cucamonga, California, and one navel orange installation at Upland, California. The initial projections were for a five-year study of both lemons and oranges.

## Results

The treatments were begun on lemons at Upland, California, in January 1962.

TABLE 27–1

*Experimental Field Installations on Citrus, Upland and Cucamonga, California*

| Tree Atmosphere | Treatment of Atmospheres | Toxicant Remaining |
|---|---|---|
| Filtered air | Activated carbon, limestone | -- |
| Filtered air + fluoride | Limestone, activated carbon, hydrogen fluoride | Fluoride |
| Low ozone air | Nitric oxide | Fluoride, NO, $NO_2$, PAN[a] |
| Low fluoride air | Limestone | Ozone, PAN |
| Low ozone, low $F^-$ air | Limestone, nitric oxide | PAN, NO, $NO_2$ |
| Ambient air | -- | Fluoride, ozone PAN |
| Check | Unenclosed trees | Fluoride, ozone PAN |

[a] Peroxyacyl nitrates

The most important economic question was the effect of the various treatments on yield of fruit. Lemons were picked at four successive times as the size or color indicated maturity. The results for five years are shown for Lemon Division 1 in Table 27–2. The yield was analyzed statistically by a randomized block procedure with the differences between

TABLE 27–2

*Average Annual Yield of Lemons, Division 1*

| | Year | Filtered Air | Filtered Air + HF | Low Ozone Air | Low Fluoride Air | Low Ozone Low Fluoride Air | Ambient Air | Check |
|---|---|---|---|---|---|---|---|---|
| Kilograms of fruit | 1962-67 | 84.2 | 93.0 | 56.1 | 57.1 | 65.1 | 63.4 | 78.6 |
| Number of fruit | 1963-67 | 841 | 940 | 597 | 600 | 679 | 688 | 970 |
| Grams per fruit [a] | 1963-67 | 104 | 103 | 97 | 98 | 101 | 97 | 91 |

[a] Adjusted for number of fruit.

the seven treatments being examined by a multiple range test. With both pounds of fruit and numbers, "filtered air" was significantly greater than all other treatments at the 1 percent level. The "filtered air" was the same as "filtered air plus fluoride." These results showed that photochemical oxidants were reducing the yield drastically.

Lemon Division 2 was a replicate of the first lemon study. By 1968 the study had progressed for five years and showed that the treatments had caused the yield of fruit, Table 27–3, to show wide differences between the different treatments. The two treatments which received carbon filtered air, "filtered air," or "filtered air plus fluoride" were significantly greater both with regard to weight and numbers of fruit than the other enclosed treatments. Outside checks were also greater than the unfiltered but enclosed yields. This could be caused by the higher

TABLE 27–3

*Average Annual Yield of Lemons, Division 2*

| | Year | Filtered Air | Filtered Air + HF | Low Ozone Air | Low Fluoride Air | Low Ozone Low Fluoride Air | Ambient Air | Check |
|---|---|---|---|---|---|---|---|---|
| Kilograms of fruit | 1963-68 | 69.0 | 62.5 | 30.2 | 26.7 | 34.2 | 30.6 | 71.1 |
| Number of fruit | 1963-68 | 641 | 583 | 292 | 279 | 316 | 289 | 747 |
| Grams per fruit [a] | 1963-68 | 111 | 110 | 106 | 110 | 108 | 109 | 103 |

[a] Adjusted for number of fruit.

temperatures which promote vegetative growth over fruit production in the lemons.

Navel Orange Division 1 was begun in May 1964. By the time the experiment was concluded in January of 1968 fruit yields were widely different, Table 27–4. The two sets of trees receiving carbon filtered

TABLE 27–4

*Average Annual Yield of Navel Oranges, Division 1*

| | Year | Filtered Air | Filtered Air + HF | Low Ozone Air | Low Fluoride Air | Low Ozone Low Fluoride Air | Ambient Air | Check |
|---|---|---|---|---|---|---|---|---|
| Kilograms of fruit | 1964-67 | 66.7 | 62.0 | 21.6 | 32.4 | 41.8 | 23.9 | 11.7 |
| Number of fruit | 1964-67 | 286 | 287 | 96 | 132 | 190 | 109 | 57 |
| Grams per fruit [a] | 1964-67 | 264 | 248 | 223 | 237 | 233 | 219 | -- |

[a] Adjusted for number of fruit.

air, "filtered air," and "filtered air plus fluoride," were higher than the other treatments, and despite much variation between replicates the differences were significant at the 1 percent level. Outside checks in this case were less than the enclosed trees indicating a response different from that in the lemons.

Fruit drop in lemons was of minor consequence, Table 27–5, averaging from 2 to 11 percent of the harvested fruit, but represents a major problem in navel oranges, Table 27–6. These results showed that during summer the trees in "ambient air" and the outside checks lost 61.8 and 76.0 percent of the total fruit set. This loss is over twice that lost by trees in "filtered air" or "filtered air plus fluoride." The fall and winter fruit drop was about the same in all treatments. Thus, the fruit drop in navel oranges accounts for the major differences in yield shown later.

The rate of leaf drop from the trees was measured by two methods. In the first method all fallen leaves were raked up, air dried, and weighed. Representative data are shown in the first three columns of Table 27–5 which are the amounts dropping during a several-month period. Trees receiving "filtered air" or "filtered air plus fluoride" dropped less leaves than the other enclosed trees.

As an alternate method of measuring leaf drop, ten representative branches were selected on trees in Lemon Division 1, were tagged, and the leaves counted. Monthly counts were made. The results, Table 27–5, columns 4 and 5, showed that after 12 and 18 months "filtered air" and "filtered air plus fluoride" had lost less leaves than all other treatments inside the houses. Why the "outside checks" retained leaves better than those receiving "ambient air" is unexplained, but perhaps the higher temperature inside the houses had this effect.

TABLE 27–5

*Effect of Air Pollutants on Leaf and Fruit Drop*

| | Kg Leaves Dropped Annually[a] | | | Percent Leaves Dropped, Lemon 1 | | Percent Fruit Drop | |
|---|---|---|---|---|---|---|---|
| | Lemon 1 | Lemon 2 | Orange 1 | 12 mo. | 18 mo. | Lemon 1 | Lemon 2 |
| Filtered air | 3.45 | 4.27 | 5.11 | 8.4 | 32.0 | 2.3 | 6.0 |
| Filtered air + HF | 3.50 | 4.20 | 4.93 | 6.9 | 23.1 | 2.7 | 6.6 |
| Low ozone air | 3.94 | 4.37 | 5.59 | 12.3 | 40.2 | 3.2 | 8.4 |
| Low Fluoride air | 4.23 | 4.62 | 5.65 | 37.3 | 93.7 | 3.3 | 11.0 |
| Low ozone, low fluoride air | 3.99 | 4.59 | 5.36 | 14.8 | 58.5 | 2.6 | 8.8 |
| Ambient air | 3.86 | 4.68 | 5.66 | 19.2 | 70.1 | 3.4 | 9.8 |
| Check | 4.31 | 5.97 | 5.81 | 12.8 | 56.9 | 6.2 | 7.9 |

[a] Lemon 1, 1962-67; Lemon 2 and Orange 1, 1963-68.

Another sensitive index of plant health is the rate of photosynthesis. This measurement was made on a series of individual trees during 1961 before the treatments outlined in Table 27–1 were begun to provide a "base line" and subsequently in 1962 after the treatments were started in January.

The results showed that the trees in "ambient air" were reduced in photosynthesis to 66 percent of the "filtered air" treatments. All other treatments were reduced except the "filtered air plus HF" which increased slightly. The significance of this last effect is unknown, but small levels of fluoride do cause increased respiration in isolated tissue preparation. The photosynthesis of isolated lemon branches was measured, also (3). These results, Table 27–7, showed that branches in "ambient air" had a numerically lower rate of photosynthesis than all other treatments except in the last period and were statistically less than "filtered air" in the first two periods.

The amount of water used by lemon trees in the different treatments

TABLE 27–6

*Fruit Drop of Navel Oranges* [a]

| Season | Filtered Air | Filtered Air + HF | Low Ozone Air | Low Fluoride Air | Low Ozone Low Fluoride Air | Ambient Air | Check |
|---|---|---|---|---|---|---|---|
| Summer | 29.7 | 30.0 | 52.7 | 47.6 | 37.7 | 61.8 | 76.0 |
| Fall + Winter | 8.5 | 11.0 | 9.8 | 10.9 | 11.0 | 7.4 | 6.5 |
| Yearly Average | 38.2 | 41.0 | 62.5 | 58.5 | 48.7 | 69.2 | 82.5 |
| Total Number of Fruit Set | 651 | 706 | 536 | 551 | 647 | 507 | 508 |

[a] Average percent of total fruit set per tree.

differed appreciably. The irrigation schedule was 15 days apart but some trees required "extra" irrigations more frequently than others. "Extra" irrigations were those required to maintain soil suction below 50 centibars. The tentative findings of 1962 were confirmed in 1963, 1964 and 1965, Table 27–8, and showed that trees which received "ambient air" used less water than those which received "filtered air." "Ozone low air" trees and "ozone, fluoride low air" trees lost less water than "filtered air" in 1964 and 1965 (5 percent level). All other treatments were not different from "filtered air" statistically but all average values were numerically less.

Fluoride accumulation in leaves was measured quarterly. The average values are shown in Table 27–9. Ambient air samples accumulated

TABLE 27–7

*Apparent Photosynthesis of Selected Lemon Branches, Av. Mg $CO_2/Dm^2/hr$*

| Dates of Treatment | Filtered | Ambient | Filtered $+F^-$ | Low $O_3$ | Low $O_3,F^-$ | Low $F^-$ |
|---|---|---|---|---|---|---|
| 8-5-63 to 10-4-63 | 6.4** | 3.7 | 4.8 | 5.3* | 4.7 | 5.2* |
| 10-7-63 to 1-31-64 | 5.6* | 3.8 | 5.8* | 4.8 | 5.2 | 4.1 |
| 6-24-64 to 8-31-64 | 2.9 | 1.5 | 2.5 | 1.9 | 2.3 | 2.1 |
| 9-14-64 to 11-8-64 | 5.0 | 3.5 | 3.6 | 3.9 | 4.4 | 2.6 |

* Greater than ambient at 5% level of significance.

** Greater than ambient at 1% level of significance.

TABLE 27–8

*Total Number of "Extra" Irrigations Required by Lemon Trees with Different Treatments During the Year*

| Treatment | 1962[a] | 1963 | 1964 | 1965 | 1966 |
|---|---|---|---|---|---|
| Filtered air | 23 | 43 | 36 | 24 | 20 |
| Ambient air | 3 | 7[b] | 5[b] | 4[c] | 6 |
| Low fluoride air | 16 | 37 | 26 | 14 | 18 |
| Low ozone air | 7 | 19 | 12[c] | 10 | 7 |
| Filtered air plus fluoride | 14 | 39 | 28 | 15 | 16 |
| Low ozone, low fluoride | 12 | 37 | 22 | 5[c] | 9 |

[a] Three months of the year.

[b] Significantly less than filtered air at 1%.

[c] Significantly less than filtered air at 5%.

more fluoride than outside check trees probably because dew and rain leached fluoride from the leaves before or after absorption. None of the accumulations were high enough to cause leaf symptoms and from the other tree responses would indicate that fluoride, while present in the atmosphere, was of minor importance to the health and performance of the trees.

TABLE 27–9

*Accumulation of Fluoride by Citrus Leaves* [a]

| Grove | Year Set | Filtered Air | Filtered Air + HF | Low Ozone Air | Low Fluoride Air | Low Ozone Low Fluoride Air | Ambient Air | Outside Check |
|---|---|---|---|---|---|---|---|---|
| Lemon 1 | 1962-1966 | 13 | 36 | 30 | 14 | 11 | 35 | 24 |
| Lemon 2 | 1963-1967 | 15 | 30 | 34 | 18 | 16 | 33 | 29 |
| Orange 1 | 1964-1967 | 5 | 23 | 18 | 8 | 8 | 18 | 13 |

[a] Parts per million of fluoride on a dry weight basis.

Both tree growth as measured by increase in girth and weights of air dried prunings were measured but no consistent differences were observed.

## Discussion

The higher rate of apparent photosynthesis and water use of trees receiving "filtered air" as compared to "ambient" could be caused by early senescence of leaves in photochemical smog (4) or other metabolic effects such as those observed by Dugger *et al.* (5) who found that ozone fumigation reduced starch and reducing sugars in lemons. The reduction in yield of lemons in "ambient air" seems to be the result of lack of set either by lack of blooming or a very early abortion of fruit. In navel oranges, the set of fruit seems normal but the "June Drop," when the fruit is about 1.5 cm diameter, is so excessive in "ambient air" that a very poor crop is left for maturity.

## Summary

Commercially producing lemons and navel orange trees were enclosed in plastic covered greenhouses and were given various fractions of the air pollutants occurring in the Los Angeles Basin. In some treatments nitric oxide was supplied to the trees to react with ozone but this formed nitrogen dioxide, another phytotoxicant. The study showed that the photochemical smog complex reduced the rate of water use, apparent photosynthesis and yield of both lemons and navel oranges. Fluoride levels in the atmosphere were too low to cause detectable effects. Leaf drop was significantly less in lemons receiving carbon filtered air than those receiving "ambient air." A similar trend occurred in navel oranges. Fruit drop is a serious problem in navel oranges. This loss was significantly less in carbon filtered air than ambient. Yield of mature fruit is reduced in some cases by as much as 50 percent.

## Acknowledgement

The authors wish to acknowledge major financial support for this study by National Air Pollution Control Administration, Grant AP 00270-06, Kaiser Steel Corporation, Sunkist Growers, San Bernardino County, Southern California Edison Company, Riverside County APCD, Los Angeles Clearing House Association, Pure Gold, Inc., Orange County APCD, Southwestern Portland Cement Company, Kaiser Cement and Gypsum Corporation (Permanente), American Cement Corporation, California Portland Cement Company, DiGiorgio Fruit Corporation, Koppers Company, Inc., American Potash and Chemical Corporation, West End Chemical Company (Stauffer), Irvine Company, Griffin Wheel Company.

## Literature Cited

1. Thompson C. Ray and O. Clifton Taylor. Plastic-covered greenhouses supply controlled atmospheres to citrus trees. *ASAE* 9(3):338, 339 and 342 (1966).
2. Richards, B. L. and O. C. Taylor. *Status and redirection of research on the atmospheric pollutants toxic to field grown crops in southern California.* Proc. 53rd Annual Meeting of APCA, May 22–26 (1960).
3. Taylor O. C., E. A. Cardiff and J. D. Mersereau. Apparent photosynthesis as a measure of air pollution damage. *APCA Jour.* 15(4):171–173 (1965).
4. Darley, Ellis F. Research on damage to vegetation from air pollution at the University of California, Riverside. *Proc. Intern. Clean Air Conf.*, London p. T28 (1959).
5. Dugger W. M., Jr., Jane Koukol, and R. L. Palmer. Physiological and biochemical effects of atmospheric oxidants on plants. *J. APCA* 16(9):467–471 (1966).

# OIL POLLUTION

# 28
# A small oil spill

*Max Blumer*
*Howard L. Sanders*
*J. Fred Grassle*
*George R. Hampson*

During the last few years the public has become increasingly aware of the presence of oil on the sea. We read about the recurring accidents in oil transport and production, such as the disaster of the *Torrey Canyon* tanker, the oil well blowout at Santa Barbara, and the oil well fires in the Gulf of Mexico. To those visiting our shores the presence of oil on rocks and sand has become an everyday experience; however, few of us realize that these spectacular accidents contribute only a small fraction of the total oil that enters the ocean. In the *Torrey Canyon* episode of 1967 about 100,000 tons of crude oil were lost. By comparison, routine discharges from tankers and other commercial vessels contribute an estimated three and one-half million tons of petroleum to the ocean every year. In addition, pollution from accidents in port and on the high seas, in exploration and production, in storage, in pipeline breaks, from spent lubricants, from incompletely burned fuels, and from untreated industrial and domestic sewage contribute an equal or larger amount of oil.

---

From Environment, 13(2):2–12. Reprinted by permission of Environment. 

The authors are members of the staff of Woods Hole Oceanographic Institution, Woods Hole, Massachusetts. Max Blumer, Ph.D., and Howard L. Sanders, Ph.D., are senior scientists. J. Fred Grassle, Ph.D., is assistant scientist, and George R. Hampson, B.S., is research associate.

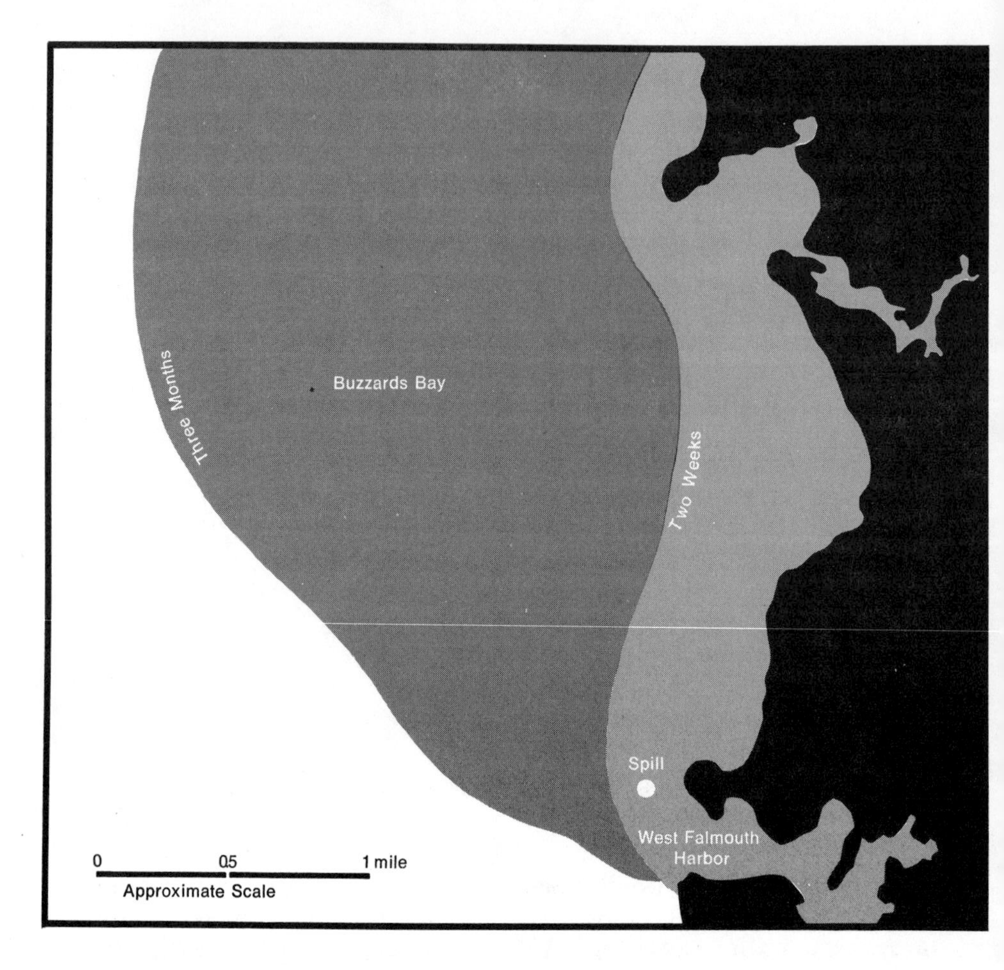


Thus, it has been estimated that the total oil influx into the ocean is between five and ten million tons per year (1).

What are the effects of oil on marine organisms and on food that we recover from the sea? Some scientists have said that the oceans in their vastness should be capable of assimilating the entire oil input. This, however, assumes that the oil is evenly distributed through the entire water profile, or water column, of the ocean. Unfortunately this assumption is not correct. Oil production, transportation, and use are heavily concentrated in the coastal regions, and pollution therefore predominantly affects the surface waters on the continental margins. J. H. Ryther has stated that the open sea is virtually a biological desert (2). Although the deeper ocean provides some fishing for tuna, bonito, skipjack, and billfish, the coastal waters produce almost the entire shellfish crop and nearly half of the total fish crop. The bulk of the remainder of the

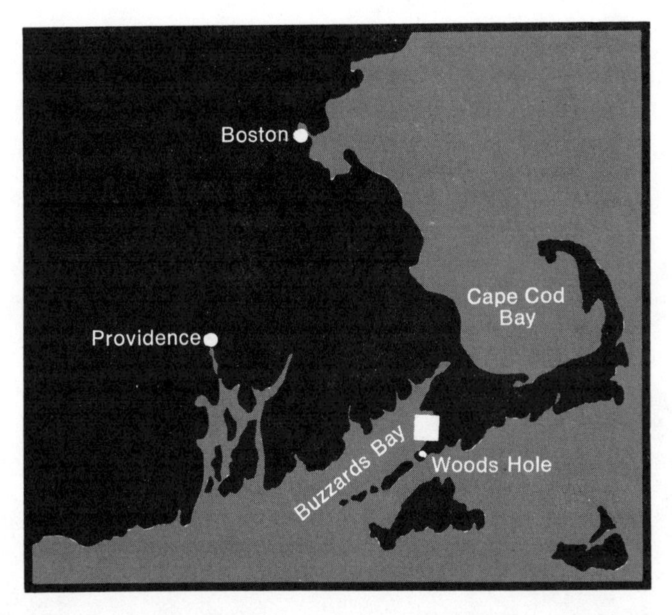


A barge loaded with fuel oil for a power plant went aground in September 1969 in Buzzards Bay, Massachusetts, shown in the map above. An interdisciplinary team of scientists from nearby Woods Hole Oceanographic Institution discovered that poisonous substances in the oil persisted in the marine environment for much longer than previously thought. Oil contamination of bottom sediment spread up into the bay with time; the map at left, an enlargement of the white area in the map above, shows the approximate boundary of the oil in the study area at two weeks and at three months after the spill.

fish crop comes from regions of upwelling water, near the continental margins, that occupy only one-tenth of one percent of the total surface area of the seas. These productive waters receive the heaviest influx of oil. They also are most affected by other activities of man, such as dredging, waste disposal, and unintentional dispersal of chemical poisons like insecticides.

Some environmentalists have expressed the belief that major oil spills such as those from the *Torrey Canyon* and the blowout at Santa Barbara have brought about little biological damage in the ocean (3). These statements are largely based on statistical measurements of the catch of adult fish. We believe that such statistics are a very insensitive measure of the ecologic damage to wide oceanic regions. Often the migratory history of the fish species studied is unknown. The fish may not have been exposed to the spill itself, or may not have suffered from a depletion of food organisms if their growth occurred in areas remote from the spill. Statistical and observational data on adult fishes will not reveal damage to the often much more sensitive juvenile forms or to intermediate members in the marine food chain. The only other studies on effects

The *Florida* with other boats in Buzzards Bay. Between 650 and 700 tons of #2 fuel oil were released into the coastal waters after the barge's accident.

of oil on marine organisms have concentrated on relatively tolerant organisms which live between the tides at the margins of affected areas. The main impact, however, would be expected in subtidal areas, and that has never been measured quantitatively.

A relatively small oil spill that occurred almost at the doorstep of the Woods Hole Oceanographic Institution at Woods Hole, Massachusetts, gave us the opportunity to study immediate and long-term ecological damage in a region for which we had extensive previous knowledge about the biology and chemistry of native marine organisms (4). On September 16, 1969, an oil barge on the way to a power plant on the Cape Cod Canal came ashore off Fassets Point, West Falmouth, in Buzzards Bay (see map, page 441). Between 650 and 700 tons of #2 fuel oil were released into the coastal waters. The oil-contaminated region in Buzzards Bay expanded steadily with time after the accident as the complex interaction of wind, waves, and bottom sediment movement spread oil from polluted to unpolluted areas. Eight months after the grounding, polluted sea bottom, marshes, and tidal rivers comprised an area many times larger than that first affected by the accident. The dispersion was much greater than expected on the basis of conventional studies of oil pollution. The situation even forced changes in our research efforts. As we shall explain later, a control point for marine surveys was established beyond the anticipated limit of the spread of oil. Within three weeks, the contamination had spread to the station. Another was established twice as far away. Three months after the accident, that too was polluted. Bottom sediment was contaminated 42 feet beneath the surface, the greatest water depth in that part of Buzzards Bay.

Ecological effects of the spreading blanket of oil beneath the surface were severe. The oil decimated offshore marine life in the immediate area of the spill during the first few days. As the oil spread out across the bottom of the bay in the following months, it retained its toxicity.

Even by May 1970, eight months after the spill, bacterial degradation (breakdown into simpler substances) of oil was not far advanced in the most polluted regions. More rapid oil deterioration in outlying, less affected areas had been reversed by a new influx of less degraded oil from the more contaminated regions.

The tidal Wild Harbor River still contained an estimated four tons of fuel oil. The contamination had ruled out commercial shellfishing for at least two years. The severe biological damage and the slow rate of biodegradation of the oil suggest that shellfish productivity will be affected for an even longer period. Furthermore, destruction of bottom plants and animals reduced the stability of marshlands and sea bottom. Resulting erosion may have promoted spread of the oil along the sea floor. Inshore, the oil penetrated to a depth of at least one to two feet in marsh sediment.

Nevertheless, compared in magnitude to other catastrophes, this was a relatively small spill; the amount of oil lost in the *Torrey Canyon* accident was 150 times larger. The interim results of our survey, coupled with research findings of other studies in this laboratory, indicate that crude oil and other petroleum products are a far more dangerous and persistent threat to the marine environment and to human food resources than we would have anticipated. Pollution from a large oil spill is very obvious and visible. It has often been thought that the eventual disappearance of this visible evidence coincides with the disappearance of any biological damage. This, however, is not true. Sensitive analytical techniques can still detect oil in marine organisms and in sediments after the visual evidence has disappeared, and biological studies reveal that this residual oil is still toxic to the marine organisms. Here we shall discuss first the general results of our study, then go more deeply into the description of the laboratory work involving biology, biochemistry, and chemistry. Our most important findings are these (4):

Crude oil and petroleum products contain many substances that are poisonous to marine life. Some of these cause immediate death; others have a slower effect. Crude oils and oil products differ in their relative composition; therefore the specific toxic effect may vary. Crude oil, in general, is less immediately toxic than some distilled products, but even crude oil that has been weathered (altered by exposure to the weather) at sea for some time still contains many of the acutely toxic hydrocarbons (5). The more persistent, slowly acting poisons (for example, the carcinogens) are more abundant in crude oil than in some of the lower boiling distillates. These poisons are quite resistant to the environmental weathering of oil.

In spite of low density, oil may mix with water, especially in a turbulent sea during storm conditions. Hydrocarbons may be dispersed through the water column in solution in the form of droplets, and the compounds may reach the sea bottom, particularly if weighted down by mineral particles. On the sea floor oil persists for long periods and can continue to damage bottom plants and animals. Thus, a single accident may result in long-term, continual pollution of the sea. This is a very important finding since biologists have long agreed that chronic pollution generally has more far-reaching effects than an accident of short duration. Hydrocarbons can be taken up by fish and shellfish. When the oil enters the fat and flesh of the animals, it is isolated from natural degradation processes. It remains essentially constant in amount and chemically intact even after the animals are transplanted into clean water for decontamination. Thus, chemicals from oil that may be poisonous to marine organisms and other animals, including man, may persist in the sea and in biological systems for many months after the spill.

By killing the bottom organisms, oil reduces cohesion of the bottom sediments and thereby accelerates transport of the sediments. Sediment movements along the sea bottom thus are a common occurrence after an oil spill. In this way contaminated sediments may be spread over great distances under the influence of tide and wave action, and the oil may be carried to areas not immediately polluted by the spill.

None of the presently available countermeasures can completely eliminate the biological damage of oil spills. The rapid removal of oil by mechanical recovery or by burning appears most promising. The use of sinking agents or detergents, on the other hand, causes the toxic and undegraded oil to spread in the ocean; the biological damage is then greater than if the spill had been left untreated. Reclamation of contaminated organisms, marshes, and offshore sediments is virtually impossible, and natural ecological recovery is slow.

With these conclusions in mind we can now turn to our experience with the West Falmouth oil spill. The effect of this relatively small spill was still acute in January 1971, almost a year and one-half after the accident. Officials in the town of Falmouth have estimated that the damage to local shellfish resources, during the first year after the accident, amounted to $118,000. This does not include the damage to other marine species and the expected damage in coming years. In addition to the loss of the oil and the barge and the cleanup expenses (estimated to be $65,000), the owner of the oil paid compensations for the losses of marine fishery resources to the town of Falmouth ($100,000) and to the Commonwealth of Massachusetts ($200,000). The actual ecological damage may far exceed this apparent cost of almost half a million dollars.

## Biological and Chemical Analysis

For our analysis (which is still continuing) bottom samples were carefully taken from the marshes and from the offshore areas. Samples for biological analysis were washed and sieved to recover living or dead organisms. These were preserved, identified, and counted. Results of counts from the affected area were compared with those from control areas that were not polluted by the spill. Some animals can be used as indicators for the presence of pollution, either because of their great sensitivity or because of their great resistance. Thus, small shrimplike animals, the amphipods of the family Ampeliscidae, are particularly vulnerable to oil pollution. Wherever the chemical analysis showed the presence of oil, these sensitive crustaceans were dying. On the other hand, the annelid worm, *Capitella capitata*, is highly resistant to oil pollution. Normally, this worm does not occur in large numbers in our area. However, after the accident it was able to benefit from the absence of other organisms which normally prey upon it and reached very high population densities. In the areas of the highest degree of pollution, however, even this worm was killed. *Capitella capitata* is well known, all over the world, as characteristic of areas heavily polluted by a variety of sources.

For chemical analysis, the sediments collected at our biological stations were extracted with a solvent that removed the hydrocarbons. The hydrocarbons were separated from other materials contained in the extracts. They were then analyzed by gas-liquid chromatography. This technique separates hydrocarbon mixtures into individual compounds, according to the boiling point and structural type. To do this, a sample is flash-evaporated in a heated tube. The vapor is swept by a constantly flowing stream of carrier gas into a long tube that is packed with a substance (substrate) that is responsible for the resolution of the mixture into its individual components. Ideally, each vaporized compound emerges from the end of the tube at a definite time and well separated from all other components. A sensitive detector and an amplifier then transmit a signal to a recorder which traces on a moving strip of chart paper a series of peaks (the chromatogram) that correspond to the individual components of the mixture. From the pattern of peaks in the gas chromatogram the chemist can learn much about the composition of the mixture. Each oil may have a characteristic fingerprint pattern by which it can be recognized in the environment for weeks, or even months, after the initial spill. Past and continuing work on the composition of those hydrocarbons that are naturally present in all marine organisms (see box, "What is Petroleum?") enabled us to distinguish easily between the natural hydrocarbons and those contained in the fuel oil. These analyses facilitated our study of the movement of the fuel oil from the West Falmouth oil spill into the bottom sediments and through the marine food chain.

## Immediate Kill

Massive, immediate destruction of marine life occurred offshore during the first few days after the accident. Affected were a wide range of fish, shellfish, worms, crabs, other crustaceans, and invertebrates. Bottom-living fish and lobsters were killed and washed up on the shores. Trawls made in ten feet of water soon after the spill showed that 95 percent of the animals recovered were dead and others were dying. The bottom sediments contained many dead snails, clams, and crustaceans. Similarly severe destruction occurred in the tidal rivers and marshes into which the oil had moved under the combined influence of tide and wind. Here again fish, crabs, shellfish, and other invertebrates were killed; in the most heavily polluted regions of the tidal marshes almost no animals survived.

The fuel oil spilled at West Falmouth was a light, transparent oil, very different from the black viscous oil associated with the *Torrey Canyon* and Santa Barbara episodes. Within days most of the dead animals had decayed and the visual evidence of the oil had almost disappeared. Casual observers were led to report to the press that the area looked as beautiful as ever. Had we discontinued our study after the visual evidence of the oil had disappeared, we might have been led to similar interpretations. From that point on, only continued, careful biological and chemical analysis revealed the extent of continuing damage.

## Persistence of Pollution

Quite recently a leading British expert on treatment of oil spills remarked that "white products, petrol, kerosene, light diesel fuel, and so forth, can be expected to be self-cleaning. In other words, given sufficient time they will evaporate and leave little or no objectionable residue." (6) Our experience shows how dangerously misleading such statements are. Chemical analyses of the oil recovered from the sediments and from the bodies of the surviving animals showed the chromatographic fingerprint of the diesel fuel, in monotonous repetition, for many months after the accident.

Dead marine animals (left) litter the Wild Harbor tidal flats polluted by toxic fuel oil after the oil barge *Florida* accident in Buzzards Bay, Massachusetts.

Bacteria normally present in the sea will attack and slowly degrade spilled oil. On the basis of visual observations it has been said that the oil spilled by the *Torrey Canyon* disappeared rapidly from the sediments. This was interpreted to mean that the action of the bacteria was "swift and complete." Our analyses, which were carried out by objective chemical, rather than by subjective observational techniques, showed the steady persistence of fuel oil that should, in principle, be even more rapidly degraded than a whole crude oil. Thus, in May 1970, eight months after the spill, oil essentially unaltered in chemical characteristics could still be recovered from the sediments of the most heavily polluted areas. By the end of the first year after the accident, bacterial degradation of the oil was noted at all locations, as evidenced by changes in the fingerprint pattern of the oil. Yet only partial detoxification of the sediments had occurred, since the bacteria attacked the least toxic hydrocarbons first. The more toxic aromatic hydrocarbons remained in the sediments.

## Spread of Pollution

For our chemical and biological work we established an unpolluted control station, outside of the area that was polluted, immediately after the accident. For a short period after the accident the sediments at this station were still clean and the organisms alive in their normal abundance and distribution. However, within three weeks, oil was found at this station and a significant number of organisms had been killed. Another control station was established twice as far from shore. Within three months fuel oil from the spill was evident at this station, and again there was a concomitant kill of bottom-living animals. This situation was repeated several times in sequence, and by spring 1970 the pollution had spread considerably from the area affected initially. At that time, the polluted area offshore was ten times larger than immediately after the accident and covered 5,000 acres (20 square kilometers) offshore and 500 acres (2 square kilometers) in the tidal river and marshes.

Another significant observation was made in the spring of 1970: Between December 1969 and April 1970, the oil content of the most heavily contaminated marine station two and one-half miles north of the original spill increased tenfold. Similar but smaller increases were observed at about the same time at other stations more distant from shore. The oil still showed the typical chromatographic fingerprint of the diesel fuel involved in the September 1969 oil spill. This and the lack of any further accident in this area suggested that oil was spreading from the most heavily contaminated inshore regions to the offshore sediments. We believe that the increase in the pollution level and the spread of oil to outlying areas are related to a transportation mechanism that we do not yet fully understand. However, the drastic kill of the animals that occurred with the arrival of oil pollution at the offshore stations showed that mortality continued for many months after the initial spill, even though no visible evidence of oil remained on the shores.

We believe these observations demonstrate that chronic oil pollution can result from a single spill, that the decimation of marine life can extend to new regions long after the initial spill, and that, once poisoned, the sea bottom may remain toxic to animals for long time periods.

## Destruction of Shellfish Resources

Our analyses showed that oysters, soft-shell clams, quahaugs (another variety of clam), and scallops took up the fuel oil. Because of the pollution, the contaminated regions had to be closed to the harvesting of shellfish. Continuing analyses revealed that the contamination of the 1970 shellfish crop was as severe as that of the 1969 crop. Blue mussels that were juveniles in the polluted area at the time of the spill generally were sexually sterile the next season—they developed almost no eggs or sperm. Furthermore, in 1970 distant areas contained shellfish contaminated by fuel oil. Therefore, harvesting prohibitions had to be maintained in 1970 and had to be extended to polluted shellfish grounds that had not been closed to the public immediately after the accident.

It has long been common to transfer shellfish polluted by human sewage into clean water to make the animals marketable again. It has been thought that a similar flushing process would remove the oil from animals exposed to oil. Indeed, taste tests showed that the objectionable oily taste disappeared from animals maintained for some period in clean water. However, we removed oysters from the contaminated areas and kept them in clean running sea water up to six months. Fuel oil was still found in the animals by chemical analysis at essentially the same concentration and in the same composition as at the beginning of the flushing period.

Thus, we discovered that hydrocarbons taken up into the fat and flesh of fish and shellfish are not removed by natural flushing or by internal metabolic processes. The substances remain in the animals for long periods of time, possibly for their entire lives. The presence or absence of an oily taste or flavor in fish products is not a measure of contamination. The reason is that only a relatively small fraction of the total petroleum product has a pronounced taste or odor. Subjective observations cannot detect the presence of the toxic but tasteless and odorless pollutants. Only objective chemical analysis measures the presence of these chemical poisons. It is important to note in this regard that state and federal laboratories in the public health sector are not generally equipped to carry out these important chemical measurements. Such tests are vital, however, for the protection of the consumer.

Thus, our investigation demonstrated that the spill produced immediate mortality, chronic pollution, persistence of oil in the sediments and in the organisms, spread of pollution with the moving sediments, destruction of fishery resources, and continued harm to fisheries for a long period after the accident. Our continuing study will assess the persistence and toxicity of the oil and the eventual ecological recovery of the area. At the present time, one and one-half years after the spill, only the pollution-resistant organisms have been able to reestablish themselves in the more heavily contaminated regions. The original animal populations there have not become reestablished. Many animals that are able to move, early in their life cycles, as free-swimming larvae reach the polluted area and are killed when they settle on the sea bottom or in the marshes at West Falmouth.

In addition, revitalization of bottom areas probably will be hampered by oxygen depletion caused by oxygen-requiring bacteria that degrade oil (7).

## The Significance of West Falmouth

Some scientists are convinced that the effects at West Falmouth are a special case and have little applicability to spills of whole, unrefined crude oils. They contend that #2 fuel oil is more toxic than petroleum and that therefore it has effects that would not be comparable to those of whole petroleum. We cannot agree with this view.

Fuel oil is a typical oil-refining product. It is frequently shipped by sea, especially along coastal routes, and it is spilled in accidents like those which occurred at West Falmouth and off Baja California following the grounding of the *Tampico Maru* in 1957 (8).

More importantly, fuel oil is a part of petroleum, and as such it is contained within the whole petroleum. Surely, hydrocarbons that are toxic when they are in fuel oil must also be toxic when they are contained in petroleum. Therefore, the effects observed in West Falmouth are typical both for that fuel oil and the whole crude oil. In terms of chemical composition, crude oils span a range of molecular weights and structures. Many light crude oils have a composition not too dissimilar from that of fuel oil, and their toxicity and effects on the environment are very similar. Other heavier crude oils, while still containing the fuel oil components, contain higher proportions of the long-lasting poisons that are much more persistent and that include, for instance, some compounds that are potent carcinogens (cancer-producing agents) in experimental animals. Such heavy crude oils can be expected to be more persistent than a fuel oil, and they will have longer lasting long-term effects. Even weathered crude oils may still contain these long-term poisons, and in many cases some of the moderately low-boiling, immediately toxic compounds. In our view, these findings differ from those of other investigators principally for two reasons: Our study is based on objective measurement and is not primarily concerned with the mobile, adult marine species—the fish whose migratory history is largely unknown—or the highly resistant intertidal forms of life. We are studying quantitatively the effects of the spill on the sessile (bottom) animals that cannot escape the spill or the polluted sediment and that are thus exposed to chronic pollution. Since all classes of bottom animals are severely affected by the oil, we believe that the effects on free-swimming animals should be just as drastic. The difficulty of measuring the total impact of oil on the marine life has led many to doubt the ecological seriousness of oil pollution. Our findings, extending far beyond the period when the visual evidence of the oil had disappeared, are based on objective chemical analyses and quantitative biological measurements, rather than on subjective visual observations. They indict oil as a pollutant with severe biological effects.

It is unfortunate that oil pollution research has been dominated so strongly by subjective, visual observations. Clearly, oil is a *chemical* that has severe *biological* effects, and therefore oil pollution research, to be fully meaningful, must combine chemical with biological studies. Those few investigators who are using objective chemical techniques find patterns in the environmental damage by oil that are similar to those demonstrated by the West Falmouth spill. Thus, R. A. Kolpack reported that oil from the blowout at Santa Barbara was carried to the sea bottom by clay minerals and that within four months after the accident the entire bottom of the Santa Barbara basin was covered with oil from the spill (9). Clearly, this is one of the most significant observations in the aftermath of that accident. A concurrent and complementary biological study would have appreciably enhanced our understanding of the ecological damage caused by the Santa Barbara oil spill.

G. S. Sidhu and co-workers, applying analytic methods similar to those used by us, showed that the mullet, an edible finfish, takes up petroleum hydrocarbons from waters containing low levels of oil pollution from refinery outflows. In their chemical structures the hydrocarbons isolated by the investigators are similar to those found in the polluted shellfish of West Falmouth. The compounds differ markedly from those hydrocarbons present as natural components in all living organisms, yet closely approximate the hydrocarbons in fossil fuels (10).

Numerous results of crude-oil toxicity tests, alone or in the presence of dispersants, have been published in the literature. However, in almost all cases such tests were performed on relatively hardy and resistant species that can be kept in the laboratory and on adult animals for short time periods under unnatural conditions or in the absence of food. At best, such tests may establish only the relative degree of the toxicity of various oils. We are convinced that the exposure of more sensitive animals, especially young ones, to oil pollution over many months would demonstrate a much greater susceptibility to the damaging effects of the oil. Such effects have been demonstrated in the studies of the West

Falmouth oil spill. These studies represent a meaningful field test in open waters.

Thus, we believe that the general toxic potential and the persistence of the West Falmouth oil are typical of most oils and oil products both at the sea bottom and in the water column.

## Conclusions

Our analysis of the aftermath of the West Falmouth oil spill suggests that oil is much more persistent and destructive to marine organisms and to man's marine food resources than scientists had thought. With the advent of objective chemical techniques, oil pollution research has entered a new stage. Earlier interpretations of the environmental effect of oil spills that were based on subjective observation, often over a short time span, have questionable validity. Crude oil and oil products are persistent poisons, resembling in their longevity DDT, PCB and other synthetic materials [which have been discussed in these pages]. Like other long-lasting poisons that, in some properties, resemble the natural fats of the organisms, hydrocarbons from oil spills enter the marine food chain and are concentrated in the fatty parts of the organisms. They can then be passed from prey to predator where they may become a hazard to marine life and even to man himself.

Natural mechanisms for the degradation of oil at sea exist—the most important of which is bacterial decomposition. Unfortunately, this is least effective for the most poisonous compounds in oil. Also, oil degrades slowly only in marine sediments, and it may be completely stable once it is taken up by organisms. It has been thought that many of the immediately toxic low-boiling aromatic hydrocarbons are volatile and evaporate rapidly from the oil spilled at sea. This has not been the case at West Falmouth, where the low-boiling hydrocarbons found their way into the sediments and organisms. We believe that the importance of evaporation has been overestimated.

Oil-laden sediments can move with bottom currents and can contaminate unpolluted areas long after the initial accident. For this reason a *single* and relatively small spill may lead to *chronic*, destructive pollution of a large area.

We have not yet discussed the low-level effects of oil pollution. However, a growing body of evidence indicates that oil as well as other pollutants may have seriously damaging biological effects at extremely low concentrations, previously considered harmless. Some of this information was presented in Rome at the December 1970 Food and Agriculture Organization's Conference on the Effects of Marine Pollution on Living Resources and Fishing. Greatly diluted pollutants affect not only the physiology but also the behavior of many animals. Many behavioral patterns which are important for the survival of marine organisms are mediated by extremely low concentrations of chemical messengers that are excreted by marine creatures. Chemical attraction and repulsion by such compounds play a key role in food finding, escape from predators, homing, finding of habitats, and sexual attraction. Possibly, oil could interfere with such processes by blocking the taste receptors of marine animals or by mimicking natural stimuli and thus eliciting false responses. Our general ignorance of such low-level effects of pollution is no excuse for neglecting research in these areas nor for complacency if such effects are not immediately obvious in gross observations of polluted areas.

Recent reports suggest an additional environmental threat from oil pollution. Oil may concentrate other fat-soluble poisons, such as many insecticides and chemical intermediates (11). Dissolved in an oil film, these poisons may reach a concentration many times higher than that which occurs in the water column. In this way other pollutants may become available to organisms that would not normally be exposed to the substances and at concentrations that could not be reached in the absence of oil.

The overall implications of oil pollution, coupled with the effects of other pollutants, are distressing. The discharge of oil, chemicals, domestic sewage, and municipal wastes, combined with overfishing, dredging, and the filling of wetlands may lead to a deterioration of the coastal ecology. The present influx of pollutants to the coastal regions of the oceans is as damaging as that which has had such a detrimental effect on many of our lakes and freshwater fishery resources. Continued and progressive damage to the coastal ecology may lead to a catastrophic deterioration of an important part of marine resources. Such a deterioration might not be reversed for many generations and could have a deep and lasting impact on the future of mankind.

Since present oil-spill countermeasures cannot completely eliminate the biological damage, it is paramount to prevent oil spills. The recent commitment by the United States to take all steps to end the intentional discharge of oil from its tankers and nontanker vessels by the mid 1970s is important. As a result of this step and of the resolution of the NATO Ocean Oil Spills Conference of the Committee on Challenges of Modern Society in Brussels, December 1970, other countries hopefully also will adopt necessary measures to halt oil pollution from ships. This would eliminate the largest single source of oceanic oil pollution. At the same time steps also must be taken to reduce oil pollution from many other, less readily obvious sources, such as petrochemical operations on shore, disposal of automotive and industrial lubricants, and release of unburned hydrocarbons from the internal combustion engine.

## Acknowledgments

The authors acknowledge the continued support of their basic and applied research efforts by the National Science Foundation, The Office of Naval Research, and the Federal Water Quality Administration.

## Notes

1. Blumer, M., "Scientific Aspects of the Oil Spill Problem," paper presented at the Oil Spills Conference, Committee on Challenges of Modern Society, NATO, Brussels, Nov. 1970.
2. Ryther, J. H., "Photosynthesis and Fish Production in the Sea," *Science*, 166:72–76. 1969,
3. McCaull, Julian, "The Black Tide," *Environment*, 11(9):10, 1969.
4. Blumer, M., G. Souza, J. Sass, "Hydrocarbon Pollution of Edible Shellfish by an Oil Spill," *Marine Biology*, 5(3):195–202, March 1970. Blumer, M., J. Sass, G. Souza, H. L. Sanders, J. F. Grassle, G. R. Hampson, "The West Falmouth Oil Spill," Reference No. 70–44, unpublished manuscript available from senior author, Woods Hole Oceanographic Institution, Woods Hole, Massachusetts, September 1970.
5. Blumer, M., G. Souza, J. Sass, "Hydrocarbon Pollution," *op. cit.*, p. 198.
6. Smith, J. Wardly, "Dealing with Oil Pollution Both on the Sea and on the Shores," paper presented to the Ocean Oil Spills Conference, Conference on Challenges of Modern Society, NATO, Brussels, November 1970.
7. Murphy, T. A., "Environmental Effects of Oil Pollution," presented at American Society of Civil Engineers, Boston; available from author at Edison Water Quality Laboratory, Edison, New Jersey, p. 14–15, July 13, 1970.
8. Jones, Laurence G., Charles T. Mitchell, Einar K. Anderson, and Wheeler J. North, "A Preliminary Evaluation of Ecological Effects of an Oil Spill in the Santa Barbara Channel," W. M. Keck Engineering Laboratories, California Institute of Technology.
9. Kolpack, R. A., "Oil Spill at Santa Barbara, California, Physical and Chemical Effects" paper presented to the FAO Technical Conference on Marine Pollution, Rome, Dec. 1970.
10. Sidhu, G. S., G. L. Vale, J. Shipton and K. E. Murray, "Nature and Effects of a Kerosene-like Taint in Mullet," paper presented to FAO Technical Conference on Marine Pollution, Rome, Dec. 1970.
11. Hartung, R., and G. W. Klinger, "Concentration of DDT by Sedimented Polluting Oils," *Environmental Science and Technology*, 4:407, 1970.
12. Gruse, W. A., and D. R. Stevens, *The Chemical Technology of Petroleum*, Mellon Institute of Industrial Research, Second Edition, McGraw-Hill Book Company, Inc., New York, p. 2, 1942.
13. For example, see: Clark, R. C. and M. Blumer, "Distribution of n-paraffins in Marine Organisms and Sediments," *Limnology and Oceanography*, 12:79–87, 1967.

# THERMAL POLLUTION

# 29

# Water temperature criteria to protect aquatic life

*J. A. Mihursky*
*V. S. Kennedy*

## Introduction

The objectives of this paper are to discuss the ecological significance of temperature in the aquatic habitat and to use the principles elucidated as standards for judgment concerning temperature regulations that may be necessary for preserving aquatic life. These objectives are difficult ones, especially in view of the physical, chemical and biological complexities involved and the lack of temperature information in many key areas. However, examination of available temperature data and the developing thermal loading problem does allow certain definitive statements to be made and should help place the entire topic in proper perspective.

## Why be Concerned with Temperature Criteria?

Although there are many causes of changed temperature regimes in the aquatic habitat, such as from dams, irrigation practices and industrial waste heat, one specific area, the steam electric station (S.E.S.) industry, appears to pose the greatest threat. This industry has the greatest non-consumptive industrial demand for water as a heat transfer medium. Picton (Figure 29–1) in 1960 reported that projected 1980 power needs

---

Reprinted with permission from *Transactions of the American Fisheries Society*, 96(Supplement):20–32 (1967).
The authors are with the Natural Resources Institute of the University of Maryland, Chesapeake Biological Laboratory, Solomons, Maryland 20688.

will use one-fifth to one-sixth of the total freshwater runoff in the United States for cooling water; discounting flood flows which occur about one-third of the year and account for two-thirds of total runoff, the S.E.S. industry will require about half of the total runoff for cooling purposes the remaining two-thirds of the year. This problem is even greater in certain heavily populated and industrialized northeastern United States watersheds where presently over 100% of flows may pass through the various power stations located within the watersheds during low flow periods (Mihursky and Cory, 1965).

More recent analysis shows a doubling of electricity needs in the United States every 10 years (Trembley, 1965) and in some areas, 6 years. Using these values and projecting from a 1960 base year reveals an increase of 30 to possibly 256 times by 2010. The report by the advisory committee for the control of stream temperatures in Pennsylvania (1962) listed many S.E.S. discharge temperatures between 100° F and 115° F and found river temperatures up to 95° F almost 5 miles downstream from a S.E.S. installation. Individual power plants may require up to 0.5 million gallons of water per minute for cooling purposes.

## S.E.S. Industry Trends

Recent decisions to construct nuclear plants as opposed to conventional fossil fuel plants will result in greater amounts of waste heat per kilowatt of electricity produced since nuclear units are less efficient than fossil fuel plants. To be economically competitive, nuclear plants will have 1,000- to 3,000-megawatt (mw) capacities compared to 100- to 700-mw ranges of present conventional utilities. These engineering changes will result in much larger amounts of waste heat production per installation, which in turn means much more heat released to local ecosystems than now occurs. The future need for larger water volumes and larger surface areas for maximum cooling rates will result in a larger percentage of S.E.S. to be located in estuaries—32% in 1980 compared to 22% in 1950 (Picton, 1960).

Because of the trend toward larger generating units and the increases in steam pressure and temperature, additional chemical releases will become commonplace from S.E.S. Cadwallader (1965) reported that higher purity standards for boiler feed water will require demineralization units which must be cleaned regularly. The need to keep internal surfaces of high pressure boilers meticulously clean will result in eventual discharge of a variety of cleaning chemicals ranging from detergents and hydrochloric acid to corrosion inhibitors.

Although industrial technological change has been rapid, no major

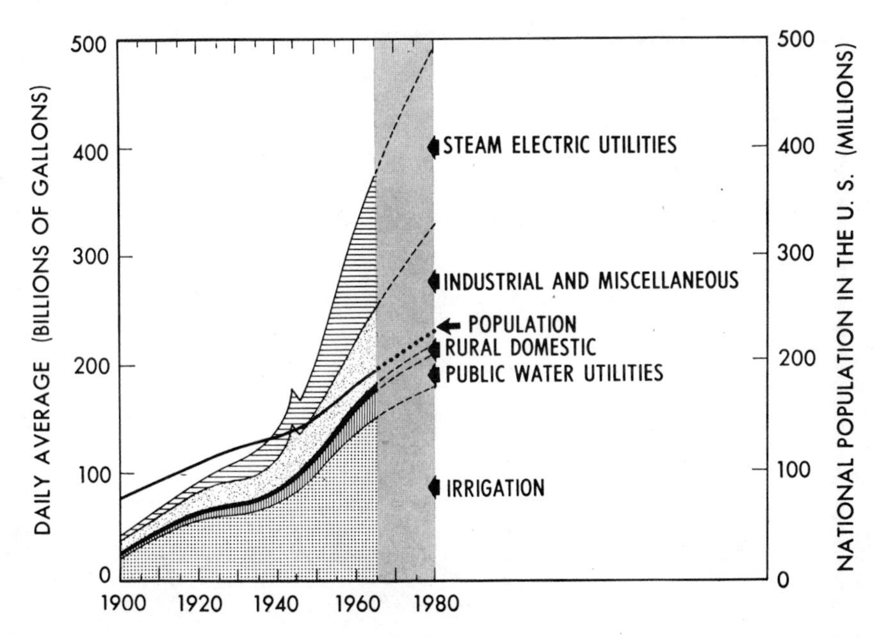


FIGURE 29–1. Water use in the United States (after Picton, 1960).

engineering design changes are envisioned which will result in appreciable reduction of waste heat to streams from steam electric stations during this century (Cootner and Löf, 1965).

## Effect of Temperature on Aquatic Life

In view of these projected trends, ecologists are becoming more concerned with the effects of thermal loading upon ecosystems. Although temperature plays a complex role as an environmental factor, certain basic principles, important to the present problem, have been elucidated in the literature and should be understood in any discussion of water quality criteria.

### *Temperature as a Lethal Factor*

A number of investigators have examined temperature as a lethal factor by use of the classic pharmacological assay method, and have demonstrated that aquatic organisms cannot live above or below certain temperature levels. The LD-50, or lethal temperature dose necessary to kill 50% of test animals, is one standard means of data reporting. Numerous experiments have shown that only a relatively small temperature

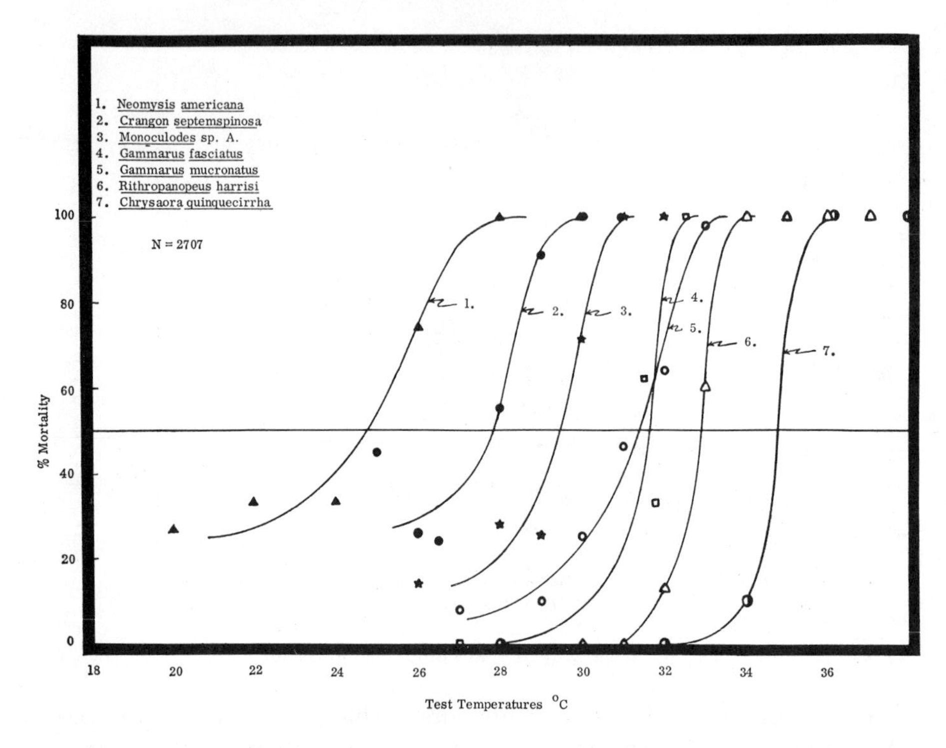


FIGURE 29–2. Comparative temperature mortality curves for seven estuarine species.

rise may cause test amimals to go from 0 to 100% mortalities. Plotted assay data typically shows a sigmoid pattern as shown in Figure 29–2.

### *Species Vary in Their Thermal Tolerances*

Figure 29–2 also illustrates another principle; namely, that temperature tolerance varies from species to species. The seven estuarine invertebrate organisms shown were acclimated between 59.0° and 60.8° F and exposed to various temperature levels for 24 hours. The opossum shrimp, *Neomysis americana* (1), is the least temperature tolerant species shown. Collected from the Patuxent estuary in Maryland, this shrimp species is near its southern limit of distribution in the United States (Tattersall, 1951). The polyp stage of *Chrysaora quinquecirrha* (7), the troublesome stinging sea nettle, is the most temperature tolerant and shows an LD-50 level 18° F above *N. americana*, even though acclimated to the same temperature. The sand shrimp, *Crangon septemspinosa* (2), is similar to *Neomysis* in its zoogeography and is relatively temperature sensitive. Three amphipod species (3, 4, 5) are grouped in the center of the graph with the deep water channel *Monoculodes* sp. (3) showing less thermal tolerance than either of the two shallow water shelf species which inhabit higher temperature regimes in the Patuxent ecosystem. These latter two species are the only ones shown of the same genus (*Gammarus*) and significantly show similar LD-50 levels. *Rhithropanopeus harrisi*, the mud crab, is a temperate-tropical form showing relatively high thermal tolerance. It has been reported to increase in abundance below S.E.S. outfalls in England (Naylor, 1965).

### Acclimation

Temperature effects are greatly modified by the past thermal history of an individual. Temperature acclimation refers to the thermal level to which an individual is physiologically adjusted. Figure 29–3 shows how LD-50 levels increase for *N. americana* with increasing acclimation levels

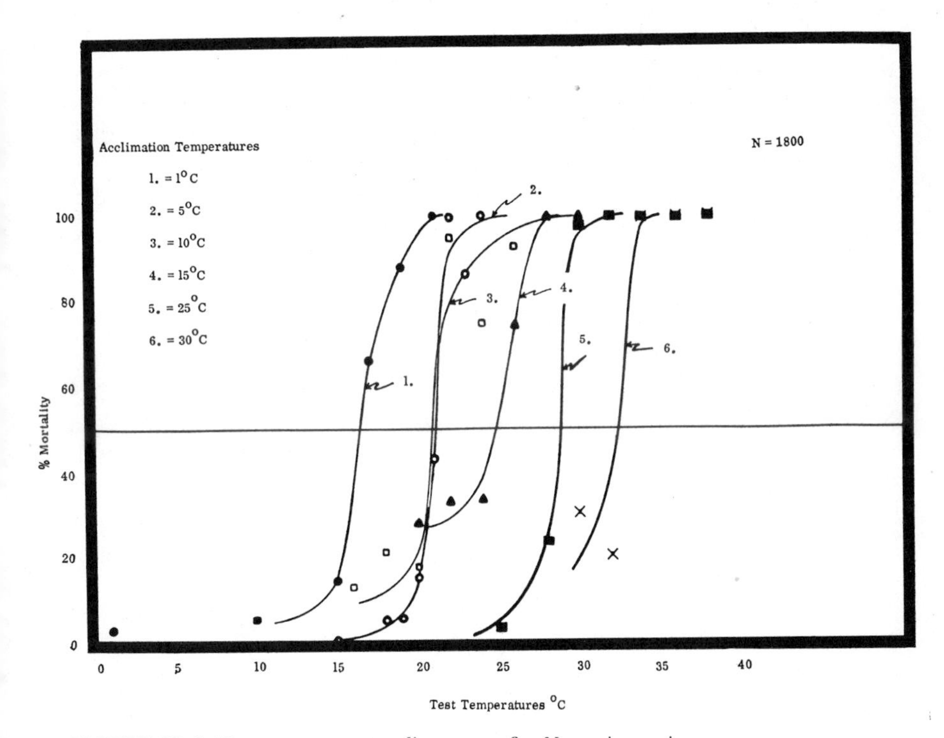


FIGURE 29–3. Temperature mortality curves for *Neomysis americana*.

(Gibson, unpublished). Ability to acclimate is limited, however, and the maximum upper or lower acclimation point has been called the ultimate incipient lethal level by Fry (1947).

It has also been shown that most organisms acclimate upward to high temperatures faster than they acclimate downward to lowered temperatures. Dickie (1958) has shown that *Placopecten magellanicus*, the giant scallop, acclimates upward rapidly (1.7 C/day), but may take up to 3 months to lose this acclimation to high temperatures. Brett (1956) reported similar findings for fish.

### Thermal Shock

Lack of acclimation to a new temperature level can produce a condition referred to as thermal shock. It can occur when no acclimation time is possible due to an abrupt change in the thermal environment.

### Critical Thermal Maximum (C.T.M.)

Disorientation and cessation of directed activities can be caused by high or low temperatures. The concept of critical thermal maximum has been especially elaborated on reptiles and amphibians (Hutchison, 1961; Lowe and Vance, 1955) but is applicable to other aquatic organisms. The C.T.M. is the thermal point at which the locomotory activity becomes disorganized and the animal loses its ability to escape from conditions that will soon cause its death. Even if temperature itself is not responsible for an animal's death, a C.T.M. condition can cause or lead to death from predation by a more active temperature resistant predator.

### Metabolism

Since temperature affects the rate of biochemical reactions, it affects metabolic rates and oxygen consumption (Prosser and Brown, 1961; Beamish, 1964); therefore, temperature can affect activity through metabolism (Fry, 1947). Figure 29–4 shows how increased temperatures result in increased oxygen consumption in four fish species studied by Beamish (1964).

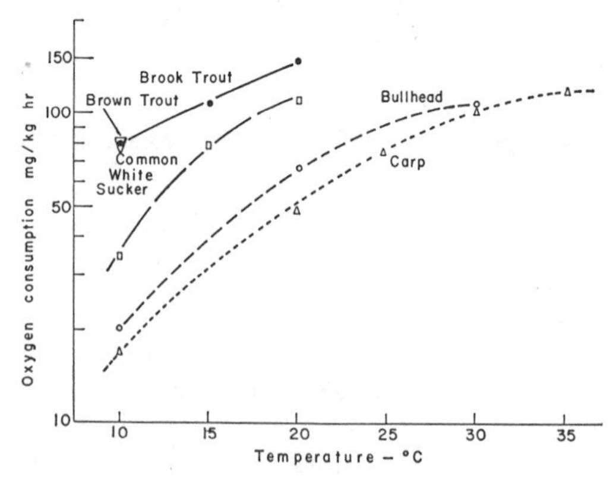


FIGURE 29–4. Standard oxygen consumption in relation to temperature for four freshwater fish species (after Beamish, 1964).

### *Reproduction*

Temperature affects reproduction, one of the most critical activities carried out by organisms. Some bivalves can be kept at temperatures that are high enough for gametogenesis, but not high enough to permit spawning (Loosanoff and Davis, 1951). Kinne (1963) in reference to marine and estuarine animals stated that given the proper physiological condition and enough space and food, the breeding season depends on temperature, rather than salinity, light, pressure or biological factors. Most aquatic species reproduce when a given temperature level is reached after a period of either increasing or decreasing temperature change (Gutsell, 1930; Thorson, 1946; Sastry, 1963).

After egg deposition, developmental rates are also often temperature dependent. For most marine invertebrates, the cleavage stages require narrower temperature ranges than do the larval stages (Kinne, 1963).

### *Behavior*

The horizontal and vertical migrations of aquatic organisms are highly dependent upon temperature. Good review papers on this subject have been given by Orton (1920), Hutchins (1947), Gunter (1957) and Kinne (1963).

### *Temperature Requirements Vary with Life History Stage*

As Brett (1960) has emphasized, temperature effects depend upon the life history stage of an organism. Effects are modified by age and size of the organism and by the season. Brett showed thermal requirements for different life processes in Pacific salmon (Figure 29–5). Egg survival range is narrower, especially during the hatching process, than it is for other life history stages. Changes in temperature requirements occur with variations in seasonal environmental temperatures. Range for growth is narrower than that for survival and shows a decrease in required temperature range for eggs and hatching similar to the survival curve. Reproduction requirements are still more restricted.

Brett (1958) has shown the variations in thermal requirements for different life processes in still another manner. Figure 29–6 is a thermal polygon formed by plotting upper and lower LD-50 levels for young sockeye salmon against acclimation temperature. Included are hypothetical polygons for an LD-5 level as well as for loading and inhibiting levels. The loading level indicates the temperature regime that requires an organism to expend more energy than it is able to replace; the inhibiting level indicates the point beyond which temperature reduces an organism's ability to execute normal functions, thus reducing chances

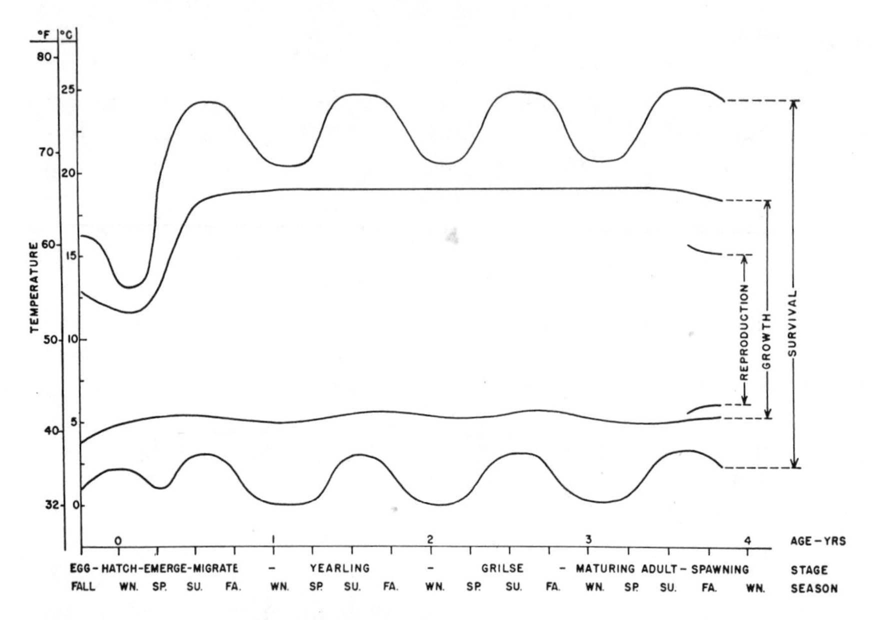


FIGURE 29–5. Schematic representation of temperature requirements for different life processes of the Pacific salmon (after Brett, 1960).

of species survival. An inhibiting effect, for example, is one that may prevent spawning because of improper temperature patterns.

## Responses to Multivariate Conditions

Normally under natural conditions, an organism is subjected to a variety of ever changing water quality patterns, some of which are ecologically significant while others are not. Laboratory experimentation in the past generally was concerned with one variable. Much of the present data on temperature effects on aquatic organisms are based on single variable designs. Although such designs are important in focusing on a single parameter and determining important principles, experiments must eventually consider other variables. Thus it is now known that temperature LD-50 levels are affected by dissolved oxygen decreases, carbon dioxide increases and increases in toxic substances.

Multivariate design work reported by Costlow, Bookhout, and Monroe (1960) has demonstrated the interaction complexities between multiple parameters (Figures 29–7 and 29–8). Percentage mortality curves were obtained using the statistical method of a fitted response surface and involve observed mortality under twelve different salinity and temperature combinations on various life history stages of the crab, *Sesarma cinereum*. They show how critical different temperature-salinity combinations are to different life history stages of this species.

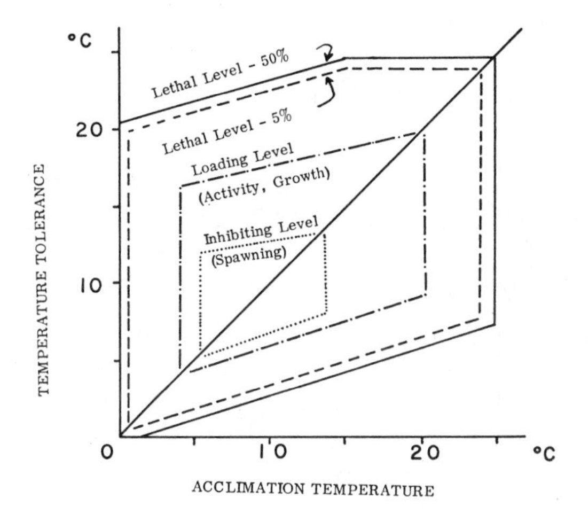


FIGURE 29–6. Thermal polygon showing measured and hypothetical temperature tolerances for the sockeye salmon acclimated to various temperature levels (after Brett, 1958).

Temperature, salinity and dissolved oxygen interactions on the lobster (*Homarus americanus*) have been studied by McLeese (1956). Figure 29–9 shows the boundary lines of lethal conditions for the three parameters. It also shows those regions where each variable operates independently and in combination with other variables. McLeese used 27 combinations of these three variables.

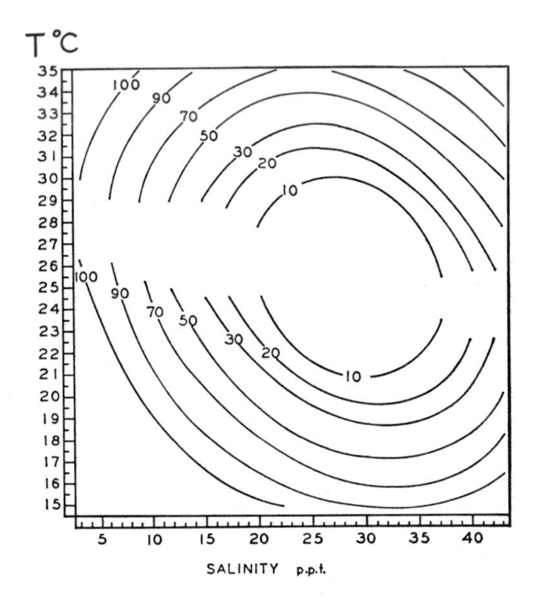


FIGURE 29–7. Estimation of percentage mortality of the first zoeal stage of *Sesarma cinereum* based on the fitted response surface to mortality under 12 different combinations of temperature and salinity (after Costlow, Bookhout, and Monroe, 1960).

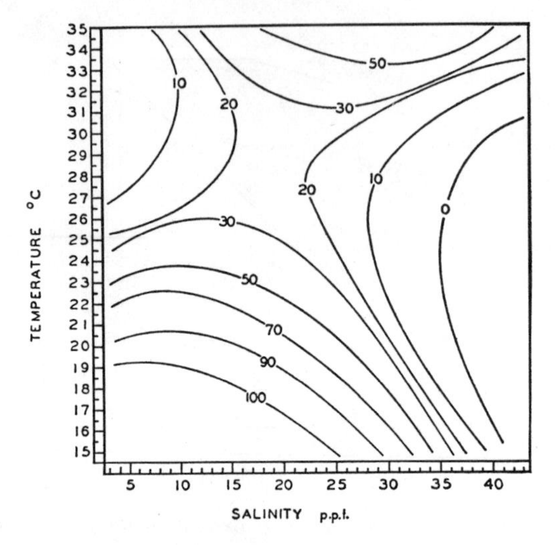


FIGURE 29–8. Estimation of percentage mortality of the megalops state of *Sesarma cinereum* based on the fitted response surface to observed mortality under 12 different combinations of temperature and salinity (after Costlow, Bookhout, and Monroe, 1960).

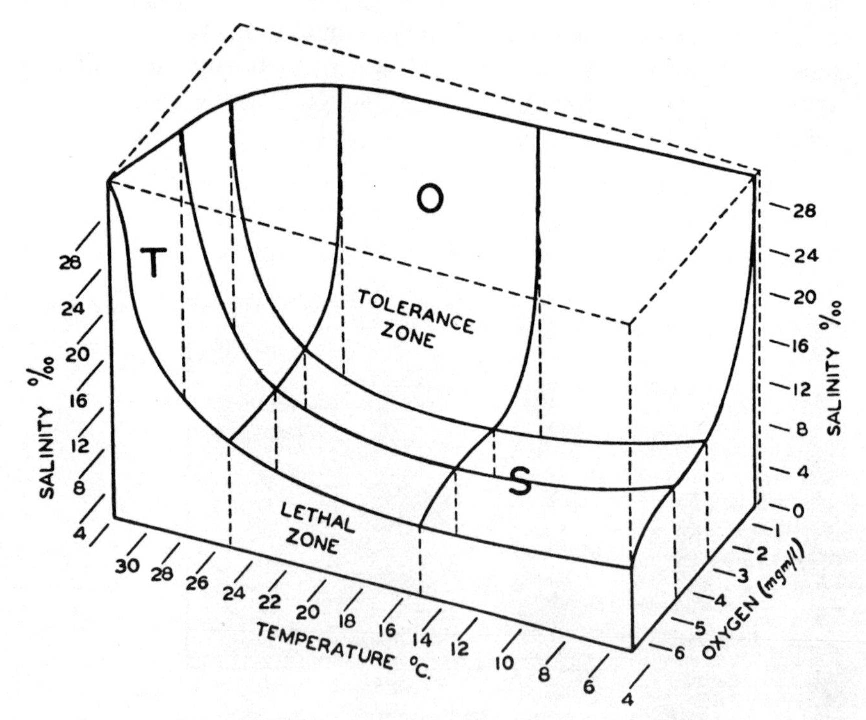


FIGURE 29–9. Diagram of the boundary of lethal conditions for lobsters in various combinations of temperature, salinity, and oxygen. T, region in which temperature alone acts as a lethal factor; S, region in which salinity alone acts as a lethal factor; O, region in which oxygen alone acts as a lethal factor (after McLeese, 1956).

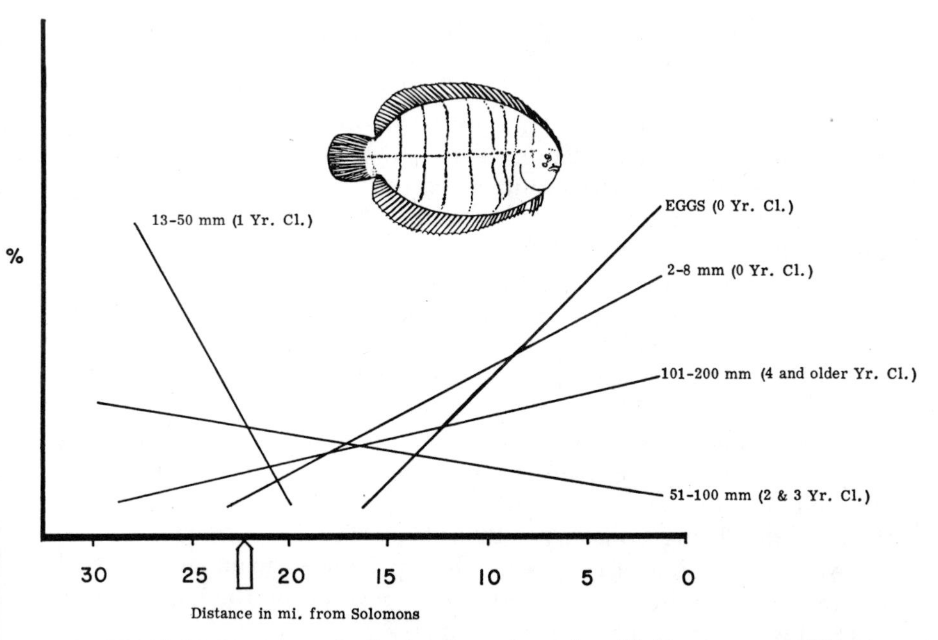


FIGURE 29–10. Percentage distribution slopes for various life history stages of *Trinectes maculatus* in the Patuxent River estuary during the spawning season May–Sept. 1963. Arrow, S.E.S. location.

## Future Research Efforts and Needs

Multivariate studies are coming closer to duplicating conditions found in nature. To do a definitive job in establishing water quality criteria, it is necessary to understand completely all the interactions between all those important water quality characteristics that can be limiting. To an extent it is meaningless to examine biological effects of only one variable to the exclusion of other possible interacting variables.

The multivariate approach is necessary not only on adult stages, but on all life history stages of important food web representatives. Before designing a laboratory program to evaluate an environmental change, however, it is essential to conduct any necessary field studies to determine important food web components. An ideal evaluation program will combine laboratory and field data, for only by interpreting laboratory data with respect to survival value for a species as determined with the aid of field work does a laboratory study obtain ecological meaning.

Information obtained from a multidisciplined program on the Patuxent River estuary in Maryland may be used to illustrate some additional specific aspects. Work has shown the flatish, *Trinectes maculatus*, to be an important estuarine species in relation to biomass. Examination of its life history pattern by combining data obtained from three collecting gears (Figure 29–10) shows it to have a lower estuary spawning area and an upper estuary nursery zone (Dovel, et al.). Eggs and larvae up to 8 mm are obtained only in the lower river, while the next largest size group collected begins at 13 mm and is found concentrated about 25 miles upriver. Thus an endemic estuarine species shows considerable movement throughout the Patuxent ecosystem. By some mechanism, the post larval 8- to 13-mm size group moves upriver and must now attempt to pass through an area at which a new 700-mw power plant is located (arrow). The plant pumps 500,000 gallons of water per minute and heats the water 11.5° F during the warmer seasons. An intensive 3-year fish egg and larvae sampling program has failed to obtain a single specimen between 8 and 12 mm from the Patuxent ecosystem. In order to aid in evaluating the ecological impact of this S.E.S. upon the Patuxent, it will be necessary to determine what time of year and in what level of the water column these post larval stages move to their upstream nursery grounds above the power plant. Of additional concern is post-larval stage rheotaxis, for this may influence their chances of being pumped through the plant and being subjected to thermal shock.

*Neomysis americana*, the opossum shrimp, is known to undergo diurnal vertical migration in the water column. It remains on the bottom as an epibenthic species during the day, but rises into the surface waters at night. It remains to be determined what effect extreme stratification would have upon vertical migration. Heated water from a power plant typically remains at the surface and bottom waters may remain normal. If *Neomysis* moves up from the normal temperature bottom water and encounters an unusually high temperature, what will be the behavior pattern? If the behavior pattern is altered, will it affect survival of the species?

Many critical questions remain to be answered concerning other invertebrates in artificially heated environments. Among them is the question of whether certain minimal winter temperatures must prevail for completion of normal life history patterns.

Insect hatching in artificially warmed waters is a similar difficult problem. If water temperatures were kept at safe levels in reference to direct mortalities, what would be the consequence of early season hatching; would adults find themselves in air temperatures too cold for normal mating behavior?

Thus many areas of research remain, among them, multivariate experiments, determination of seasonal cycles and factors that govern them, and factors that control behavior patterns of non-migratory and migratory species. It is perhaps superfluous to indicate temperature research is necessary on many more species. To focus on this need is the fact that no single species has been subjected to multivariate studies for all life history stages. Also, of the almost 1,900 fish species listed in the American Fisheries Society's list of fishes from the United States and

Canada (1960) less than 5% have been specifically examined for their response to temperature—quite a small figure upon which to base water temperature requirements for the entire United States.

## Regulatory Considerations

As is obvious from the previous brief consideration of the role of temperature upon living systems, the problem is complex and has been recognized as such by many other workers. However, by using some present state regulations and existing temperature data obtained from various sources on a variety of species, let us examine and project the effect of certain levels of heated water upon three ecosystems: (1) cold-water salmonid streams, (2) warm-water centrarchid environments, and (3) estuaries.

### *Cold-water Salmonid Streams*

A typical cold-water fish species in North America, the brook trout (*Salvelinus fontinalis*), does not usually occur in bodies of water where temperature rises much above 68 F (Henderson, 1963). Graham (1949) gave 57 to 61 F as the final temperature preferendum, which compares well to the 59 F given by Beamish (1964) as the temperature of maximum spontaneous activity. Normal gametogenesis rate under natural illumination does not appear to be restricted by temperature between 47.3° F and 60.8° F (Henderson, 1963). Fry reported 66.2° F as the temperature for peak of activity in reference to oxygen consumption. Fisher and Elson (1950) gave 50° F as a preferred temperature level of *S. fontinalis* specimens previously acclimated to 39.2° F. Fry (1951) reported 77.5° F as an upper lethal level for adult fish acclimated to 77° F and reported that brook trout fry tolerated 74.3° F for 12 to 14 hours.

Gaufin (1965) in his paper on environmental requirements of Plecoptera, one of the most important groups of macroinvertebrates in trout diets, reported 50 to 60° F as the optimum temperature range for stonefly species tested. When temperatures rose above 60° F stoneflies became sensitive to low dissolved oxygen levels.

Thus, if the objective is to protect salmonid streams, there is ample justification for ORSANCO's 1956 statement that temperature of receiving water "shall not be raised in streams suitable for trout propagation" and Pennsylvania's present ruling that no heat is to be added to trout waters, except that, during colder seasons between October and May when water temperatures are below 58 F, effluent discharge temperatures shall not exceed 58 F.

Lack of specific regulations in many states and vague rulings in others, in view of known requirements of cold water forms could mean the loss of many miles of valuable trout streams due to future thermal loading.

### Warm-water Centrarchid Environments

Concerning so-called warm water species, Bennett (1965) in discussing environmental requirements of centrarchids reported that nest building in largemouth bass (*Micropterus salmoides*) begins at 56° F and spawning begins at 66° F. He quotes Ferguson (1958) as listing field observations on preferred temperature for largemouth bass at 80 to 82° F. Brett (1956) listed upper lethal temperature for largemouth bass acclimated to 86° F to be 97.5° F. Limited data on smallmouth bass (*Micropterus dolomieui*) show it to be less temperature tolerant than largemouth bass. For example, spawning begins at 62° F, 4° F lower than the largemouth. Field observations listed 68.5 to 70.3° F as preferred temperatures, while laboratory data gave 82.4° F as a preferred temperature (Bennett, 1965).

Data on typical macroinvertebrate food organisms of warm water streams have been given for Trichoptera by Roback (1965), for Tendipedidae by Curry (1965), and for mixed riffle populations by Trembley (1961) and Coutant (1962). Trichoptera genera were reported well distributed up to 95° F with greatest diversity, however, reported at 82.4° F. The majority of Tendipedidae were reported to have an upper limit of 86 to 91.4° F, with some genera found at 100 to 103° F. Trembley (1960) and Coutant (1962) showed a tolerance limit less than 90° F for a normal population structure of riffle macroinvertebrates from the Delaware River. An extensive loss in numbers, diversity and biomass occurred with further temperature elevations above 90° F (Figure 29–11).

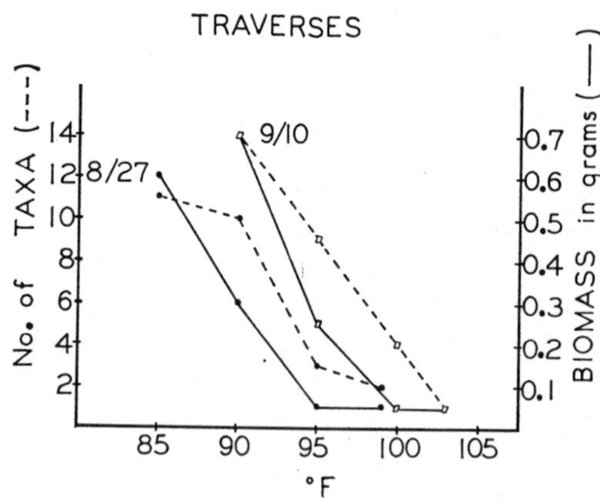


FIGURE 29–11. Number of taxa and biomass per square foot versus temperature for macroinvertebrates collected in transect fashion from a maximally heated area below a power plant discharge and progressing laterally across a riffle area towards less thermally influenced (after Coutant, 1962).

Thus, state regulations such as in Maryland where discharge temperatures are allowed up to 99.9 F if beyond 50 feet from the discharge structure can expect almost complete elimination of most vertebrate and macroinvertebrate species in river sections reaching 95 to 99.9 F.

Pennsylvania's ruling of a 93 F maximum river temperature would appear to cause a reduction of species number. It is yet to be demonstrated whether all life history stages of important game fish species such as largemouth and smallmouth bass could tolerate such temperature levels.

### *Estuaries*

The Patuxent River estuary in Maryland may be used as an estuarine example. A power plant which has recently begun full operation in the Patuxent is designed to produce an 11.5° F differential across the condensers during the warmer seasons and a 23° F differential during the colder seasons. It pumps 250,000 gallons of water per minute (gal/min) during cold season operations and 500,000 gal/min during warm season operations (Mihursky, 1963).

Recent LD-50 work at our laboratory on arthropods, pelecypods, and fish shows that under winter and summer power plant operating conditions, lethal levels will be reached for a number of species studied if the organisms are subjected to 24-hour shock exposures to the discharge temperatures. One mud crab, one shrimp, and one fish species when acclimated to 95° F were only able to withstand about a 4 to 7° F temperature elevation for 24 hours before 50% mortality occurred.

The problem of thermal shock to pelagic life history stages is extremely critical in marine environments. Unlike organisms from nontidal freshwater rivers, most marine species have pelagic stages during some phase of their life history, even benthic forms, and consequently risk being pumped through a power station where abrupt thermal changes occur.

Of primary practical concern to the Patuxent ecosystem is the valuable striped bass (*Roccus saxatilis*) which has a spawning and nursery area just upstream from the power plant. In addition to considerations of adult, juvenile, larval, and egg thermal tolerances is the fact that juveniles are dependent upon the opossum shrimp (*Neomysis americana*) in their diet. As stated earlier, *Neomysis* is a north temperate latitude species and the Chesapeake system is near its southern limit of distribution. Laboratory work, as reported earlier, has shown *Neomysis* to be the least temperature tolerant estuarine species examined in our laboratory to date. As Figure 29–3 shows, it does not appear to survive above 87.8° F, a surface water temperature occasionally reached and surpassed in the Chesapeake system during extended hot, humid periods. Disregarding any temperature effect on striped bass itself, the possible loss of a major food item of juveniles in their nursery area may be extremely detrimental.

In the previously described Patuxent estuary situation it appears certain that estuarine temperatures between 90 and 99.9° F, which are allowable under present Maryland laws, are going to suppress or eliminate certain species. The overall biological impact on local and migratory species in the Patuxent will depend on the volume and the vertical and horizontal distribution of heated water between 90 and 99.9° F.

## Conclusions

Thus, using a limited amount of field and laboratory data and applying them to practical considerations of thermal discharges in relation to some present state laws, and especially when viewed in the light of present and impending thermal discharges, it is obvious a very serious problem will exist in the near future. Present heat loads tend to be found in localized situations; however, projected engineering designs of larger power plants and the need to stagger power plants along the length of a river will result in much larger areas of fresh and marine waters receiving significant temperature elevations. Indeed, engineering proposals have been made to utilize 100% of river flows at a single installation during low flow periods.

Lessons from past environmental changes indicate environmental repair is difficult to economically or politically justify once the damage has occurred. The task concerning temperature considerations is to indicate exactly the ecological alterations to be expected under given temperature regimes and to determine exactly the temperature requirements of the species involved. Undoubtedly, at the present level of understanding in this complex problem, it is imperative that a conservative point of view be maintained when establishing regulations.

## Acknowledgments

We would like to acknowledge use of unpublished data developed with Mr. C. I Gibson on *Neomysis americana*, and use of unpublished data developed with Mr. W. Dovel and Mr. A. J. McErlean on *Trinectes maculatus*.

## Literature Cited

American Fisheries Society. 1960. A list of common and scientific names of fishes from the United States and Canada, *Spec. Publ.* No. 2, 102 p.

Aquatic Life Advisory Committee of the Ohio River Valley Water Sanitation Commission (ORSANCO). 1956. Aquatic life water quality criteria, *2nd Prog. Rep. Sewage and Industrial Wastes*, 28(5): 678–690.

Beamish, F. W. H. 1964. Respiration of fishes with special emphasis on standard oxygen consumption II. Influence of weight and temperature on respiration of several species. *Canadian J. Zool.* 42: 177–188.

Bennett, G. W. 1965. "The environmental requirements of centrarchids with special reference to largemouth bass, smallmouth bass, and spotted bass." In *Biological Problems in Water Pollution.* 3rd Seminar, 13–17 August 1962. P.H.S. No. 999-WP-25, p. 156–160.

Brett, J. R. 1956. Some principles in the thermal requirements of fishes. *Quart. Rev. Biol.* 31 (2): 75–87.

———. 1958. "Implication and assessments of environmental stress." In: *The Investigation of Fish-Power Problems.* University of British Columbia, p. 69–83.

———. 1960. "Thermal requirements of fish—three decades of study, 1940–1970."In: Biological Problems in Water Pollution, 2nd Seminar, 1959. *Robert A. Taft Sanitary Engineering Center. Tech. Rept.* W60-3, p. 110–117.

Cadwallader, L. W. 1965. "Power". In *Industrial Waste Water Control.* S. C. Gurnham (Ed.). Academic Press New York, p. 413–427.

Cootner, P. H., and G. O. Löf. 1965. *Water demands for steam generation* Resources for the Future, Inc., Washington, D. C.

Costlow, J. D., Jr., C. G. Bookhout, and R. Monroe. 1960. The effect of salinity and temperature on larval development of *Sesarma cinereum* (Bosc) reared in the laboratory. *Biol. Bull.* 118: 183–202.

Coutant, C. C. 1962. The effect of a heated water effluent upon the macroinvertebrate riffle fauna of the Delaware River. *Proc. Penn. Acad. Sci.* 36: 58–71.

Curry, L. L. 1965. "A survey of the environmental requirements for the midge (Diptera: Tendipedidae)" In: *Biological Problems in Water Pollution*, 3rd Seminar, 13–17 August 1962, P.H.S. Publ. No. 999-WP-25, p. 127–141.

Dickie, L. M. 1958. Effects of high temperatures on survival of the giant scallop. *J. Fish. Res. Bd. Canada* 15(6): 1189–1211.

Dovel, W. L., J. A. Mihursky, and A. J. McErlean. Life history aspects of the hogchoker, *Trinectes maculatus*, in the Patuxent River estuary, Maryland. Unpublished.

Ferguson, R. C. 1958. The preferred temperature of fish and their mid-summer distribution in temperate lakes and streams. *J. Fish. Res. Bd. Canada* 15(4): 607–624.

Fisher, K. C., and P. F. Elson. 1950. The selected temperature of Atlantic salmon and speckled trout and the effect of temperature on the response to electrical stimulus. *Physiol. Zool.* 23(1): 27–34.

Fry, F. E. J. 1947. Effects of the environment on animal activity. *Univ. Toronto Stud., Biol. Ser.* 55 (Pub. Ontario Fish. Res. Lab. 68). 62 p.

———. 1951. Some environmental relations of the speckled trout (*Salvelinus fontinalis*). *Rept. Proc. N. E. Atlantic Fish. Conf.*

Gaufen, A. R. 1965. "Environmental requirements of plecoptera". In *Biological Problems in Water Pollution*, 3rd Seminar, 13–17 August 1962. P.H.S. Publ. No. 999-WP-25, p. 105–110.

Gibson, C. I. Thermal tolerance studies on the opossum shrimp, *Neomysis americana*, from the Patuxent River estuary, Maryland. Unpublished.

Graham, J. M. 1949. Some effects of temperature and oxygen pressure on the metabolism and activity of the speckled trout, *Salvelinus fontinalis*. *Canadian J. Res.*, D 27:270–288.

Gunter, G. 1957. Temperature. *In:* Treatise on Marine Ecology and Paleoecology. I. Edited by J. W. Hedgpeth, *Geol. Soc. Amer. Mem.*, No. 67, p. 159–184.

Gutsell, J. S. 1930. Natural history of the bay scallop. *Bull. U.S. Bur. Fish.*46:569–632.

Henderson, N. E. 1963. Influence of light and temperature on the reproductive cycle of the eastern brook trout, *Salvelinus fontinalis* (Mitchill). *J. Fish. Res. Bd. Canada* 20(4): 859–897.

Hutchins, L. W. 1947. The bases for temperature zonation in geographical distribution. *Ecol. Monogr.* 17(3): 325–335.

Hutchinson, V. H. 1961. Critical thermal maxima in salamanders. *Physiol. Zool.* 34:92–125.

Kinne, O. 1963. The effects of temperature and salinity on marine and brackish water animals. I. Temperature. *Oceanogr. Mar. Biol. Ann. Rev.* 1: 300–340.

Loosanoff, V. L., and H. C. Davis. 1950. Temperature requirements for maturation of gonads of northern oysters. *Biol. Bull.* 103: 80–96.

Lowe, C. H., and V. J. Vance, 1955. Acclimation of the critical thermal maximum of the reptile, *Urasaurus ornatus. Science* 122: 73–74.

McLeese, D. W. 1956. Effects of temperature, salinity and oxygen on the survival of the American lobster. *J. Fish. Res. Bd. Canada* 13: 247–272.

Naylor, E. 1965. Effects of heated effuents upon marine and estuarine organisms. *Adv. Mar. Biol.* 3: 63–103.

Mihursky, J. A. 1963, Patuxent River estuary study with special reference to the effect of heated steam electric station water upon estuarine ecology. *Univ. of Maryland Nat. Res. Inst. Ref.* No. 63–66.

———. and R. L. Cory. 1965. Thermal loading and the aquatic environment: An approach to understanding on estuarine ecosystem. *International Conference on Industrial Electronics/Control Instrumentation*, 8–10 September 1965. Philadelphia, Pennsylvania.

Pennsylvania Department of Health. 1962. *Heated discharges—their effect on streams*. Report by the Advisory Committee for the control of stream temperatures to the Pennsylvania Sanitary Water Board. Pub. No. 3, 107 p. Division of Sanitary Engineering, Bureau of Environmental Health, Pennsylvania Department of Health, Harrisburg, Pennsylvania.

Orton, J. H. 1920. Sea-temperature, breeding, and distribution in marine animals. *J. Mar. Biol. Ass.* 12: 339–366.

Picton, W. L. 1960. *Water use in the United States*, 1900–1980. Water and Sewerage Division, U. S. Dept. of Commerce.

Prosser, C. L., and F. A. Brown, Jr. 1961. *Comparative animal physiology*, 2d ed. W. B. Saunders Co., Philadelphia and London. 688 p.

Roback, S. S. 1965. "Environmental requirements of Trichoptera." In: *Biological Problems in Water Pollution*, 3rd Seminar, 13–17 August 1962. P.H.S. Publ. No. 999-WP-25, p. 118–126.

Sastry, A. N. 1963. Reproduction of the bay scallop, *Aequipectin irradians* Lamarck. Influence of temperature on maturation and spawning. *Biol. Bull.* 125(1): 146–153.

Tattersall, W. M. 1951. A review of the Mysidacea of the United States National Museum. *U. S. Nat. Mus. Bulletin* No. 201, 292 p.

Thorson, G. 1946. Reproduction and larval development of Danish marine bottom invertebrates, with special reference to the planktonic larvae in the sound (Öresund). *Medd. Komm. Danmarks Fisk. Havundersog. Ser. Plank* 4 (1): 1–523.

Trembley, F. J. (Ed.) 1961. *Research project on effects of condenser discharge water on aquatic life*. Progress Report 1960. Institute of Research. Lehigh University. 80 p.

———. 1965. "Effects of cooling water from steam-electric power plants in stream biota." In: *Biological problems in water pollution*, 3rd Seminar, 13–17 August 1962. P.H.S. Publ. No. 999-WP-25, p. 334–345.

# VIII

# Fertility and productivity of world ecosystems

# AQUATIC ECOSYSTEMS

# 30

# Cultural eutrophication is reversible

*Arthur D. Hasler*

Many lakes the world over are becoming less desirable places on which to live because of nutrient wastes pouring into them from a man-changed environment. Human activities, through population and industrial growth, intensified agriculture, river-basin development, recreational use of waters, and domestic and industrial exploitation of shore properties, are contributing to excessive nutrient enrichment of lakes, streams, and estuaries. This accelerated process of enrichment (cultural eutrophication) causes undesirable changes in plant and animal life, reduces the aesthetic qualities and economic value of the body of water, and threatens the destruction of precious water resources. Overwhelming excessive scums of blue-green algae and aquatic plants choke the open water, rendering the water turbid and nonpotable. The algae and aquatic plants die and rot, yielding a repugnant odor, and the organic matter from this crop sinks and consumes the deep-water oxygen vital for fish and other animal life.

Under natural conditions lakes proceed toward geological extinction at varying rates through eutrophication or bog formation. Many lakes, in unpopulated temperate zones, and lying in sandy granite drainage basins are still pristine and clear (oligotrophic) even though 10,000 years have elapsed since their formation. Other lakes in the same area, such

Hasler, Arthur D. 1969. Cultural eutrophication is reversible. *BioScience*, 19(5):425–431. Reprinted with permission.

The author is with the Laboratory of Limnology, University of Wisconsin, Madison 53706.

as shallow bog lakes which were also formed during the same glacial epoch, are already extinct. They are grown over with mats of sphagnum moss interspersed with orchids and pitcher plants. Brown-colored water lies below the mat which deteriorates and slowly fills in the basin. In some, the terminal stages of bog formation are evident because these former lakes are now covered with shrubs, tamarack, and black spruce forests, but this type of extinction is not eutrophication. How this succession or continuum proceeds from open lake to forest is too complex to be developed in this brief essay.

Archeological studies by G. E. Hutchinson and R. Patrick of cores of lake sediments of the Italian lake, *Lago di Monterosi*, reveal that the Romans, by constructing roads, inadvertently increased the nutrient drainage of a landscape by cutting the trees and exposing limestone strata. The erosion from these nutrient-richer strata was followed by a eutrophic period in the lake's history as recognized by the kinds of diatoms found in the cores. E. S. Deevey, Yale University, also recognized prehistoric changes of climate and rate of eutrophication in Linsley Pond, Conn., which are correlated with the abundance and variety of fossil organisms in the different strata.

The rate of eutrophication of lakes in geological time can often be predicted by examining the soil and vegetation of their drainage basins. If the drainage area is large, the vegetation pristine, and the soil rich and erodible, the lake water will be rich in algae and fish; if poor, the water will produce little and will retain its clarity.

## The Algal Community (Phytoplankton)

Algae are microscopic one-celled plants which require for their growth the same nutrients as do garden flowers and lawns. If fertilized richly in spring and summer, they flourish; if impoverished, they grow sparsely. A community of free-floating algae is richer and more diverse than is any garden.

Pure cultures of algae grown in the laboratory multiply and grow rapidly if nitrogen and phosphorus-bearing chemicals are added. Algae in identical cultures will grow even more luxuriantly when small amounts of sewage effluent are added demonstrating that in addition to P and N, there are ingredients (probably vitamins and growth hormones) in sewage which promotes growth.

A nutrient-poor, temperate zone lake will be clear; hence, one might collect thousands of minuscule algae cells at depths of 150 ft and more. In a nutrient rich lake, on the other hand, the high numbers of algae will lend a greenish cast to the surface water, restrict the penetration of sunlight, and therefore limit photosynthesizing algae and rooted aquatic plants to the shallower depths.

The onset of eutrophicated (nutrient-enriched) conditions has adverse ramifications. When the enriched conditions are due to man-made effluents, the algae grow so profusely that the water fleas (the basic food of all larval fishes) cannot consume the algae fast enough to reduce their numbers significantly; hence, abnormal amounts die uneaten.

The biological communities of a lake become upset when bacteria are unable to convert dead organic matter into plant and animal food.

Not only is oxygen in the deep, cool water exhausted by organic products, but hydrogen sulfide (rotten egg gas) accumulates to poisonous levels.

The finale in these despoiled depths is the demise of all noble fishes, e.g., whitefish, trout, and cisco, which require oxygen-rich water depths for life. Moreover, some noble fishes such as cisco spawn in the fall—their eggs must incubate throughout winter—but an enriched lake, having lost its oxygen in the deep layers (under the ice), cannot nourish the eggs for hatching in the spring.

## Sources of Nutrients

Phosphate additions appear to be one of the major factors in pollution of European and North American lakes although the rate at which nutrients pass through chemical and biological cycles is also important. Sources of plant nutrients are principally from human sewage and industrial wastes, including the phosphate-rich detergents (Table 30–1).

TABLE 30–1

*Summary of estimated nitrogen and phosphorus reaching Wisconsin surface waters*

| Source | N | P | N | P |
|---|---|---|---|---|
| | Lbs. per year | | (% of total) | |
| Municipal treatment facilities | 20,000,000 | 7,000,000 | 24.5 | 55.7 |
| Private sewage systems | 4,800,000 | 280,000 | 5.9 | 2.2 |
| Industrial wastes[a] | 1,500,000 | 100,000 | 1.8 | 0.8 |
| Rural sources | | | | |
| Manured lands | 8,110,000 | 2,700,000 | 9.9 | 21.5 |
| Other cropland | 576,000 | 384,000 | 0.7 | 3.1 |
| Forest land | 435,000 | 43,500 | 0.5 | 0.3 |
| Pasture, woodlot & other lands | 540,000 | 360,000 | 0.7 | 2.9 |
| Ground water | 34,300,000 | 285,000 | 42.0 | 2.3 |
| Urban runoff | 4,450,000 | 1,250,000 | 5.5 | 10.0 |
| Precipitation on water areas | 6,950,000 | 155,000 | 8.5 | 1.2 |
| Total | 81,661,000 | 12,557,500 | 100.0 | 100.0 |

[a] Excludes industrial wastes that discharge to municipal systems. Table does not include contributions from aquatic nitrogen fixation, waterfowl, chemical deicers and wetland drainage.

Drainage from farmland is second in importance as a nutrient source in temperate zones, where farm manure, spread on frozen ground in winter, is flushed into streams during spring thaws and rains. A shocking statistic which points up the gravity of our contemporary situation is that farm animals in the Midwest alone provide unsewered and untreated excrement which is equivalent to that from a population of 350 million people. Also, it is surprising that substantial quantities of nitrates of combustion engine and smokestack origin augment these sources (see rain and groundwater, Table 30–1). City streets also provide sources of phosphates and nitrates that must be dealt with.

*In toto*, the results of man-induced eutrophication are catastrophic as noted in the case history of Lake Zürich, Switzerland, where all noble, deep-water fishes, which had provided gourmet specimens for generations, disappeared within 20 years after sewage disposal in surrounding villages was changed from the "Chick Sale" type to flush toilets.

## A Case History

The Zürichsee, a lake in the foothills of the Alps, offers a sad example of the effects of sewage effluent. It is composed of two distinct basins, the Obersee (50 m) and the Untersee (141 m), separated only by a narrow passage. In the past five decades the deeper of the two, at one time a decidedly clear and oligotrophic lake, has become strongly eutrophic, owing to urban effluents originating from a group of small communities totaling about 110,000 people. The shallower of the two received no major urban drainage and retained its oligotrophic characteristics for a longer period. Thus we have an experimental and reference lake side by side.

## History of Fishing

Dr. L. Minder (Zürich, Switzerland) observed that, hand in hand with domestic fertilization, the Zürichsee changed from a whitefish (coregonid) lake to a coarse fish lake. In fact, the trout, *Salmo salvelinus*, and a whitefish, *Coregonus exignus*, disappeared from the Untersee and are no longer common in the Obersee; restocking has not been successful. An upsurge of cyprinid fishes (minnow-carp family), chiefly *Abramis brama* and *Leuciscus rutilus*, have also become abundant with the progressive eutrophication.

## History of Plankton Succession

Minder is convinced that the decided increase of plankton is not an expression of a natural ripening process, but is due to plant nutrients, principally P and N, from domestic sources. The diatom *Tabellaria fenestra* appeared explosively in 1896. Two years later an eruption of the blue-green alga *Oscillatoria rubescens* occurred. The latter had been known from the eutrophic Murtenersee for 70 years, but otherwise had been recorded only from the Baldeggersee in spring and winter 1894. It had not been seen in the Zürichsee plankton until the 1896 eruption when it replaced the usually dominant *Fragilaria capucina.* When Minder studied the lake during 1920–24, *Oscillatoria rubescens* appeared in quantities in the surface plankton, with a maximum in fall and winter, while flourishing in the deeper water of the lake in summer. In 1936 Minder observed a red scum over most of the lake. An odor of fish oil is frequently noticeable in summer. There were 1.75 g wet weight of algae per liter, chiefly *Oscillatoria*, on 5 May 1899.

Further evidence for a recent sudden increase in biological productivity can be found in bottom sediment studies. These demonstrate that *Tabellaria* occurs in only the most recent layers. Moreover, the modern layers are laminated, at least in the deeper parts of the lake, and everywhere are darker than the underlying sediment. The dark, laminated character of the sediments is especially pronounced from 1896 onward, the date being determined by counting the seasonal laminae.

Minder cites some comparative plankton analyses on the Untersee and Obersee: first, in no series were the biocoenoses of the two sections identical; second, *Oscillatoria rubescens* was never found in the Obersee and *Tabellaria*, very infrequently; third, quantitatively the entire plankton of the Untersee was vastly richer; and fourth, plankton quantities are greater in the Untersee downstream from the town of Rapperswil where most of the sewage enters. Minder also observed the rotifer *Keratella quadrata* as appearing first in 1900. *Bosmina longirostris* largely replaced *B. coregoni* after 1911. Since 1920 one of the Ulotricales has become common in summer.

## History of Other Limnological Factors

Chemical analyses by Minder have shown that certain elements of domestic sewage origin, notably chloride ion, have increased gradually over a relatively few decades. Analyses in 1888 showed the water contained 1.3 mg Cl/liter, by 1916 it had risen to 4.9 mg/liter. The organic matter as measured by loss on ignition also rose from 9 mg/liter in 1880 to 20 in 1914. Vollenweider gives as a rule of thumb 0.2–0.5 $g/m^2/yr$ of P and 5–10 $g/m^2/yr$ of N as the levels of these nutrients which are associated with nuisance blooms of algae in European lakes.

Minder gives comparisons of the changes in transparency (average of 100 readings)

| | Maximum disc reading | Minimum disc reading |
|---|---|---|
| Before 1910 | 16.8 | 3.1 |
| 1905-10 | 10.0 | 2.1 |
| 1914-28 | 10.0 | 1.4 |

It is significant also that $O_2$ values in the deep water have decreased in the last four decades. Midsummer values at 100 m were nearly 100% saturation from 1910–30; from 1930–42, however, they averaged about 50% saturation but did occur as low as 9%.

## Man-Made Lakes

Lakes are more adversely affected by sewage effluent than are flowing streams, chiefly because rapidly flowing water is not conducive to the growth and attachment of algae and rooted aquatic plants. Hence, the diversion of sewage around lakes and into streams is the lesser evil. Nevertheless, while alleviating the lake problem, it places an increasing burden upon the stream's biological system and upon the communities downstream which must purify it. In modern times most large streams have man-made dams for impounding the water and the outflow provides energy for hydroelectric power. If such a reservoir receives sewage, the quality of the water deteriorates as does a lake, and the cost of water purification and odor control rises for downstream communities. There is therefore a finite limit. Permissible levels of impurities will have to

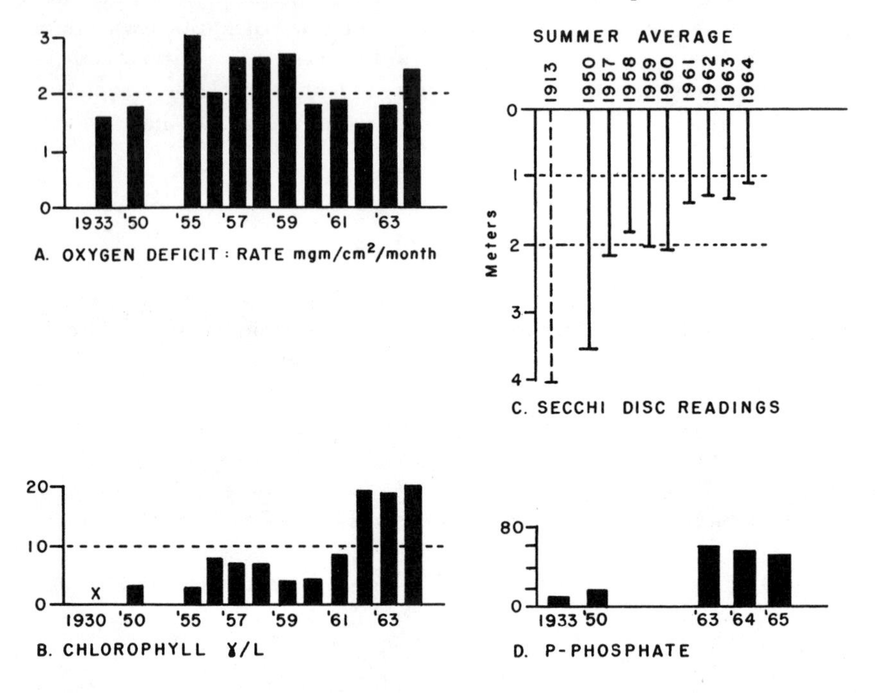


FIGURE 30–1. Changes that have occurred in Lake Washington. (A) Rate of development of relative areal oxygen deficit below 20 m depth between 20 June and 20 September. (B) Mean concentration of chlorophyll in top 10 m of lake during same period of time. (C) Mean Secchi disc transparency, June–September. (D) Maximum concentration of phosphate phosphorus in surface water during winter. (After Edmondson, 1968)

decrease, hence the technological improvement to obtain more complete nutrient removal must become more efficient.

## Limnological Features of Eutrophicated Lakes

Figure 30–1 shows the data obtained by Edmondson and his associates during their continuing long-term investigation of Lake Washington at Seattle. Changes in chlorophyll concentration in the epilimnion correspond with the trend of phosphate accumulation, reduced transparency, and increase in the $O_2$ deficit, and are all therefore expressions of racing eutrophication.

Findenegg (Fig. 30–2) has used the $^{14}C$ method (radioactive carbon) to evaluate the degree of eutrophication. A eutrophic lake fixes carbon principally in the near-surface water because its turbidity prevents light from supplying essential photosynthetic energy to the deeper layers. Higher nutrient levels in a eutrophic lake also serve to stimulate higher rates of $^{14}C$ fixation in the surface waters (see example No. 4, Fig. 30–2).

Dr. Findenegg's example demonstrates a general principle in biology that in senility the older animal often consumes as much food as when he was young, but utilizes it less efficiently. In jargon we say, "He spins his wheels." With increasing levels of eutrophication there is a steady

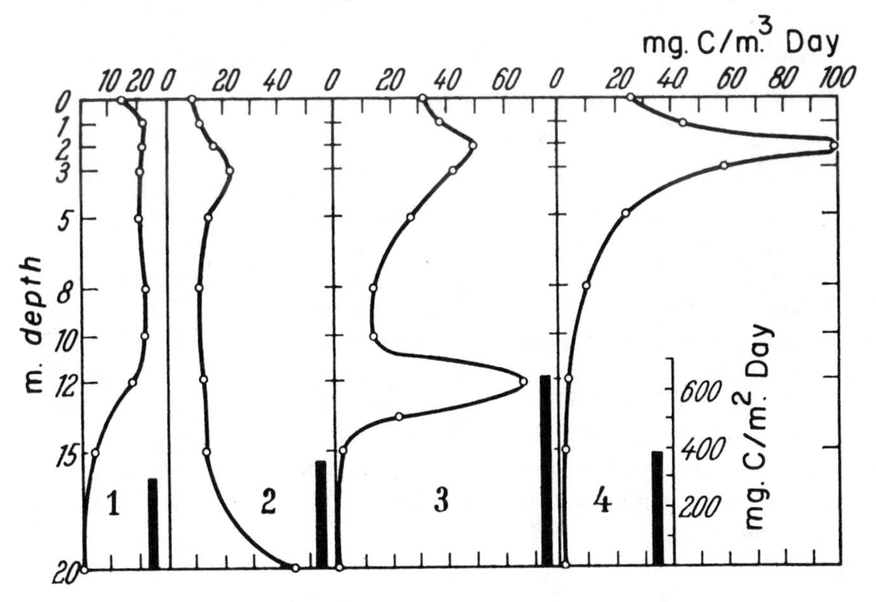


FIGURE 30–2. The production of carbon by photosynthesizing algae in four alpine lakes (1. Millstatter; 2. Klopeiner; 3. Worther; and 4. Lower part of the Lake of Constance) expressed as mg C per m³ per day. The columns give the total production below 1 m² in mg C per m² per day. Note the restriction of production to the surface waters in the most eutrophic lake, No. 4.

increase in carbon fixed, but as turbidity rises, photosynthesis is restricted to the surface waters, hence less depth can be used for production and the utilization drops off—frankly the lake is overfed and obese, perhaps also physiologically senile, and it is spinning its wheels.

Reduction in transparency (Secchi Disc) of Lakes Washington and Zürich appear to be characteristic for rapidly eutrophicated oligotrophic lakes. Hypereutrophication of a natural eutrophic lake, Lake Mendota, has not changed the average transparency nor the average hypolimnetic deficit although other characteristics such as loss of cisco, increase in macrophytes, and increase in algal blooms are conspicuous features.

## Phytoplankton

Comparisons of algal diversity of oligotrophic Trout Lake and eutrophic Lake Mendota show the latter to have fewer species, but the size of the organisms is considerably larger indicating higher levels of production than in the oligotrophic lake.

Often the low species diversity of the phytoplankton in eutrophic lakes is a result of high populations of blue-green algae such as *Aphanizomenon flosaquae* and *Anabaena spiroides*. In some seasons *Fragilaria crotonensis* and *Stephanodiscus astrae* become dominant.

In many eutrophicated northeastern U.S. lakes, rooted aquatic plants *Myriophyllum* and *Ceratophyllum* become festooned with the filamentous alga *Cladophora* and form dense mats in shallow areas.

## Great Lakes

Until recently it was thought that eutrophication would not be a major problem in large lakes because of the vast diluting effect of their size. However, evidence is accumulating that indicates eutrophication is occurring in the lower Great Lakes. Furthermore, the undesirable changes in the biota appear to have been initiated in relatively recent years. Charles C. Davis, utilizing long-term records from Lake Erie, has observed both qualitative and quantitative changes in the phytoplankton owing to cultural eutrophication. Total numbers of phytoplankton have increased more than threefold since 1920, while the dominant genera have changed from *Asterionella* and *Synedra* to *Melosira*, *Fragilaria*, and *Stephanodiscus*.

Other biological changes usually associated with the eutrophication process in small lakes have also been observed in the Great Lakes. Alfred M. Beeton recently summarized the literature pertaining to the trophic status of the Great Lakes in terms of their biological and physiochemical characteristics and indicated that of the five lakes Lake Erie has undergone the most noticeable changes due to eutrophication. In terms of the annual harvests, commercially valuable species of fish such as the lake herring or cisco, sauger, walleye, and blue pike have been replaced by less desirable species such as the freshwater drum or sheepshead, carp, and smelt. Similarly, in the organisms living in the bottom sediments, drastic changes in species composition have been observed. Where formerly the mayfly nymph *Hexagenia* was abundant to the extent of 500 organisms per square meter, it presently occurs at levels of five and less per square meter. Chironomid midges and tubificid worms now are dominant members in this community.

## What Can Be Done to Reduce the Galloping Rate of Eutrophication?

The deterioration of our lakes proceeds at such a galloping pace that there is insufficient time to raise an enlightened younger generation which could cope with the causes of eutrophication. Every effort must be undertaken to convince government officials and voters that action, even though expensive, must be taken immediately to avoid catastrophe. In attempting to obtain positive returns, time operates negatively against delay. To insure a brighter future, universities, colleges, churches, service clubs, the press, radio, and television must acquire a knowledge of the causes, prevention, and cure and begin without delay to help disseminate factual information.

Provided they are given the facts, preachers and rabbis could preach sin against the environment as convincingly as they preach sin against the soul.

Decision makers such as legislators, state, county, and village officials, decision planners such as architects, decision formulators such as lawyers and judges, decision executors such as realtors, engineers, and contractors must all receive enlightenment about the implication of possible perturbations of the landscape whose environmental health influences the well-being of the lake into which the land's effluents flow.

I urge that every educational body, in every community, organize week-long intensive clinics, seminars, or working groups to which experienced limnologists and ecologists are invited as teachers, lecturers, demonstrators, and guides. I urge journalists and editors, television and radio directors to send their personnel to clinics, or to meet with experienced ecologists in order to obtain facts and illustrations, to provide bases on which readable and effective articles and programs can be prepared. They should be taken on field trips to areas where the reality of a eutrophied lake can be demonstrated, in order to capture their enthusiasm and stimulate their originality toward preparation of dynamic and imaginative programs. I would urge legislative lawyers to draft critical legislation for water usage in regions where the legal procedures are inadequate and encourage them to draft laws which will provide adequate protection of a lake or reservoir from perturbations.

The processes of eutrophication are too rapid to risk delay in taking legal action. In applying new concepts of water law to the alleviation of eutrophication, there is a need for proper zoning ordinances and forthright public initiative in modernizing the law when the scientific data, even if not complete, suggest action. Public planners recommend the formation of a County or Drainage Basin Authority, which can act for the towns, counties, and municipalities to deal with all problems of water quality. It would have authority to make water and sewage assessments, control erosion, create zoning ordinances, conduct studies toward evaluation of problems and evolve improvements.

In Wisconsin the late Jacob Beuscher, through his association with ecologists and landscape architects, drafted unique zoning legislation for Wisconsin which now has been passed (Wisconsin Water Resources Law, 1965).

If a dwelling or resort is planned, it must meet exact specifications for sewage disposal so that no effluent can seep into the water. Some soils are less able to absorb effluents than others, hence the setback of a planned hotel or dwelling might need to be at the extreme end of the 1000-ft maximum setback on lakes, 300 on rivers. Beuscher legislation also specified beauty for this zoned corridor as a quality to be preserved. Vilas County, rich in lakes, prohibits cutting the natural vegetation from more than 10% of the shoreline fronting a property.

## Social, Legal, and Economic Aspects

In the preceding discussion emphasis was placed upon the effectiveness of various management procedures. Of equal importance are comprehensive economic analyses of new approaches to management. Included should be studies to develop methods to quantify costs and benefits and to analyze public opinion so that the management programs developed are acceptable to society.

Beuscher writes:

Since resource management requires not only scientific knowledge and techniques but also governmental and legal structures by which desired management can be achieved, the entire field of legal and governmental structure is a necessary research area. Wisconsin's assertion that it is trustee of all navigable waters of the state is one of the strongest examples of a state's assertion of its right and duty to protect public interests in natural resources, and this could form a basis for a variety of strong regulatory policies. Potential conflicts between the asserted trusteeship and the rights of private littoral and riparian owners exist, however, and should be investigated as a guide both to the potentials for regulation and to the limitations on regulation without compensation.

Zoning is one type of regulatory action which is likely to be of significant value in attacking the problems associated with inland lakes. As for other water resources, the Wisconsin Legislature has acted, and the counties are presently required by law to enact river and lakeshore zoning ordinances. Research is needed to review the powers of the counties under present statutes, especially noting where powers which seem necessary to accomplish desired regulation are lacking or unclear. Creative proposals and careful analyses are needed concerning the present procedures for administration of zoning by the counties. Projects in various areas of scientific research could be undertaken with a view toward producing facts and testing procedures which the counties might employ to guide and defend their regulation of stream and lakeshore lands.

Owing to the traditional lack of compensation for regulations imposed, zoning is a limited device for the control of lakeshore lands. Imaginative legal research is needed on a broad range of new control devices, such as compensable regulation and partial condemnation, some of which are being tried in some parts of the country. Finally, the powers of all levels of government and the potential powers of private groups and resource control and management corporations should be analyzed as means toward proposing more systematic and creative methods of management than exist presently.

## Examples of Success

Our knowledge of what causes eutrophication is sufficiently good that firm and effective precautions can be recommended. They may be expensive to achieve, but the predictive facts are at hand. With improvement, methods can be made more economical, but it is not a lack of knowledge which prevents us from action. Three case histories are at hand.

## Lake Monona

Complaints about the unpleasant odors arising from Madison's Lake Monona were published in the newspapers as early as 1850. Sewage effluents were impugned as the villain in 1885 when a consultant J. Nader advised:

> . . .that the lakes were not properly used as receptacles for sewage in the crude state.

In 1895 a 1½-mill sewer tax was imposed on assessed valuation, but the sewage treatment plant built from these funds failed in 1898. Septic tanks and cinder filter beds were then constructed but reached capacity in 1906, and it was not until 1914 that a modern sewage treatment plant was constructed. Its effluent entered Lake Monona and continued to feed the algae and weeds. The process of eutrophication accelerated in Lake Monona and in 1920 the city council minutes read:

Winds ... drive detached masses of putrefying algae onto shores ... if stirred with a stick, look like human excrement and smell exactly like odors from a foul and neglected pig sty.

In 1921 consideration was given to piping the effluent to the Wisconsin River, but another plant was built below Lake Monona in 1928. One-half of Madison's effluent then first passed into Lake Waubesa; later (1936), all of it. In spite of heavy applications of the algal poison $CuSo_4$ to the lake, the build-up of offensive and obnoxious odors in Waubesa and Kegonsa worsened. Anti-pollution legislation was introduced in 1941, but was vetoed because of conflicting opinions on whether sewage or rural runoff was the culprit.

In 1942 the Burke Plant, which had been discontinued in the '30's, was reopened to accommodate the military needs of Truax Field in World War II. The need for more copper sulfate during this period is obvious from the graph (Fig. 30–3). In 1943 the Lewis Anti-Pollution law was passed, to take effect one year after the war.

In actuality, the effluent did not bypass the lakes until 1950. During this century of time, buck passing, economy measures, false information from communications, inconclusive action, and lack of cooperation between government and citizens hampered progress.

The fact that copper sulfate treatment of the lake dropped from carload quantities to minimal local treatments (Fig. 30–3) is proof that even though agricultural drainage unfortunately continues, the diversion of city sewage produced a change for the better.

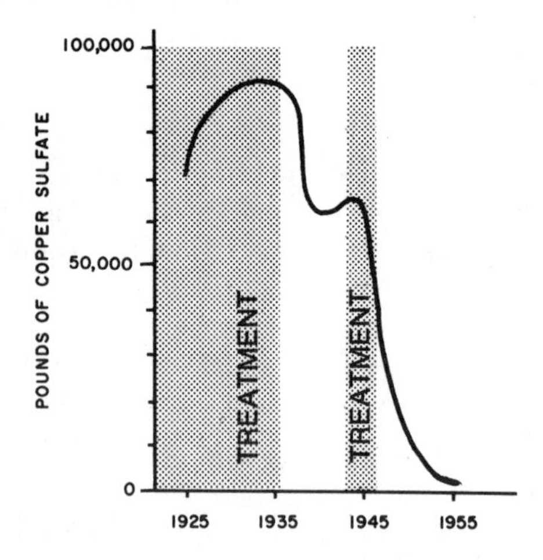


FIGURE 30–3. The amount of copper sulfate required to control alga nuisances in Lake Monona near Madison, Wisconsin, in the period 1925–56. The amount used in a function both of the abundance of algae and the degree of public disturbance; thus small differences are probably unimportant, but large changes are significant. The main trend is shown by the solid line.

## Lake Washington

Lake Washington was used first for the disposal of raw sewage and later treated sewage from the City of Seattle. About 1930, the last major source of raw sewage was removed from the lake, but up until 1959 untreated sewage still entered the lake in relatively small quantities through storm sewer overflows and seepage from septic tanks. In 1959 there were 10 sewage treatment plants, serving 64,000 people, putting treated effluent into Lake Washington.

While biologists and engineers warned the community of the impending doom of the lake, it took dense blooms of *Oscillatoria rubescens*, the same lavender-colored alga that produced nuisances in Lake Zürich, to awaken the citizens to the reality of these warnings. While some argued as they did in Madison that the runoff from fertile land was causing the nuisances, other contended that the major source (city sewage) could be diverted. A powerful combination of radio and television debates, citizens' group meetings, and door-to-door campaigning culminated in arousing sufficient public support for a bond issue. This amounted to an expenditure of $121 million for diversion of 50 million gallons of sewage from Lake Washington to Puget Sound. While the diversion is not yet entirely complete, a major part of the sewage has now (1968) been diverted, with subsequent improvement of the quality and clarity of water in Lake Washington.

## Lake Tahoe

To protect the pristine beauty and crystal clarity of Lake Tahoe, several sanitary engineers (P. McGauhey, G. Rohlich, and others) made a study of the sewage disposal problems of Lake Tahoe and published a comprehensive study in 1963. They described the problem and projected the rate of growth of the communities and tourist facilities whose sewage from treatment plants and septic tanks enters the lake. The lake's great size and immense depth meant that it could absorb some sewage without showing general signs of deterioration. However, objectionable accumulations of green algae at sewage outlets indicated that it was only a matter of time before the increased sewage load from a skyrocketing population would change this clearest of all North American lakes to one of lesser esthetic value.

In spite of the 70-odd governmental units in Nevada and California which surround the lake, this initial limnological and engineering evaluation inspired the creation of citizen-governmental action committees and associations which went into action. South Tahoe (1968) has built a $19 million sewage treatment plant in which the treated effluent is pumped over a 7000 ft pass to a reservoir, where the fertile water will be used for irrigation. Some $10 million of the total budget was in federal grants acquired by the South Tahoe Public Utility District to help offset this cost—a demonstration of cooperative action by federal and local government in solving a local and national problem. Acknowledging the imminent danger of despoiling this esthetic and financial resource, other communities are facing reality in an action program. The small community of Round Hill in Nevada with only 42 voting citizens, albeit many are owners of gaming houses, has bonded itself for $5.8 million to treat its sewage and pump it out of the basin. We can only hope that the hotel and residential sewage can be similarly diverted from other parts of this once gin-clear lake in time to avoid the certain despoliation of Sierra Nevada's most magnificent landscape gem.

Two European lakes, Schliersee and Tegernsee in Germany's Bavarian Alps, were eutrophicated by hotel sewage, but are now slowly reverting to more tolerable conditions following diversion of sewage to the outlets. Lake d'Annecy, France (near Lake Geneva) is following suit. All lakes from which sewage has been diverted have shown improvement (see case for Lake Monona, Fig. 30–3), thereby demonstrating that ingredients in sewage contribute greatly to eutrophication. These facts negate the argument "Why divert sewage at great cost if rainwater and rural drainage is so rich in nitrogen?" The "healing" is, of course, more rapid in lakes in oligotrophic and high rainfall landscapes. The Madison, Wis., lakes are naturally eutrophic; nevertheless the hypereutrophic, repulsive conditions have been alleviated following diversion, as demonstrated by a shift from single-species to multi-species blooms in lakes Waubesa and Kegonsa.

## Harvesting and Utilization of Excess Crops of Plants and Fish

More machinery is needed for harvesting large aquatic plants. Removal of this crop, which contains significant quantities of phosphorus and nitrogen, will aid in impoverishing the water. It also improves esthetic quality, opens the water area to boating and swimming, and creates better shoreline sanitation. Development of new and more effective harvesting machines is needed, and research is also needed to find a commercial outlet for the harvested vegetation. Eutrophicated lakes produce large crops of fish which should also be harvested more intensively, by commercial fishermen if necessary, because the unharvested fish die and decompose, adding nutrients to the already over-rich environment.

In Lake Mendota in late 1966, 40 lb./acre (44 kg/ha) of carp alone were harvested in a single seine haul. Yields of 250/lb./acre/yr (280 kg/ha/yr) of fish of all species could be harvested easily from this lake without damaging the fishery. In terms of nitrogen and phosphorus, 1000 lb. of rotting fish would yield to the lake 25 lb. (11.3 kg) of N and 2 lb. (0.9 kg) of P.

## Chemical Control of Nuisance Growths

Chemicals which poison unwanted aquatic plants and algae have deleterious effects in and around treated areas. Moreover, the killed weeds rot and add to the nutrient supply. This is a bad conservation practice because no good is accomplished. Chemicals distort the structure of multi-species aquatic communities and hence are less useful in lakes than they are in agriculture where weeds are to be eradicated from a single-species crop such as wheat. Herbicide usage cannot be justified in a lake ecosystem. In addition, chemicals are more difficult to manage than on land, for they are soon drifted to other areas. The toxic actions of these chemicals on other species in the lake have not been tested, nor have the possible insidious side effects of sublethal actions over longer periods. At present, the use of chemicals to combat algal blooms or rooted aquatic vegetation can be no more than a palliative and should be used only as a last resort. What is eradicated is sure to be replaced by something else that may be more difficult to poison.

## Utilization of Sewage and Farm Manure

Each human produces 1.5 to 4 lb./yr (0.7–1.8 kg/yr) of phosphorus; Chicago's effluent amounted to 30 tons (27,000 kg) of phosphorus per day in 1960. After sewage has undergone secondary treatment it still contains phosphorus. The average P content of sewage after secondary treatment is 8 mg/1. Hormones, vitamins, and growth substances are also fertilizing ingredients. Moreover, phosphate-rich detergents now added are not entirely removed. In fact, in most treatment plants secondary treatment removes only about 80% of the P and high costs deter removing more. An increasing human population adds to the total residual left in sewage effluent after treatment. "We seem to be on a treadmill," comments G. A. Rohlich, an eminent engineer, who states further that in spite of new advances we are not much further ahead of the problem than at the turn of the century.

The price of clean water may rise to a point where we may have to insist upon and want to afford evaporation of the effluents in order to obtain a dry solid for use as a farm fertilizer. Secondary treatment of sewage does not remove organic growth factors but probably produces them; hence, evaporation looms as a likely though expensive treatment.

In temperate climates of North America it is customary to scatter farm manure on the frozen land. Large amounts of valuable fertilizer are flushed into streams and lakes during early spring thaws and spring rains. Because a cow produces 6 lb. (3 kg) nitrogen and 1.5 lb. (0.7 kg) phosphorus per year, it is clear that this source is important. Modernization of the European method of fluidizing dairy cow manure, storing it in huge tanks, and distributing it with a "honey wagon" as soon as the soil can absorb it is now being recommended. However, economic limitations inhibit progress in converting to a more efficient method of manuring. Fortunately, forest and agricultural soils have a remarkable tenacity for phosphorus. Agricultural and forest crops could profit from the fertilizers from our domestic wastes, but the technology for processing and distributing them is still very expensive when compared with artificial fertilizers.

The volume and weight of dried sewage to be disposed of will have staggering proportions. Settlings from primary treatment of sewage abound near every city, but a farmer can buy and distribute sacked fertilizer cheaper than he can haul dried sewage sludge which is available free of charge. We are too affluent to be able to afford the use of our "night soil."

The City of Milwaukee markets a dried sludge called Milorganite from its primary settlings which is rich in organic matter. Every city could do this, but instead it piles up and presents a disposal problem because Milwaukee's product satisfies the available market for this product.

## Benefits of Guided Eutrophication

All eutrophication is not necessarily bad. Well-planned enrichment could increase the production of food organisms for fish and hence raise the protein productivity of a natural or man-made lake. Many lakes in Canada, Alaska, and USSR are candidates for this potential. Because of the complexity of interactions at various depths and seasons, more knowledge than we now have is needed before we can guide eutrophication and harvest the increase in fish produced without exceeding the fertility levels that destroy the esthetics of a lake.

## Predictions

Predicting the consequences of eutrophication would be highly desirable for decision-makers. Systems analysis offers new techniques in multifactorial analysis which may be employed in constructing models of drainage basins in order to evaluate the impact of various eutrophicating factors.

## In Summary

It is now of the greatest urgency to prevent further damage to water resources and to take corrective steps to reverse present damages. Suggested preventive and corrective measures include removing nutrients from municipal, industrial, and agricultural wastes; diversion of treated effluents from lakes; harvesting algae, aquatic plants, and fish from lakes in order to help impoverish the water and to improve esthetic qualities; and establishing regulations for shoreland corridors in order to protect lakes from further damage.

## General References (Not referenced in text)


Edmondson, W. T. 1968. Water-quality management and lake eutrophication: The Lake Washington case. Reprinted from *Water Resources Management and Public Policy*, Thomas H. Campbell and Robert O. Sylvester (eds.), University of Washington Press, Seattle, p. 139–178.

Findenegg, I., 1964. Bestimmung des trophiegrades von seen nach der radiocarbonmethode. *Naturwissenschaften*, **51:** 368–369.

Hasler, Arthur D. 1947. Eutrophication of lakes by domestic drainage. *Ecology*, **28** (4): 383–395.

International Symposium on Eutrophication. 1968. National Academy of Sciences-National Research Council. (In press)

McGauhey, P. H., R. Eliassen, G. A. Rohlich, H. F. Ludwig, and E. A. Pearson. 1963. Comprehensive study of protection of water resources of Lake Tahoe. To Lake Tahoe Area Council Engineering-Sciences, Inc., Arcadia, Calif.

Minder, Leo. 1938. De Zürichsee als entrophierungsphänomen. Summarische ergebnisse aus fünfzig jahren zürichseeforschung. *Geol. Meere Binnengewässer*, **2** (2): 284–299.

Rohlich, G. A., and K. Stewart. 1967. Eutrophication—a review. California State Water Quality Control Board. Publ. No. 34. 188 p.

# 31

# Eutrophication of the St. Lawrence Great Lakes

*Alfred M. Beeton*

## Introduction

Evidence of appreciable change in the biota and physicochemical conditions in Lake Erie (Beeton 1961) and speculation on possible changes in the other lakes, stemming from increases in total dissolved solids (Ayers 1962; Rawson 1951), have directed attention to eutrophication of the Great Lakes. The question does not concern the existence of eutrophication, because all lakes are aging and there is no reason to believe that the Great Lakes are exceptional. Of greatest consequence is the possibility of detection and perhaps even measurement of the rate of eutrophication of these large lakes. Important also is the effect of mankind on the normal rate of eutrophication.

Accelerated rates of aging due to man's activity have been detected in a number of lakes (Hasler 1947). The classic example is the Untersee of Lake Zürich, which urban effluents have changed from an oligotrophic to a eutrophic lake in a relatively short time. Recently, studies of Lake Washington at Seattle (Edmondson, Anderson, and Peterson 1956), and Fure Lake, Denmark (Berg *et al.* 1958) have demonstrated that these relatively large lakes are undergoing accelerated eutrophication due to man's influence. None of these lakes, however, is large in comparison with any of the St. Lawrence Great Lakes.

Reprinted with permission from *Limnology and Oceanography*, 10:240–254 (1965).

## Present Trophic Nature of the Great Lakes

The meaning of eutrophication seems to vary according to the special interests of the individual. Limnologists agree that eutrophication is part of the aging of a body of water and implies an increase in the nutrient content of the waters. Most lakes change gradually from a nutrient-poor, oligotrophic, to a nutrient-rich, eutrophic, condition. At this point agreement ends in arguments on lake classification. Although the terms eutrophic, mesotrophic, and oligotrophic are used freely by limnologists and are well established in the literature, it is difficult to determine precisely what is meant by them. Various investigators have attached considerable significance to one or more of the following criteria in classifying lakes: abundance or species of plankton or both; benthic organisms; chemical characteristics; sediment types; distribution of dissolved oxygen; productivity; fish populations; and morphometry and morphology of the lake basin. Lake classification has been closely related to regional limnology and has been developed primarily from observations on small lakes. It is not surprising, then, to find that the Great Lakes do not fit readily into the various classification schemes that have been proposed. Rawson (1955, 1956) reviewed the problem of classifying large lakes and considered certain characteristics of the St. Lawrence Great Lakes.

Despite the troublesome problems of lake classification, an attempt should be made to classify the Great Lakes for discussion of their eutrophication and to facilitate comparison with other lakes. A better classification surely is needed since Lake Erie, for example, has been called oligotrophic, mesotrophic, and eutrophic by various investigators over the past 30 years.

### *Physicochemical conditions*

Morphometry, transparency, total dissolved solids, conductivity, and dissolved oxygen content of the water appear to be useful in lake classification. On the basis of these five physical and chemical factors, two of the lakes would be classified as oligotrophic, one as tending toward the mesotrophic, and two as eutrophic. The low specific conductance and total dissolved solids, and the high transparency and dissolved oxygen of Lakes Huron and Superior agree with the commonly accepted characteristics of oligotrophy (Table 31–1). Lakes Erie and Ontario have the high specific conductance and total dissolved solids and, of special significance, the low concentration of dissolved oxygen in the hypolimnion characteristic of eutrophic lakes. Data were not available on the

TABLE 31–1

*Physical and chemical characteristics of the Great Lakes**

| Lake | Mean depth (m) | Transparency (average Secchi disc depth, m) | Total dissolved solids (ppm) | Specific conductance (μmhos at 18C) | Dissolved oxygen |
|---|---|---|---|---|---|
| **Oligotrophic** | >20 | High | Low: around 100 ppm or less | <200 | High, all depths all year |
| Superior | 148.4 | 10 | 60 | 78.7 | Saturated, all depths |
| Huron | 59.4 | 9.5 | 110 | 168.3 | Saturated, all depths |
| Michigan | 84.1 | 6 | 150 | 225.8 | Near saturation, all depths |
| **Eutrophic** | <20 | Low | High: >100 | >200 | Depletion in hypolimnion: <70% saturation |
| Ontario | 86.3 | 5.5 | 185 | 272.3 | 50 to 60% saturation in deep water in winter |
| Erie, average for lake | 17.7 | 4.5 | 180 | 241.8 | |
| Central basin | 18.5 | 5.0 | — | — | <10% saturation, hypolimnion |
| Eastern basin | 24.4 | 5.7 | — | — | 40 to 50% saturation, hypolimnion |

*Criteria designating lake types are based primarily on factors considered important by Rawson (1960); specific conductance limits from Dunn (1954); dissolved oxygen criteria, Thienemann (1928). Data from Bureau of Commercial Fisheries except transparency and dissolved oxygen for Lake Ontario (Rodgers 1962).

oxygen content of the hypolimnetic waters of Lake Ontario, but it is even more significant that the deep waters had a low percentage saturation of dissolved oxygen under essentially isothermal conditions. The transparencies of Lakes Erie and Ontario are not especially low in comparison with many small lakes, but in comparison to Lakes Huron and Superior, their average transparencies are indeed low. Lake Michigan falls between the other lakes from the standpoint of specific conductance, total dissolved solids, and transparency. The dissolved oxygen content of Lake Michigan water is near saturation, however, at all depths. On occasion, water in the deeper areas has a saturation between 70 and 80 percent, but these low values are infrequent.

### *Biological characteristics*

The biological characteristics of all the Great Lakes, except Lake Erie, may place them in the oligotrophic category (Table 31–2). This classification surely holds for Lakes Huron, Michigan, and Superior, but is uncertain for Lake Ontario, where recent and detailed information is lacking on benthos and plankton. The types of bottom fauna, especially the dominant midge larvae, have been recognized for many years as good criteria for classifying lakes (Elster 1958); Brundin (1958) held this basis had world-wide application. Lakes Huron, Michigan, and Superior are all of the *Orthocladius-Tanytarsus* (*Hydrobaenus-Calopsectra*) type. *Hydrobaenus* and related genera are the only midges in the deeper parts of these lakes. *Calopsectra* occurs in shallower areas with a variety of other midges. In Lake Erie, *Procladius* and *Tendipes* spp. (*T. plumosus* group) are the dominant midges. *Cryptochironomus* and *Coelotanypus* also are abundant in the western basin. *Calopsectra* and a variety of midge larvae become progressively more abundant toward the eastern end of the lake and only a few *Tendipes* appear in samples from the deep (maximum, 64 m) eastern basin.

The crustaceans, *Mysis relicta* and *Pontoporeia affinis*, are characteristic of oligotrophic lakes, although they occur in lakes classed as mesotrophic. Both organisms require fairly high dissolved oxygen concentrations and cold water. Consequently, they are absent from highly eutrophic lakes. Both crustaceans are found in all of the Great Lakes, but in Lake Erie the major population is restricted to the eastern basin.

Oligotrophic lakes of the north have been considered salmonid lakes. It is not implied that salmonids do not occur in eutrophic lakes, but, when present, they are not the dominant fishes and usually they have a restricted bathymetric distribution. Salmonids dominate the fish populations in all of the lakes except Lake Erie, although in Lake Ontario ictalurids and percids are almost as important in the commercial fishery as salmonids (Table 31–2). Three native species (formerly four) of salmonids occur in Lake Erie, but during most of the year they are restricted to the eastern basin. Warmwater fishes dominate the fish fauna of Lake Erie now, including the eastern basin.

Considerable controversy exists over the value of plankton in lake classification. Rawson (1956) pointed out that the number of species present, as well as the ecological dominants, should be considered in characterizing plankton. He stressed that the dominant species of plankters in the oligotrophic lakes of Canada, as well as the Great Lakes, were not the species usually associated with oligotrophic lakes. Furthermore, a number of plankters commonly accepted as indicative of eutrophy are dominant species in these lakes. These observations are supported by the more recent plankton data from the Great Lakes. *Fragilaria crotonensis* is a dominant diatom species in Lakes Huron and Michigan as well as Lake Erie, and *Melosira granulata* is a dominant plankter in Lake Superior. Both species are considered to be indicators of eutrophy by Teiling (1955). On the other hand, plankters that are accepted as indicators of oligotrophy (*Dinobryon*, *Tabellaria*, *Cyclotella*) are also dominant in the Great Lakes (Table 31–2). It is agreed, nevertheless, that the Cyanophyceae, *Aphanizomenon*, *Anabaena*, and *Microcystis*, which are among the dominant plankters in Lake Erie, are good indicators of eutrophy. Little purpose could be served here, however, by further discussion of the problem of plankton indicators of lake types. Consequently, Rawson's (1956) ranking of algae in order of their occurrence from oligotrophy to eutrophy has been used, in part, in Table 31–2, since he worked on lakes similar in many ways to the St. Lawrence Great Lakes. On the basis of plankton abundance and the dominant species of phytoplankton, Lakes Huron, Michigan, and Superior would be considered oligotrophic and Lake Erie eutrophic.

TABLE 31–2

*Biological characteristics of the Great Lakes**

| Lake | Bottom fauna and dominant midges | Dominant fishes | Plankton abundance | Dominant phytoplankton† |
|---|---|---|---|---|
| Oligotrophic | *Orthocladius-Tanytarsus* type (*Hydrobaenus-Calopsectra*) | Salmonids | Low | *Asterionella formosa*<br>*Melosira islandica*<br>*Tabellaria fenestrata*<br>*Tabellaria flocculosa*<br>*Dinobryon divergens*<br>*Fragilaria capucina* |
| Superior | *Pontoporeia affinis*<br>*Mysis relicta*<br>*Hydrobaenus* | Salmonids | Very low | *Asterionella formosa*<br>*Dinobryon*<br>*Synedra acus*<br>*Cyclotella*<br>*Tabellaria fenestrata*<br>*Melosira granulata* |
| Huron | *Pontoporeia affinis*‡<br>*Mysis relicta*<br>*Hydrobaenus*<br>*Calopsectra* | Salmonids | Low | *Fragilaria crotonensis*<br>*Tabellaria fenestrata*<br>*Fragilaria construens*<br>*Fragilaria pinnata*<br>*Cyclotella kutzingiana*<br>*Fragilaria capucina* |
| Michigan | *Pontoporeia affinis*§<br>*Mysis relicta*<br>*Hydrobaenus* | Salmonids | Low | *Fragilaria crotonensis*<br>*Melosira islandica*‖<br>*Tabellaria fenestrata*<br>*Asterionella formosa*<br>*Fragilaria capucina* |
| Ontario | *Pontoporeia affinis*<br>*Mysis relicta* | Salmonids, ictalurids, percids | — | — |
| Eutrophic | *Tendipes plumosus* type | Yellow perch, pike, black bass | High | *Microcystis aeruginosa*<br>*Aphanizomenon*<br>*Anabaena* |
| Erie<br>Central basin | *Tendipes plumosus* | Yellow perch, smelt, freshwater drum | High | *Melosira binderana*<br>*Stephanodiscus*<br>*Cyclotella*<br>*Fragilaria crotonensis* |
| Eastern basin | *Pontoporeia affinis*<br>few *Calopsectra* | Yellow perch, smelt, few salmonids | High | *Microcystis*<br>*Aphanizomenon* |

*Criteria for lake types as follows: tendipedid larvae, Brudin (1958); fish and plankton abundance, Welch (1952); dominant phytoplankton, Rawson (1956).

†Data for Lake Michigan from Bureau of Commercial Fisheries; Lake Superior from Putnam and Olson (1961); Lake Erie from Davis (1962); Lake Huron from Williams (1962).

‡Data from Teter (1960).

§Data from Merna (1960).

‖Refers to *Melosira islandica-ambigua* type.

The combined biological, chemical, and physical characteristics of Lakes Huron and Superior clearly are those of oligotrophy. The biota and the high dissolved oxygen content of the deep hypolimnetic waters characterize Lake Michigan as oligotrophic but contrariwise, the high content of total dissolved solids indicates a trend toward mesotrophy. Lake Ontario, as a mesotrophic lake, retains the biota of an oligotrophic lake, but the physicochemical characteristics are those of eutrophy. The chemical content of the waters of Lake Ontario is closely similar to that of Lake Erie, since the main inflow to Lake Ontario is from Lake Erie via the Niagara River. The trophic nature of Lake Ontario has been determined to a large extent by the chemical history of Lake Erie waters. Lake Ontario, and perhaps Lake Michigan, would be eutrophic except for the large volumes of deep waters. Even in Lake Erie, the eastern basin has components of a fauna associated with oligotrophy and sufficient deep, cold, oxygenated water to maintain this fauna. These conditions exist despite the highly eutrophic nature of the central basin (flow through the lake is from west to east). The evident ability of the total dissolved oxygen content of the deep hypolimnetic waters to meet the oxygen demand of the organic production of the epilimnetic waters, as well as the oxygen demand of allochthonous materials, makes Lakes Michigan, Ontario, and eastern Lake Erie in some measure oligotrophic (or mesotrophic) because of their morphometry.

## Evidence of Eutrophication

The present trophic nature of the Great Lakes is to a considerable degree the result of their gradual aging since formation. Evidence is accumulating, however, which indicates that human activity is greatly accelerating the eutrophication of all of the lakes but Lake Superior. This evidence is most spectacular for Lake Erie. A difficult problem is one of finding acceptable indices of change.

Various criteria have been used by different investigators to demonstrate eutrophication. Hasler (1947) compiled information on 37 lakes affected by enrichment from domestic and agricultural drainage. Among the changes in many of these lakes were: the dramatic decline and disappearance of salmonid fishes and increases in populations of coarse fish; changes in the species composition of plankton; and blooms of blue-green algae. Special significance has been attached to blooms of *Oscillatoria rubescens*. As the Untersee of Lake Zürich changed from a salmonid to a coarse-fish lake, plankton abundance increased, different species became dominant in the plankton, transparency decreased, and the dissolved oxygen content of the deep waters decreased. At the same time, the concentrations of chlorides and organic matter increased

(Minder 1918, 1938, 1943). Minder (1938) attributed the increase and changes in the plankton to the growing amount of phosphorus and nitrogen from domestic sources. *Oscillatoria rubescens* appeared explosively in 1898 and replaced the formerly dominant *Fragilaria capucina*. The cladoceran *Bosmina longirostris* replaced *B. coregoni* after 1911. Blooms of *Oscillatoria rubescens*, declines in the hypolimnetic oxygen, decrease in transparency, and increases in the abundance of plankton were cited by Edmondson, Anderson, and Peterson (1956) as evidence of eutrophication of Lake Washington. They held this increased productivity to be the result of growing discharges of treated sewage into the lake. Similar changes were observed in Fure Lake by Berg et al. (1958). Species composition of the phytoplankton changed, transparency decreased, dissolved oxygen concentrations became low in the hypolimnion, and conductivity rose. These changes have occurred during the last 40 to 50 years. Berg (Berg et al. 1958, p. 176) stated, "The cause is an increased introduction of material with the sewage."

Our knowledge of the limnology of the Great Lakes in earlier years is seriously deficient. Observations useful for tracing changes in the Great Lakes are mostly limited to water-quality data, commercial fishing records, and a few observations on plankton. The fishing records and chemical data have the longest history and are the most reliable.

### *Chemical characteristics*

Chemical data representative of the lakes proper were compiled from many sources (Table 31–3) ranging from isolated samplings and water-intake data to extensive lake-wide sampling. Consequently, records for a particular year may represent an average of hundreds, thousands, or only a few determinations. Data on magnesium were available from most of the sources plotted in Figures 31–2 and 31–3, but they are not included because no significant change in concentration could be detected in any of the lakes. For example, magnesium concentrations in Lake Erie averaged 7.6 p.p.m. in 1907 (Dole 1909), 8.0 p.p.m. in 1934 (Mangan, Van Tuyl, and White 1952), and 8.0 p.p.m. in recent years (Bureau of Commercial Fisheries data). Broadly speaking, there has been no significant change in Lake Superior. Other lakes in order of increasing chemical change are Huron, Michigan, Erie, and Ontario.

*Lake Superior* The indicated slight downward trend in total dissolved solids in Lake Superior is not significant and concentrations have remained at approximately 60 p.p.m. throughout the years (Fig. 31–1). Calcium, chloride, and sulfate concentrations also have remained the same since 1886 (Fig. 31–2). The close agreement among analyses of Lake Superior water by various individuals using different methods and techniques is unusual. The slight decrease in the sodium-plus-potassium content of the water probably is not real, because present analytical methods differ substantially from former ones. The uniformity in the chemical analyses here lends confidence to the reliability of the chemical data for the other lakes.

*Lake Huron* The slight increase in total dissolved solids in this lake is probably real, since about 30 percent of the inflow to Lake Huron is from Lake Michigan, where dissolved solids have risen significantly (Fig. 31–1). The sodium-plus-potassium content has remained about the same over the years of record. Some rather low values were reported for these ions during the 1930's, but the 1890–1910 data agree with recent determinations (Fig. 31–2). An increase of 3 p.p.m. in chloride appears to have occurred during the past 30 years. The major source of chlorides within the Lake Huron watershed is in the Saginaw Valley, where considerable quantities of brine are pumped to the surface in the oil fields and for use in the chemical industry. The increased influx of brine during the past 30 years may account for most of the increase in chloride in the lake. Sulfate concentrations have increased 7.5 p.p.m. in the past 54 years.

*Lake Michigan* Total dissolved solids have increased about 20 p.p.m. since 1895 (Fig. 31–1). Calcium has remained constant (Fig. 31–2). The greatest increase in any ion in Lake Michigan has been that of sulfate, which has risen 12 p.p.m. since 1877. The chloride concentrations have risen slowly but steadily by 4 p.p.m. The sodium-plus-potassium content has not changed since 1907 but it exceeds that extant in 1877–1900. The determinations before 1907 may be too low, since they are below those reported for Lakes Huron and Superior. If, however, these early determinations are reasonably accurate, the increase that occurred between 1877 and 1907 may be attributed to population growth in the Chicago area. The population of Chicago exceeded 1 million in 1890 and the Chicago Sanitary Canal, to divert sewage from the lake, was not completed until 1900. Consequently, during these early years considerable amounts of raw sewage entered the lake at Chicago.

*Lake Erie* Total dissolved solids, calcium, chloride, sodium-plus-potassium, and sulfate all increased significantly in Lake Erie during the past 50 years. Total dissolved solids have risen by almost 50 p.p.m. (Fig. 31–1). Increases of approximately 8, 16, 5, and 11 p.p.m. have taken place in the concentrations of calcium, chloride, sodium-plus-potassium, and sulfate, respectively (Fig. 31–2).

*Lake Ontario* The rate of increase in total dissolved solids in Lake Ontario has been the same as in Lake Erie. This rate was similar to that occurring in Lake Michigan prior to the late 1920's but has been higher than in Lake Michigan since about 1930 (Fig. 31-1). Close agreement of chemical data for Lake Ontario for 1854 and 1884 with those for 1970

TABLE 31-3

*Sources of data used in preparing Figs. 31–1, 31–2, and 31–3*

| Source and date | Data |
|---|---|
| Allen (1964) | Major ions, south-central Lake Huron, 1956. |
| Bading (1909) | Chlorides, Lake Michigan, 1909. |
| Barnard and Brewster (1909) | Chlorides, Lake Michigan, 16 Sept 1908, table 29. |
| Bartow and Birdsall (1911) | Major ions, average concentrations in 10 open Lake Michigan samples collected about 1910. |
| Beeton, Johnson, and Smith (1959) | Major ions, Lake Superior, 1953. |
| Birge and Juday (no date) | Total dissolved solids, Lake Erie, 1928–1930; data on analyses for total dissolved solids by L. A. Youtz for Lakes Erie, Huron, and Superior, 1928. |
| Bowles (1909) | Total dissolved solids, Lake Michigan, April 1908. |
| Collins (1910) | Major ions, Lake Michigan, 1885, from J. H. Long; total dissolved solids and chlorides, Chicago, average of weekly values 1897–1900. |
| Clarke (1924) | Major ions, Lake Michigan, Milwaukee, 1877; Kenosha and Racine, Wis., 1911. |
| Dole (1909) | Major ions and total dissolved solids, Lakes Erie, Huron, Michigan, Ontario, and Superior, averages of monthly determinations 1906–07. |
| Eddy (1943) | Major ions and total dissolved solids, Lake Superior, 1934. |
| Erie, Pa., Bur. Water (1956, 1957, 1959) | Major ions and total dissolved solids, Lake Erie, 1956, 1957, 1959. |
| Fish (1960) | Chlorides, Lake Erie, 1929. |
| Hunt (1857) | Major ions in water collected at Pointe des Cascades, Vandreuil, Que., Lake Ontario, 1854. |
| International Joint Commission (1951) | Chlorides and total dissolved solids, open Lake Erie near Detroit River mouth, sampling ranges P-1-W and LC, table N-17; open Lake Huron above Port Huron; open Lake Ontario, sampling locations 4 miles (6.4 km) or more from Niagara River, 1946–48. |
| Jackson (1912) | Chlorides and total dissolved solids, Lake Erie, 1910, tables 69, 70; 1911, tables 71, 72; 1912, table 68. |
| Kramer (1961) | Major ions, except potassium, Lake Erie, 1961. |
| Kramer (1962) | Major cations, western Lake Ontario, 1959. |
| Lake Michigan Water Commission (1909) | Chlorides and total dissolved solids, Lake Michigan, October 1908. |
| Lane (1899) | Sodium chloride, Lake Huron, 12 miles (19.3 km) above Port Huron and Alpena, Mich., 1895; Lake Michigan, 5 miles (8 km) off Milwaukee, Wis., 1895–99; total solids, Chicago, Ill., 1895; major ions and total dissolved solids, open Lake Superior 50 miles (80 km) from Keweenaw Pt., 1886. |
| Lenhardt (1955) | Major ions and total dissolved solids, Lake Michigan, 1954. |
| Leverin (1942) | Major ions and total dissolved solids, Lake Huron, Pt. Edward, Ont., 1934–37; Lake Ontario, average of 6 analyses, Kingston, Ont., 1934–38 and 1940. |
| Leverin (1947) | Major ions and total dissolved solids, Lake Erie, average of 6 determinations, Fort Erie, Ont., 1934–38; Lake Ontario, average of 7 determinations, Toronto, Ont., 1934–38; Lake Superior, average of 5 determinations, Sault Ste. Marie, Ont., 1936–38, one sample taken in open lake midway between Fort William and Sault Ste. Marie, 1942. |
| Lewis (1906) | Chlorides and total dissolved solids, Lake Erie, average of 6 analyses, Erie, Pa., 1901–1903. |
| Mangan, Van Tuyl, and White (1952) | Major ions and total dissolved solids, Lake Erie, 1934, 1945, and 1951. |
| Michigan Water Resources Commission (1954) | Major ions and total dissolved solids, Lake Huron at Alpena, East Tawas, and Harbor Beach, Mich.; Lake Michigan at Muskegon, St. Joseph, and Traverse City, Mich.; Lake Superior at Calumet, Mich. |
| Ohio, State of (1953) | Major ions, Lake Erie, Lorain, O., average values, 1950–52. |
| Reade (1903) | Major ions, Lake Ontario, water sample collected in the St. Lawrence River opposite Montreal, Que., 1884. |
| Thomas (1954) | Major ions and total dissolved solids, Lake Erie, at Chippawa, Ont.; Lake Huron at Goderich and Sarnia, Ont.; Lake Ontario at Gananoque and Port Hope, Ont.; Lake Superior at Sault Ste. Marie, Ont., averages of monthly analyses, 1948–49. |
| U.S. Geological Survey (1960) | Major ions and total dissolved solids, Lake Erie, Niagara River at Buffalo, N.Y.; Lake Ontario, St. Lawrence River at Cape Vincent, N.Y., analyses for August 1957. |
| U.S. Public Health Service (1961) | Chlorides, sulfates, and total dissolved solids, Lake Erie at Buffalo, N.Y.; Lake Huron at Port Huron, Mich.; Lake Michigan at Milwaukee, Wis.; Lake Ontario at Massena, N.Y.; Lake Superior at Duluth, Minn.; average values for Oct 1960–30 Sept 1961. |
| Wright (1955) | Chlorides, western Lake Erie, 1930. |

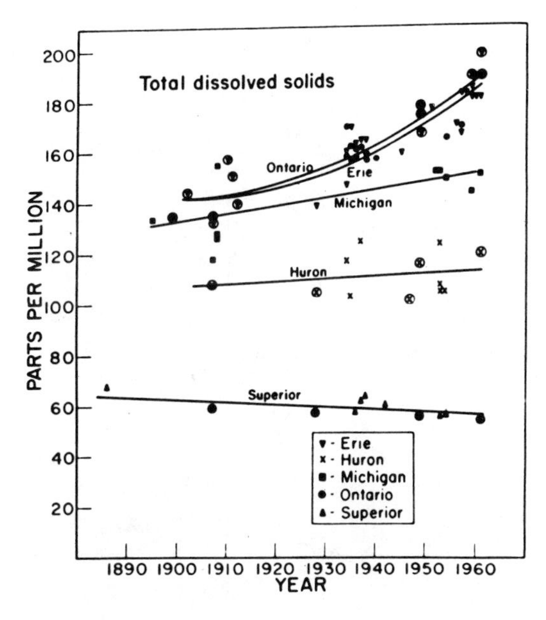


FIGURE 31–1. Concentrations of total dissolved solids in the Great Lakes. Circled points are averages of 12 or more determinations. Data are from sources presented in the bibliography and Table 31–3.

indicates that the chemical characteristics of the water were altered little during this period (Fig. 31–2). Calcium, chloride, and sodium-plus-potassium increased to the same extent as noted in Lake Erie since 1910. Sulfate concentrations increased by 13 p.p.m., which is somewhat higher than in Lake Erie.

*Summary of chemical changes* The extent of change in total dissolved solids in Lakes Erie, Michigan, and Ontario has not been as great as that indicated by Ayers (1962) for Lake Michigan or by Rawson (1951) for Lakes Erie and Ontario; both used the observations of Dole (1909) in 1906–1907 as their base. Dole's estimates of total dissolved solids (and of various ions as well) for these lakes were 9 to 16 p.p.m. lower than those of several other investigators during this period. The 1907 data for Lake Michigan on sulfate, chloride, and especially calcium probably were all low because Dole collected his samples in the Straits of Mackinac where Lake Michigan water enters Lake Huron and the mixing of the water from these two lakes, as well as the occasional inflow of Lake Superior water into this area, could produce low concentrations of ions.

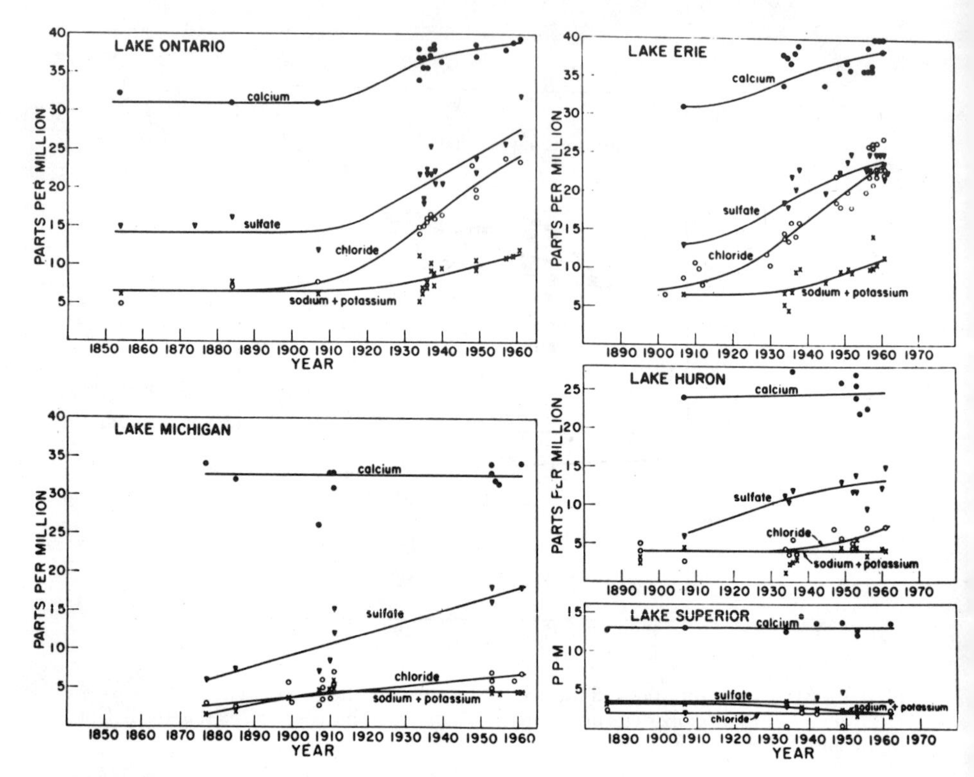


FIGURE 31–2. Changes in the chemical characteristics of Great Lakes waters. Data for Lake Erie, 1958; Lake Huron, 1956; Lake Michigan, 1954, 1955, 1961; Lake Ontario, 1961; Lake Superior, 1952, 1953, 1961, 1962 are from the Ann Arbor Biological Laboratory, U.S. Bureau of Commercial Fisheries. Other data are from sources presented in the bibliography and Table 31–3.

Changes in the chemical characteristics of Lake Ontario have closely paralleled those in Lake Erie (Fig. 31–3). Prior to 1910 the chemical characteristics of the two lakes were similar and conditions in Lake Erie were probably the same as indicated by the 1854 and 1884 analyses of Lake Ontario water. Concentrations of calcium, chloride, sodium-plus-potassium, and sulfate have been somewhat higher in Lake Ontario than in Lake Erie during the past 50 years. The greater concentrations of salts in Lake Ontario probably can be attributed to growth of the Toronto, Hamilton, and Rochester metropolitan areas and the industrial expansion along the upper Niagara River.

Lakes Erie and Ontario are the only lakes in which calcium increased materially (Fig. 31–3). Increases in sulfate have been significant in all of the lakes except Superior. The 11 p.p.m. and 13 p.p.m. increases in Lakes Erie and Ontario have taken place in 30 years, whereas the rise of 12 p.p.m. in Lake Michigan has been more gradual over a period of 84 years. The sulfate change in Lake Huron parallels that in Lake Michigan and may have resulted largely from the inflow of Lake Michigan waters. The degree of change in the chloride content of Lakes Erie and Ontario is similar to that for sulfate, but chloride has not increased as much as sulfate in Lakes Huron and Michigan.

### *Plankton*

Few plankton data are available for Lakes Huron, Ontario, and Superior, especially for earlier years. Some rather extensive plankton data do exist, however, for Lakes Erie and Michigan.

*Lake Michigan* Studies of Lake Michigan phytoplankton by Briggs (1872, Thomas and Chase (1886), Eddy (1927), Ahlstrom (1936), Damann (1945), and Williams (1962) show that the diatom species dominant 90 years ago have maintained their importance. The relative abundance and occurrence of individual species give no evidence of change in the phytoplankton. (Some confusion exists over the identification of certain species and some species are listed by one investigator and not by another.)

The best information of change of plankton abundance probably

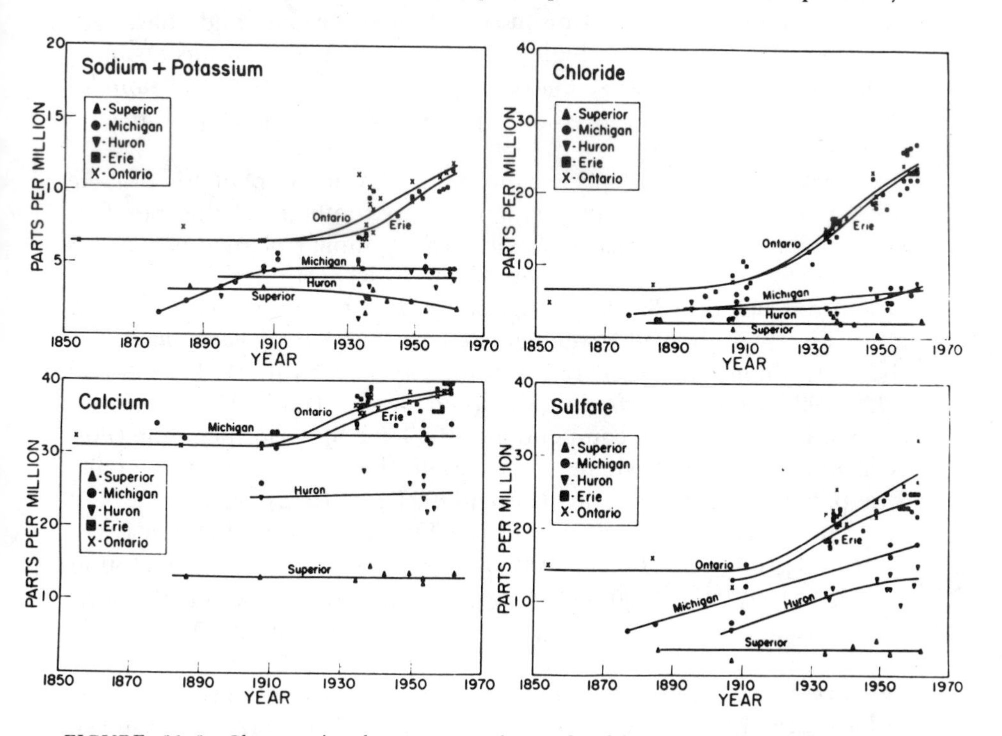


FIGURE 31–3. Changes in the concentrations of calcium, chloride, sodium-plus-potassium, and sulfate in each of the Great Lakes. Sources of data same as for Fig. 31–2.

comes from Damann's (1960) publication on 33 years of plankton data from the Chicago water intake which showed an average increase of 13 organisms per ml per year in the standing crop of the total plankton. Damann's work, and indeed most of the past work on plankton, has been in the extreme southern end of the lake except for the open-lake data of Ahlstrom (1936). Damann (1945, p. 771) pointed out that the data he published "...are an expression of the plankton activity in only a small portion of the southwestern corner of Lake Michigan."

Recent plankton collections during July 1960 to July 1961 from Gary, Indiana, at the southern end of the lake, and Milwaukee, Wisconsin, near the middle of the lake (Williams 1962), show considerable differences between these two localities. The average phytoplankton count at Milwaukee was 975 plankters/ml. This figure is close to the annual average of 952/ml for the Chicago water intake 1926–1942 (Damann 1960). The average phytoplankton count at Gary was 1,914 plankters/ml, well above the annual average of 1,222/ml reported at Chicago for 1943–1958 (Damann 1960). Local differences in the relative importance of the major diatom species at Gary and Milwaukee also are apparent. Sampling by the Bureau of Commercial Fisheries, in 1960, yielded further evidence of wide variability. The relative abundance of the plankton species was different from that at the Gary and Milwaukee water intakes, and a much higher abundance of plankton was obtained with nets towed vertically. Plankton counts in the open-lake samples ranged from about 450 to 12,000 plankters/ml and averaged around 4,500/ml.

Two changes in the species composition of the zooplankton of Lake Michigan may be of some consequence. Apparently the cladoceran, *Bosmina longirostris*, has replaced *B. coregoni.* A similar change was noted as evidence of eutrophication in Lake Zürich (Minder 1938). Wells (1960) did not find *B. coregoni*, although *B. longirostris* was present in all of his 1954 and 1955 samples. *Bosmina coregoni* (*B. longispina*) was the most abundant cladoceran in samples collected in 1887–1888 and 1926–1927, and *B. longirostris* was listed as rare (Eddy 1927). *Bosmina coregoni* is an important component of the Lake Superior plankton (Putnam and Olson 1961), whereas *B. longirostris* has been important in Lake Erie plankton for many years (Fish 1960; Davis 1962).

The other possible change in the Lake Michigan plankton is the increased prominence of *Diaptomus oregonensis.* This species was not found by Eddy (1927), but Wells (1960) reported it to be present on all collection dates and it was a dominant diaptomid in the fall of 1955.

*Lake Erie* Some significant changes have been observed in the plankton of Lake Erie. Evidently copepods and especially cladocerans have shown a marked increase in abundance since 1939 (Bradshaw 1964). Bradshaw attached some importance to the "recent" occurrence of the cladocerans *Eurycerus lamellatus*, *Chydorus sphaericus*, and *Ilyocryptus* sp., since they had not been found by Chandler (1940) in 1938–1939. *Eurycerus lamellatus*, *Chydorus sphaericus*, and two species of *Ilyocryptus* were present, however, in 1929 (Fish 1960). A copepod, *Diaptomus siciloides*, which was reported as "incidental" in Lake Erie plankton in 1929 and 1930 (Wright 1955), is now one of the two most abundant diaptomids (Davis 1962). Marsh's (unpublished manuscript) account of plankton in 1929 and 1930 included the statement, "The occurrence of *Diaptomus siciloides* in Lake Erie is a matter of decided interest, as it has never before been found in any of the Great Lakes." Later he continued, "*D. siciloides* may be considered an accidental intruder in the lake plankton . . ."

## Changes in Lake Erie

Several important changes in Lake Erie have not been detected in the other lakes; all indicate an accelerated rate of eutrophication.

### *Fish populations*

The abundance of several commercially important fishes in Lake Erie has changed markedly during the past 40 years.[1] The fish populations of all of the lakes have changed but most changes, except in Lake Erie, have been the direct or indirect consequence of the buildup in sea lamprey populations (Smith 1964). The sea lamprey has not been important in Lake Erie, where few of the tributaries offer suitable spawning conditions.

The lake herring or cisco contributed around 20 million pounds (9 million kg) annually and as much as 48.8 million pounds (22.1 million kg) to the commercial catch (U.S. and Canada) prior to 1925. In 1925, the production declined to 5.8 million pounds (2.6 million kg) and continued to decrease. The take has amounted to a fraction of that since the early 1930's, except for landings of more than 2 million pounds (about 1 million kg) in 1936 and 1937 and a production of 16.2 million pounds (7.4 million kg) in 1946. The total production was only 7,000 pounds (3,200 kg) in 1962.

The whitefish fishery has been at an all-time low since 1948. The 1962 catch was 13,000 pounds (6,000 kg), whereas production had been 2 million pounds (1 million kg) or more for many years.

The sauger was contributing 1 million pounds (500,000 kg) or more

---

Statistical data cited are from Baldwin and Saalfeld (1962).

to the commercial production prior to 1946. The catch has not reached 0.5 million pounds (0.22 million kg) since 1945 and has declined progressively since 1953. Production has been only 1 to 4 thousand pounds (450–1,800 kg) in recent years.

The walleye production increased during the 1940's and 1950's to reach 15.4 million pounds (7 million kg) in 1956. Production has decreased to less than 1 million pounds (450,000 kg).

The commercial catch of blue pike has dropped disastrously. The production fluctuated around an average of about 15 million pounds (6.8 million kg) for many years. The landings dropped to 1.4 million pounds (640,000 kg) in 1958, to 79,000 pounds (36,000 kg) in 1959, and to only 1,000 pounds (450 kg) in 1962. Of the few blue pike caught in 1963, most were more than 10 years old.

The total production of all species in Lake Erie continues to be around 50 million pounds (22.7 million kg), but only because more freshwater drum (sheepshead), carp, yellow perch, and smelt are being caught than in the past. The major factor in the decline in the commercial catch of the more desirable species has been their failure to reproduce.

### Bottom fauna

The changes in the species composition of the bottom fauna of the area west of the islands in Lake Erie have been sweeping. Carr and Hiltunen (unpublished manuscript)[2] have shown that few of the formerly abundant mayfly nymphs (*Hexagenia*) now inhabit this area and that tubificids are far more abundant now than 30 years ago (Fig. 31–4). Midge larvae and tubificids have increased and mayflies have decreased also among the islands and in the western part of the central basin (Beeton 1961).

### Dissolved oxygen

Synoptic surveys of Lake Erie in 1959 and 1960 have demonstrated that low dissolved oxygen concentrations (3 p.p.m. or less) appear in about 70 percent of the hypolimnetic waters of the central basin during late summer (Beeton 1963). Scattered observations of some relatively low dissolved oxygen concentrations have been made during the past 33 years. The information we have indicates that the severity of depletion is more frequent and greater now than in the past and probably affects a more extensive area (Carr 1962).

[2]Carr, J. F., and J. K. Hiltunen. Changes in the bottom fauna of Lake Erie, west of the islands, 1930–1961. Paper presented at the 11th annual meeting Midwest Benthological Society, Murfresboro, Tenn., 18 April 1963.

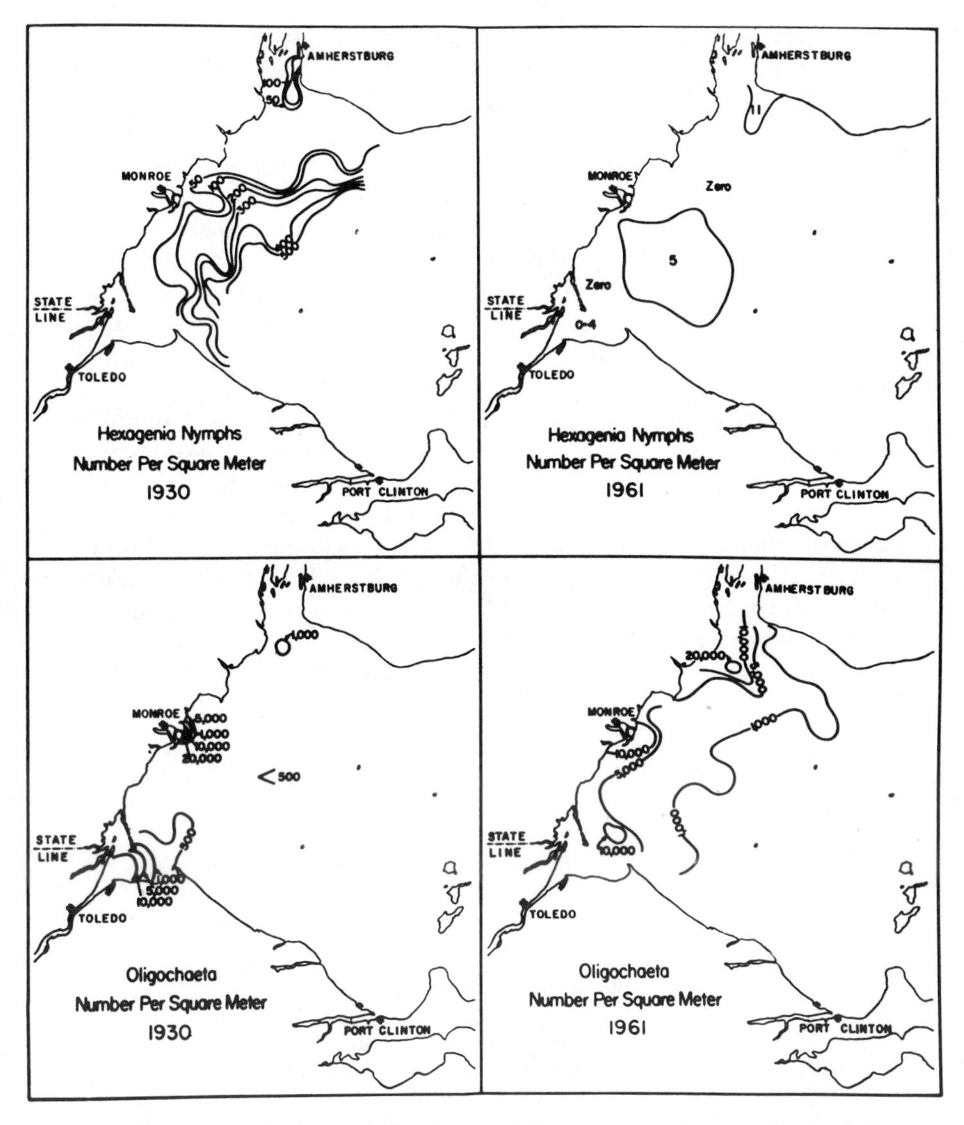


FIGURE 31–4. Distribution and abundance of *Hexagenia* nymphs and oligochaetes in western Lake Erie, 1930 and 1961.

## Conclusion

The chemical content of the water in all of the Great Lakes except Lake Superior has changed in some measure. The biota also has changed in Lake Michigan and especially Lake Erie. These changes, remarkable for such large lakes, are those characteristic of eutrophication in smaller lakes and have come about over the relatively short time of 50 to 60 years. Man's activities clearly have accelerated the rate of eutrophication. This rate has been greatest in Lakes Erie, Ontario, and Michigan and these lakes have had the largest population growth within their drainage areas. An indication of the growth of population comes from census data for the northeast central states; the population there increased from 4.5 million to 16 million between 1850 and 1900, and by 1960 the population was 36.3 million. The rate of population growth increased sharply after 1910. The substantial increases in the chemical content of the waters of Lakes Erie and Ontario also have appeared since 1910. The increases have been greatest for chloride and sulfate, both of which are conspicuous in domestic and industrial wastes, whereas magnesium concentrations have not changed measurably. Most of the magnesium comes from the limestones in the Lake Michigan basin; the stability of magnesium concentrations, therefore, indicates no appreciable change in the erosion of these deposits. The population along Lake Superior always has been sparse. The population along Lake Huron has been far less than on Lakes Michigan, Erie, and Ontario. Most changes in the open-lake waters of Lake Huron have resulted from the inflow of Lake Michigan water. Undoubtedly, Saginaw Bay, Lake Huron, has been changed appreciably by the extensive growth of industry and agriculture within the Saginaw River valley during the past 50 years. High chloride concentrations have been measured in the bay (Adams 1937); no attempt has been made, however, to assess the extent of change.

## References


Adams, M. P. 1937. *Saginaw Valley report*. Mich. Stream Control Commission. 156 pp.

Ahlstrom, E. H. 1936. The deep-water plankton of Lake Michigan, exclusive of the Crustacea. *Trans. Am. Microscop. Soc.* 55:286–299.

Allen, H. E. 1964. Chemical characteristics of south-central Lake Huron. *Great Lakes Res. Div., Inst. Sci. and Tech., Univ. Mich., Publ.* No. 11, pp. 45–53.

Ayers, J. C. 1962. Great Lakes waters, their circulation and physical and chemical characteristics. *Am. Assoc. Advan. Sci., Publ.* No. 71, pp. 71–89.

Bading, G. A. 1909. *Water conditions at Milwaukee*. Lake Michigan Water Comm., Rept. No. 1, pp. 36–39.

Baldwin, N. S., and R. W. Saalfeld. 1962. Commercial fish production in the Great Lakes 1867–1960. Great Lakes Fish. Comm., Tech. Rept. No. 3. 166 pp.

Barnhard, H. E., and J. H. Brewster. 1909. *The sanitary condition of the southern end of Lake Michigan, bordering Lake County, Indiana*. Lake Michigan Water Comm., Rept. No. 1, pp. 193–266.

Bartow, E., and L. I. Birdsall. 1911. *Composition and treatment of Lake Michigan water*. Lake Michigan Water Comm., Rept. No. 2, pp. 69–86.

Beeton, A. M. 1961. Environmental changes in Lake Erie. *Trans. Am. Fisheries Soc.*, 90:153–159.

———. 1963. *Limnological survey of Lake Erie 1959 and 1960*. Great Lakes Fish. Comm., Tech. Rept. No. 6. 32 pp.

———. J. H. Johnson, and S. H. Smith. 1959. Lake Superior limnological data. *U.S. Fish Wildlife Serv., Spec. Sci. Rept. Fisheries* No. 297. 177 pp.

Berg, K., K. Andersen, T. Christensen, F. Ebert, E. Fjerdingstad, C. Holmquist, K. Korsgaard, G. Lange, J. M. Lyshede, H. Mathiesen, G. Nygaard, S. Olsen, C. V. Otterstrøem, A. Skadhsuge, E. Steemann Nielsen. 1958. Investigations on Fure Lake 1950–54. Limnological studies on cultural influences. *Folia Limnol. Scandinavica*, 10(1958). 189 pp.

Birge, E. A., and C. Juday. No date. The organic content of the water of Lake Erie. Supplemental data to "A limnological survey of western Lake Erie with special reference to pollution," by Stillman Wright. Ohio Div. Wildlife. Unpublished manuscript. 281 pp.

Bowles, J. T-B. 1909. *Investigation of typhoid fever epidemic at Sheboygan, Wisconsin*. Lake Michigan Water Comm., Rept. No. 1, pp. 90–95.

Bradshaw, A. S. 1964. The crustacean zooplankton picture: Lake Erie 1939–49–59; Cayuga 1910–51–61. *Verhandl. Intern. Ver. Limnol.*, 15:700–708.

Briggs, S. A. 1872. The Diatomaceae of Lake Michigan. The Lens, 1:41–44.

Brudin, L. 1958. The bottom faunistical lake type system. *Verhandl. Intern. Ver. Limnol.*, 13:288–297.

Carr, J. F. 1962. Dissolved oxygen in Lake Erie, past and present. *Great Lakes Res. Div., Inst. Sci. and Tech., Univ. Mich., Publ.* No. 9, pp. 1–14.

Chandler, D. C. 1940. Limnological studies of western Lake Erie. I. Plankton and certain physical-chemical data of the Bass Islands Region, from September, 1938, to November, 1939. *Ohio J. Sci.*, 40:291–336.

Collins, W. D. 1910. The quality of the surface waters of Illinois. *U.S. Geol. Surv., Water Supply Papers*, 239. 94 pp.

Clarke, F. W. 1924. The composition of the river and lake waters of the United States. *U.S. Geol. Surv., Profess. Papers*, No. 135. 199 pp.

Damann, K. E. 1945. Plankton studies of Lake Michigan. I. Seventeen years of plankton data collected at Chicago, Illinois. *Am. Midland Naturalist*, 34:769–796.

———. 1960. Plankton studies of Lake Michigan. II. Thirty-three years of continuous plankton and coliform bacteria data collected from Lake Michigan at Chicago, Illinois. *Trans. Am. Microscop. Soc.*, 79:397–404.

Davis, C. C. 1962. The plankton of the Cleveland Harbor area of Lake Erie in 1956–1957. *Ecol. Monographs*, 32:209–247.

Dole, R. B. 1909. The quality of surface waters in the United States. Part I. Analyses of waters east of the one hundredth meridian. *U.S. eol. Surv., Water Supply Papers,* 236. 123 pp.

Dunn, D. R. 1954. Notes on the bottom fauna of twelve Danish lakes. *Vidensk. Medd. Dansk. Naturhist. Foren.*, 116:251–268.

Eddy, S. 1927. The plankton of Lake Michigan. *Illinois Nat. Hist. Surv. Bull.*, 17:203–232.

———. 1943. Limnological notes on Lake Superior. *Proc. Minn. Acad. Sci.*, 11:34–39.

Edmondson, W. T., G. C. Anderson, and D. R. Peterson. 1956. Artificial eutrophication of Lake Washington. *Limnol. Oceanog.*, 1:47–53.

Elster, H.-J. 1958. Das limnologische Seetypensystem, Rückblick und Ausblick. Verhandl. *Intern. Ver. Limnol*, 13:101–120.

Erie, Pennsylvania, Bureau of Water. 1956. Ninetieth annual report, 1956. 63 pp.

———. 1957. Ninety-first annual report, 1957, 63 pp.

———. 1959. Ninety-third annual report, 1959. 56 pp.

Fish, C. J. 1960. Limnological survey of eastern and central Lake Erie, 1928–1929. *U.S. Fish Wildlife Serv. Spec. Sci. Rept. Fisheries*, No. 334. 198 pp.

Hasler, A. D. 1947. Eutrophication of lakes by domestic drainage. *Ecology*, 28:383–395.

Hunt, T. S. 1857. The chemical composition of the waters of the St. Lawrence and Ottawa Rivers. *Phil. Mag.* Ser. 4, 13:239–245.

International Joint Commission. 1951. *Report of the International Joint Commission United States and Canada on pollution of boundary waters*. Washington and Ottawa. 312 pp.

Jackson, D. D. 1912. *Report on the sanitary conditions of the Cleveland water supply*. Cleveland. 148 pp.

Kramer, J. R. 1961. Chemistry of Lake Erie. Great Lakes Res. Div., *Inst. Sci. and Tech., Univ. Mich., Publ.* No. 7, pp. 27–56.

———. 1962. Chemistry of western Lake Ontario. Great Lakes Res. Div., *Inst. Sci. and Tech., Univ. Mich., Publ.* No. 9, pp. 21–28.

Lake Michigan Water Commission. 1909. *Comparative analysis of samples of water from Lake Michigan*. Rept. No. 1, pp. 103–105.

Lane, A. C. 1899. Lower Michigan waters: a study into the connection between their chemical composition and mode of occurrence. *U.S. Geol. Surv. Water-supply Irrigation Papers*, 31. 97 pp.

Lenhardt, L. G. 1955. *Water quality and water usage of the Great Lakes public water supplies*. The Great Lakes and Michigan. Great Lakes Res. Inst., Univ. Mich., pp. 13–15.

Leverin, H. A. 1942. *Industrial waters of Canada*. Can. Dept. Mines Resources, Mines Geol. Branch, Rept. 807. 112 pp.

———. 1947. *Industrial waters of Canada*. Can. Dept. Mines Resources, Mines Geol. Branch, Rept. 819. 109 pp.

Lewis, S. J. 1906. Quality of water in the upper Ohio River basin and at Erie, Pa. *U.S. Geol. Surv. Water-supply Irrigation Papers*, 161. 114 pp.

Mangan, J. W., D. W. Van Tuyl, and W. F. White, Jr. 1952. Water resources of the Lake Erie shore region in Pennsylvania. *U.S. Geol. Surv. Circ.* 174. 36 pp.

Marsh, C. D. No date. The Crustacea of the plankton of western Lake Erie. Supplemental data to "A limnological survey of western Lake Erie with special reference to pollution," by Stillman Wright. Ohio Div. Wildlife. Unpublished manuscript. 31 pp.

Merna, J. 1960. A benthological investigation of Lake Michigan. M.S. Thesis, Michigan State Univ. 74 pp.

Michigan Water Resources Commission. 1954. Great Lakes water temperatures. Unpublished manuscript. 50 pp.

Minder, Leo. 1918. Zur Hydrophysik des Zürich u. Walensees, nebst Beitrag zur Hydrochemie u. Hydrobakteriologie des Zürichsees. *Arch. Hydrobiol.*, 12:122–194.

———. 1938. Der Zürichsee als Eutrophierungsphänomen. Summerische Ergebnisse aus fünfzig Jahren Zürichseeforschung. *Geol. Meere Binnengewä*sser, 2:284–299.

———. 1943. Neuere Untersuchungen über den Sauerstoffgehalt and die Eutrophie des Zürichsees. *Arch. Hydrobiol*, 40:279–301.

Ohio, State of. 1953. *Lake Erie pollution survey, supplement*. Ohio Div. Water, Final Rept., Columbus, Ohio. 39 tables, 125 pp.

Putnam, H. E., and T. A. Olson. 1961. *Studies on the productivity and plankton of Lake Superior*. School Public Health, Univ. Minn., Rept. No. 5. 58 pp.

Rawson, D. S. 1951. The total mineral content of lake waters. *Ecology*, 32:669–672.

———. 1955. Morphometry as a dominant factor in the productivity of large lakes. *Verhandl. Intern. Ver. Limnol.*, 12:164–174.

———. 1956. Algal indicators of trophic lake types. *Limnol. Oceanog.*, 1:18–25.

———. 1960. A limnological comparison of twelve large lakes in northern Saskatchewan. *Limnol. Oceanog.*, 5:195–211.

Reade, T. M. 1903. *The evolution of earth structure*. Longmans, Green, New York, 342 pp.

Rodgers, G. K. 1962. *Lake Ontario data report*. Great Lakes Inst., Univ. Toronto, Prelim. Rept. No. 7. 102 pp.

Smith, S. H. 1964. Status of the deepwater cisco population of Lake Michigan. *Trans. Am. Fisheries Soc.*, 93:209–230.

Teiling, E. 1955. Some mesotrophic phytoplankton indicators. *Verhandl. Intern. Ver. Limnol.*, 12:212–215.

Teter, H. E. 1960. The bottom fauna of Lake Huron. *Trans. Am. Fisheries Soc.*, 89:193–197.

Thienemann, A. 1928. Der Sauerstoff im eutrophen und oligotrophen See. Die Binnengewässer, Band 4, Schweizerbart, Stuttgart, 75 pp.

Thomas, F. J. F. 1954. *Industrial water resources of Canada*. Upper St. Lawrence River-central Great Lakes drainage basin in Canada. Can. Dept. Mines Tech. Surv., Water Surv. Rept. No. 3, Mines Branch Rept. 837. 212 pp.

Thomas, B. W., and H. H. Chase. 1887. Diatomaceae of Lake Michigan as collected during the last sixteen years from the water supply of the city of Chicago. *Notarisia, Commetarium Phycologium*, 2:328–330.

U.S. Geological Survey. 1960. Quality of the surface waters of the United States. *U.S. Geol. Surv., Water Supply Papers*, 1520. 641 pp.

U.S. Public Health Service. 1961. National water quality network. Annual compilation of data October 1, 1960–September 30, 1961. *U.S. Public Health Serv. Publ.* 663. 545 pp.

Welch, P. S. 1952. *Limnology*. McGraw-Hill, New York, 538 pp.

Wells, LaRue. 1960. Seasonal abundance and vertical movements of planktonic Crustacea in Lake Michigan. *U.S. Fish Wildlife Serv., Fishery Bull.*, 60(172):343–369.

Williams, L. G. 1962. Plankton population dynamics. *U.S. Public Health Serv. Publ.*, 663. 90 pp.

Wright, Stillman. 1955. Limnological survey of western Lake Erie. *U.S. Fish Wildlife Serv., Spec. Sci. Rept. Fisheries*, 139. 341 pp.

# 32

# Nitrogen, phosphorus, and eutrophication in the coastal marine environment

*John H. Ryther*
*William M. Dunston*

The photosynthetic production of organic matter by unicellular algae (phytoplankton) in the surface layers of the sea is accompanied by, is indeed made possible by, the assimilation of inorganic nutrients from the surrounding water. Most of these substances are present at concentrations greatly in excess of the plants' needs, but some, like nitrogen and phosphorus, occur at no more than micromolar levels and may be utilized almost to the point of exhaustion by the algae. It is, in fact, the availability of these nutrients that most frequently controls and limits the rate of organic production in the sea.

Harvey (*1*) was among the first to point out that phytoplankton growth caused the simultaneous depletion of both nitrate and phosphate from the ambient seawater. Much has since been written about the interesting coincidence that these elements are present in seawater in very nearly the same proportions as they occur in the plankton (*2–4*). For example, Redfield (*3*) reported atomic ratios of available nitrogen to phosphorus of 15 : 1 in seawater, depletion of nitrogen and phosphorus in the ratio of 15 : 1 during phytoplankton growth, and ratios of 16 : 1

---

Reprinted with permission from *Science*, 171:1008–1013(12 March 1971).  The authors are affiliated with Woods Hole Oceanographic Institution, Woods Hole, Massachusetts 02543.

for laboratory analyses of phytoplankton. This relationship may have resulted from adaptation of the organisms to the environment in which they live, but Redfield suggested a mechanism, the microbial fixation of elementary nitrogen, which could regulate the level of fixed nitrogen in the sea relative to phosphorus to the same ratio as these elements occur in the plankton. In other words, any deficiency of nitrogen could be made up by nitrogen fixation.

Such a process could, in times past, have adjusted the oceanic ratio of nitrogen to phosphorus to its present value, and it may be important in regulating the level or balance of nutrients in the ocean as a whole and over geological time. It is certainly not effective locally or in the short run. As analytical methods have improved and as the subject has been studied more intensively, it has become increasingly clear that the concept of a fixed nitrogen to phosphorus ratio of approximately 15 : 1, either in the plankton or in the water in which it has grown, has little if any validity.

As early as 1949 Ketchum and Redfield (*5*) showed that deficiencies of either element in culture mediums may drastically alter their ratios in the algae. They reported (*5*) nitrogen to phosphorus ratios by atoms in cultures of *Chlorella pyrenoidosa* of 5.6 : 1 for normal cells, 30.9 : 1 for phosphorus-deficient cells, and 2.9 : 1 for nitrogen-deficient cells. A number of subsequent studies of both algal cultures (*6*, *7*) and oceanic particulate matter (*8*, *9*) have reported highly variable ratios of nitrogen to phosphorus. These ratios are somewhat difficult to interpret in oceanic particulate matter, since living algae may comprise a very small fraction of the total particulate organic matter collected by the usual sampling methods, and the origin and nature of the remaining material are largely unknown. On the other hand, the chemical composition of algae grown in the usual culture mediums may differ significantly from that of naturally occurring organisms. Despite these uncertainties, the following generalizations may be made: (i) ratios of nitrogen to phosphorus from less than 3 : 1 to over 30 : 1 (by atoms) may occur in unicellular marine algae; (ii) the ratio varies according to the kind of algae grown and the availability of both nutrients; and (iii) although there is no indication of any "normal" or "optimum" nitrogen to phosphorus ratio in algae, values between 5 : 1 and 15 : 1 are most commonly encountered and an average ratio of 10 : 1 is therefore a reasonable working value.

In seawater, a 15 : 1 atomic ratio may be typical of the ocean as a whole. But since 98 percent of its volume lies below the depth of photosynthesis and plant growth, such mean values have little relevance to the present discussion. If one considers only the remaining 2 percent of the ocean's volume, the so-called euphotic layer, high ratios approaching 15 : 1 occur only at the few times and places where relatively deep water is mixed or upwelled into the euphotic layer (*9*). Over the greater part of the sea surface, the two elements appear to bear no constancy in their interrelationship (*5*, *9–12*).

Detailed examination of the nutrient data from the sea surface reveals that, as the two elements are utilized, nitrogen compounds become depleted more rapidly and more completely than does phosphate. This is particularly true when only nitrate-nitrogen is considered. Both Vaccaro (*9*) and Thomas (*13*) have pointed out, however, that ammonia may often be quantitatively a more important nitrogen source than is nitrate in surface ocean waters, particularly when nitrogen levels in general become reduced through plant growth. But even when all known forms of available nitrogen are considered together, they are often found to be reduced to levels that are undetectable in the euphotic layer. In this event, almost invariably a significant amount of phosphate remains in solution. There is, in short, an excess of phosphate, small but persistent and apparently ubiquitous, in the surface water of the ocean, relative to the amount of nitrogen available to phytoplankton nutrition. This is true in both the Atlantic (*9*, *10*, *12*) and Pacific oceans (*11*, *13*, *14*). Thus, the ratio of nitrogen to phosphorus in surface seawater may range from 15 : 1, where subeuphotic water has recently been mixed or upwelled to the surface, to essentially zero when all detectable nitrogen has been assimilated. Since most of the surface waters of the ocean are nutrient deficient most of the time, nitrogen to phosphorus ratios appreciably less than 15 : 1 are the common rule.

A puzzling question remains to be answered. If the ultimate source of nutrients is deep ocean water containing nitrate and phosphate at an atomic ratio of 15 : 1 and if the average phytoplankton cell contains these elements at a ratio of about 10 : 1, why is it that nitrogen compounds are exhausted first from the water and that a surplus of phosphate is left behind? How can nitrogen rather than phosphorus be the limiting factor? Before turning to this question, we will present two examples of nitrogen as a limiting factor to phytoplankton growth.

## Long Island Bays

Great South Bay and Moriches Bay are contiguous and connected embayments on the south shore of Long Island, New York, formed by the barrier beach that extends along much of the East Coast of the United States. They are shallow, averaging 1 to 2 m in depth, have a hard sandy bottom, and traditionally have supported a productive fishery of oysters and hard clams. Introduction and growth of the Long Island duckling industry, centered along the tributary streams of Moriches Bay, resulted in the organic pollution of the two bays and the subsequent development of dense algal blooms in the bay waters, to the detriment of the shellfisheries.

TABLE 32–1

*Regeneration of nitrogen and phosphorus accompanying the decomposition of mixed plankton [after Vaccaro (18)]. Excess phosphorus (last column) was calculated on the assumption that all nitrogen was assimilated as produced and that phosphorus was assimilated at a nitrogen to phosphorus ratio of 10 : 1*

| Days | $NH_3$ + $NO_2$ + $NO_3$ (μg atoms N/ liter) | $PO_4$ (μg atoms P/ liter) | N : P (by atoms) | Excess P (μg atoms P/ liter) |
|---|---|---|---|---|
| 0 | 0.00 | 0.80 | 0.0 | 0.80 |
| 7 | 0.86 | 0.79 | 1.1 | 0.70 |
| 17 | 2.81 | 0.84 | 3.3 | 0.56 |
| 30 | 3.08 | 1.05 | 2.8 | 0.74 |
| 48 | 3.86 | 0.98 | 3.9 | 0.59 |
| 87 | 4.14 | 1.04 | 4.1 | 0.63 |

As a result of studies by the Woods Hole Oceanographic Institution during the period 1950–55, the ecology of the region and the etiology of its plankton blooms were described in some detail (*15*). The situation has since changed, but certain of the unpublished results of the study are especially pertinent to this discussion and will be reviewed here.

During the period of dense phytoplankton blooms, the peak in the abundance of phytoplankton occurred in Moriches Bay in the region nearest the tributaries, where most of the duck farms were located. The algal populations decreased on either side of this peak in a manner that suggested dilution from tidal exchange via Great South Bay to the west and Shinnecock Bay to the east (Fig. 32–1). Further study suggested that growth of the phytoplankton was actually confined to the tributaries themselves and that the algae in the bays represented a nongrowing population that was able to persist for long periods of time, during which they became distributed in much the same way that a conservative oceanographic property (for instance, freshwater) would behave.

Roughly coincident with the distribution of the phytoplankton was that of phosphate, which reached a maximum concentration of 7.0 μmole per liter in eastern Moriches Bay and fell to levels of about 0.25 μmole per liter at the eastern and western ends of the region (Fig. 32–1). Phosphate, in fact, was used throughout the study as the most convenient and diagnostic index of pollution from the duck farms.

Analyses were also made for nitrogen compounds, including nitrate, nitrite, ammonia, and uric acid (uric acid is the nitrogenous excretory product of ducks). Except in the tributaries that were in direct receipt of the effluent from the duck farms, no trace of nitrogen in any of

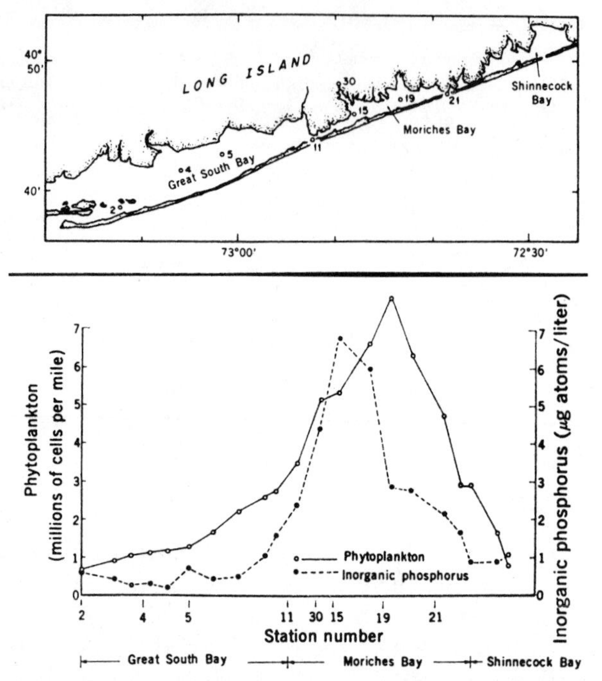


FIGURE 32–1. The distribution of phytoplankton and inorganic phosphorus in Great South Bay, Moriches Bay, and Shinnecock Bay, Long Island, in the summer of 1952. Station numbers on the map (above) correspond to station numbers on the abscissa of the figure (right).

the above forms was found throughout the region studied. It was tentatively concluded that growth of the phytoplankton was nitrogen-limited and that the algae quickly assimilated nitrogen in whatever form it left the duck farms, exhausting the element from the water well up in the tributaries before it could reach the bay.

To confirm this theory, water samples were collected from a series of stations (Nos. 2, 4, 5, 11, 15, 30, 19, and 21) in Great South Bay and Moriches Bay and in the Forge River, one of the tributaries of Moriches Bay on which several duck farms were located. These station locations are indicated by number in Fig. 32–1. The water samples were Millipore-filtered, and each was then separated into three 50-ml portions. The first of these served as a control while the other two received separately $NH_4Cl$ and $Na_2HPO_4 \cdot 12H_2O$ at concentrations of 100 and 10 μmole per liter, respectively. All flasks received an inoculum of *Nannochloris atomus*, the small green alga that was the dominant species in the blooms. The cultures were then incubated for 1 week at 20 C and approximately 11,000 lu/m² of illumination, after which the cells were counted (Fig. 32–2).

The algae in the unenriched controls increased in number by roughly two- to fourfold, the best growth occurring in the water collected from within (station 30) or near (stations 11 and 15) the Forge River. No growth occurred in any of the samples enriched with phosphate, with the exception of that taken from the Forge River (station 30), in which the cell count increased about threefold. In fact, the addition of phosphate seemed to inhibit the growth of the algae relative to that observed in the unenriched controls. In contrast, all the water samples to which ammonia-nitrogen were added supported a heavy growth of *Nannochloris*, resulting in cell counts an order of magnitude greater than were attained in the control cultures. About twice as many cells were

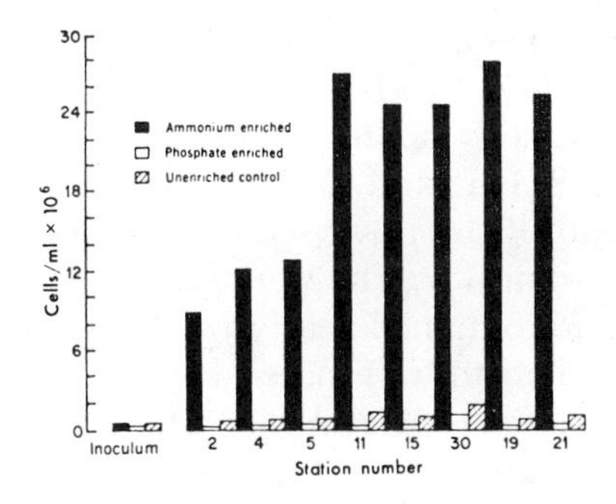


FIGURE 32–2. *Nannochloris atomus* in unenriched, ammonium-enriched, and phosphate-enriched water collected from Great South Bay and Moriches Bay at the station locations indicated by number in Fig. 32–1

produced in the samples from Moriches Bay as were produced in samples from Great South Bay, which suggests that some other nutrient became limiting in the latter series. One might surmise, from the distribution of phosphate in the two bays (Fig. 32–1), that phosphorus was the secondary limiting factor in Great South Bay, but this possibility was not investigated. There can be little doubt, however, that nitrogen, not phosphorus, was the primary limiting factor to algal growth throughout the region.

## New York Bight and the Eastern Seaboard

In September 1969, an oceanographic cruise (R.V. *Atlantis II*, cruise 52) was undertaken along the continental shelf of the eastern United States between Cape Cod and Cape Hatteras. The primary objective of the cruise was to study the effects of pollution of various kinds from the population centers of the East Coast upon the productivity and cycles of organic matter in the contiguous coastal waters. To obtain input data, samples were collected inside the New York bight, at the locations where sewage sludge and dredging spoils from New York City are routinely dumped, as well as from up the Raritan and Hudson rivers.

Of the 52 stations occupied during the cruise, 16 will be discussed here. These stations constitute three sections originating in New York Harbor, one extending eastward along the continental shelf south of Long Island and the New England coast, one extending southeasterly along the axis of Hudson Canyon and terminating at the edge of Gulf Stream, and one running southerly along the coast of New Jersey to the mouth of Delaware Bay (Fig. 32–3A). The nearshore nontidal currents of the region are predominantly to the south (Fig. 32–3B), so that pollution emanating from New York Harbor would be expected to spread in that direction roughly along the axis of section 3. Section 1 and the inshore stations of section 2 might be considered as typical of unpolluted or moderately polluted coastal waters, whereas the two distal stations of section 2 (1513 and 1547) should be oceanic in character.

At the inshore end of the three sections, station 1507 was located in heavily polluted Raritan Bay. The water at that location was a bright apple-green in color and contained nearly a pure culture of a small green alga identified in an earlier study of the area by McCarthy (*16*) as *Didymocystis* sp. Karshikov. Stations 1505 and 1504 are the respective dumping sites for dredging spoils and sewage sludge for the city of New York.

The distribution of total particulate organic carbon at the surface is shown for the three sections in Fig. 32–4. The measurements were made by the method of Menzel and Vaccaro (*17*). It is clear from the two sets of measurements that living algal cells made up only a small fraction of the particulate organic content of the water. On the basis of either criteria, one can see that the high content of particulate organic matter characteristic of the New York bight extends seaward for less than 80 km to the east and southeast, whereas evidence of pollution occurs at least 240 km to the south (section 3), along the New Jersey coast to Delaware Bay, presumably the direction of flow of the water flushed out of the bight.

The distribution of inorganic nitrogen, as combined nitrate, nitrite, and ammonia, in the surface water of the three sections is shown in Fig. 32–5. If one considers that the two terminal stations of section 2 (1513 and 1547) are oceanic in character, it is clear that the level of inorganic nitrogen immediately outside New York Harbor, even at the two dumping sites, is as low as, if not lower than, those found in the open sea and that such low values are, in fact, characteristic of the entire continental shelf. Phosphate (Fig. 32–6), however, presents a quite different picture. Its surface concentrations throughout the shelf area and particularly in the water south of the New York bight are appre-

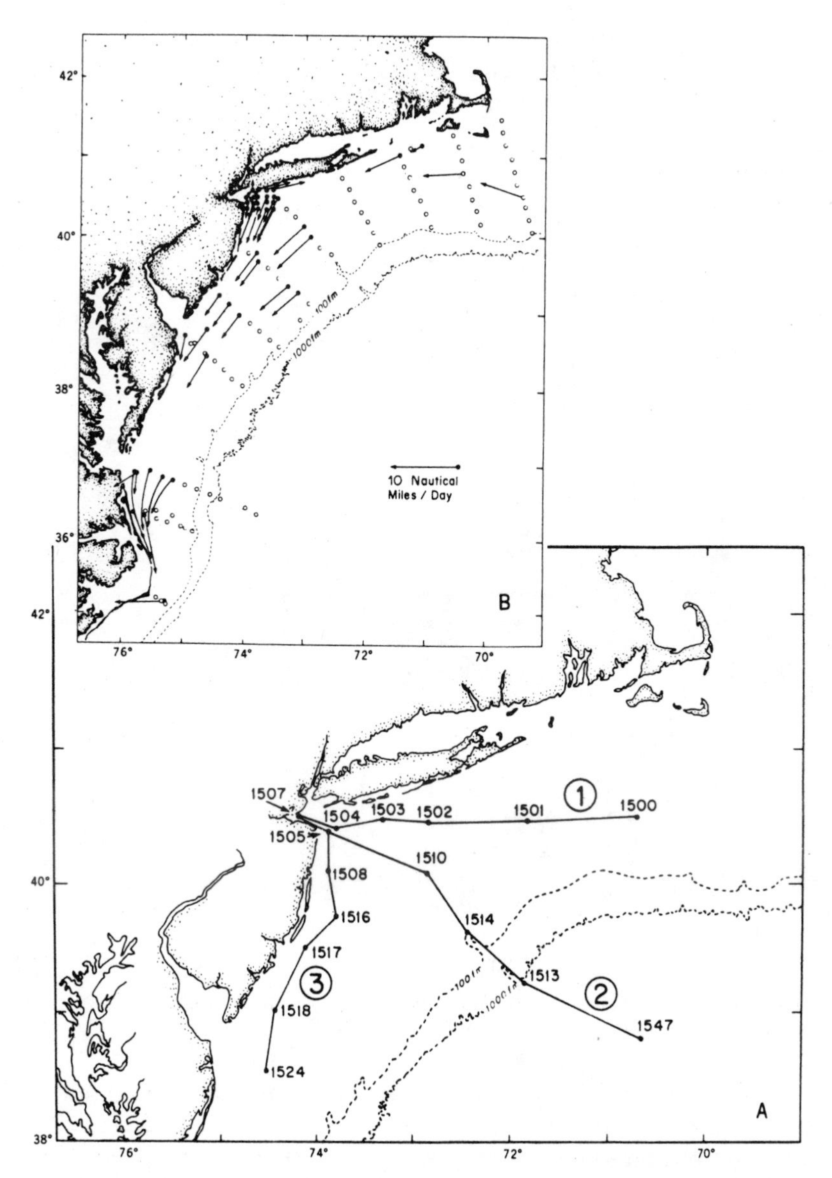


FIGURE 32–3. (A) Three oceanographic sections occupied on cruise 52 of *Atlantis II*, with station locations indicated. (B) Current speed and direction during August (10 nautical miles = 18.5 km) [from unpublished data of D. F. Bumpus, Woods Hole Oceanographic Institution].

ciably higher than those observed at the two oceanic stations. As in the Great South Bay–Moriches Bay situation, the available nitrogen from the center of high pollution within the New York bight seems to be utilized by microorganisms as quickly as it becomes available, but there is a surplus of phosphate, which is carried seaward and is distributed throughout the continental shelf.

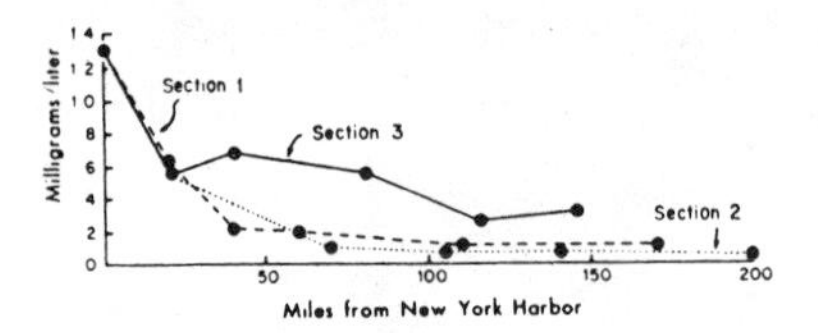


FIGURE 32–4. The distribution of particulate organic carbon along the three sections shown in Fig. 32–3 (50 miles = 80 km).

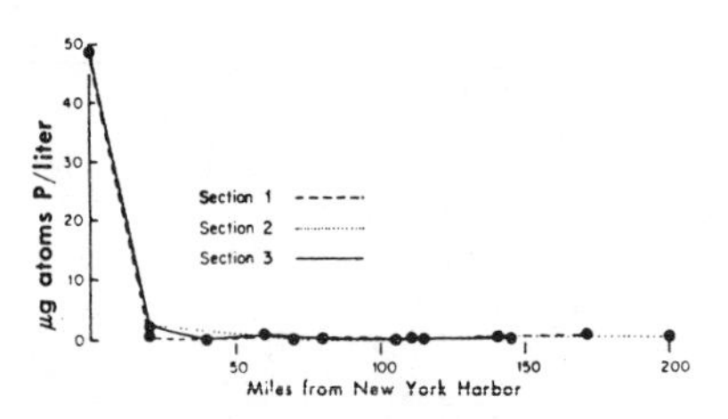


FIGURE 32–5. The distribution of inorganic nitrogen ($NO_2$ + $NO_3$ + $NH_3$) along the three sections shown in Fig. 32–3 (50 miles = 80 km).

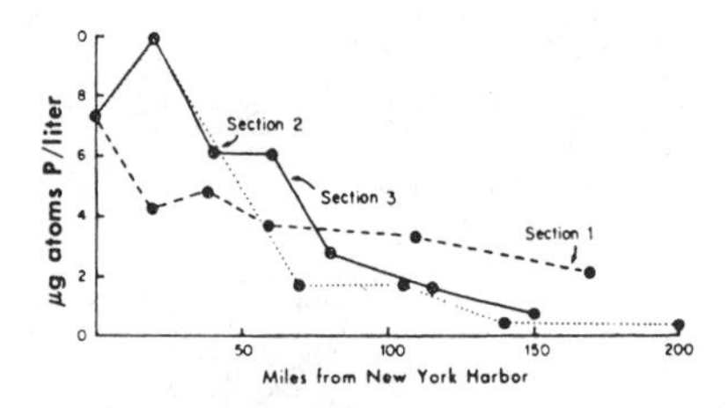


FIGURE 32–6. The distribution of phosphate along the three sections shown in Fig. 32–3 (50 miles = 80 km).

Experiments similar to those described above were carried out with surface water collected from the 16 stations of sections 1 to 3. This water was Millipore-filtered immediately after its collection and was stored frozen in polycarbonate bottles until used. In the laboratory, the water from each station was divided into three portions, as in the experiments described earlier; one served as a control, one received 10 $\mu$mole of sodium phosphate, and one received 100 $\mu$mole of ammonium chloride. In this instance, the mediums were inoculated with the common coastal diatom *Skeletonema costatum*, the cells of which had been washed and nutrient-starved in sterile, unenriched Sargasso Sea water for 2 days prior to their use. Growth of the cultures at 20° C and 11,000 lu/m$^2$

of illumination proceeded for 5 days; the results are shown in Fig. 32–7.

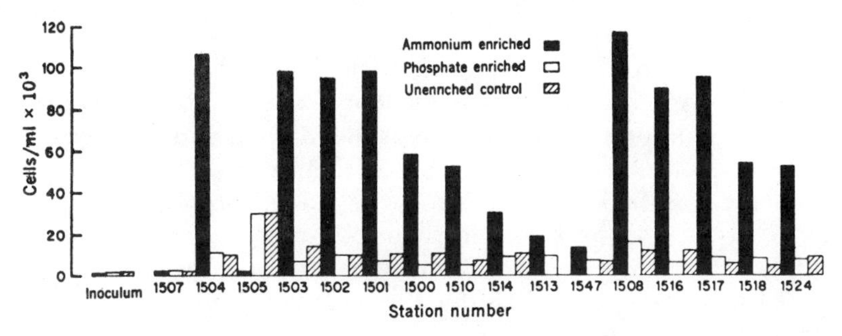


FIGURE 32–7. Growth of *Skeletonema costatum* in unenriched, ammonium-enriched, and phosphate-enriched water from the New York bight collected from the stations shown in Figure 32–3.

As in the earlier experiment, there was some growth in most of the unenriched controls, and this varied from station to station. Growth in the samples enriched with phosphate was no better and, in several cases, not so good as growth in the control cultures. In contrast, and again as in the bay experiments, heavy growth of *Skeletonema* occurred in most of the samples enriched with ammonia; in several cases growth was ten times or more the growth in the controls and phosphate-enriched samples.

Variability of the growth in the $NH_4$-enriched samples from station to station again reflects the differential development of secondary nutrient deficiencies, perhaps of phosphate but possibly of one or more other nutrients. This would be expected to occur first in the more offshore stations (see Fig. 32–6), and such an explanation is consistent with the experimental results. Currently unexplained, however, is the poor growth in the ammonia-enriched samples from stations 1507 and 1505 (two of the heavily polluted stations from within the New York bight), particularly at station 1505, where the best growth occurred in the unenriched and phosphate-enriched series. With the exception of the anomalous results from those two stations, the generalization can be made, consistent with the nutrient distribution picture, that also in these waters nitrogen, not phosphorus, is the primary limiting factor to algal growth.

## Sources and Mechanisms

To return to the question of why and how nitrogen can limit the growth of phytoplankton when the amount of phosphorus relative to nitrogen in the plants is greater than it is in seawater, there are probably two explanations. One explanation applies to the ocean in general; the other, to coastal waters and estuaries specifically.

Seasonally or aperiodically, as a result of surface cooling, wind mixing, or other processes leading to vertical instability, the surface layers of the ocean are recharged with nutrients from subeuphotic depths. This mechanism, important though it is as the ultimate source of enrichment of the open sea, probably occurs infrequently. Most of the time, in the thermally stratified, nutrient-impoverished surface waters of the open ocean, organic production is maintained largely through recycling. The supply of nutrients by vertical transport from beneath the thermocline is relatively insignificant. Under these circumstances plant production is limited by the rate of regeneration of the nutrient that is mineralized most slowly. Table 32–1 lists the relative rates of mineralization of inorganic nitrogen compounds and of phosphate from a mixed plankton tow (*18*). The excess phosphate left in the water is also shown; the amount was calculated on the assumption that all the nitrogen is assimilated as quickly as it is formed and that phosphate is used at a ratio of one atom of phosphorus for each ten atoms of nitrogen assimilated (see above). Even if nitrogen and phosphorus were assimilated at a ratio of 5 : 1, an appreciable amount of phosphate would still be left unassimilated. This mechanism is probably responsible for the small but persistent supply of dissolved phosphate observed in surface waters throughout most of the open ocean environment.

The situation is quite different in coastal waters and estuaries. Here the surplus of phosphate may be quite large, as we have seen, and its source is unquestionably the land.

In Great South Bay and Moriches Bay it was pointed out that phosphate could be used as a tracer of the pollution originating from the duck farms located on the tributaries to Moriches Bay. Nitrogen and phosphorus are contained in duck feces in the ratio of 3.3 : 1 by atoms. Total nitrogen and phosphorus analyses of dissolved and suspended matter in the tributaries and in Moriches Bay itself gave nitrogen to phosphorus ratios of 2.3 : 1 to 4.4 : 1, consistent with the presumed origin of this material. About half of the total phosphorus was present as dissolved, inorganic phosphate, with the remainder being tied up in the algae and other particulate matter. All the nitrogen occurred in the latter form. As mentioned earlier, no inorganic nitrogen in any form and no uric acid could be detected anywhere in the water (*15*).

According to the above data, the ratio of nitrogen to phosphorus in the plankton would be about 6.6 : 1. The population of algae in the area consisted of an almost pure culture of two species of green algae, which were identified at the time as *Nannochloris atomus* and *Stichococcus* sp. The green algae (Chlorophyceae) are characterized by a low nitrogen to phosphorus ratio (*5*, *6*). This fact was believed to be partly responsible for their presence in the bay waters, though other factors such as low salinity and high temperature were also shown to be important selective mechanisms (*15*).

In the New York bight and the contiguous coastal waters, a high level of phosphate was again measured (Fig. 32–6). From its distributional pattern there can be little doubt that this material originated in New York Harbor and its tributaries. In Raritan Bay, as mentioned, the phytoplankton consisted almost exclusively of a small green alga, owing presumably to a combination of ecological conditions similar to those that obtained in Great South Bay and Moriches Bay. In the New York bight and the waters farther offshore, conditions more typical of the marine environment prevailed, and the plankton flora consisted of a mixture of diatoms, flagellates, and other forms (*19*). What is the origin of the surplus phosphate in this case?

From data provided by Pearson *et al.* (*20*) one can calculate that the nitrogen to phosphorus ratio by atoms in domestic wastes that have been subjected to primary sewage treatment is 5.8 : 1. In wastes that have undergone secondary treatment, the ratio, according to Weinberger et al. *(21)*, is 5.4 : 1. On the assumption that the 4 billion kl per day of domestic wastes entering the New York bight from the New York-New Jersey megalopolis have been subjected to something intermediate between primary and secondary treatment, some 90 metric tons of nitrogen and 36 metric tons of phosphorus are discharged into these waters each day. If the phytoplankton that inhabit the area assimilate nitrogen and phosphorus in the ratio of 10 : 1 by atoms (4.5 : 1 by weight), nearly half the phosphate entering the system is in excess of the amount that can be used by the plants.

## Eutrophication

As we have seen, phosphate is a convenient index or tracer of organic pollution. Its analysis by conventional colorimetric techniques is quick, accurate, and highly sensitive and is far easier than analysis of other chemical nutrients. Furthermore, it persists when other products of organic decomposition, such as nitrogenous compounds, have disappeared from solution. Thus, domestic wastes can be tracked longer and farther from their source of input by looking at the distribution and concentration of phosphate than by using almost any other criteria. From this fact, it is a short and easy step to the conclusion that phosphate is the causative agent of algal growth, eutrophication, and the other adverse effects associated with organic pollution. In the sea, such is far from true.

There is the possibility, alluded to briefly above, that blue-green algae, and possibly other microorganisms capable of fixing atmospheric nitrogen, may by this process bring enough nitrogen into the biological cycle to balance the surplus of phosphate. Filamentous blue-green algae are common in freshwater lakes, and their ability to fix nitrogen is well demonstrated (*22*). For this reason, or simply because of a high natural ratio of nitrogen to phosphorus, there is probably, as Edmondson (*23*) suggests, "A large class of lakes in which phosphorus is the dominating element," a hypothesis that he has well documented for Lake Washington. As Edmondson has also shown, however, such is true only in the relatively unpolluted condition. During the period when Lake Washington received sewage effluent, phosphate was present in excess quantities relative to the available nitrogen.

In the open tropical ocean, there are also filamentous blue-green algae, of the genus *Trichodesmium*, that are capable of fixing nitrogen, though the process is so slow and inefficient as to be almost undetectable (*24*). In the more eutrophic coastal waters and estuaries, such algae are almost unknown, and nitrogen fixation has not been demonstrated. Here, as we have shown, it is unquestionably nitrogen that limits and controls algal growth and eutrophication.

Much of the phosphate in domestic waste has its origin in detergents. The fraction of the land-derived phosphate in our coastal waters that can be attributed to this source is difficult to assess but has been estimated to be 25 to 50 percent of the total (*25*). The total land-derived phosphate also includes human excreta, agricultural runoff, industrial wastes, and other material, all of which vary greatly from place to place. As shown earlier, the nitrogen to phosphorus ratio in domestic waste is slightly higher than 5 : 1 by atoms. Even if as much as half of the phosphate in sewage came from detergents and if all of the phosphate from this source could be eliminated by its complete replacement with other compounds, which is a most unlikely possibility (*26*), the amounts of nitrogen and phosphorus entering the environment would still be in the atomic ratio of 10 : 1, and no reduction of algal growth or eutrophication could be expected.

If, in fact, the phosphate in detergents is replaced with nitrilotriacetic acid (NTA), as is the current trend in the industry (*26*), the net effect could be an acceleration and enhancement of the eutrophication process. In sewage treatment (and presumably in nature, if more slowly), NTA undergoes biodegradation and probably yields glycine and glycolic acid as intermediate decomposition products (*27*). These compounds may be used directly as a nitrogen source by at least some species of unicellular algae (*15*), or they may be deaminated to ammonia, which is universally available to phytoplankton.

Coastal waters already receive the sewage of roughly half the population of the United States. To replace a portion of the phosphate in this sewage with a nitrogenous compound and to then discharge it into an environment in which eutrophication is nitrogen-limited may be simply adding fuel to the fire.

## References and Notes


1. H. W. Harvey, *J. Mar. Biol. Ass. U.K.* **14**, 71 (1926).
2. A. C. Redfield, *James Johnston Memorial Volume* (Univ. of Liverpool Press, Liverpool, 1934), p. 176.
3. ———, *Amer. Sci.* **46**, 205 (1958).
4. H. U. Sverdrup, M. W. Johnson, R. H. Fleming, *The Oceans* (Prentice-Hall, New York, 1942), p. 236; L. H. N. Cooper, *J. Mar. Biol. Ass. U.K.* **22**, 177 (1937).
5. B. H. Ketchum and A. C. Redfield, *J. Cell. Comp. Physiol.* **33**, 281 (1949).
6. J. D. H. Strickland, *Bull. Fish. Res. Bd. Can. No. 122* (1960).
7. T. R. Parsons, K. Stephens, J. D. H. Strickland. *J. Fish. Res. Bd. Can.* **18**, 1001 (1961); C. D. McAllister, T. R. Parsons, K. Stephens, J. D. H. Strickland, *Limnol. Oceanogr.* **6**, 237 (1961).
8. D. W. Menzel and J. H. Ryther, *Limnol. Oceanogr.* **9**, 179 (1964).
9. R. F. Vaccaro, *J. Mar. Res.* **21**, 284 (1963).
10. E. Harris and G. A. Riley, *Bull. Bingham Oceanogr. Collect. Yale Univ.* **15**, 315 (1956).
11. U. Stafánsson and F. A. Richards, *Limnol. Oceanogr.* **8**, 394 (1963).
12. B. H. Ketchum, R. F. Vaccaro, N. Corwin, *J. Mar. Res.* **17**, 282 (1958).
13. W. H. Thomas, *Limnol. Oceanogr.* **11**, 393 (1966).
14. ——— and A. N. Dodson, *Biol. Bull.* **134**, 199 (1968); W. H. Thomas, *J. Fish. Res. Bd. Can.* **26**, 1133 (1969).
15. J. H. Ryther, *Biol. Bull.* **106**, 198 (1954).
16. A. J. McCarthy, thesis, Fordham University (1965).
17. D. W. Menzel and R. F. Vaccaro, *Limnol. Oceanogr.* **9**, 138 (1964).
18. R. F. Vaccaro, in *Chemical Oceanography*, J. P. Riley and G. Skirrow, Eds. (Academic Press, New York, 1965), vol. 1, p. 356.
19. E. M. Hulbert, unpublished data.
20. E. A. Pearson, P. N. Storrs, R. E. Selleck, *Ser. Rep. 67–3* (Sanitary Engineering Research Laboratory, Univ. of California, Berkeley, 1969).
21. L. W. Weinberger, D. G. Stephan, F. M. Middleton, *Ann. N.Y. Acad. Sci.* **136**, 131 (1966).
22. R. C. Dugdale, V. A. Dugdale, J. C. Neess, J. J. Goering, *Science* **130**, 859 (1959); D. L. Howard, J. I. Frea, R. M. Pfister, P. R. Dugan, ibid. **169**, 61 (1970).
23. W. T. Edmondson, ibid. **169**, 690 (1970).
24. R. C. Dugdale, D. W. Menzel, J. H. Ryther, *Deep-Sea Res.* **7**, 298 (1961).
25. F. A. Ferguson, *Environ. Sci Technol.* **2**, 188 (1968).
26. *Chem. Eng. News* **1970**, 18 (17 August 1970).
27. R.D. Swisher, M.M. Crutchfield, D. W. Caldwell, *Environ. Sci. Technol.* **1**, 820 (1967).


28. Supported by the Atomic Energy Commission, contract AT(30-1)-3862, ref. NYO-3862-40, and by NSF grant GB 15103. Contribution 2537 from the Woods Hole Oceanographic Institution.

# 33

# Effects of decreased river flow on estuarine ecology

B. J. Copeland

With rapidly expanding industrialization along the waterways of the world, water usage has become a legal and economic problem. The phenomenon of low flow occurs naturally in times of drought and many studies during drought conditions have shown that the effects of reduced river flow are diverse and far-reaching. The practice of decreasing river flow by the installation of reservoirs is frequently followed without consideration or prior knowledge of the effects downstream, particularly in estuaries.

Estuaries are important in many ways. Many important commercial and sport fishes and invertebrates use estuaries during some stage of their development. Estuaries probably are the greatest producing areas accessible to man for available protein. Many people use these waters for recreation, and industry uses them as factory sites because water is available for transportation, processing, and waste disposal.

As used in this paper, the word "estuary" means a body of water between the mouth of a river and the sea. Since lagoons are really "old estuaries," the term estuary, as used here, includes lagoons, bays, and sounds.

---

Reprinted with permission from *Journal Water Pollution Control Federation*, Vol. 38, pages 1831–1839 (November, 1966), Washington, D.C. 20016. The author is with the Zoology Department, North Carolina State University, Raleigh, North Carolina.

## Currents and Hydrography

An intricate current system within estuaries is produced by the balance of river discharge and the contribution of the sea. With decreased river flow, the current system can be so altered that shoaling and scouring can set up complete foreign physical conditions.

The most important hydrobiological parameter is salinity. Under normal conditions, salinity is variable in some estuaries. Collier and Hedgpeth (1) showed a direct correlation between river flow and estuarine salinity in Corpus Christi Bay, Texas. If river flow is restricted by upstream reservoirs, the salinity level in the receiving estuary may increase to the detriment of estuarine biological communities.

Closely connected to river flow is the maintenance of natural inlets between barrier islands, connecting estuaries to the sea. Simmons and Hoese (12) suggested the necessity of river flow for maintaining discharge through Cedar Bayou, a natural inlet on the Central Texas coast. Lack of river flow during the early 1950's resulted in the closing of Cedar Bayou. Passes can be kept open only at great expense if river flow is reduced below the level required to maintain them. The passes are important passageways for fish and invertebrates between nursery grounds and feeding areas (2) (3) (4) (5) (6) (7).

## Oyster Production

Oysters flourish in a wide range of salinity, and the production of oyster is exclusively estuarine. Oyster production in 1962 amounted to about $16,000,000 in Chesapeake Bay and $6,000,000 on the Gulf Coast. Production by weight was about the same, however, amounting to $20 \times 10^6$ and $19 \times {}^6$ lb ($8.6 \times 10^6$ and $9.1 \times 10^6$ kg) respectively (8).

Under normal estuarine conditions, occasional flushing with fresh water from rivers helps rid oyster populations of damaging parasites and eliminates species that compete with oysters for food (9) (10). With reduced river flow, however, salinities may remain near or above that of seawater with the result that damaging parasites and competing species thrive.

Wells (11) reported a decrease in number of species of oyster parasites and scavengers in the Beaufort, N.C., area as salinity decreased. The abundance of boring sponges, major oyster parasites, decreases as salinity decreases (12) (13) (14). Most of the boring sponges in the high salinity areas of Newport River, N.C., were killed when the salinity was lowered during heavy flooding of the river (14).

*Dermocystidium marinum*, a parasitic fungus, is often a serious threat to oyster populations and may be controlled by dilution with fresh water (15) (16) (17) (18) (19) (20) (21) and (many others). *D. marinum* is common on oysters in moderate salinities, nearly always absent in salinities below 10–15 ppt, and only occasionally found in salinities approaching 35 ppt (18). Although the average salinities may be similar in various estuaries, areas with the greatest fluctuation in freshwater inflow (salinity fluctuations) have the least *D. marinum* infestation. A sudden drop in salinity as a result of flooding with fresh water may be enough to check the fungal growth. Apparently it is necessary that freshwater inflow occur in spring to check the infestation effectively.

Another widespread and destructive oyster parasite is the oyster drill, *Thais haemastoma*, which also is sensitive to salinity changes and is usually eliminated by moderate to low salinities (22). The Atlantic oyster drills, *Urosalpinx* and *Eupleura*, are sometimes quite destructive to oysters (23). They, too, are controlled in moderate to low salinities and/or with periodic flushing with fresh water.

Franklin (24), in a well-documented newspaper article, discussed the new, yet unnamed, oyster disease MSX. Apparently, MSX is limited to waters with over 15 to 20 ppt salinity, thus controlled by freshwater contributions.

In spite of the tremendous need for fresh water by the oyster industry, an overabundance will upset the balance. Andrews *et al.* (25) reported oyster kills when the salinity was lowered in the James River, Va. Butler (26) and Gunter (27) reported oyster kills over large areas of the Mississippi Sound during floods of fresh water. However, if flooding occurs for only a short period, oysters are capable of remaining closed for several days and avoiding the harmful effects.

Galtsoff (10) discussed the decrease in the oyster population of Texas bays during the drought of the 1950's. He suggested that the resulting increase in salinity in coastal waters during that time allowed the influx of parasites and diseases and the higher salinities (above 40 ppt) restricted gonad development. Surveys by the Texas Parks and Wildlife Department (28) (29) indicate little or no oyster harvest in Texas bays that receive little fresh water.

## Shrimp

Commercially important shrimp (*Peneidae*) are spawned offshore and the young migrate into estuarine nursery grounds, generally as postlarvae, to complete the life cycle. There, utilizing the higher productivity and abundant dissolved organic material, they grow rapidly, sometimes more than one millimeter per day. Along the Gulf coast shrimp is the most valuable of all fishery products, amounting to more than $60 million in 1962 (8).

The entrance of postlarval penaeid shrimp through the Aransas Pass Inlet, Texas, corresponds to high flow of the rivers of the area in spring and fall (Fig. 33–1) (30). Undoubtedly, this coupling of peak migration and increased river flow (accomplished through years of natural selection) is essential for the propagation of penaeid shrimp.

Important fluxes occur in estuarine ecosystems during the high flows of spring and fall, including flows of vitamins and other dissolved organic compounds, nutrients, lowered salinity by the addition of fresh water, and flushing and mixing influences. Burkholder and Burkholder (31) reported a greater concentration of vitamin $B_{12}$ in the mud and estuarine waters of Georgia than in the adjacent seawater. Starr and Sanders (32) found similar results in other areas, and postulated that productivity of the nearshore sea was greatly dependent on suspended solids brought into the estuary by river flow. In attempts at raising shrimp from the egg to juvenile stage in laboratory experiments, it was found that it was necessary to add vitamin $B_{12}$ to the sea water

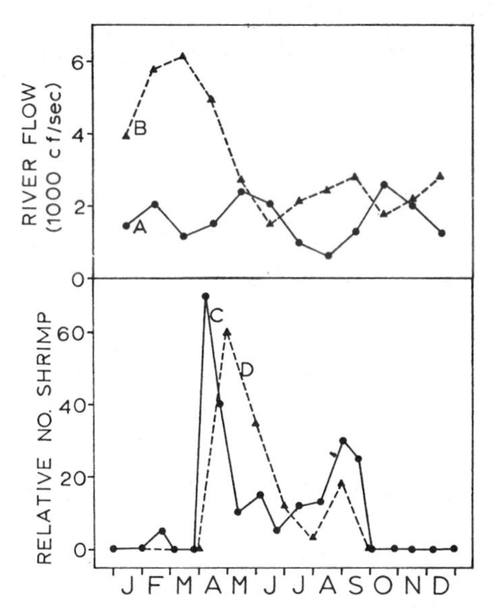


FIGURE 33–1. Relative number of postlarval penaeid shrimp entering the bays, and river flow of coastal rivers. A. River flow for Guadelupe River near Victoria, Tex.; 10-yr. avg. 1953–1962; B. river flow for Neuse River, near Kingston, N. C.; 10-yr. avg. 1955–1964; C. postlarval *Penaeus aztecus* and *P. duorarum* entering Texas bays through the Aransas Pass Inlet (30); D. postlarval *P. aztecus* and *P. duorarum* entering the Brunswick-Onslow bay area of North Carolina (59).

aquarium, according to a personal communication from the Bureau of Commercial Fisheries, Galveston, Texas.

Gunter (33) found that postlarval penaeid shrimp were most abundant in waters of moderate salinities, although other features were the same. At least 2 species of *Penaeus* are hypoosmotic to seawater and hyperosmotic to low-salinity waters, with isotonicity occurring between 25 and 30 ppt (34). At lowered temperatures, however, the ability to regulate osmosis is slightly impaired, which may account for movement to deeper (and more saline) waters.

Gunter and Hildebrand (35) showed a correlation between catch of white shrimp on the Texas coast and average rainfall for the state. Their correlation coefficients were significant to the one-percent level when rainfall of the two previous years was correlated with an annual catch. This method of analysis was considered valid since the reproductive cycle of shrimp is one year or more, large bodies of water change their salinity slowly, and dry land absorbs more water (less runoff) following droughts.

A plot of shrimp catch and average rainfall for Texas is shown in Figure 33–2. An increase or decrease in rainfall is followed by similar fluctuations in shrimp catch, generally after a two-year period. After the drought during the early 1950's, white shrimp populations never fully recovered, in spite of technological advances and increased fishing

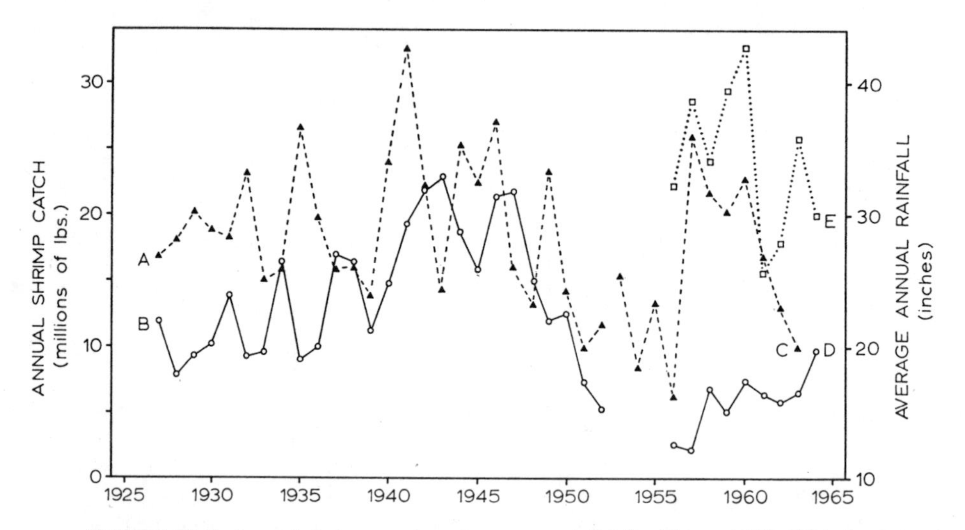


FIGURE 33–2. Annual shrimp catch and average rainfall for Texas, 1927–1964. A. Avg. rainfall for Texas, 1927–1952 (35); B. annual catch of white shrimp (*Penaeus setiferus*) from Texas waters, 1927–1952 (35); C. avg. rainfall for Texas, 1953–1964 (U. S. Weather Bureau, Climatological Data, Annual Summary, Vol. 58–70); D. annual catch of white shrimp (*P. setiferus*) from Texas waters, 1956–1964 (from U. S. Bureau of Commercial Fisheries, Gulf Coast Landings, 1956–1964). E. annual catch of brown shrimp (*P. aztecus*) from Texas waters, 1956–1964 (from U. S. Bureau of Commercial Fisheries, Gulf Coast Landings, 1956–1964).

effort. Presumably, the addition of several reservoirs in Texas during this time prevented river runoff from approaching the level of previous years.

Gunter (36) reported a general decline of white shrimp along the Texas coast and suggested that it be attributed to the drought. He further postulated that the appearance of larger white shrimp populations in Galveston Bay and smaller populations in the bays to the south be attributed to the decrease in freshwater contribution to Texas bays in a southerly direction.

## Blue Crab

The blue crab (*Callinectes* sp.) fisheries ranked sixth in value in the United States, with a value of $1.5 million on the Gulf coast in 1962 (8). The life cycle of blue crabs is closely connected to estuaries, and estuarine ecology is a vital factor in the maintenance of a continuous crab fishery.

There have been pronounced fluctuations in crab production through the years, according to U.S. Fisheries statistics. There is a direct correlation between crab reproduction and the salt content of estuarine waters (37). The optimum range of salinity for hatching of blue crab eggs was between 23 and 28 ppt. Hoese (38) noted a decrease in the crab population in Mesquite Bay, Texas, during the Texas drought, when salinities were greater than that of seawater. After the drought was broken and salinities were moderately low, the blue crab population increased to its normal level.

## Primary Production

Estuaries are among the most productive ecosystems known (39). The maintenance of high productivity in estuaries is due to the "nutrient trap" created by the mixing of river waters with ocean waters. The inflow of nutrient sulfates, carbonates, phosphorus, and nitrogenous compounds via rivers contributes greatly to estuarine productivity.

Considerable amounts of organic detritus are brought into the estuary from adjacent marshes and from upstream (40). Detritus is the principal food of many estuarine organisms (39). This material is decomposed by bacteria and fungi, releasing large amounts of organic and inorganic substances which are absorbed by estuarine organisms.

Estuarine phytoplankton growth is related to terrigenous nutrients in Patuxent River and Chesapeake Bay region (41). Plankton blooms occurred just after peak flows down the Patuxent. Similarly, nutrient concentrations were higher following the peak river flow and prior to the plankton blooms. Hutner and Provasoli (42), in their review of research in algal nutrition, have verified that abundant growth of marine algae requires vitamins and nutrients in greater concentration than those found in ordinary seawater.

Estuaries in the southeastern United States are more productive near the mouths of contributing rivers (43) (44) (45) (46). It is not clear, however, whether higher productivity is because of the contribution of rivers or is because the greater mixing regenerates nutrients in that area.

On the other hand, accumulated nutrients tend to be flushed out of estuarine areas receiving large amounts of flood waters of low nutrient concentration. Such floods also kill many of the larger organisms so that their stored nutrients also are lost (38).

The productivity of Texas estuaries was lower just after heavy floods following prolonged droughts (47) (48). During times of normal river discharge, however, the algal productivity was higher than that in most estuarine areas. Texas estuaries are unique in that they are very shallow and have minimal tidal fluctuations; thus, freshwater flow is important in mixing processes.

## General Discussion

The influx of fresh water is one of the principal sources of dissolved nutrients in estuaries. The relationship of growth in estuarine phytoplankton to terrigenous nutrients is shown by the increase following periods of heavy rainfall and runoff. Organic materials, which are decomposed by bacteria and fungi to release large amounts of organic and inorganic substances that are absorbed by estuarine organisms via the aquatic medium, are brought into the estuaries by river inflow (40).

Estuaries that have proved to be important nursery areas possess a well-defined salinity gradient between river mouth and tidal pass, accommodating a large variety of species. River inflow maintains the salinity gradient to a large extent and without it the entire estuary could become hypersaline, as in the Texas Laguna Madre. On the other hand, too much freshwater inflow may cause the entire estuary to become fresh or near-fresh (Sabine Lake, in east Texas) and destroy the salinity gradient.

The estuarine systems of Texas lie in a broad arc of approximately 375 miles (600) km) and pass through a variety of climatic regions, ranging from almost fresh water in Sabine Lake to hypersaline water in the Laguna Madre. Characteristics of the five estuaries other than Sabine Lake and Laguna Madre show the effects of freshwater contributions. Since 1959, the average annual freshwater contribution to these estuaries has been 20.5 acre-ft per surface acre ($62.5 \times 10^3$ cu m/ha) for Galveston Bay, 4.4 ($13.4 \times 10^3$) for Matagorda Bay, 11.5 ($35.0 \times 10^3$) for San Antonio Bay, 1.4 ($4.3 \times 10^3$) for Aransas Bay, and 1.9 ($5.8 \times 10^3$) for Corpus Christi Bay. The large inflow in Galveston and San Antonio Bays, plus rainfall, has contributed to the maintenance of thousands of acres of marshes and bayous that provide habitat for the young of many important animals.

Galveston Bay has continually produced more oysters than other Texas bays, presumably because of the more favorable freshwater conditions. The San Antonio Bay system is the only other bay approaching Galveston Bay in producing oyster reefs. Matagorda, Aransas, and Corpus Christi Bays produce few or no oysters at the present time (27) (28). Oyster production corresponds very closely to the amount of fresh water received by each estuary.

Blue crab production was highest in the Galveston and San Antonio Bay areas, and lowest in the bays receiving the least amount of fresh water (49).

Shrimp are the most important fishery product of the Texas coast. Galveston and Matagorda Bay support the largest shrimp populations and Corpus Christi Bay the smallest. White shrimp are now almost nonexistent in Aransas and Corpus Christi Bays (receiving very little fresh water), where in years past they were quite abundant (33) (50) (51).

The data available for comparison of freshwater input and commercial fishery output are scant. However, in Texas, where freshwater input is at a critical level in most estuarine areas, some startling conclusions can be drawn. The minimum freshwater contribution required to maintain the present commercial fishery is not reached in some years in Matagorda, Aransas, and Corpus Christi Bays (Fig. 33–3). In Galveston and San Antonio Bays, larger commercial fishery yields have been harvested during years of intermediate or below-average freshwater input.

The data presented in Figure 33–3 do not complete the real picture. Perhaps more important than the total commercial fisheries output is the change in species composition that makes up the total. During the year following an above-average freshwater input, shrimp and oysters make up a larger percentage of the total. These products are more valuable economically than are fish products, which make up a larger percentage of the total during years of below-average freshwater input.

Many fish species whose juveniles reside in low-salinity waters depend on estuaries for the completion of their life cycles (52) (53). Freshwater input is important to the propagation of fishes because of its salinity control.

Many of the same effects that have been observed in North American estuaries have been observed in estuaries in other areas. In the St. Lucia estuary system on the east coast of South Africa, serious silting and variations in rainfall, along with man's increasing demand for water, has affected salinity gradient (54) (55). Most of the formerly rich fauna was lost because of the practices of civilization. On the other hand, Richard's Bay, just 36 miles (58 km) south, has not suffered the effects seen

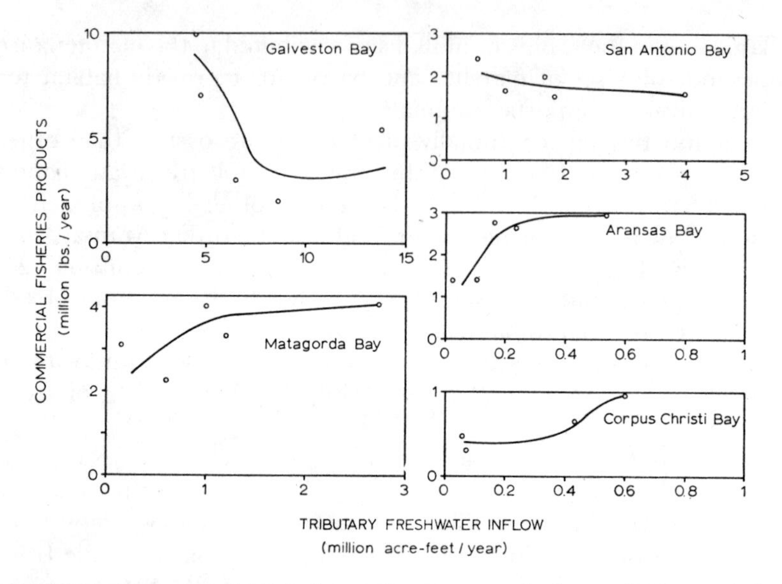


FIGURE 33–3. Commercial fisheries products vs. annual tributary freshwater inflow for Texas bays. 1959–1964. One year's freshwater contribution was plotted against the following year's commercial harvest. Solid lines indicate trends only. Data on annual commercial fishery yields were obtained from the Bureau of Commercial Fisheries, Biological Laboratories, Galveston, Texas. Data on tributary freshwater contribution to each bay were obtained from the Texas Water Commission.

so clearly in the St. Lucia system (56). Presumably, this is because the river flow into Richard's Bay has not been changed by man and his enterprises.

As shown by several studies in Russia, the regulation of flow of rivers by the construction of dams led to a considerable accumulation of nutrient salts in reservoirs and to a sharp reduction in the amounts of nutrient salts flowing to the estuaries (57). Following the construction of the Tsimliansk hydroelectric dam, the biomass of the phytoplankton in the Sea of Azov was reduced by 2 to 3 times, and the zooplankton from about 600 mg/cu m to about 50 mg/cu m. Ultimately, this had an adverse effect on the feeding conditions of fishes and their populations.

An important question regarding low river flow is whether the loss of river input and the resulting higher salinities actually result in lessened productivity in the estuary of simply in changes in the productivity channels. Galtsoff (10) reported a general change in the species of oysters in bays of the central Texas coast as salinities increased during the drought of the early 1950's. With the increase in salinity there was a gradual replacement of *Crassostrea virginica*, the commercial oyster, by

*Ostrea equestris*, the noncommercial oyster. In 1952 over half of the young oysters (spat) were *O. equestris*, whereas in years of normal salinity the reefs were comprised almost entirely of *C. virginica*. Gunter (58) has discussed this subject and concluded that salinity and freshwater inflow are important factors in limiting the species composition of estuaries.

As has been shown in the previous discussion, freshwater input to estuaries is an important factor. Without it, estuaries become hypersaline and species composition can be altered drastically. With continuation of man's activities in allowing less and less fresh water downstream to the estuary, man may have to pave the estuarine areas and sell them for real estate.

## Acknowledgments

The author acknowledges the courtesy of C. H. Chapman and R. A. Diener of the Bureau of Commercial Fisheries and of F. Masch and C. Urban of the Texas Water Commission in permitting the use of data included in Figure 33–3.

## References

1. Collier, A., and J. W. Hedgpeth. 1950. "An Introduction to the Hydrography of Tidal Waters of Texas." *Publ. Inst. Marine Sci. Univ. Texas*, 1:2, 121.
2. Simmons, E. G., and H. D. Hoese. 1959. "Studies on the Hydrography and Fish Migration of Cedar Bayou, a Natural Tidal Inlet on the Central Texas Coast." *Publ. Inst. Marine Sci. Univ. Texas*, 6:56.
3. Copeland, B. J. 1965. "Fauna of the Aransas Pass Inlet, Texas. I. Emigration as Shown by Tide Trap Collections." *Publ. Inst. Marine Sci. Univ. Texas*, 10:9.
4. Daugherty, F. M. 1952. "The Blue Crab Investigation, 1949–50." *Texas Jour. Sci.*, 4:77.
5. Hoese, H. D. 1958. "The Case of the Pass." *Texas Game Fish.*, 16(6): 18 and 30.
6. Reid, G. K., Jr. 1955. "A Summer Study of the Biology and Ecology of East Bay, Texas." *Texas Jour. Sci.*, 7:316.
7. Reid, G. K., Jr. 1957. "Biological and Hydrographic Adjustment in a Disturbed Gulf Coast Estuary." *Limnol. Oceanogr.*, 2:198.
8. U. S. Fish and Wildlife Service, "United States Fisheries." Annual Summary, Bureau of Commercial Fisheries, Commercial Fisheries Statistics, No. 3471.
9. Collier, A., S. M. Ray, A. W. Magnitzky, and J. O. Bell. 1953. "Effect of Dissolved Organic Substances on Oysters." *Fishery Bull.*, U.S. Fish and Wildlife Serv., 54:167.
10. Galtsoff, P. S. 1964. "The American Oyster *Crassostrea virginica* Gmelin." *Fish. Bull.*, U.S. Fish and Wildlife Serv., 64:1.
11. Wells, H. K. 1961. "The Fauna of Oyster Beds, with Special Reference to the Salinity Factor." *Ecol. Monogr.*, 31:239.
12. Hopkins, S. H. 1956. "Notes on the Boring Sponges in Gulf Coast Estuaries and Their Relation to Salinity." *Bull. Marine Sci. Gulf Caribb.*, 6:44.

13. Hopkins, S. H. 1962. "Distribution of Species of *Cliona* (Boring Sponge) on the Eastern Shore of Virginia in Relation to Salinity." *Chesapeake Sci.*, 3:121.
14. Wells, H. W. 1959. "Boring Sponges (Clionidae) of Newport River, North Carolina." *Jour. Elisha Mitchell Sci. Soc.*, 75:168.
15. Andrews, J. D. 1955. "Notes on Fungus Parasites of Bivalve Mollusks in Chesapeake Bay." *Proc. Natl. Shellfish. Assn.*, 45:157.
16. Andrews, J. D., and W. G. Hewatt. 1957. "Oyster Mortality Studies in Virginia. II. The Fungus Disease Caused by *Dermocystidium marinum* in Oysters of Chesapeake Bay." *Ecol. Monogr.*, 27:1.
17. Hewatt, W. G., and J. D. Andrews. 1954. "Oyster Mortality Studies in Virginia. I. Mortalities in Oysters in Trays at Gloucester Point, York River." *Texas Jour. Sci.*, 6:121.
18. Hoese, H. D. 1963. "Absence of *Dermocystidium marinum* at Port Aransas, Texas, with Notes on an Apparent Inhibitor. *Texas Jour. Sci.*, 15:98.
19. Mackin, J. G. 1951. "Histopathology of Infection of *Crassostrea virginica* (Gmelin) by *Dermocystidium marinum* Mackin, Owen, and Collier." *Bull. Mar. Sci. Gulf Caribb.*, 1:72.
20. Mackin, J. G. 1955. "*Dermocystidium marinum* and Salinity." *Proc. Natl. Shellfish. Assn.*, 46:116.
21. Mackin, J. G. 1962. "Oyster Disease Caused by *Dermocystidium marinum* and Other Microorganisms in Louisiana." *Publ. Inst. Marine Sci. Univ. Texas*, 7:132.
22. Butler, P. A. 1953. "The Southern Oyster Drill." *Proc. Natl. Shellfish. Assn.*, 44:67.
23. Carriker, M. R. 1955. "Critical Review of Biology and Control of Oyster Drills *Urosalpinx* and *Eupleura*." Spec. Sci. Rept. U.S. Fish and Wildlife Serv. 148:1.
24. Franklin, B. A. (October 31, 1965). "Once-Plentiful Eastern Oyster Has Become Victim of Drought and Disease." *The N.Y. Times.*
25. Andrews, J. D., D. Haven, and D. B. Quayle. 1959. "Fresh-Water Kill of Oysters (*Crassostrea virginica*) in James River, Virginia, 1958." *Proc. Natl. Shellfish. Assn.*, 49:29.
26. Butler, P. A. 1952. "Effect of Floodwaters on Oysters in Mississippi Sound in 1950." Res. Rept. U.S. Fish and Wildlife Serv. 31:1.
27. Gunter, G. 1953. "The Relationship of the Bonnet Carre Spillway to Oyster Beds in Mississippi Sound and the 'Louisiana Marsh', with a Report on the 1950 Opening." *Publ. Inst. Marine Sci. Univ. Texas*, 3:1, 17.
28. Hefferman, T. L. (mimeo). "Computation, Analysis and Preparation of Coastwide Oyster Population Data." Marine Fisheries Projects Reports for 1961–62, Texas Game and Fish Comm., Austin, Texas.
29. Hofstetter, R. (mimeo). "A Summary of Oyster Studies along the Texas Coast." Coastal Fisheries Projects Reports for 1963, Texas Parks and Wildlife Dept., Austin, Texas, 163.
30. Copeland, B. J., and M. V. Truitt. 1966. "Fauna of the Aransas Pass Inlet, Texas. II. Penaeid Postlarvae." *Texas Jour. Sci.*, 18:65.
31. Burkholder, P. R., and L. M. Burkholder. 1956. "Vitamin $B_{12}$ *in Suspended Solids and Marsh Muds Collected Along the coast of Georgia.*" *Limnol. Oceanog.*, 1:202.
32. Starr, T. J., F. Sanders. 1959. "Some Ecological Aspects of Vitamin $B_{12}$-active Substances." *Texas Rept. Biol. Med.*, 17:49.
33. Gunter, G. 1950. "Seasonal Population Changes and Distributions as Related to Salinity, of Certain Invertebrates of the Texas Coast, Including the Commercial Shrimp." *Publ. Inst. Marine Sci. Univ. Texas*, 1:2, 7.
34. Williams, A. B. 1960. "The Influence of Temperature on Osmotic Regulation in Two Species of Estuarine Shrimps (*Penaeus*)," *Biol. Bull. Marine Biol. Lab., Woods Hole*, 119:560.
35. Gunter, G., H. H. Hildebrand. 1954. "The Relation of Total Rainfall of the State and Catch of the Marine Shrimp (*Penaeus setiferus*) in Texas Waters." *Bull. Marine Sci. Gulf Caribb.*, 4:95.

36. Gunter, G. 1962. "Shrimp landings and production of the state of Texas for the period 1956–1959, with a comparison with other gulf states." *Publ. Inst. Marine Sci. Univ. Tex.*, 8:216.
37. Sandoz, M., and R. Rogers. 1944. "The Effect of Environmental Factors on Hatching, Moulting, and Survival of Zoea Larvae of the Blue Crab *Callinectes sapidus* Rathbun." *Ecology*, 25:216.
38. Hoese, H. D. 1960. "Biotic Changes in a Bay Associated with the End of a Drought." *Limnol. Oceanog.*, 5:326.
39. Odum, E. P. 1959. "Fundamentals of Ecology." 2nd Ed., W. B. Saunders Co., Philadelphia, Pa.
40. Diener, R. A. 1964. "Texas Estuaries and Water Resource Development Projects." *Proc. 9th Conf. Water for Texas*, 9:25.
41. Nash, C. B. 1947. "Environmental Characteristics of a River Estuary." *Jour. Marine Res.*, 6:147.
42. Hutner, S. H., L. Provasoli. 1964. "Nutrition of Algae." *Amer. Rev. Plant Physiol.*, 15:37.
43. Marshal, N. 1956. "Chlorophyll *a* in the Phytoplankton in Coastal Waters of the Eastern Gulf of Mexico." *Jour. Marine Res.*, 15:14.
44. Pomeroy, L. R. 1959. "Algal Productivity in Salt Marshes of Georgia." *Limnol. Oceanog.*, 4:386.
45. Pomeroy, L. R., H. H. Haskin, and R. A. Ragotskie. 1956. "Observations on Dinoflagellate Blooms." *Limnol. Oceanog.*, 1:54.
46. Ragotskie, R. A. 1959. "Plankton Productivity in Estuarine Waters of Georgia." *Publ. Inst. Marine Sci. Univ. Texas*, 6:146.
47. Odum, H. T., and C. M. Hoskin. 1958. "Comparative Studies of the Metabolism of Marine Waters." *Publ. Inst. Marine Sci. Univ. Texas*, 5:16.
48. Odum, H.T., and R. F. Wilson. 1962. "Further Studies on Reaeration and Metabolism of Texas Bays, 1958–1960." *Publ. Inst. Marine Sci. Univ. Texas*, 8:23.
49. Childress, U. R. (mimeo). "Coordination of the Blue Crab Studies of the Texas Coast." Marine Fisheries Projects Reports for 1963, Texas Parks and Wildlife Dept., Austin, Texas.
50. Compton, H. (mimeo). "A Study of the Bay Populations of Juvenile Shrimp, *Penaeus aztecus* and *Penaeus setiferus*." Marine Fisheries Projects Reports for 1960–61, Texas Game and Fish Comm., Austin, Texas.
51. Moffett, A. (mimeo). "A Study of the Texas Bay Populations of Juvenile Shrimp." Marine Fisheries Project Reports for 1963, Texas Parks and Wildlife Dept.
52. Gunter, G. 1938. "Seasonal Variations in Abundance of Certain Estuarine and Marine Fishes in Louisiana, with Particular Reference to Life Histories." *Ecol. Monogr.*, 8:313.
53. Springer, V. G., and K. D. Woodburn. 1960. "An Ecological Study of the Fishes of the Tampa Bay Area." Florida State Bd. Conservation Professional Paper Ser. No. 1, 1.
54. Day, J. H. 1951. "The Ecology of South African Estuaries, Part I: A Review of Estuarine Conditions in General." *Trans. Roy. Soc. S. Africa*, 33:53.
55. Day, J. H., N. A. H. Millard, and G. J. Broekhuysen. 1954. "The Ecology of South African Estuaries. Part IV: The St. Lucia System." *Trans. Roy. Soc. S. Africa*, 34:129.
56. Millard, N. A. H., and A. D. Harrison. 1954. "The Ecology of South African Estuaries. Part V: Richard's Bay." *Trans. Roy. Soc. S. Africa*, 34:157.
57. Nikolsky, G. V. 1963. "The Ecology of Fishes." Academic Press, New York.
58. Gunter, G. 1961. "Some Relations of Estuarine Organisms to Salinity." *Limnol. Oceanog.*, 6:182.
59. Williams, A. B. 1955. "A Survey of North Carolina Shrimp Nursery Grounds." *Jour. Elisha Mitchell Sci. Soc.*, 71:200.

# TERRESTRIAL ECOSYSTEMS

# 34

# The ecosystem concept and the rational management of natural resources

*F. Herbert Bormann*
*Gene E. Likens*

Natural resource managers have traditionally emphasized strategies that maximize productivity of some product or service with little or no regard to secondary effects of their strategies. Thus, we export food surpluses, while natural food chains become increasingly contaminated with pesticides and surface and ground waters carry ever larger burdens of pollutants derived from fertilizers and farm wastes. We cut forests with inadequate perception of the effect on regional water supplies, wildlife, recreation, and aesthetic values. We preside over the conversion of wetlands to profitable commercial purposes, with little concern over important hydrologic, biologic, aesthetic and commercial values lost in the conversion.

It has long been apparent to ecologically oriented resource managers that a new conceptual approach to resource management is needed, and it has been widely suggested that the ecological system or ecosystem concept provides the appropriate stage for the play of management roles. The ecosystem is the next higher integrative level beyond that of organisms, and considered to be the basic functional unit of nature. It includes both organisms and their non-living environment intimately linked

Reprinted with permission from *Yale Scientific*, 1971, 45 (7):2–8. The authors are Professor, School of Forestry, Yale University, and Associate Professor, Section of Ecology and Systematics, Cornell University, respectively.

together by a variety of biological, chemical, and physical processes. The biotic and abiotic components interact among themselves and with each other, they influence each other's properties, and both are essential for the maintenance and development of the ecosystem. Knowledge of the structure and function of ecosystems allows us to see more clearly how nature works and how managerial practices affect the working of nature; it spotlights ecological realities which in the long run determine whether or not new technology or economic policy is wise.

The study of the structure and function of ecological systems can be approached from several directions. Many ecologists have emphasized the flow of energy through food webs within the ecosystem; generally they have concentrated their analyses on the biological fraction of the system. Several years ago, we devised an analytical approach based on the measurement of nutrient movement into and out of the ecosystem and their circulation within it. This approach integrates methods data, and concepts from the fields of biology, soils, geology, hydrology and meteorology; it also allows consideration of energy flow through food chains but as a component of the total system.

In the following paragraphs, we discuss the model we use for the study of nutrient cycles, how the parameters of the model are made more measurable when small watersheds meeting certain specifications are used as the basic units of study, and the results of studies on natural and experimentally modified forests. At appropriate points along the way, we introduce managerial considerations made more apparent by the application of the ecosystem concept, and we conclude by reiterating a set of observations on the structure and function of ecosystems that have direct applicability to the management of a wide variety of wildland and man dominated ecosystems.

Our discussion will focus on the deciduous forest since we are most familiar with its structure, function, and development. To gain dimension, the reader might visualize our ecosystem as a 1000 hectare (247 acres) stand of mature deciduous forest. The lateral boundaries of the system might be the biological edge of the stand or some arbitrarily determined line. Vertical boundaries of the ecosystem are defined as the tops of the trees and the deepest depths of soil at which significant biological activity occurs. Thus, our hypothetical forest ecosystem occupies a three-dimensional space including the soil-air interface.

In our conceptual model of the deciduous forest ecosystem, nutrients may be thought of as occurring in any of four basic compartments intimately linked together by an array of natural processes (Fig. 34–2). Compartments are: 1) the organic compartment composed of the biota and its organic debris (there are probably more than 2500 species of plants and animals in our 1000 hectare system); 2) the available nutrient

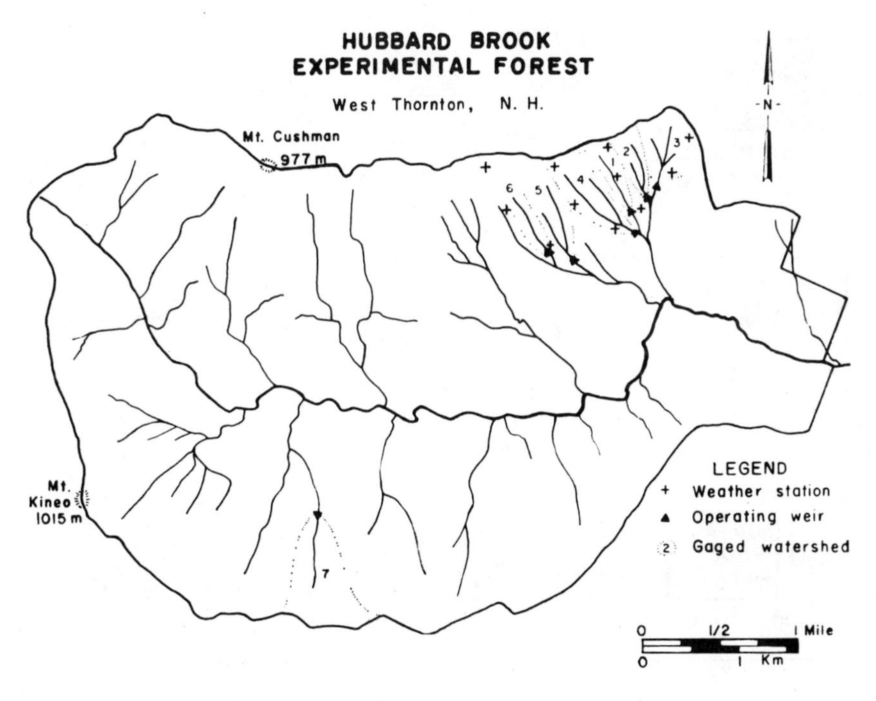


FIGURE 34–1. Outline map of the Hubbard Brook Experimental Forest showing the various drainage streams that are tributary to Hubbard Brook (after Likens et al. 1967).

compartment composed of nutrients held on exchange sites of the clay-humus complex in the soil or in the soil solution; 3) the soil and rock compartment containing nutrients in forms temporarily unavailable to organisms, and 4) the atmospheric compartment made up of gases of the ecosystem both above and below ground. The quantity of any nutrient in any compartment at any time can be estimated by appropriate sampling and analytical procedures.

Nutrients may flux between these compartments along a variety of pathways most of which are powered directly or indirectly by solar energy. Available nutrients may be taken up and assimilated by vegetation and microorganisms, they may circulate in highly complex food webs within the organic compartment, and then be made available again through decomposition or leaching from living or dead organic matter. Soil and rock minerals may be decomposed by weathering and nutrients made available. Available nutrients may be returned to the soil and rock compartment through the formation of new minerals such as clays or insoluble compounds. Nutrients, particularly those without gaseous phase, tend to cycle among the organic, available nutrient, and soil and rock compartments forming an intrasystem cycle (Fig. 34–2). Nu-

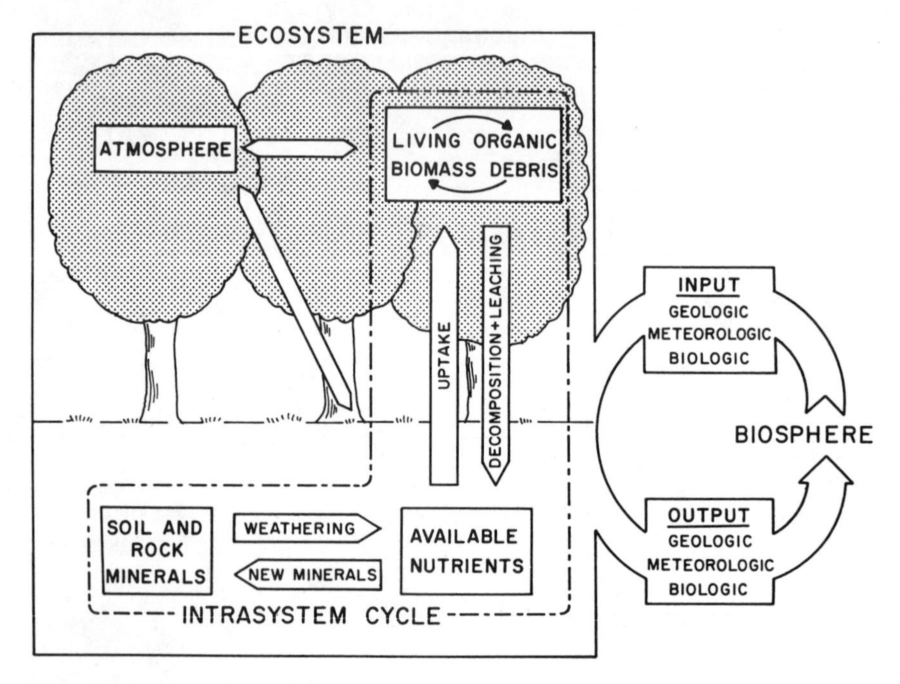


FIGURE 34–2. Nutrient relationships of a terrestrial ecosystem. Inputs and outputs for the ecosystem are moved by meteorologic, geologic, accumulation and exchange pathways within the ecosystem are shown (modified from Bormann and Likens 1967).

trients in gaseous form are continually being transferred to and from other compartments by inorganic chemical reactions such as oxidation and reduction or organic reactions related to such processes as photosynthesis, respiration, nitrogen fixation and nitrogen volatilization.

The forest ecosystem is not an independent entity but is connected to surrounding ecosystems—forest, fields, streams and lakes—and to the biosphere of the earth in general by a system of inputs and outputs. Nutrient inputs and outputs in both organic and inorganic forms are moved through the ecosystem boundary by meteorologic, geologic, and biologic vectors. Meteorologic vectors are phenomena such as solar radiation, wind and precipitation. Geologic input and output is carried by vectors such as moving water or soil creep, while biologic vectors are animals moving into and out of the ecosystem. Activities dependent on human vectors such as fertilization or harvesting are here considered as a specialized biologic vector.

From a managerial point of view, it is important to understand that the structure and function of an ecosystem can be strongly influenced by nutrient inputs, both natural and man-made originating outside the system. The effects of sewage, agricultural, and industrial wastes on lakes and streams are well known. For forest ecosystems, we have only to look at the forest devastation around Ducktown, Tennessee or Sudbury, Ontario to see the effects of $SO_2$ input resulting from metal smelting (Scheffer and Hedgcock 1955) or the dying forest around Los Angeles to see the effect of ozone and PAN (peroxacetyl nitrate) resulting from the internal combustion engine (Cobb and Stark 1970). More subtle, but potentially more devastating effects of air pollution may result from the input of acidified rain (Odén 1968). This will be discussed in somewhat greater detail later.

Equally, the manager (and the consumer) must realize the intimate way his ecosystem is linked to others and that he can produce effects beyond his system that will not appear in his ledgers but will appear in the overall ecologic ledgers. And, of course, it is by the ecological ledgers that we ultimately judge the viability of economic systems. Examples of land management that may extend well beyond the managed ecosystem are practices that accelerate erosion, pest control techniques that result in export of persistent pesticides or residues to regional or world-wide food webs, controlled burning methods that make significant contributions to regional air pollution, and fertilization practices that result in significant leakages of nutrients to surface and ground waters.

The last point should be one of particular concern to forest managers. During the last three decades, the world use of nitrogen in manufactured fertilizer has risen from 4 million metric tons in 1948–52 to use of about 20 million metric tons in 1966–67 (Byerly 1970). Coupled with this is widespread and steadily increasing nitrate pollution of our surface and ground waters, part of which is directly attributable to increasing leakage of nitrate from agricultural ecosystems (Commoner 1970; Byerly 1970). Nitrate pollution of our water supplies is becoming a major national problem and probably will get worse. Forest managers would be wise to take this into account *before* embarking on policies of widespread forest fertilization with nitrogen compounds.

For the past seven years, we have been quantifying the nutrient relationships shown in Figure 34–2 in both undisturbed and experimentally manipulated northern hardwood forest ecosystems in central New Hampshire. To do this, we use small forested watersheds as fundamental units for study and experimental manipulation. Here, the lateral boundaries of the ecosystem are clearly identified as topographic divides and as a consequence nutrient inputs are restricted to meteorologic and biologic vectors since by definition there can be no transfer of geologic inputs between adjacent watersheds (Bormann and Likens 1967). Furthermore, if the watershed-ecosystem 1) is a part of a larger homogeneous biotic unit, 2) has an impermeable geologic substrate, and 3) is characterized by relatively humid conditions, the input-output budget for nutrients without a gaseous phase may be *simply* determined from the difference between meteorological input (nutrients contained in rain and snow) and geological output (nutrients contained in dissolved substances and particulate matter in streams draining the watershed).

Our study is being carried out in cooperation with the Northeastern Forest Experiment Station at the Hubbard Brook Experimental Forest in central New Hampshire (Fig. 34–1). This is the major installation of the U.S. Forest Service in New England for measuring aspects of the hydrologic cycle as they occur in small, forested watersheds. Six watersheds, tributary to Hubbard Brook and 12 to 43 hectares in size, are being studied. Each watershed is forested by a well-developed second-growth stand of sugar maple, beech and yellow birch. The area was heavily cut over in 1919, but since that time there has been no disturbance by cutting or fire.

Precipitation entering these watersheds is measured by a network of gauging stations (Fig. 34–1) and drainage water is measured by weirs anchored on the bedrock across the stream draining each watershed. Since there is no deep seepage, the loss of water by evapotranspiration is calculated by subtracting hydrologic output from hydrologic input.

Water budgets for 1955 to 1968 indicate average precipitation, runoff, and evapotranspiration of 123 cm, 72 cm, and 51 cm, respectively (Likens *et al.* 1970). Precipitation is distributed rather evenly throughout the year whereas runoff is not. Most of the runoff (57%) occurs during the snowmelt period of March, April, and May. In fact, 35% of the total runoff occurs in April. In marked contrast, only 0.7% of the yearly runoff occurs in August.

To measure chemical parameters of the ecosystem, weekly water samples of input (rain and snow) and output (stream water) are collected and analyzed for calcium, magnesium, potassium, sodium, aluminum, ammonia, nitrate, sulfate, chloride, bicarbonate and hydrogen ions and dissolved silica. The concentrations of these elements in precipitation and in stream water are entered into a computing system where weekly concentrations are multiplied by the weekly volume of water entering or leaving the ecosystem and the input and output of chemicals are computed in terms of kilograms of an element per hectare of watershed.

The concentrations of dissolved chemicals in precipitation are very variable over short periods of time. In contrast, concentrations of dissolved chemicals in stream water are largely constant and regression analyses indicate predictable relationships between concentrations of chemical elements and discharge rates of the stream (Johnson *et al.* 1969). Thus, sodium and silica concentrations are found to be inversely related to discharge rates, while aluminum, hydrogen ion, and nitrate concentrations increase as discharge rates increase. Magnesium, calcium, sulfate,

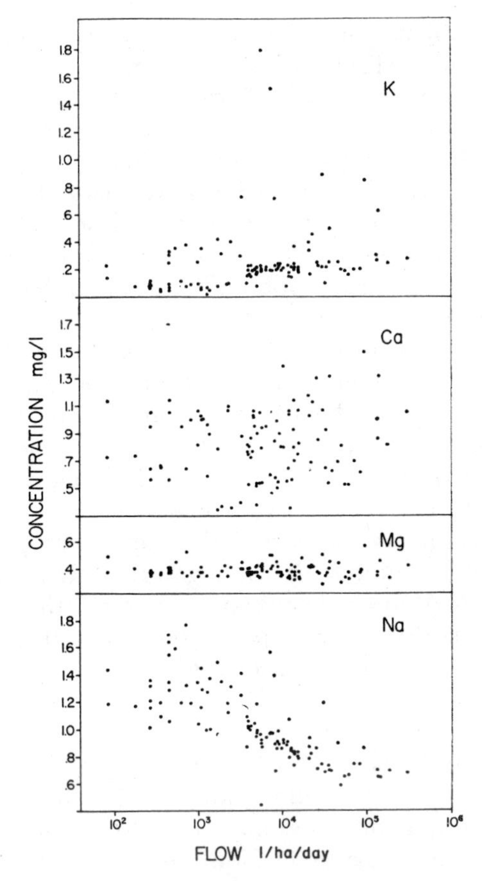


FIGURE 34–3. Relationship between cation concentration and volume of flow for stream in watershed No. 4 during 1963–1965 (after Likens et al. 1967).

chloride and potassium are relatively independent of discharge rate. These systematic and predictable variations in chemical concentrations show the remarkable degree to which streamwater chemistry is under the control of processes inherent in the undisturbed forest ecosystem.

Although there are highly predictable changes in streamwater concentrations as discharge rates change, the magnitudes of concentration change are relatively small (Fig. 34–3). For example, concentrations of potassium, calcium, and magnesium hardly change, while the concentration of sodium decreases only 3X as discharge rate increases by four orders of magnitude, from summer trickles to spring torrents. These results were unexpected since during the spring melt period, when the bulk of liquid water is discharged from our ecosystems, we thought there would be considerable dilution and that concentrations of elements in stream water would be very low. This was not the case.

The chemical stability of the stream water drainage of the mature forest ecosystem at Hubbard Brook is very likely due to the mature highly permeable podzolic soils. Because of microtopography, loose soil structure, and absence of frozen ground during the winter, virtually all of the drainage water must pass through the soil. The intimate contact afforded by this passage, plus the relatively large buffering capacity of the soil materials (relative to the quantities of chemicals lost to percolating water) and the relatively small range of temperatures within the soil mass (10°C at a depth of 91 cm) apparently succeed in buffering the chemistry of transient waters.

Because of the relative constancy of ionic concentrations, total output is strongly dependent on volume of stream flow. Consequently based on a knowledge of hydrologic output alone, it is now possible to predict with a fair degree of accuracy not only chemical concentrations but also total chemical output in the stream water draining from the mature forested ecosystem. This relationship would seem to have considerable value for regional planners concerned with water quality.

A particularly interesting finding is that almost the whole loss of positively charged cations in stream water from the undisturbed forest ecosystem is balanced by the input of positively charged hydrogen ions in precipitation. The occurrence of hydrogen ions, in turn, is related to the amount of sulfate in the precipitation. It is estimated that 50% of the sulfate in precipitation results from industrial air pollution with sulfur dioxide and other sulfur products. These sulfur compounds may ultimately form ionized sulfuric acid, that is, hydrogen ions and sulfate ions. When the precipitation enters the ecosystem, the hydrogen ions replace the positively charged cations on the negatively charged soil exchange sites and the cations are washed out of the system in stream water. Thus, air pollution is directly related to a small but continuous loss of fertility from the terrestrial ecosystem and a small but continuous chemical enrichment of streams and lakes. This may be an extremely important hidden cost that has been made apparent in ecosystem analyses.

Knowing the input and output of chemicals, we have constructed nutrient budgets for nine elements. An average of 2.6, 1.5, 0.7, and 1.1 kg/hectare of calcium, sodium, magnesium, and potassium, respectively entered the ecosystem, while 11.8, 6.9, 2.9, and 1.7 kg/ha were flushed out of the system. These inputs and outputs represent connections of the forest ecosystem with world-wide biogeochemical cycles. These data, from undisturbed forest ecosystems, also provide comparative data whereby we may judge the effect of managerial practices on biogeochemical cycles.

Net losses of calcium, sodium, and magnesium occurred in every year spanning a range of wet, dry, and average years. Potassium, a major component of the bedrock, showed small net gains in two years and a smaller 6-year average net loss. This suggests that potassium is a particularly sensitive element and that it is accumulating in the ecosystem relative to the other cations. Several factors may account for this. Part of the potassium may be retained in the structure of illitic clays developing within the ecosystem and/or potassium may be retained in proportionally greater amounts than other cations in the slowly increasing biomass of the system.

Thus far, we have mentioned only chemical losses occurring as dissolved substances. Losses also occur when chemicals locked up in particulate matter such as rock or soil particles and in organic matter like leaves and twigs, are washed out of the ecosystem by the stream. We have also measured these outputs and developed equations expressing particulate matter losses as a function of discharge rate of the stream. (Bormann, Likens and Eaton 1969).

Briefly, 1) losses of dissolved substances account for the great bulk of chemical losses from our undisturbed ecosystem, and 2) while dissolved substance losses are largely independent of discharge rate, particulate matter losses are highly dependent on discharge rate. The matter point is of special interest since forest management practices can either increase or decrease discharge rates, and thus shift the balance between dissolved substance and particulate matter losses.

One of the principal parameters of any terrestrial ecosystem is weathering, or the rate at which elements bound in primary minerals are made available for biological use. Based on net losses of elements from our ecosystems, a relatively uniform geology in the watershed and a knowledge of the bulk chemistry of the underlying rock and till, we estimate that the nutrients contained in 800 ± 70 kg/ha of bouldery till are made available each year by weathering (Johnson et al. 1968).

We now have for the undisturbed northern hardwood ecosystem of Hubbard Brook estimates of chemical input in precipitation, output in stream water, and the rates at which ions are generated by the weathering of minerals within the system. To complete the picture of nutrient cycling according to the model shown in Figure 34–2, it is also necessary to measure nutrient content of the compartments (biomass-organic debris, available nutrient, primary-secondary mineral, and atmosphere) and the other flux rates between them (uptake, decomposition-leaching, and the formation of new minerals). These parameters, with the exception of new minerals, are shown for the calcium cycle (Fig. 34–4).

These data suggest a remarkable ability of these undisturbed northern hardwood forests to hold and circulate nutrients, for the net annual loss of 9 kg/ha represents only about 0.3% of the calcium in the available nutrient and organic compartments of the system of 1.3% of available nutrients alone.

Knowledge of the nutrient relationships shown in Figure 34–4

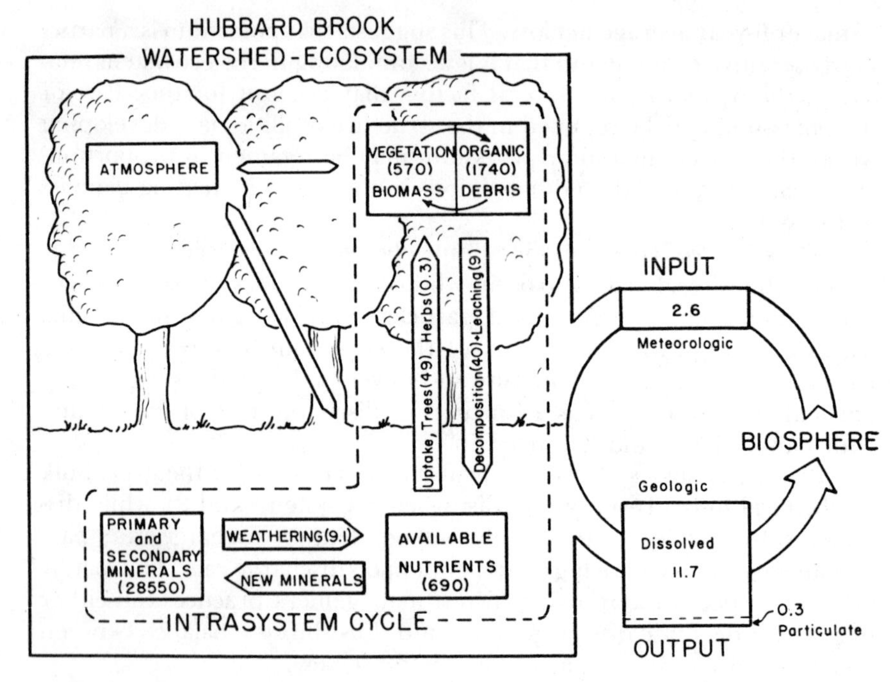


FIGURE 34–4. Major parameters of the calcium cycle in the Hubbard Brook watershed ecosystems. All data in Kg/ha and Kg/ha-yr (after Likens and Bormann 1972).

has considerable value for long term forest management. These data, coupled with information on the quantity of nutrients removed from the ecosystem in forest products, provide a means to evaluate the nutrient drain of harvesting on the nutrient capital of the ecosystem, and provide a rational basis for the development of forest fertilization practices.

The small watershed approach provides a means by which we can conduct experiments at the ecosystem level. Using a watershed ecosystem calibrated in terms of hydrological-nutrient cycling parameters it is possible to impose treatments and to determine treatment effects either by comparison with control watersheds or with predicted behavior had the watershed not been treated. This is, of course, the longstanding method employed by the U.S. Forest Service in its study of the hydrological parameters of forests. This approach makes it possible to evaluate managerial practices in terms of the whole ecosystem rather than isolated parts and to test the effects of various land-management practices (cutting, controlled burning, grazing, etc.) or to determine the effect of potential environmental pollutants (pesticides, herbicides, fertilizers, etc.) on the behavior of nutrients, water, and energy in the system.

In 1965, the forest of one watershed at Hubbard Brook was completely cut in an experiment designed (1) to determine the effect of deforestation on stream flow, (2) to examine some of the fundamental chemical relationships of the forest ecosystem, and (3) to evaluate the effects of forest manipulation on nutrient relationships and eutrophication of stream water. This type of research is particularly relevant in view of the increasingly serious water shortages occurring in the megalopolis belt bordering the inland forested regions of the northeastern United States.

The experiment was begun in the winter of 1965–66 when the forest of one watershed, 15.6 ha in size, was completely leveled by the U.S. Forest Service. All trees, saplings, and shrubs were cut, dropped in place, and limbed so that no slash was more than 1.5 m above ground. No products were removed from the forest and great care was exercised to prevent disturbance of the soil surface thereby minimizing soil erosion. The following summer, June 23, 1966, regrowth of the vegetation was inhibited by an aerial application of 28 kg/ha of the herbicide, Bromacil. The succeeding two years an additional herbicide 2, 4, 5-T was sprayed on the regrowth.

One of the objectives of this severe treatment was to block a major ecosystem pathway, nutrient uptake by the higher plants (Fig. 34–2), while the ultimate decomposition pathway continued to function. Under these circumstances, we questioned whether or not the ecosystem had the capacity to hold the nutrients accumulating in the available nutrient compartment.

Streamwater samples were collected weekly and analyzed, as they had been for a two-year period preceding the treatment. Similar measurements on adjacent undisturbed watersheds provided comparative information.

The treatment had a pronounced effect on runoff. Beginning in May of 1966, runoff from the cutover watershed began to increase over values expected had the watershed not been cut. The cumulative runoff value for 1966 exceeded the expected value by 40%. The greatest difference occurred during the months of June through September, when actual runoff values were 418% greater than expected values (Bormann *et al.* 1968). The difference is directly attributable to changes in the hydrologic cycle resulting from the removal of the transpiring surface. Accelerated runoff has continued through the summers of 1967, 1968, and 1969 (Pierce *et al.* 1970). The marked increase in water output is of particular interest in view of the developing water shortage in southern New England. However, other factors, yet to be discussed, show that the water increase following our drastic treatment is bought at the expense of other valuable ecosystem functions.

In the deforested ecosystem, our treatment resulted in a fundamental alteration of the nitrogen cycle which in turn caused extraordinary losses of soil fertility. This is best understood by a brief consideration of the nitrogen cycle and patterns of nitrogen output in the undisturbed ecosystem. Nitrogen fixation and volatilization are not discussed because, as yet, we have no data on these processes.

Nitrogen incorporated in organic compounds is ultimately decomposed in a number of steps to ammonium-nitrogen, $NH_4+$ (Fig. 34–5). Ammonium is a positively charged cation and can be held fairly tightly within the soil on negatively charged exchange sites. Ammonium ions may be taken up directly by green plants and used in

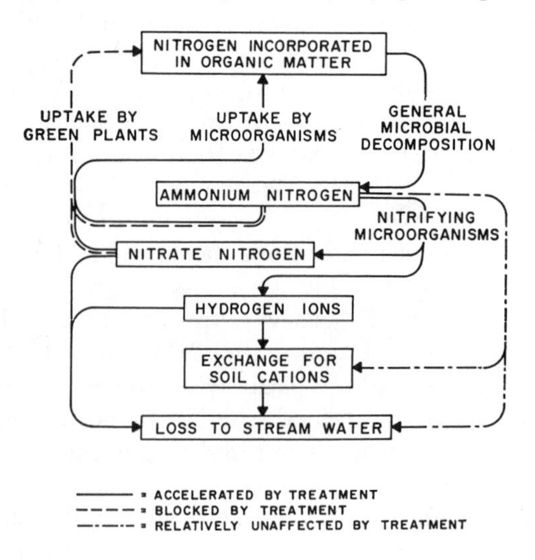


FIGURE 34–5. Effects of deforestation on the nitrogen cycle.

the fabrication of nitrogen-containing organic compounds, or they might be used as the substrate for the nitrification process. Two genera of soil bacteria, *Nitrosomonas* and *Nitrobacter* oxidize ammonium to nitrate, $NO_3^-$. In the process of nitrification, two hydrogen ions, $H^+$, are produced for every ion of ammonium oxidized to nitrate. As we have already mentioned, $H^+$ ions may play a vital role in the release of cations. Nitrate, a negatively charged ion, is highly leachable, and if it is not taken up by higher plants it can be easily removed from the ecosystem in drainage water.

As yet we have been unable to quantify the amount of nitrification that goes on in the podzol soil underlying our undisturbed forests. A fairly high rate may occur, but this process may be so closely coupled to higher plant uptake that we only see minor amounts in the drainage water. Also, we have some evidence that heterotrophic nitrification occurs. On the whole, however, certain aspects of our data plus the general findings of soil scientists working with podzol soils under mature forests suggest that the process of nitrification is of minor importance in our undisturbed forest. Hence, our working hypothesis is that the bulk of the nitrogen is decomposed to ammonium, which is taken up directly by green plants rather than converted to nitrate-nitrogen.

Nitrate concentrations in stream water are generally quite low (<1mg/l, except during late winter and spring) and the output of nitrate from the undisturbed ecosystem can be accounted for by its input in precipitation. In fact, according to budgetary analyses, undisturbed ecosystems are accumulating nitrogen at the rate of about 2 kg/ha-yr. Data from the undisturbed forests indicate a strong and reproducible seasonal cycle of nitrate concentrations in stream water (Fig. 34–6). Higher concentrations are associated with the winter period from November through April, while low concentrations persist from April through November.

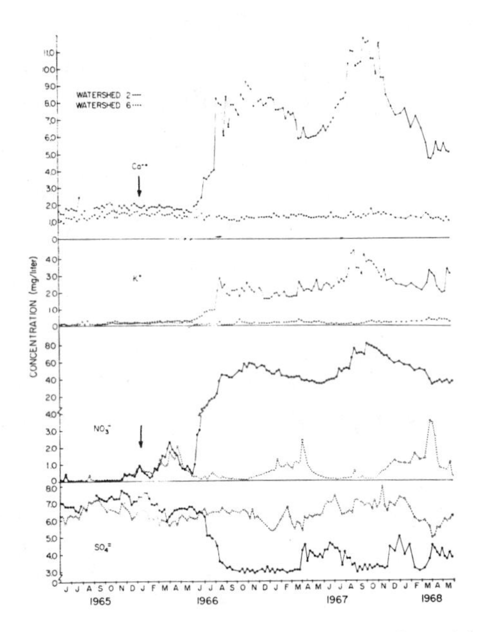


FIGURE 34–6. Measured streamwater concentrations for calcium, potassium, nitrate and sulfate ions in watersheds 2 (deforested) and 6. Note the change in scale for the nitrate concentration. The arrow indicates the completion of cutting in watershed 2 (modified from Likens *et al.* 1970).

The decline of nitrate concentration during May and low concentrations throughout the summer correlate with heavy nutrient demands by the vegetation and generally increased biologic activity associated with warming of the soil. The winter pattern of nitrate concentration may be explained primarily in physical terms, since input of nitrate in precipitation from November through May largely accounts for nitrate lost in stream water during this period.

Beginning on June 7, 1966, 16 days *before* the herbicide application, nitrate concentrations in the deforested watershed showed a precipitous rise while the undisturbed ecosystem showed the normal spring decline (Fig. 34–6). High concentrations of nitrate have continued for three years in the stream water draining from the deforested system. The magnitude of nitrate losses is a clear indication of the acceleration of nitrification in the deforested ecosystem. There is no doubt that cutting and repeated applications of herbicides drastically altered conditions controlling the nitrification process.

The action of the herbicide in the cutover watershed is one of reinforcing the already well-established trend of nitrate loss induced by forest cutting alone (Fig. 34–6). This is probably effected through the destruction of the remaining vegetation, herbaceous plants and root sprouts, by the herbicide. Even in the event of rapid transformation of all the nitrogen in the Bromacil, this source could at best contribute less than 1% of the total nitrogen lost as nitrate during 3 years and the herbicide 2, 4, 5-T does not contain nitrogen.

Average net losses of nitrate nitrogen were 120 kg/ha during the period 1966 through 1968 (Likens *et al.* 1970). We estimate that the annual nitrogen turnover in our undisturbed forests is approximately 60 kg/ha. Consequently, an amount of elemental nitrogen equivalent to double the amount normally taken up by the forest has been lost each year since cutting.

In the process of nitrification, hydrogen ions are produced and these replace metallic cations on the exchange surfaces of the soil. This is precisely what is seen in the deforested ecosystem. Calcium, magnesium, sodium, and potassium concentrations in the stream water increased almost simultaneously with the increase in nitrate. This was followed about one month later by a sharp rise in the concentrations of aluminum.

Net losses of potassium, calcium, aluminum, magnesium, and sodium were 21, 10, 9, 7, and 3 times greater, respectively, than those for an undisturbed watershed for the water years 1966–67 and 1967–68 (Likens et al. 1970). This represents a major loss of nutrients from the ecosystem, and the data from this experiment suggest that commercial forestry should focus more attention on the effect of harvesting practices on the loss of solubilized nutrients in drainage water. However, it should be emphasized that regrowth of vegetation following a normal cutting operation would act to decelerate nutrient losses.

Our results indicate that the capacity of the ecosystem to retain nutrients is dependent on the maintenance of intrasystem nutrient cycling (Fig. 34–2). When the cycle is broken, as by vegetation destruction, losses of nutrients are greatly accelerated. The accelerated rate is related both to the cessation of nutrient uptake by plants and to the larger quantities of drainage water passing through the ecosystem. Accelerated losses may be also related to increased decomposition rates resulting from changes in the physical environment, e.g. increased soil temperature, somewhat moister soil or a modest increase in organic matter available for decomposition.

However, the direct effect of the vegetation on the process of nitrification cannot be overlooked. There is a body of evidence from other regions which suggests that various types of climax vegetation have the capacity to chemically inhibit the process of nitrification. Under these conditions, the positively charged ammonium ion may be held on the exchange sites, while the production of the highly leachable nitrate ion is inhibited. Disturbance of these ecosystems may lead to a loss of inhibition and an acceleration of nitrate production. If this applies to Hubbard Brook, release of the process of nitrification as a result of cutting the vegetation and inhibiting regrowth by herbicide applications may account for the major losses of nutrients from the cutover ecosystem.

Results of the particulate matter study also indicate a basic change in the pattern of losses with 3 years of data indicating approximately a nine-fold increase over a comparable undisturbed ecosystem. After an initial surge, losses of particulate organic matter in the stream water draining from the deforested watershed have declined as a result of the virtual elimination of primary production of organic matter within the ecosystem. In contrast, loss of inorganic material from the stream bed has accelerated because of the greater erosive capacity of the now augmented streamflow and because several biological barriers to surface soil erosion and stream bank erosion have been greatly diminished. The extensive network of fine roots that tended to stabilize the bank is now dead as a result of cutting and herbicide treatment, and the dead leaves that tended to plaster over exposed banks are now gone. The continuous layer of litter that once protected the soil surface is now discontinuous. With continued denudation of the watershed, we would expect an exponential increase in output of inorganic particulate matter.

Finally, the export of nitrate to the small stream draining the completely cut watershed has resulted in nitrate concentrations exceeding established pollution levels for more than three years. In general, the continued deforestation treatment has resulted in eutrophication of the stream ecosystem and in the occurrence of algal blooms. This latter finding indicates that under some circumstances forest management practices can make significant contributions to the eutrophication of our streams.

Our comparative study of undisturbed and completely cut watersheds illustrates, very clearly, that homeostasis of the ecosystem is linked to orderly flow of nutrients between the biotic and abiotic fractions of ecosystem and the production and decomposition of biomass. These processes, integrated with the climatic cycle, result in relatively tight intra-system nutrient cycles, minimum output of nutrients and water, and maximum stability of the ecosystem in terms of its capacity to resist erosion. Destruction of the vegetation sets off a complex chain of interactions (Fig. 34–7) the net effect of which is an increase in amount and flow rates of water and the breakdown of biological barriers to erosion and transportation coupled with an increase in the export of nu-

HUBBARD BROOK ECOSYSTEM STUDY BASED ON TWO YEARS OF DATA 1966 - 1968

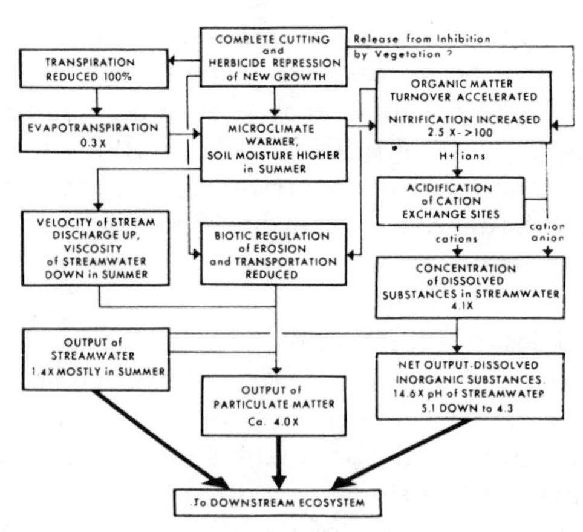


FIGURE 34–7. A summary of some of the ecological effects of deforestation of watershed No. 2 in the Hubbard Brook Experimental Forest (modified from Bormann and Likens 1970).

trient capital (available nutrients and organic matter) and inorganic particulate matter. The rate at which the ecosystem can return to the more stable condition after imposed denudation ceases is dependent on the degree to which the ecosystem can support vegetative regrowth. If this capacity is seriously diminished long term successional processes will be required to restore the ecosystem to its original level of productivity.

Our study at Hubbard Brook not only illustrates the structure and function of the northern hardwood ecosystems, it also reinforces the suggestion of many ecologists that the ecosystem concept is a powerful analytical tool useful in assessing management strategies. It provides the best means for seeing nature whole and for a rational scheme of ecologic bookkeeping.

Three points, inherent in the ecosystem concept and emphasized by the Hubbard Brook study, should be recognized as fundamental knowledge for any wise scheme of land-use management. A) The ecosystem is a highly complex natural unit composed of all the organisms (plants, animals, including man, and microbes) and their inorganic environment (air, water, soil, and rock). B) All parts of an ecosystem are intimately linked together by natural processes that are part of the ecosystem, such as uptake of nutrients, fixation of energy, movement of nutrients and energy through food webs, release of nutrients by decomposition of organic matter, weathering of rock and soil minerals to release nutrients, and the formation of new minerals. C) Individual ecosystems are linked to surrounding land and water ecosystems and to the biosphere in general by input and output connections with world-wide circulation of air and water and with food webs.

Among some engineers and technologists concerned with natural resource management, there has been a tendency to leap over the basic complexity of the ecosystem, particularly biologic complexity, and to deal with a simplified model containing only obvious physical and biological components. The flaw in this strategy is evident in the many instances where discounted organisms or physical factors have become major disturbing factors in environmental management schemes. For example, the increasingly important role of humid zone pests in desert areas with overhead irrigation; the rising levels of schistosomiasis in tropical countries with expanding ditch type irrigation systems; the rise in importance of new pests following pesticide control of old pests; and the short life of many reservoirs due to unanticipated sedimentation.

Management practices applied to any single aspect of the ecosystem will be transmitted to other parts of the ecosystem by existing pathways and beyond the ecosystem by its output links with the surrounding biosphere. This principle is illustrated in Figure 34–7 where we have portrayed an array of biotic and abiotic interactions resulting from the removal and suppression of new forest growth at Hubbard Brook.

The complex linkages that exist in any ecosystem bring under question the wisdom of any management plan that does not consider them as fully as possible. Although a manager may limit utilization to one good or service and enter into his ledgers only money values related to that good or service, in actuality, by the very nature of the ecosystem, he is affecting all values of the system in some degree or other. Whether or not he chooses to valuate these changes is a management decision, nevertheless changes occur and are ecologic realities. Consequently, good environmental management, i.e. good for society, requires that managerial practices be imposed only after a careful analysis and valuation of all the ramifications the proposed practice might have on the natural ecosystem.

Failures in environmental management often result because of a widespread lack of appreciation of the complexity of nature and the assumption that we can manage one part of nature alone, or from an acceptance of the naive assumption that somehow or other nature has the capacity to absorb all the types of minipulations we throw at her and still serve man in ways most desirable to him. Technologists have provided many "solutions" to environmental problems only to find in the long run that they have merely shifted the problem from one segment of the ecosystem to another. The polluted condition of our air and water, the tragedy of Lake Erie, the increasing burden of thermal pollution and of persistent pesticides on the world ecosystem are, in large part, the accumulated result of many short term "solutions" developed without due regard to all of the ecological factors involved.

It is necessary, if we are to find real solutions, that the talents of technologists be put to use within the holistic framework of ecological systems, where interrelations of all matter, living and dead, and the limits of growth and development are recognized.

## References

Bormann, F.H. and G.E. Likens 1967. Nutrient cycling. *Science* 155:424–429.

——— 1970. The nutrient cycles of an ecosystem. *Scientific American* 223(4):92–101.

Bormann, F.H., G.E. Likens, and J.S. Eaton 1969. Biotic regulation of particulate and solution losses from a forest ecosystem. *BioScience* 19(7):600–610.

Bormann, F.H., G.E. Likens, D.W. Fisher, and R.S. Pierce 1968. Nutrient losses accelerated by clear-cutting of a forest ecosystem. *Science* 159:882–884.

Byerly, T.C. 1970. Environment and agriculture; issues and answers. *Agr. Sci. Rev.* 8(1):1–8.

Cobb, F. W., Jr. and R. W. Stark 1970. Decline and mortality of smog-injured Ponderosa Pine. *J. of Forestry* 68:147–149.

Commoner, B. 1970. Soil and fresh water: damaged global fabric. *Environment* 12(4):4–11,

Johnson, N.M., G.E. Likens, F.H. Bormann, and R.S. Pierce 1968. Rate of chemical weathering of silicate minerals in New Hampshire. *Geochim. Cosmochim. Acta.* 32:531–545.

Johnson, N.M., G.E. Likens, F.H. Bormann, D.W. Fisher, and R.S. Pierce 1969. A working model for the variation in stream water chemistry at the Hubbard Brook Experimental Forest, New Hampshire. *Water Resources Research* 5(6):1353–1363.

Likens, G.E. and F.H. Bormann 1972. Nutrient cycling. *In* J. Wiens (Ed.), *Ecosystems, structure and function*, Oregon State Univ. Press, Corvallis, pp. 25–67.

Likens, G.E., F.H. Bormann, N.M. Johnson, and R.S. Pierce 1967. The calcium, magnesium, potassium, and sodium budgets of a small forested ecosystem. *Ecology* 48:772–785.

Likens, G.E., F.H. Bormann, R.S. Pierce, and D.W. Fisher 1970. Nutrient-hydrologic cycle interaction in small forested watershed-ecosystem. Proc. Colloque sur la Productivite des Ecosystems Forestiers dans la Monde. UNESCO, Brussels.

Odén, S. 1968. Nederbordens och Luftens Forsurning-dess Orsaker, Forlopp och I Olida Miljoer, Statens Naturvetenskapliga Forskningsgrad, Stockholm, Bull. No. 1, 86 pp.

Pierce, R.S., J.W. Hornbeck, G.E. Likens, and F.H. Bormann 1970. Effects of elimination of vegetation on stream water quantity and quality. *Int. Symp. on the Results of Research on Representative and Experimental Basins*. Int. Assoc. Sci. Hydrol., New Zealand, pp. 311–328.

Scheffer, T.C. and G.G. Hedgcock 1955. Injury to northwestern forest trees by sulfur dioxide from smelters. *Tech Bull. U.S. Dept. Agr.* No. 1117, 49 pp.

# 35

# The unexploited tropics

*Daniel H. Janzen*

I question statements to the effect that we can solve the food problems of underdeveloped tropical countries with better methods of farming. Aside from the obvious fact that, even if successful, this would be only a most temporary solution to the problem of tropical overpopulation, there are sound ecological reasons why such a grand experiment should not be tried at all. The lowland tropics are already overexploited for the most part and we are in danger of hurting future generations of tropical peoples when we suggest that the solution of the world's food problem lies in converting more wild tropical vegetation to permanent cropland. In the following paragraphs, I will explore some of the distinctive aspects of tropical agriculture, and ask the significance of this distinctiveness to tropical agricultural peoples.

A plant has three major external defenses against herbivores. There is weather inimical to insects, mites, rodents and other plant pests. There is the large complex of carnivorous predators and parasites, each of which does its part in slowing the rate at which plant pests increase their populations once they have located host plants. Finally, the individuals of the host plant population are scattered through the community, making it difficult for the host specific plant pests to locate them.

When a *temperate zone* forest or prairie is converted to a monoculture field or orchard, only the latter two defenses of the plants remaining in the community are destroyed. The winter works for the farmer, not

Reprinted with permission from the *Bulletin of the Ecological Society of America*, 51:4–7, 1970.

against him. The two missing defenses may be replaced by integrated biological and chemical control methods, with chemical control being viewed as either augmenting the internal chemical defenses of the plant, or an exceptionally general predator with no reproductive ability.

When a *tropical zone* community is converted to a monoculture, all of the primary external defenses of the plant are destroyed. The more the weather regime for the field approaches that which we think of as lowland humid tropical, (1) the more the plant must escape from herbivores through its distance from other individuals of the same species and (2) the less competent the parasite and predator complex will be at retarding the growth rates of its increasingly diffuse prey populations. If any given combination of (1) host plant spatial heterogeneity and (2) predator-parasite complex is adequate for survival of the plant in a temperate zone community, it will probably be less than adequate in a tropical one. The lack of a winter will probably have to be made up for with superior internal chemical defenses by the tropical plant (or with increased pesticide application).

Small wonder, then, that farming success in the lowland tropics is directly proportional to the length and severity of the dry season, with the limit being the point where there is not enough wet season to grow a crop.

We are confronted now with two questions: What is the prognosis for attempts to increase the yield of tropical lands already under cultivation, and should more uncultivated land be converted to monocultures?

When tropical vegetation is cleared, high soil temperatures and rainfall lead to almost immediate removal of the inorganic nutrients released from the dead plants. At best, the remedy of fertilizers, intelligent methods of clearing, and promotion of micorrhizal associations makes the new field an adequate substrate for growing crops. It does not guarantee a crop sufficient to compensate for the total cost of its production. Tropical fields in so-called primitive slash-burn agriculture systems are allowed to go fallow after 1 to 3 years of use, because yield declines drastically after that time. The reasons for the decline are often obvious on close inspection, and well known to Central American farmers that live close to their crops. More fertilizer only aggravates the problem. The fields are left fallow because, per unit volume of vegetation, they have more herbivorous insects than does any other kind of tropical habitat.

These insect populations have been steadily building up on host plants that are much more closely spaced than under natural circumstances. They have been relatively free of the predator-parasite complex, which is largely made up of species that have not evolved a behavior and physiology adequate to harvest such immense quantities of prey made available so suddenly. Further, the plants under this herbivore pressure are undergoing extreme root and canopy competition from wild perennial weeds that do not lose their entire vegetative investment on an annual basis, or more often, as do most crop plants. Finally, man wishes to harvest seeds, fruits, or storage tubers. However, when the plant is placed under severe herbivore pressure and weed competition, these are generally the last part of the energy budget to be attended to. The plant must grow vegetatively or lose status in the field.

As the human population density rises in such an ecosystem, the shorter must be the fallow period and hence the closer the fields in space and time. However, the closer the fields, the larger and more constant will be the weed seed and insect inoculum from other fields. The shortening fallow period will be progressively less effective at eliminating the offending pest populations through normal succession. As the speed and magnitude of inoculation from outside and inside the site increase, we can expect an exponential rate of pest increase to the point where yield does not warrant the effort of planting and harvesting. It is of interest to note that if the plant species that used to occupy the older stages in fallow successions have been made locally extinct by shortening the fallow period, the situation cannot immediately be alleviated by simply increasing the fallow time. Plant communities do not go through successional changes unless seed trees of the replacement species are present. It is tempting to ask how important such an irreversible perturbation of the ecosystem may be to tropical peoples that plan to clear large areas of tropical forest that appear underexploited by contemporary methods of slash-burn agriculture.

For the insect component of the problem of increasing tropical agricultural yield, the cures appear to the (1) judicious use of insecticides coupled with maintenance of habitat heterogeneity by crop rotation, (2) promotion of biological control agents, (3) growing plants far from their native pests, (4) use of tree crops, and (5) intensive hand care of crops. Lest we imagine that such a repertoire of techniques will revolutionize tropical agriculture, I would like to explore some of their more salient features.

Insect resistance to insecticides is a simple manifestation of the fact that insects have been countering the chemical defenses of plants for at least 75 million years. There is no reason to believe that tropical insects are less competent at this game than temperate ones. In fact, since insect-plant biochemical coevolution is a much larger part of the structural glue of tropical communities than temperate ones, I would expect tropical insects to be better at it than temperate insects. It is surely no coincidence that the tropics are well known for plants containing large amounts of defensive chemicals (rubber-tree latex, spices, rotenone, caffeine, cannabinol, strychnine, cortisone, etc.). A second major factor that may lead to faster development of insecticide resistance in tropical insects than temperate ones is that the insecticide may be the major mortality factor that the plant pest has to deal with. The tropical insect's entire physiology can evolve in the direction of a resistant phenotype without conflicting selective forces such as winter or drought-hardiness.

The speed of this evolution of resistance may be further augmented by the fact that food plants are generally bred to have few defensive chemicals, structures, and behaviors. For the crop plant drawn from a tropical community, a plant whose defense against insects was greatly dependent on secondary compounds, the problem may be especially bad. The pest insect can divert its biochemical machinery from dealing with the plant's internal defenses to the pesticide. The absence of a great variety of vegetables in lowland tropical markets, a fact often lamented by visitors from temperate zones, is a direct manifestation of the crop plant's inability to defend itself in an ecosystem where chemical defenses are paramount.

In short, as we move into the tropics we grossly increase the need for the use of insecticides, yet maximize the chances of resistance developing. Where this leads us has already been demonstrated by cotton growers in Central America. They are using up to 20 to 30 parathion applications per crop, and have managed to exterminate most wild vertebrates and non-pest insects in areas of extensive lowland agriculture. The outcome has been to increase the number of pest species by 2 to 10 fold and reduce the resource base of large areas such as lowland El Salvádor to a level probably far below that of pre-Columbian agriculture.

Crop rotation is pest control through escape of the crop in time. The effectiveness of this escape is measured by the number of pests that survive from one crop to the next. As the duration and intensity of the inimical seasons between crops decline (as we move from the dry tropics to the wet tropics, or as we move from temperate to tropical zones), and the availability of wild alternate plant hosts during the off season increases over the same gradient, the time between successive crops of the same species must be increased. If we are to increase agricultural yield by crop rotation in the tropics, we must have more kinds of crops. However, if there is anything that characterizes successful tropical herbaceous crops, it is their low species diversity. The situation is further aggravated by the reduction in chemical diversity among the crop species. This is brought about through selective breeding for (1) more palatable crops, (2) more rapidly growing plants, and (3) more varieties for ease of post-harvest processing.

The control of pest populations by self-reproducing populations of predators and parasites may be expected to work least well in large monocultures of tropical vegetation. Tropical entomophagous insects often appear to be more host-specific than temperate ones. When pesticides are used in the field, the predator-parasite complex is not only directly poisoned, but also suffers from a lack of its particular hosts. This is likely to make the return to initial population sizes a slower process than in temperate fields where pesticides and biological control agents are being used in concert. Every pest individual, on the other hand, is guaranteed to be sitting in a large patch of host plants with almost no intraspecific competition, whether it is a tropical or temperate zone field.

Harvest of the crop, followed by the usual migration of the pest out of the field or population crash within the field, has an effect on the predator-parasite complex similar to the application of a pesticide. In temperate areas, entomophagous insects are physiologically and behaviorally adapted to dealing with major fluctuations in their host density to a much greater degree than are their tropical counterparts.

There are two other reasons to be pessimistic about the future of biological control in large tropical monocultures. First, a tremendous number of tropical predaceous and parasitic insects pass the dry season (or other time of prey scarcity) as active adults in vegetation other than that inhabited by the majority of their usual herbivorous prey. As cultivation intensity increases, the volume, diversity, and ubiquity of such vegetation declines rapidly. Second, much of the success of biological control methods in temperate areas depends on there being a major amount of crop growth in the spring before the herbivorous prey has attained the density characteristic of a predator-prey equilibrium in the field. The equilibrium density of entomophage and prey is likely to be attained much more rapidly in a tropical crop than a temperate one, since the inoculum of predator and prey is both early and large. In fact, in many tropical fields, the prey never leave; they live on scattered plants growing in the field during the "off" season. Having the equilibrium density of herbivorous prey reached early in the life of the crop may well be associated with a higher level of damage than is acceptable in the economics of the crop.

For some tropical crop plants, growing them in new geographic areas and thereby leaving their pests behind is promising, as evidenced by the success of rubber, cacao, cotton, coffee, bananas, and sugar cane, when grown as introduced plants. However, these plants are primarily cash crops and will not feed large numbers of people; worse, they tend to subjugate their growers to the economic whims of the developed countries. When plants that yield a crop suitable for direct consumption by the grower are introduced, the local insects likewise find them worth harvesting.

Woody plants characterize the tropics. Many successful tropical agricultural monocultures have been based on perennial plants such as coconuts, rubber, cacao, coffee, tea, sugar cane, bananas, pineapples, tree fruits and nuts, etc. Such crops require a long-term investment of land, capital, and labor; they do not in themselves support large human populations. They cannot be readily rotated with other crops. They have a very low caloric yield per acre, for the reason that large plants use most of their photosynthate for maintenance and defense metabolism. For this same reason, the harvest of foliage from tropical tree crops is likely to be especially detrimental, and should yield a crop richer in plant poisons and lower in caloric content than temperate vegetation. Genetic strains of trees with the foliage palatability comparable to herbaceous crop plants will take many more years to develop than herbaceous plants, yet will have the same susceptibility to insect pests as herbaceous plants. Because of insect outbreaks, tropical foresters have generally encountered extreme difficulty in attempting to produce pure stands of native lumber trees, yet here there is not even a requirement of leaf palatability.

It is clear that intensive hand-care can overcome many of the difficulties mentioned to this point. Rice can be babied to harvest in the face of almost any competitive and herbivorous community, especially if pesticide applications are at levels that would never be accepted in developed countries. With truly excessive doses of fertilizers, and subsequent unknown pollution effects on stream runoff, high yield strains of some plants can be produced; here, the field has been turned into a hydroponic garden. If the goal of developed countries is to see the underdeveloped countries swell their populations of people tied to intensive and mindless hand-care of the crops that will feed them, then I suppose it is legitimate to continue to encourage them to further exploit the small semblance of semi-undisturbed vegetation remaining. But have you ever tried to engage in intellectual activity after cutting weeds with a machete all day in the hot sun? I may note that we are in no hurry to turn our pitifully small forest reserves, refuges, and parks into corn fields to feed the expanding populations of the world. Obviously the total yield of the tropics could be increased by some ten to twenty percent with essentially total destruction of the tropics as a place to live. We could turn the tropics into one giant cattle pasture and rice paddy. I do not, however, note a mass immigration of people to live amongst Iowa corn fields, Kansas wheat fields, or Texas cattle pastures.

I suggest that those who are so ready to use the tropics to feed the hungry masses of the world start with cultivation of the underexploited lands of the developed countries. The temperate zones' productivity could be increased by at least 50 percent. We would then be able to offer the underdeveloped countries a graphic example of how their countries will appear if they continue to allow their populations to expand on the assumption that the warm tropical sun will provide an infinite source of food.

# 36

## On living in the biosphere

G. E. Hutchinson

In discussing the subject of "The World's Natural Resources," I want first to make a number of general observations that will provide an intellectual framework into which our developing knowledge, both academic and practical, may be fitted. We live in a rather restricted zone of our planet, at the base of its gaseous envelope and on the surface of its solid phase, with temporary excursions upwards, downwards, or sideways onto or into the oceans. These regions in which we can live and which we can explore are characterized by their temperature, which does not depart far from that at which water is a liquid, and by their closeness to regions on which solar radiation is being delivered. This zone of life is spoken of as the biosphere. Within it, certain natural products can be utilized in both biological and cultural life. It is customary to consider these resources as either material or energetic, but the two categories are not easily separable; contemporary solar radiation is an energetic resource, coal and oil are to be regarded as material resources valuable for their high energy content, which we may call, epigrammatically, fossil solar radiation. There is a third very important though inseparable aspect, namely, the pattern of distribution. Most fossil sunlight, or chemical energy of carbonaceous matter, is diffused through sedimentary rocks in such a way as to be useless to us. Schrödinger says that we feed on negative entropy, and I am almost tempted to regard pattern as being as fundamental a gift of nature as sunlight or the chemical elements.

Reprinted with permission from *Scientific Monthly*, 67:393–397 (December 1948). Also reprinted in *Itinerant Ivory Tower*, Yale University Press, 1953.

The first major function of the sunlight falling on the earth's surface is as the energy of circulation of the oceans and atmosphere. The second is to increase the mobility of water molecules, to become latent heat of evaporation, and so to keep the water cycle operating. The third major function is photosynthesis. Apart from atomic energy, and a little volcanic heat which presumably is actually of radioactive origin, all industrial energy is solar and due to one or the other of these three processes.

The material requirements of life are extremely varied. Between thirty and forty chemical elements appear to be normally involved. Industrially, some use appears to be found for nearly all the natural elements, and some of the new synthetic ones also. Looking at man from a strictly geochemical standpoint, his most striking character is that he demands so much—not merely thirty or forty elements for physiological activity, but nearly all the others for cultural activity. What we may call the anthropogeochemistry of cultural life is worth examining. We find man scurrying about the planet looking for places where certain substances are abundant; then removing them elsewhere, often producing local artificial concentrations far greater than are known in nature. Such concentrations, whether a cube of sodium in a bottle in the laboratory, or the George Washington Bridge, have usually been brought into being by chemical changes, most frequently reductions, of such a kind that the product is unstable under the conditions in which accumulation takes place. Most artifacts are made to be used, and during use the strains to which they are submitted distort them, and they become worn-out or broken. This results in a very great quantity of the materials that are laboriously collected being lost again in city dumps and automobile cemeteries. The final fate of an object may depend on many factors, but it is probable that in most cases a very large quantity of any noncombustible, useful material is fated to be carried, either in solution or as sediment, into the sea. Modern man, then, is a very effective agent of zoogenous erosion, but the erosion is highly specific, affecting most powerfully arable soils, forests, accessible mineral deposits, and other parts of the biosphere which provide the things that *Homo sapiens* as a mammal and as an educable social organism needs or thinks he needs. The process is continuously increasing in intensity, as populations expand and as the most easily eroded loci have added their quotas to the air, the garbage can, the city dump, and the sea.

The most important general consideration to bear in mind in discussing the dynamics of the biosphere and its inhabitants is that some of the processes of significance are acyclical, and others, to a greater or less degree, cyclical. By an acyclical process will be meant one in which a permanent change in geochemical distribution is introduced into the system; usually a concentrated element tends to become dispersed. By a cyclical process will be meant one in which the changes involved introduce no permanent alteration in the large-scale geochemical pattern, concentration alternating with dispersion. The cylical processes are not necessarily reversible in a thermodynamic sense; in fact, they are in general no more and no less reversible than the acyclical. Most of the cyclical processes operate because a continuous supply of solar energy is led into them, and sunlight will provide no problems for the conservationist for a very long time. It is important to realize that most of the acyclical processes are so slow that man appears as an active intruder into a passive pattern of distribution. They are safer to disturb because we know what the result of the disturbance will be. If we mine the copper in a given region sufficiently assiduously, we know that ultimately there will not be any more copper available there. Cyclical processes involve complex circular paths, regenerative circuits, feedback mechanisms, and the like. Small disturbances of such processes may merely result in small temporary changes, with a rapid return to the previous steady state. This has been beautifully demonstrated in the experiments of Einsele, who added single massive doses of phosphate to a lake, changing for a time, but only for a time, its entire chemistry and biology. This stability does not imply that if large disturbances strain the mechanism beyond certain critical limits very profound disruption will not follow; in fact, the very self-regulatory mechanisms that give the system stability against small disturbances are likely to accentuate the disruption when the critical limits are transcended. In disturbing cylical processes we usually do not know what we are doing.

The most nearly perfect cyclical processes are those involving water and nitrogen. Some losses to the sediments of the deep oceanic basins must occur, but they are very small and are doubtless fully balanced or more than balanced by juvenile water and perhaps by molecular nitrogen and ammonia of volcanic origin.

The least cyclical processes are those in which material is removed from the continents and deposited in the permanent basins of the ocean. With one or two exceptions, the delivery to the deep-water sediments of the ocean is of little significance. Most of the mechanical and chemical sedimentation in the oceans takes place in relatively shallow water. The uplifting of shallow water sediments constitutes an important method of completing cycles. One, and perhaps two, exceedingly important elements are, however, sedimented less economically. Calcium, during the Paleozoic, was mainly precipitated in shallow water, but since the rise of the pelagic foraminifera in the Mesozoic, a great deal of calcium, along with an equivalent amount of carbon and oxygen, has been continually diverted to regions from which it is unlikely ever to be removed. At present the sedimentary rocks of the world are an adequate biological and commercial source of calcium, but, with progressive orogenic cycles, less and less of the element will be uplifted (Kuenen), and whatever organisms inherit the earth in that remote future will have to face the problem of the biosphere "going sour on them." For phosphorus, the case is less well established, but there is probably a slow loss in the form of sharks' teeth and the ear bones of whales (Conway), which are very resistant and which are known to be littered about on the floor of the abysses of the ocean.

It is desirable to consider two of the main geochemical cycles in order to gain an idea of the effect of man upon them. It must be admitted that we are ignorant of many matters of importance here. In the cycle of carbon we have a remarkable, possibly a unique, case in which man, the miner, increases the cyclicity of the geochemical process. It is generally admitted that our available store of carbon is ultimately of volcanic origin. A steady stream of carbon dioxide and lesser amounts of methane and carbon monoxide are entering the atmosphere from volcanic vents. Part, probably a major part, of this carbon dioxide is ultimately lost to the marine sediments as limestones; a very small part of it is then returned to the air, wherever lime kilns are in operation. Another part of the $CO_2$ entering the atmosphere is reduced in photosynthesis. A part of the organic matter so formed is fossilized, and a small part of this fossilized organic carbon is available as fuel, in the form of coal and oil. At the present time it appears that the combustion of coal and oil actually returns carbon to the atmosphere as $CO_2$ at a rate at least a hundred times greater than the rate of loss of all forms of carbon, oxidized and reduced, to the sediments (Goldschmidt). This particular process obviously cannot go on indefinitely. It concerns only the reduced carbon; to complete the cycle in the case of oxidized carbon, a great deal of energy would have to be supplied. It concerns only the reduced carbon which is aggregated. The poorest sources would be the poorest exploitable oil shales. Most of the reduced carbon is much more dispersed than this; in making an estimate of the total reduced carbon of the sediments, the commercially usable fuels constitute a negligible fraction that need not be considered.

Although the rate at which carbon dioxide is returned to the air by the human utilization of fossil fuels is so very much greater than is the primary production of carbon dioxide from volcanic sources, the rate is evidently a very small fraction—of the order of 1 percent—of the rate of photosynthetic fixation and subsequent respiratory liberation of $CO_2$ by the organisms of the earth. Since about 1890 a slight increase in $CO_2$ content in the air, at least at low altitudes over the land surfaces of the Northern Hemisphere, has been noted. This has been attributed to the accumulation of industrially produced $CO_2$, as the quantity of $CO_2$ that has appeared in the atmosphere is of the same order of magnitude as the total combustion of fuel (Callendar). In view of the small fraction of the total $CO_2$ production that industrial output represents, it seems very unlikely that merely adding an extra percent to the natural biological production should overload the cyclic process, so that it rejects quantitatively the additional load. It is known that the air at high altitudes still shows nineteenth-century values. The most reasonable explanation of the observed increase is that the photosynthetic machinery of the biosphere has been slightly impaired, probably by deforestation. It is clear that, in any intelligent long-term planning of the utilization of the biosphere, an extended study of atmospheric gases is desirable, even though for the moment it seems unlikely that the observed change is a particularly serious symptom.

The only other cycle that can be considered in any great detail is that of phosphorus. The chief event in the geochemical cycle of phosphorus is the leaching of the element from the rocks of the continents, and its transport by rivers to the sea. At the present time the rate of this transportation is of the order of 20,000,000 tons of phosphorus per year for the entire earth. Part of this phosphorus, when it enters the sea, will ultimately be deposited in the sediments of the depths of the ocean. Such phosphorus will probably be largely lost to the geochemical cycle, as has just been indicated. The sedimentary rocks of the continents, therefore, will gradually lose phosphorus; there is some evidence that this has actually occurred (Conway). The main return path is by the uplifting of sediments formed in continental seas, which then undergo renewed chemical erosion. Of particular interest are methods by which concentrated phosphorus can be returned to the land surfaces. As far as is known, there are two such methods: The first is the formation of phosphatic nodules and other forms of phosphate rock in regions of upwelling in which water at a low $pH$, rich in phosphate, is brought up to the surface of the sea. The $pH$ falls and an apatite-like phosphate is deposited. When the sea floor is later elevated, a commercial deposit may result. The second method is by the activity of sea birds, such as the guano birds of the Peruvian coast. There is little or no unequivocal evidence that guano deposits of great extent were formed prior to the late Pliocene or Pleistocene. Some of the well-known occurrences of rock phosphate, such as that of Quercy, have been explained in this way, but they are certainly not typical guano deposits. During the late Tertiary and Pleistocene, an extraordinary amount of phosphate was deposited on raised coral islands throughout the world, and bird colonies seem to provide the only reasonable agencies of deposition. The great deposits of Nauru, Ocean Island, Makatea, Angaur, the Daito Islands, Christmas Island south of Java, Curaçao, and some of the other West Indies all seem to have been formed in this way. This process is as characteristic of the time as is glaciation, though less grandiose. Its meaning is not clear, but it is probably connected with changes in vertical circulation of the ocean as glaciopluvial periods gave place to interpluvials. Today in certain regions massive amounts of guano are deposited, and it is probable that the oceanic birds of the world as a whole bring out from several tens to several hundreds of thousands of tons of phosphorus and deposit it on land. Only about 10,000 tons of the element are delivered in places where it is not washed away and where it can be carried by man to fertilize his fields.

The main processes that tend to reverse the phosphorus depletion of the continents are, therefore, the deposition of marine phosphorites on the continental shelves and subsequent elevation, and the formation of guano deposits. Both processes are evidently intermittent, and are quantitatively inadequate to arrest deflection of the element into the permanent ocean basins. Man contributes both to the loss and to the gain of phosphorus by land surfaces. He quarries phosphorite, makes superphosphate of it, and spreads it on his fields. Most of the phosphorus so laboriously acquired ultimately reaches the sea. At present the world's production of phosphate rock is about 10,000,000 tons per annum. This contains from 1,000,000 to 2,000,000 tons of elementary phosphorus. Human activity probably, therefore, accounts for from 5 to 10 percent of the loss of phosphorus from the land to the sea. Man also contributes to the processes bringing phosphorus from the sea to the land. This is done by fisheries. The total catch for the marine fisheries of the earth is of the order of $25–30 \cdot 10^6$ tons of fish, which corresponds to about 60,000 tons of elementary phosphorus. The human, no less than the nonhuman, processes tending to complete the cycle seem miserably inadequate. It is quite certain that ultimately man, if he is to avoid famine, will have to go about completing the phosphorus cycle on a large scale. It will be a harder task than that of solving the nitrogen problem, which would have loomed large in any symposium on "The World's Natural Resources" fifty years ago, but possibly an easier problem than some of the others that must be solved if we are to survive and really become the glory of the earth.

The population of the world is increasing, its available resources are dwindling. Apart from the ordinary biological processes involved in producing population saturation already known to Malthus, the current disharmony is accentuated by the effect of medical science, which has decreased death rates without altering birth rates, and by modern wars, which one may suspect put greater drains on resources than on populations. Terrible as these conclusions may appear, they have to be faced. The results of the interaction between population pressure and decline of resource potential are further partly expressed in such wars, which are pathological expressions of attempts to cope with these and other problems and which now invariably aggravate the situation. It is evident that the fundamental causes of war lie in those psychological properties of populations which make them attempt solutions in a warlike manner, and *not* in the existence of the problems themselves. The two problems of war and of resources are, however, at present very closely interrelated; it is probably impossible to find a solution for one without progress in the solution of the other. Any kind of reasonable use of the world's resources involves better international relations than now exist. It is otherwise impossible to operate on a planetary basis, or to avoid the fearful material and spiritual expense of living in a world divided into two armed camps. The lack of international trust is the first difficulty in achieving rational utilization of the resources of the world; the second difficulty for the United States is what may be called the problem of the transition from the pioneering to the old, settled community. For a pioneer, life may be hard, but in good country there is "plenty more where these came from," whether lumber or buffalo tongues or copper. This attitude is incorporated into popular thought very deeply; in a crude form, it is now completely destructive, but it seems possible that attitudes might be developed that could utilize such a point of view. There is at least one thing of which we have plenty more than we use, whether we call it Yankee ingenuity, American know-how, or the human intellect. In some industrial fields there has been a notable series of triumphs of the kind that really will give plenty more of a number of things, and give them in a cyclic manner. The rise of the magnesium industry, and the utilization of the magnesium of sea water as a source of the metal are interesting examples. The production of plastics, though it is probably at the moment not as geochemically economical as it should be, is another. These, along with the development of silicate building materials not involving any particularly uncommon elements, all point the way to the kind of material culture that permits a reasonably high living standard, at least in certain directions, without devastation of the earth.

The future outlook for the world, particularly in food resources, has been put before the public in several recent books, notably those of Fairfield Osborn and William Vogt. Anyone with any technical knowledge understands that the dangers described in these books are real enough, more real and more dangerous perhaps than the threat of an atomic world war. The problem of getting action to forestall such dangers in a culture that has developed under conditions of potentially unlimited abundance just around the corner, is obviously extremely difficult. I do not think it is impossible. The first requirement is a faith that the job can be done. The very difficulty of the task, its apparent impossibility, may here and now prove a challenge that brings the desired response. The difficult we do at once, the impossible takes a little longer. I doubt that a direct appeal to fear will produce any results except a disbelief in the prophets of doom. Cassandra seems even more unpopular in modern America than in ancient Ilium. There would seem to be forces operating in society which tend to reverse the destructive processes, or which could be made to do so. One of the most immediate needs is to find out what they are and do everything possible to strengthen them. I will give one example. A number of industrial concerns—particularly in the chemical and pharmaceutical field, but also some engineering and publishing firms—have used in their advertising a legitimate pride in the learning, skillful research, and development that have gone into the manufacture of their product. It is quite certain that there are many cases in which one particular product, the result of considerable research, and of taking risks in development, actually reduces the drain on the natural resources of the country and the world. I should like to see a small systematic experiment, on the part of some such concern, in advertising in which it is pointed out that by buying this product one is letting the industrial skill behind the product operate for the benefit of one's children. I am fully aware that if this point of view were worked up skillfully enough by a few responsible, public-spirited corporations and put into a form that would pay the corporation, as well as the country and the world, there would be other less responsible concerns that would use the method when their product is actually not one sparing natural resources. This, however, seems to be a lesser evil than the total neglect of the commercial advertising field, which is one of the most potent in determining the values of the public and which at present is largely disruptive. I do not doubt that a professional cultural anthropologist could pick out a great many other fields that could be used to promote the idea of an expanding economy based on an abundance of human ingenuity rather than on an excess of raw materials.

There is one further point that I should like to develop. Though the pursuit of happiness is embedded deeply in the constitutional foundations of the United States, we do not know much about it. It is fairly certain that no metric exists that can be applied directly to happiness, but intuitively we may proceed a little way by arguing as if such a metric could exist. It is obvious that only a very few people, with a genius for sanctity, can be happy if half-frozen and starving. If the temperature be raised, and the food supply and other amenities increased, the possibility of happiness is obviously at first also increased, but beyond a certain increase in the environmental resources available no further increase in happiness would be expected and we might begin to look for a decline. The image of an overheated kitchen, filled with too much electrical equipment and catering to overfed people, will, if adequately evoked, have a nightmarish quality. In more formal language, if we could find a function of the environmental resources that expressed the relationship of happiness to those resources, it is reasonably certain that the function would not be monotonic. For every resource there seems likely to be an optimum level of consumption, but we do not know if this optimum level is, in any particular case, widely exceeded, so that gross overutilization is actually producing avoidable distress.

The problem is not by any means as simple as that of determining discrete optima. In any given society all the cultural values are probably interrelated to form a coherent system, so that the existence of one set of values may greatly modify the others. If, as seems possible, our attitude toward food leads a considerable section of our population to be definitely overweight, it is legitimate to inquire to what extent, by changes in the upbringing of our children, the psychological needs filled by food can be satisfied in other ways so that the psychologically optimum intake falls to a level nearer the psychological optimum for the individual and the moral optimum for the world. We might ask, for instance, how we can substitute the delights of ballet and Mozartian opera, which are geochemically very cheap, for part of those provided by hot dogs or apple pie and ice cream, which may in the long run prove too expensive to use except as a source of energy and essential nutrients. This example is chosen with a view to indicating that the kind of substitutions that might be considered need not be in the least puritanical. It may appear overintellectualized, but that at least is a guarantee that it is not inhuman. Indeed it raises the very interesting problem that those of us in the educational world have to face, namely, why we are raising a generation in the belief that the majority of constructive, complicated, difficult activities are boring duties, when the same generation shows us that in certain specific fields, mainly concerned with electronic amplifiers and with the explosive combustion of hydrocarbons, complicated activity can be very entertaining. What we have to do is to show by example that a very large number of diversified, complicated, and often extremely difficult constructive activities are capable of giving enormous pleasure. This is, in fact, the reason why it is essential that the teachers in our colleges and universities should be enthusiastic investigators in the fields of scholarship or practitioners and critics in their arts. It ought to be possible to show that it is as much fun to repair the biosphere and the human societies within it as it is to mend the radio or the family car.

# 37

# Man-made climatic changes

*Helmut E. Landsberg*

Climate, the totality of weather conditions over a given area, is variable. Although it is not as fickle as weather, it fluctuates globally as well as locally in irregular pulsations. In recent years some people have voiced the suspicion that human activities have altered the global climate, in addition to having demonstrated effects on local microclimates. There have also been a number of proposals advocating various schemes for deliberately changing global climate, and a number of actual small-scale experiments have been carried out. For most of the larger proposals, aside from considerations of feasibility and cost, one can raise the objection that a beneficial effect in one part of the earth could well be accompanied by deterioration elsewhere, aside from the inevitable disturbances of the delicate ecological balances.

But the question "Has man inadvertently changed the global climate, or is he about to do so?" is quite legitimate. It has been widely discussed publicly—unfortunately with more zeal than insight. Like so many technical questions fought out in the forum of popular magazines and the daily press, the debate has been characterized by misunderstandings, exaggerations, and distortions. There have been dire predictions of imminent catastrophe by heat death, by another ice age, or by acute oxygen

---

Reprinted with permission from *Science*, 170:1265–1274 (18 December 1970).  A revision of this article is to be published under the title "Man-made climatic changes (1971 revision)" by the World Meteorological Organization, Geneva, in 1972. The author is research professor at the Institute of Fluid Dynamics and Applied Mathematics, University of Maryland, College Park 20742.

deprivation. The events foreseen in these contradictory prophecies will obviously not all come to pass at the same time, if they come to pass at all. It seems desirable to make an attempt to sort fact from fiction and separate substantive knowledge from speculation.

## Natural Climatic Fluctuations

In order to assess man's influence, we must first take a look at nature's processes.

The earth's atmosphere has been in a state of continuous slow evolution since the formation of the planet. Because of differences in the absorptive properties of different atmospheric constituents, the energy balance near the surface has been undergoing parallel evolution. Undoubtedly the greatest event in this evolution has been the emergence of substantial amounts of oxygen, photosynthetically produced by plants (*1*). The photochemical development of ozone in the upper atmosphere, where it forms an absorbing layer for the shortwave ultraviolet radiation and creates a warm stratum, is climatically also very important, especially for the forms of organic life now in existence. But for the heat balance of the earth, carbon dioxide ($CO_2$) and water vapor, with major absorption bands in the infrared, are essential constituents. They absorb a substantial amount of the dark radiation emitted by the earth's surface. The condensed or sublimated parts of the atmospheric water vapor also enter prominently into the energy balance. In the form of clouds they reflect incoming short-wave radiation from the sun, and hence play a major role in determining the planetary albedo. At night, clouds also intercept outgoing radiation and radiate it back to the earth's surface (*2*).

Over the past two decades Budyko (*3*) has gradually evolved models of the global climate, using an energy balance approach. These models incorporate, among other important factors, the incoming solar radiation, the albedo, and the outgoing radiation. Admittedly they neglect, as yet, nonlinear effects which might affect surface temperatures (*4*) but it seems unlikely that, over a substantial period, the nonlinear effects of the atmosphere-ocean system will change the basic results, though they may well introduce lags and superimpose rhythms. Budyko's calculations suggest that a 1.6 percent decrease in incoming radiation or a 5 or 10 percent increase in the albedo of the earth could bring about renewed major glaciation.

The theory that changes in the incoming radiation are a principal factor governing the terrestrial climate has found its major advocate in Milankovitch (*5*). He formulated a comprehensive mathematical model of the time variations of the earth's position in space with respect to the sun. This included the periodic fluctuations of the inclination of the earth's axis, its precession, and the eccentricity of its orbit. From these elements he calculated an insolation curve back into time and the corresponding surface temperature of the earth. He tried to correlate minima with the Pleistocene glaciations. These views have found considerable support in isotope investigations, especially of the $^{18}O/^{16}O$ ratio in marine shells (*6*) deposited during the Pleistocene. Lower $^{18}O$ amounts correspond to lower temperatures. Budyko and others (*7*) raise some doubts that Milankovitch's theory can explain glaciations but admit that it explains some temperature fluctuations. For the last 1700 years there is also evidence that the $^{18}O$ content of Greenland glacier ice is inversely correlated to a solar activity index based on auroral frequencies (*8*). Again, low values of $^{18}O$ reflect the temperature at which the precipitation that formed the firn fell.

The fluctuations of externally received energy are influenced not only by the earth's position with respect to the sun but also by changes in energy emitted by the sun. Extraterrestrial solar radiation fluctuates with respect to spectral composition, but no major changes in total intensity have yet been measured outside the atmosphere. The occurrence of such fluctuations is indicated by a large number of statistical studies (*9*), but ironclad proof is still lacking. Such fluctuations are of either long or short duration. They have been tied to the solar activity cycle. Inasmuch as details are yet unknown, their effect on climate is at present one factor in the observed "noise" pattern.

In the specific context of this discussion, we are not concerned with the major terrestrial influences on climate, such as orogenesis, continental drift, and pole wanderings. But other, somewhat lesser, terrestrial influences are also powerful controllers of climate. They include volcanic eruptions that bring large quantities of dust and $CO_2$ into the air, and natural changes of albedo such as may be caused by changes in snow and ice cover, in cloudiness, or in vegetation cover (*10*). The fact that we have not yet succeeded in disentangling all the cause-and-effect relations of natural climatic changes considerably complicates the analysis of possible man-made changes.

## The Climatic Seesaw

It was only a relatively short time ago that instrumental records of climate first became available. Although broad-scale assessments of climate can be made from natural sources, such as tree rings (*11*) or pollen associations, and, in historical times, from chronicles that list crop conditions or river freezes, this is tenuous evidence. But a considerable number of instrumental observations of temperatures and precipitation are available for the period from the early 18th century to the present, at least for the Northern Hemisphere. These observations give a reasonably objective view of climatic fluctuations for the last two and a half centuries. This is, of course, the interval in which man and his activities have multiplied rapidly. These long climatic series are mostly from western Europe (*12*), but recently a series for the eastern seaboard of the United States has been reconstructed from all available data sources. In this series Philadelphia is used as an index location, since it is centrally located with respect to all the earlier available records (*13*). Figures 37–1 and 37–2 show the annual values for temperature and precipitation for a 230-year span; there are some minor gaps where the data were inadequate. These curves are characteristic of those for other regions, too.

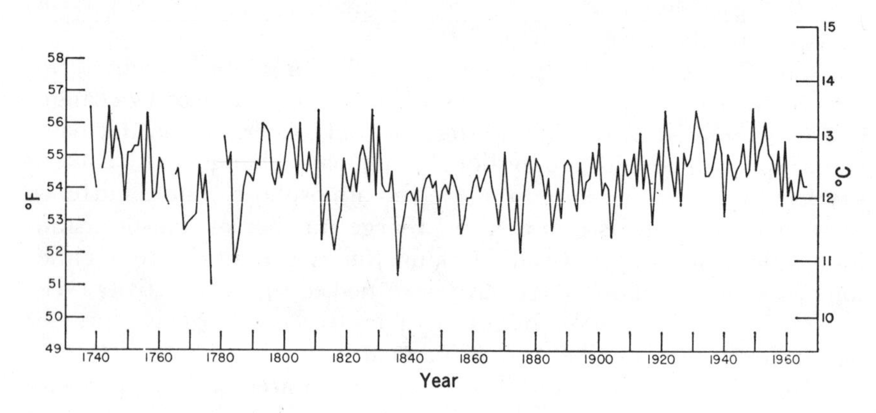


FIGURE 37–1. Annual temperatures for the eastern seaboard of the United States for the period 1738 to 1967—a representative, reconstructed synthetic series centered on Philadelphia.

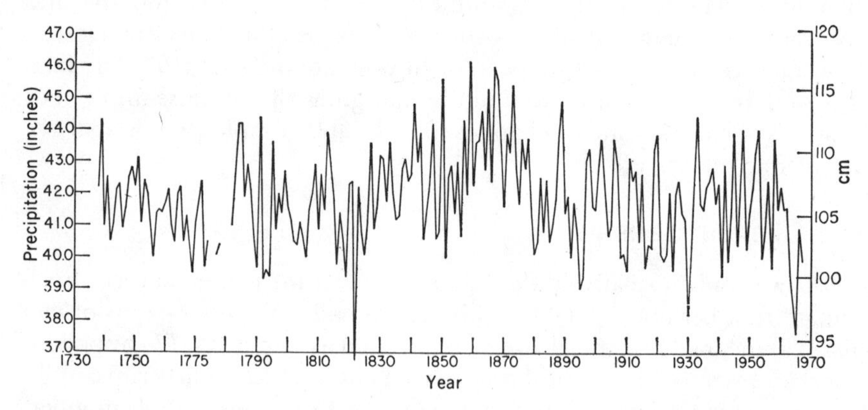


FIGURE 37–2. Annual precipitation totals for the eastern seaboard of the United States for the period 1738 to 1967—a representative, reconstructed synthetic series centered on Philadelphia.

In particular they reflect the restlessness of the atmosphere. Many analysts have simply considered the variations to be quasi-random. Here I need only say that they do not reflect any pronounced one-sided trends. However, there are definite long or short intervals in which considerable one-sided departures from a mean are notable. On corresponding curves representing data for a larger area that encompasses most of the regions bordering the Atlantic, the major segments are those for the late 18th century, which was warm; the 19th century, which was cool; and the first half of the 20th century, in which there was a notable rising trend. This trend was followed by some cooling in the past 2 decades.

In the precipitation patterns, "noise" masks all trends, but we know that during a period in the middle of the last century there was considerably more precipitation than there is now. For shorter intervals, spells of drought alternate with high precipitation. Sometimes, for small areas, these can be quite spectacular. An example is the seasonal snowfall on Mount Washington, in New Hampshire; there the snowfall increased from an average of 4.5 meters in the winters of 1933–34 to 1949–50 to an annual average of 6 meters in the period 1951–52 to 1966–67 (*14*). Yet these values should not be taken as general climatic trends for the globe, or even for the hemisphere. Even if we take indices that integrate various climatic influences, we still cannot make categorical statements. Glacier conditions are typical in this group of indices. For example, the glaciers on the west coast of Greenland have been repeatedly surveyed since 1850. In consonance with temperature trends for lower latitudes, they showed their farthest advances in the 7th decade of the 19th century and have been retreating ever since (*15*). This pattern fits the temperature curves to the 1950 turning point, but, although glaciers in some regions of the world have been advancing since then, this is by no means true of all glaciers. The question of whether these changes reflect (i) relatively short-term temperature fluctuations, or (ii) alterations in the alimenting precipitation, or (iii) a combination of these two factors remains unanswered.

Many of the shorter fluctuations are likely to be only an expression of atmospheric interaction with the oceans. Even if external or terrestrial impulses affect the energy budget and cause an initial change in atmospheric circulation, notable lag and feedback mechanisms involving the oceans produce pulsations which, in turn, affect the atmosphere (*16*). The oceans have a very large thermal inertia, and their horizontal motions and vertical exchanges are slow. Namias (*17*) has investigated many of the fluctuations of a few years' duration. He concluded, for example, that drought conditions on the eastern seaboard of the United States in the 1960's were directly affected by the prevailing wind system and by sea-surface temperatures in the vicinity but that the real dominant factor was a wind-system change in the North Pacific. Such teleconnections (relations among conditions in distant parts of the globe) complicate interpretations of local or even regional data tremendously. The worldwide effect of changes in the Pacific wind system is obvious from Namias's estimate that accelerations and decelerations cause large-scale breaks in the regime of sea-surface temperatures. These seem to occur in sequences of approximately 5 years and may cause temperature changes of 0.5°C over the whole North Pacific. Namias estimates that this can cause differences of $8 \times 10^{18}$ grams in the annual amounts of water evaporated from the surface. The consequences for worldwide cloud and rain formation are evident. It is against this background that we have to weigh climatic changes allegedly wrought by man.

## Carbon Dioxide

The fact that the atmospheric gases play an important role in the energy budget of the earth was recognized early. Fourier, and then Pouillet and Tyndall, first expressed the idea that these gases acted as a "greenhouse" (*18*). After the spectrally selective absorption of gases was recognized, their role as climatic controls became a subject of wide debate. The capability of $CO_2$ to intercept long-wave radiation emitted by the earth was put forward as a convenient explanation for climatic changes. Arrhenius (*19*) made the first quantitative estimates of the magnitude of the effect, which he mainly attributed to fluctuating volcanic activity, although he also mentioned the burning of coal as a minor source of $CO_2$. The possibility that man-made $CO_2$ could be an important factor in the earth's heat balance was not seriously considered until Callendar (*20*), in 1938, showed evidence of a gradual increase in $CO_2$ concentration in the earth's atmosphere. But it was Plass (*21*) who initiated the modern debate on the subject, based on his detailed study of the $CO_2$ absorption spectrum. The crucial question is, How much has $CO_2$ increased as a result of the burning of fossil fuels? It is quite difficult to ascertain even the mean amount of $CO_2$ in the surface layers of the atmosphere, especially near vegetation. There are large diurnal and annual variations. Various agriculturists have reported concentrations ranging from 210 to 500 parts per million. The daily amplitudes during the growing season are about 70 parts per million (*22*). Nearly all early measurements were made in environments where such fluctuations took place. This, together with the lack of precision of the measurements, means that our baseline—atmospheric $CO_2$ concentrations prior to the spectacular rise in fossil fuel consumption of this century—is very shaky. Only since the International Geophysical Year have there been some regularly operating measuring points in polar regions and on high mountains and reliable data from the oceans which give some firm information on the actual increase (*23*).

The best present estimate places the increase in atmospheric $CO_2$ since 1860 at 10 to 15 percent. This is hardly a spectacular change, but the rate of increase has been rising, and various bold extrapolations have been made into the 21st century. Much depends on the sinks for $CO_2$ which at present are not completely known. At present concentrations, atmospheric $CO_2$ and $CO_2$ stay in approximate equilibrium, through the photosynthetic process in plants. It is estimated that $150 \times 10^9$ tons of $CO_2$ per year are used in photosynthesis (*24*). A corresponding amount is returned to the atmosphere by decay, unless the total volume of plant material increases (*25*). This volume is one of the unknowns in the estimates of $CO_2$ balance. Perhaps satellite sensors can give some bulk information on that point in the future. The oceans are a major sink for $CO_2$. The equilibrium with the bicarbonates dissolved in seawater determines the amount of $CO_2$ in the atmosphere. In the exchange between atmosphere and ocean, the temperature of the surface water enters as a factor. More $CO_2$ is absorbed at lower surface-water temperatures than at higher temperatures. I have already pointed out the fact that surface-water temperatures fluctuate over long or short intervals; most of these ups and downs are governed by the wind conditions. The interchange of the cold deep water and the warm surface water through downward mixing and upwelling, in itself an exceedingly irregular process, controls, therefore, much of the $CO_2$ exchange (*26*). Also, the recently suggested role of an enzyme in the ocean that facilitates absorption of $CO_2$ has yet to be explored. Hence it is quite difficult to make long-range estimates of how much atmospheric $CO_2$ will disappear in the oceanic sink. Most extrapolators assume essentially a constant rate of removal. Even the remaining question of how much the earth's temperature will change with a sharp increase in the $CO_2$ content of the atmosphere cannot be unambiguously answered. The answer depends on other variables, such as atmospheric humidity and cloudiness. But the calculations have been made on the basis of various assumptions. The model most widely used is that of Manabe and Wetherald (*27*). They calculate, for example, that, with the present value for average cloudiness, an increase of atmospheric $CO_2$ from 300 to 600 parts per million would lead to an increase of 2°C in the mean temperature of the earth at the surface. At the same time the lower stratosphere would cool by 15°C. At the present rate of accumulation of $CO_2$ in the atmosphere, this doubling of the $CO_2$ would take about 400 years. The envisaged 2°C rise can hardly be called cataclysmic. There have been such worldwide changes within historical times. Any change attributable to the rise in $CO_2$ in the last century has certainly been submerged in the climatic "noise." Besides, our estimates of $CO_2$ production by natural causes, such as volcanic exhalations and organic decay, are very inaccurate; hence the ratio of these natural effects to anthropogenic effects remains to be established.

## Dust

The influence on climate of suspended dust in the atmosphere was first recognized in relation to volcanic eruptions. Observations of solar radiation at the earth's surface following the spectacular eruption of Krakatoa in 1883 showed measurable attenuation. The particles stayed in the atmosphere for 5 years (*28*). There was also some suspicion that summers in the Northern Hemisphere were cooler after the eruption. The inadequacy and unevenness of the observations make this conclusion somewhat doubtful. The main exponent of the hypothesis that volcanic dust is a major controller of terrestial climate was W. J. Humphreys (*29*). In recent years the injection into the atmosphere of a large amount of dust by an eruption of Mount Agung has renewed interest in the subject, not only because of the spectacular sunsets but also because there appears to have been a cooling trend since (*30*). The Mount Agung eruption was followed, in the 1960's, by at least three others from which volcanic constituents reached stratospheric levels: those of Mount Taal, in 1965; Mount Mayon, in 1968; and Fernandina, in 1968. Not only did small dust particles reach the stratosphere but it seems likely that gaseous constituents reaching these levels caused the formation of ammonium sulfate particles through chemical and photochemical reactions (*31*). The elimination of small particulates from the stratosphere is relatively slow, and some backscattering of solar radiation is likely to occur.

As yet man cannot compete in dust production with the major volcanic eruptions, but he is making a good try. However, most of his solid products that get into the atmosphere stay near the ground, where they are fairly rapidly eliminated by fallout and washout. Yet there is some evidence that there has been some increase in the atmospheric content of particles less than $10^{-4}$ centimeter in diameter (*32*). The question is simply, What is the effect of the man-made aerosol? There is general agreement that it depletes the direct solar radiation and increases radiation from the sky. Measurements of the former clearly show a gradual increase in turbidity (*33*), and the same increase in turbidity has been documented by observations from the top of Mauna Loa, which is above the level of local contamination (*34*). From these observations the conclusion has been drawn that the attenuation of direct solar radiation is, in part at least, caused by backscattering of incoming solar radiation to space. This is equivalent to an increase in the earth's albedo and hence is being interpreted as a cause of heat loss and lowered temperatures (*35*). But things are never that categorical and simple in the atmosphere. The optical effects of an aerosol depend on its size distribution, its height in the atmosphere, and its absorptivity. These properties have been studied in detail by a number of authors (*36*). It is quite clear that most man-made particulates stay close to the ground. Temperature inversions attend to that. And there is no evidence that they penetrate the stratosphere in any large quantities, especially since the ban, by most of the nuclear powers, of nuclear testing in the atmosphere. The optical analyses show, first of all, that the backscatter of the particles is outweighed at least 9 to 1 by forward scattering. Besides, there is a notable absorption of radiation by the aerosol. This absorption applies not only to the incoming but also to the outgoing terrestrial radiation. The effectiveness of this interception depends greatly on the overlapping effect of the water vapor of the atmosphere. Yet the net effect of the man-made particulates seems to be that they lead to heating of the atmospheric layer in which they abound. This is usually the stratum hugging the ground. All evidence points to temperature rises in this layer, the opposite of the popular interpretations of the dust effect. The aerosol and its fallout have other, perhaps much more far-reaching, effects, which I discuss below. Suffice it to say, here, that man-made dust has not yet had an effect on global climate beyond the "noise" level. Its effect is puny as compared with that of volcanic eruptions, whose dust reaches the high stratosphere, where its optical effect, also, can be appreciable. No documented case has been made for the view that dust storms from deserts or blowing soil have had more than local or regional effects.

Dust that has settled may have a more important effect than dust in suspension. Dust fallen on snow and ice surfaces radically changes the albedo and can lead to melting (*37*). Davitaya (*38*) has shown that the glaciers of the high Caucasus have an increased dust content which parallels the development of industry in eastern Europe. Up until 1920 the dust content of the glacier was about 10 milligrams per liter. In the 1950's this content increased more than 20-fold, to 235 milligrams per liter. So long as the dust stays near the surface, it should have an appreciable effect on the heat balance of the glacier. There is fairly good evidence, based on tracers such as lead, that dusts from human activities have penetrated the polar regions. Conceivably they might change the albedo of the ice, cause melting, and thus pave the way for a rather radical climatic change—and for a notable rise in sea level. There has been some speculation along this line (*39*), but, while these dusts have affected microclimates, there is no evidence of their having had, so far, any measurable influence on the earth's climate. The possibility of deliberately causing changes in albedo by spreading dust on the arctic sea ice has figured prominently in discussions of artificial modification of climate. This seems technologically feasible (*40*). The consequences for the mosaic of climates in the lower latitudes have not yet been assessed. Present computer models of world climate and the general circulation are far too crude to permit assessment in the detail necessary for ecological judgments.

All of the foregoing discussion applies to the large-scale problems of global climate. On that scale the natural influences definitely have the upper hand. Although monitoring and vigilance is indicated, the evidence for man's effects on global climate is flimsy at best. This does not apply to the local scale, as we shall presently see.

## Extraurban Effects

For nearly two centuries it has been said that man has affected the rural climates simply by changing vast areas from forest to agricultural lands. In fact, Thomas Jefferson suggested repetitive climatic surveys to measure the effects of this change in land use in the virgin area of the United States (*41*). Geiger has succinctly stated that man is the greatest destroyer of natural microclimates (*42*). The changeover from forest to field locally changes the heat balance. This leads to greater temperature extremes at the soil surface and to altered heat flux into and out of the soil. Cultivation may even accentuate this. Perhaps most drastically changed is the low-level wind speed profile because of the radical alteration in aerodynamic roughness. This change leads to increased evaporation and, occasionally, to wind erosion. One might note here that man has reversed to some extent the detrimental climatic effects of deforestation in agricultural sectors, by planting hedges and shelter belts of trees. Special tactics have been developed to reduce evaporation, collect snow, and ameliorate temperature ranges by suitable arrangements of sheltering trees and shrubs (*43*).

The classical case of a local man-made climatic change is the conversion of a forest stand to pasture, followed by overgrazing and soil erosion, so that ultimately nothing will grow again. The extremes of temperature to which the exposed surface is subjected are very often detrimental to seedlings, so that they do not become established. Geiger pointed this out years ago. But not all grazing lands follow the cycle outlined above. Sometimes it is a change in the macroclimate that tilts the balance one way or another (*44*).

Since ancient times man has compensated for vagaries of the natural climates by means of various systems of irrigation. Irrigation not only offsets temporary deficiencies in rainfall but, again, affects the heat balance. It decreases the diurnal temperature ranges, raises relative humidities, and creates the so-called "oasis effect." Thornthwaite, only a decade and a half ago, categorically stated that man is incapable of deliberately causing any significant change in the climatic patterns of the earth. Changes in microclimate seemed to him so local and trivial that special instrumentation was needed to detect them. However, "Through changes in the water balance and sometimes inadvertently, he exercises his greatest influence on climate" (*45*).

What happens when vast areas come under irrigation? This has taken place over $62 \times 10^3$ square kilometers of Oklahoma, Kansas, Colorado, and Nebraska since the 1930's. Some meteorologists have maintained that about a 10 percent increase in rainfall occurs in the area during early summer, allegedly attributable to moisture reevaporated from the irrigated lands (*46*). Synoptic meteorologists have generally made a good case for the importation, through precipitation, of moisture from marine sources, especially the Gulf of Mexico. Yet $^3H$ determinations have shown, at least for the Mississippi valley area, that two-thirds of the precipitated water derives from locally evaporated surface waters. Anyone who has ever analyzed trends in rainfall records will be very cautious about accepting apparent changes as real until many decades have passed. For monthly rainfall totals, 40 to 50 years may be needed to establish trends because of the large natural variations (*47*).

This century has seen, also, the construction of very large reservoirs. Very soon after these fill they have measurable influences on the immediate shore vicinity. These are the typical lake effects. They include reduction in temperature extremes, an increase in humidity, and small-scale circulations of the land- and lake-breeze type, if the reservoir is large enough. Rarely do we have long records as a basis for comparing conditions before and after establishment of the reservoir. Recently, Zych and Dubaniewicz (*48*) published such a study for the 30-year-old reservoir of the Nysa Klodzka river in Poland, about 30 square kilometers in area. At the town of Otmuchow, about 1 kilometer below the newly created lake, a 50-year temperature normal was available (for the years 1881 to 1930). In the absence of a regional trend there has been an increase in the annual temperature of 0.7°C at the town near the reservoir. It is now warmer below the dam than above it, whereas, before, the higher stations were warmer because of the temperature inversions that used to form before the water surface exerted its moderating influence. It is estimated that precipitation has decreased, because of the stabilizing effect of the large body of cool water. Here, as elsewhere, the influence of a large reservoir does not extend more than 1 to 3 kilometers from the shore. Another form of deliberate man-controlled interference with microclimate, with potentially large local benefits, is suppression of evaporation by monomolecular films. Where wind speeds are low, this has been a highly effective technique for conserving water. The reduction of evaporation has led to higher water surface temperature, and this may be beneficial for some crops, such as rice (*49*).The reduction of fog at airports by seeding of the water droplets also belongs in this category of man-controlled local changes. In the case of supercooled droplets, injection of suitable freezing nuclei into the fog will cause freezing of some drops, which grow at the expense of the remaining droplets and fall out, thus gradually dissipating the fog. For warm fogs, substances promoting the growth or coalescence of droplets are used. In many cases dispersal of fog or an increase in visual range sufficient to permit flight operations can be achieved (*50*). Gratifying though this achievement is for air traffic, it barely qualifies as even a microclimatic change because of the small area and brief time scale involved. Similarly, the changes produced by artificial heating in orchards and vineyards to combat frosts hardly qualify as microclimatic changes.

Finally, a brief note on general weather modification is in order. Most of the past effort in this field has been devoted to attempts to augment rainfall and suppress hail. The results have been equivocal and variously appraised (*51*). The technique, in all cases, has been cloud seeding by various agents. This produces undoubted physical results in the cloud, but the procedures are too crude to permit prediction of the outcome. Thus, precipitation at the ground has been both increased and decreased (*52*). The most reliable results of attempts to induce rainfall have been achieved through seeding clouds forming in up-slope motions of winds across mountains and cap clouds (*53*). Elsewhere targeting of precipitation is difficult, and the effects of seeding downwind from the target area are not well known. No analysis has ever satisfactorily shown whether cloud seeding has actually caused a net increase in precipitation or only a redistribution. In any case, if persistently practiced, cloud seeding could bring about local climatic changes. But an ecological question arises: If we can do it, should we? This point remains controversial.

Attempts to suppress hail by means of cloud seeding are also still in their infancy. Here the seeding is supposed to achieve the production of many small ice particles in the cloud, to prevent any of them from growing to a size large enough to be damaging when they reach the ground. The seeding agent is introduced into the hail-producing zone of cumulonimbus—for example, by ground-fired projectiles. Some successes have been claimed, but much has yet to be learned before one would acclaim seeding as a dependable technology for eliminating this climatic hazard (*54*).

Hurricane modification has also been attempted. The objective is reduction of damage caused by wind and storm surges. Seeding of the outer-wall clouds around the eye of the storm is designed to accomplish this. The single controlled experiment that has been performed, albeit successfully in the predicted sense, provides too tenuous a basis for appraising the potential of this technique (*55*). Here again we have to raise the warning flag because of the possibility of simultaneous change in the pattern of rainfall accompanying the storm. In many regions tropical storm rain is essential for water supply and agriculture. If storms are diverted or dissipated as a result of modification, the economic losses resulting from altered rainfall patterns may outweigh the advantages gained by wind reduction (*56*). As yet such climatic modifications are only glimpses on the horizon.

## General Urban Effects

By far the most pronounced and locally far-reaching effects of man's activities on microclimate have been in cities. In fact, many of these effects might well be classified as mesoclimatic. Some of them were recognized during the last century in the incipient metropolitan areas. Currently the sharply accelerated trend toward urbanization has led to an accentuation of the effects. The problem first simply intrigued meteorologists, but in recent years some of its aspects have become alarming. Consequently the literature in this field has grown rapidly and includes several reviews summarizing the facts (*57*).

We are on the verge of having a satisfactory quantitative physical model of the effect of cities on the climate. It combines two major features introduced by the process of urbanization.They concern the heat and water balance and the turbulence conditions. To take changes in turbulence first, the major contributory change is an increase in surface roughness. This affects the wind field and, in particular, causes a major adjustment in the vertical wind profile so that wind speeds near the surface are reduced. The structural features of cities also increase the number of small-scale eddies and thus affect the turbulence spectrum.

The change in the heat balance is considerably more radical. Here, when we change a rural area to an urban one, we convert an essentially spongy surface of low heat conductivity into an impermeable layer with high capacity for absorbing and conducting heat. Also, the albedo is usually lowered. These radical changes in surface that accompany the change from rural to urban conditions lead to rapid runoff of precipitation and consequently to a reduction in local evaporation. This is, of course, equivalent to a heat gain—one which is amplified by radiative heat gain resulting from the lowering of the albedo. This heat is effectively stored in the stone, concrete, asphalt, and deeper compacted soil layers of the city. In vegetated rural areas usually more incoming radiation is reflected and less is stored than in the city. Therefore structural features alone favor a strongly positive heat balance for the city. To this, local heat production is added. The end result is what has been called the urban heat island, which leads to increased convection over the city and to a city-induced wind field that dominates when weather patterns favor weak general air flow.

Most of the features of the near-the-surface climatic conditions implied by this model have, over the years, been documented by comparisons of measurements made within the confines of cities and in their rural surroundings, mostly at airports. Such comparisons gave reasonably quantitative data on the urban effect, but some doubts remained. These stemmed from the fact that many cities were located in special topographic settings which favored the establishment of a city—such as a river valley, a natural harbor, or an orographic trough. They would by nature have a microclimate different from that of the surroundings. Similarly, airport sites were often chosen for microclimatic features favorable for aviation. Some of the uncertainties can be removed by observing atmosheric changes as a town grows. An experiment along this line was initiated 3 years ago in the new town of Columbia, Maryland. The results so far support earlier findings and have refined them (*58*).

Perhaps of most interest is the fact that a single block of buildings will start the process of heat island formation. This is demonstrated by air and infrared surface temperature measurements. An example is given in Fig. 37–3. The observations represented by the curves of Fig. 37–3 were made in a paved court enclosed by low-level structures which were surrounded by grass and vegetated surfaces. On clear, relatively calm evenings the heat island develops in the court, fed by heat stored in the daytime under the asphalted parking space of the court and the building walls. This slows down the radiative cooling process, relative

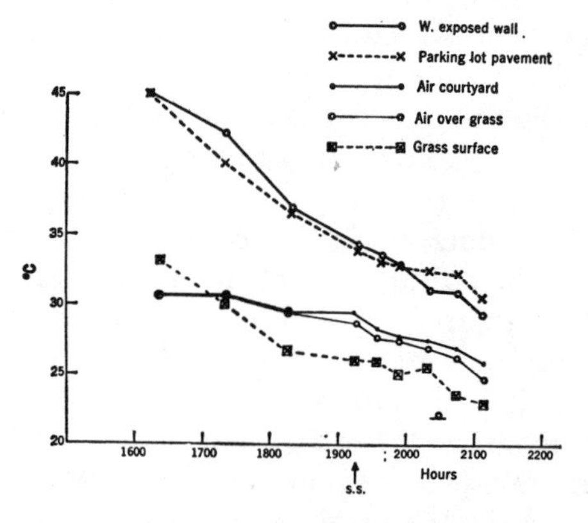


FIGURE 37–3. A typical example of microclimatic heat island formation in incipient urbanization. The top two curves show radiative temperatures of wall and parking lot pavement on a clear summer evening (6 August 1968). The two middle curves show air temperatures (at elevation of 2 meters) in the paved courtyard and over an adjacent grass surface; from sunset (*s.s.*) onward, the courtyard is warmer than the air over the grass. The bottom (dashed) curve gives the radiative temperature of grass. The symbol at 2030 hours indicates the start of dew formation.

to cooling from a grass surface, and keeps the air that is in contact with the surface warmer than that over the grass (*59*).

The heat island expands and intensifies as a city grows, and stronger and stronger winds are needed to overcome it (*60*). And although it is most pronounced on calm, clear nights, the effect is still evident in the long-term mean values. Figure 37–4 shows the isotherms in the Paris region, which is topographically relatively simple and without appreciable differences in elevation. A pronounced metropolitan heat island of about 1.6°C in the mean value can be seen. This is typical of major cities. In the early hours of calm, clear nights the city may be 6° to 8°C warmer than its surroundings. The Paris example is noteworthy because it has been demonstrated that the rise in temperature is not confined to the air but also affects the soil. It has been observed in a deep cave under the city, where temperatures have been measured for two centuries (*61*). Curiously enough, the cave temperature was once considered so invariant that the cave in question was proposed as one of the fixed points for thermometer scales. This artificially introduced trend in temperatures also plays havoc with the long-term temperature records from cities. They become suspect as guides for gaging the slow, natural climatic fluctuations.

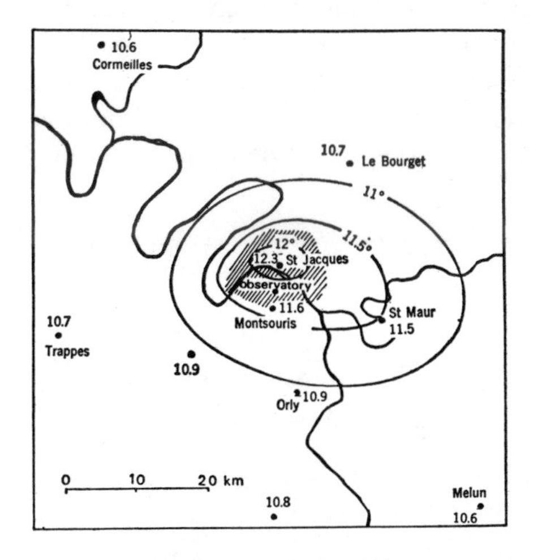


FIGURE 37–4. The urban heat island of Paris, shown by mean annual isotherms in degrees Celsius. The region is characterized by minimal orographic complexity. [After Dettwiller (*61*)].

Part of the rise in temperature must be attributed to heat rejection from human and animal metabolism, combustion processes, and air-conditioning units. Energy production of various types certainly accounts for a large part of it. In the urbanized areas the rejected energy has already become a measurable fraction of the energy received from the sun at the surface of the earth. Projection of this energy rejection into the next decades leads to values we should ponder. One estimate indicates that in the year 2000 the Boston-to-Washington megalopolis will have 56 million people living within an area of 30,000 square kilometers. The heat rejection will be about 65 calories per square centimeter per day. In winter this is about 50 percent, and in summer 15 percent, of the heat received by solar radiation on a horizontal surface (*62*). The eminent French geophysicist J. Coulomb has discussed the implications of doubling the energy consumption in France every 10 years; this would lead to unbearable temperatures (*63*). It is one of a large number of reasons for achieving, as rapidly as possible, a steady state in population and in power needs.

An immediate consequence of the heat island of cities is increased convection over cities, especially in the daytime. That has been beautifully demonstrated by the lift given to constant-volume balloons launched across cities (*64*). The updraft leads, together with the large amount of water vapor released by combustion processes and steam power, to increased cloudiness over cities. It is also a potent factor in the increased rainfall reported from cities, discussed below in conjunction with air pollution problems. Even at night the heating from below will counteract the radiative cooling and produce a positive temperature lapse rate, while at the same time inversions form over the undisturbed countryside. This, together with the surface temperature gradient, creates a pressure field which will set a concentric country breeze into motion (*65*). A schematic circulation system of this type is shown in Fig. 37–5.

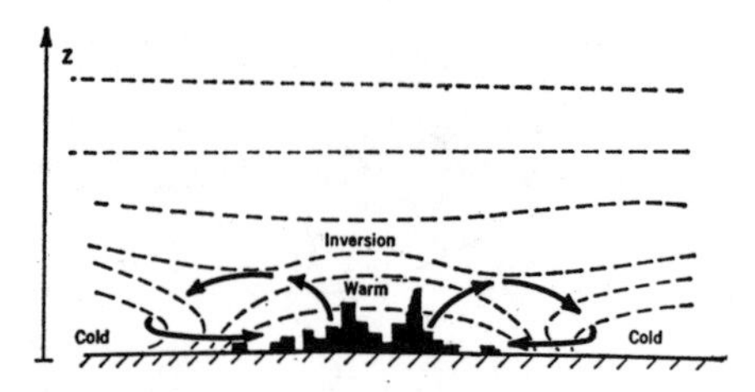


FIGURE 37–5. Idealized scheme of nocturnal atmospheric circulation above a city in clear, calm weather. The diagram shows the urban heat island and the radiative ground inversions in the rural areas, a situation that causes a "country breeze" with an upper return current. (Dashed lines) Isotherms; (arrows) wind; $Z$, vertical coordinate.

The rapid runoff of rainfall caused by the imperviousness of the surfaces of roads and roofs, as well as by the drainage system, is another major effect of cities. In minor rainfalls this has probably only the limited consequence of reducing the evaporation from the built-up area and thus eliminating much of the heat loss by the vaporization that is common in rural areas. But let there be a major rainstorm and the rapid runoff will immediately lead to a rapid rise of the draining streams and rivers. That can cause flooding and, with the unwise land use of flood plains in urban areas, lead to major damage. The flood height is linearly related to the amount of impervious area. For the 1- to 10-year recurrence intervals, flood heights will be increased by 75 percent for an area that has become 50 percent impervious, a value not at all uncommon in the usual urban setting. Observations in Hempstead, Long Island, have shown, for example, that, for a storm rainfall of 50 millimeters, direct runoff has increased from 3 millimeters in the interval from 1937 to 1943 to 7 millimeters in the interval from 1964 to 1966. This covers the time when the area changed from open fields to an urban community (*66*).

It is very difficult to document the decrease of wind speed over cities. Long records obtained with unchanged anemometer exposures at representative heights are scarce. Reasonable interpretations of available records suggest a decrease of about 25 percent from the rural equivalents. This is not unreasonable in the light of measurable increases in aerodynamic roughness. These are around 10 to 30 centimeters for meadows and cultivated fields and around 100 centimeters for woodland. There are several estimates for urban areas. I will give here a value calculated from the unique wind measurements on the Eiffel Tower at a height of 316 meters, and from other wind records in the Paris region (*67*). These data yield values around 500 centimeters. They also suggest a decrease in wind at the top of the Eiffel Tower from the interval 1890–1909 to the interval 1951–1960 of 0.4 meter per second, or 5 percent of the mean wind speed. In view of the height of this anemometer, this is quite a notable adjustment of the wind profile to the increase in terrain roughness.

## Air Pollution Effects

Most spectacular among the effects of the city upon the atmospheric environment are those caused by air pollution. The catalogue of pollutants put into the air by man is long and has been commented upon in so many contexts that the reference to the literature will have to suffice (*68*). Nor shall I dwell here on the special interactions of pollutants with the atmosphere in climatically and topographically specialized instances, such as the much investigated case of Los Angeles (*69*). I shall concentrate, instead, on the rather universal effects of pollutants on local climates.

Among these is the attenuation of solar radiation by suspended particulates. Although this affects the whole spectrum, it is most pronounced in the short wavelengths. The total direct radiation over most major cities is weakened by about 15 percent, sometimes more in winter and less in summer. The ultraviolet is reduced by 30 percent, on an average, and in winter often no radiation of wavelengths below 390 nanometers is received. The extinction takes place in a very shallow layer, as simultaneous measurements taken at the surface and from a tall steeple have shown (*70*).

Horizontally, the particulate haze interferes with visibility in cities. When shallow temperature inversions are present, the accumulation of aerosols can cause 80- or 90-percent reduction of the visual range as compared with the range for the general uncontaminated environment. The haze effect is accentuated by the formation of water droplets around hygroscopic nuclei, even below the saturation point. This is the more noteworthy because relative humidities near the surface are generally lower in cities than in the countryside. This is attributable partially to the higher temperatures and partially to the reduced evaporation. Nonetheless, fog occurs from two to five times as often in the city as in the surroundings. Fortunately, this seems to be a reversible process. Recent clean-up campaigns have shown that, through the use of smokeless fuels, considerable lessening of the concentration of particulates, and hence of fog and of the attenuation of light, can be achieved. In London, for example, with the change in heating practices, winter sunshine has increased by 70 percent in the last decade, and the winter visibilities have improved by a factor of 3 since the improvements were introduced (*71*).

I have alluded above to the increase in cloudiness over cities. It is likely that the enormous number of condensation nuclei produced by human activities in and around cities contributes to this phenomenon. Every set of measurements made has confirmed early assessments that these constituents are more numerous by one or two orders of magnitude in urbanized regions than in the country (*72*). Every domestic or industrial combustion process, principally motor vehicle exhaust, contributes to this part of the particulate. Independent evidence suggests that there is more rainfall over cities than over the surrounding countryside. But the evidence that pollutants are involved is tenuous. There is little doubt that the convection induced by the heat island can induce or intensify showers. This has been demonstrated for London, where apparently thundershowers yield 30 percent more rain than in the surrounding area (*73*). Orographic conditions would lead one to expect more showers in hilly terrain. This is not the case. Although this buoyancy effect is certainly at work, it does not stand alone: in some towns there are observations of precipitation increases from supercooled winter stratus clouds over urban areas. Some well-documented isolated cases of snow over highly industrialized towns suggest a cloud-seeding effect by some pollutants that may act as freezing nuclei (*74*). Also the rather startling variation of urban precipitation in accordance with the pattern of the human work week argues for at least a residual effect of nucleating agents produced in cities. The week is such an arbitrary subdivision of time that artificial forces must be at work. Observations over various intervals and in various regions indicate increased precipitation for the days from Monday through Friday as compared with values for Saturday and Sunday. These increases usually parallel the increase in industrialization, and, again, there is evidence for a more pronounced effect in the cool season (*75*).

Although most studies indicate that the increase in precipitation in urban areas is around 10 percent—that is, close to the limit of what could still be in the realm of sampling errors—some analyses have shown considerably larger increases in isolated cases. These instances have not yet been lifted out of the umbra of scientific controversy (*76*). But we should note here that some industrial activities and internal combustion engines produce nuclei that can have nucleating effects, at least on supercooled cloud particles. In the State of Washington in some regions that have become industrialized there is evidence of a 30-percent increase in precipitation in areas near the pulp mills over an interval of four decades (*77*). There are also incontrovertible observations of cloud banks forming for tens of kilometers in the plumes of power plants and industrial stacks. This is not necessarily associated with increased precipitation but raises the question of how far downwind man's activities have caused atmospheric modifications.

In the absence of systematic three-dimensional observations, we have to rely on surface data. A recent study by Band (*78*) throws some light on the conditions. He found that, for a heat island 3°C warmer than its surroundings, a small but measurable temperature effect was still notable 3 kilometers to leeward of the town. Similarly, a substantial increase in the number of condensation nuclei was noted 3 kilometers downwind from a small town. In the case of a major traffic artery, an increased concentration of nuclei was measurable to 10 kilometers downwind. For a major city, radiation measurements have suggested that the smoke pall affects an area 50 times that of the built-up region. These values, which are probably conservative, definitely indicate that man's urbanized complexes are beginning to modify the mesoclimate.

As yet it is very difficult to demonstrate that any far-reaching climatic effects are the results of man's activities. If man-made effects on this scale already exist or are likely to exist in the future, they will probably be a result of the vast numbers of anthropogenic condensation and freezing nuclei. Among the latter are effective nucleating agents resulting from lead particles in automobile exhaust. These particles have become ubiquitous, and if they combine with iodine or bromine they are apt to act as freezing nuclei. Schaefer and others have pointed out that this could have effects on precipitation far downwind (*79*). These inadvertent results would lead either to local increases in precipitation or to a redistribution of natural precipitation patterns. They are, however, among the reversible man-made influences. As soon as lead is no longer used as a gasoline additive—which, hopefully, will be soon—the supply of these nucleating agents will stop and the influence, whatever its importance, should vanish promptly because of the relatively short lifetime of these nuclei.

Perhaps more serious, and much more difficult to combat, is the oversupply of condensation nuclei. Gunn and Phillips pointed out years ago that, if too many hygroscopic particles compete for the available moisture, cloud droplets will be small and the coalescence processes will become inhibited (*80*). This could lead to decreases in precipitation, a view that has recently been confirmed (*81*).

There remains one final area of concern: pollution caused by jet aircraft. These aircraft often leave persistent condensation trails. According to one school of thought, these artificial clouds might increase the earth's albedo and thus cause cooling. Although on satellite pictures one can occasionally see cloud tracks that might have originated from these vapor trails, they seem to be sufficiently confined, with respect to space and time, to constitute a very minute fraction of the earth's cloud cover. The other view of the effect of these vapor trails, which change into cirriform clouds, is that ice crystals falling from them may nucleate other cloud systems below them and cause precipitation. Any actual evidence of such events is lacking. And then we have the vivid speculations concerning weather modifications by the prospective supersonic transport planes. For some time military planes have operated at the altitudes projected for the supersonic transports. The ozone layer has not been destroyed, and no exceptional cloud formations have been reported. The water vapor added by any probable commercial fleet would be less than $10^{-9}$ of the atmospheric water vapor; thus, no direct influence on the earth's heat budget can be expected. At any rate, it seems that the sonic boom is a much more direct and immediate effect of the supersonic transport than any possible impact it may have on climate (*82*).

There is little need to comment on the multitude of schemes that have been proposed to "ameliorate" the earth's climate. Most of them are either technologically or economically unfeasible. All of them would have side effects that the originators did not consider. The new trend toward thinking in ecological terms would lead us to require that much more thoroughgoing analyses of the implications of these schemes be made than have been made so far before any steps are taken toward their implementation (*83*).

## Summary

Natural climatic fluctuations, even those of recent years, cover a considerable range. They can be characterized as a "noise" spectrum which masks possible global effects of man-caused increases of atmospheric $CO_2$ and particulates. Local modifications, either deliberate or inadvertent, measurably affect the microclimate. Some artificial alterations of the microclimate are beneficial in agriculture. Among the unplanned effects, those produced by urbanization on local temperature and on wind field are quite pronounced. The influences on rainfall are still somewhat controversial, but effects may extend considerably beyond the confines of metropolitan areas. They are the result of water vapor released by human activity and of the influence of condensation and freezing nuclei produced in overabundance by motor vehicles and other combustion processes. Therefore it appears that on the local scale man-made influences on climate are substantial but that on the global scale natural forces still prevail. Obviously this should not lead to complacency. The potential for anthropogenic changes of climate on a larger and even a global scale is real. At this stage activation of an adequate worldwide monitoring system to permit early assessment of these changes is urgent. This statement applies particularly to the surveillance of atmospheric composition and radiation balance at sites remote from concentrations of population, which is now entirely inadequate. In my opinion, man-made aerosols, because of their optical properties and possible influences on cloud and precipitation processes, constitute a more acute problem than $CO_2$. Many of their effects are promptly reversible; hence, one should strive for elimination at the source. Over longer intervals, energy added to the atmosphere by heat rejection and $CO_2$ absorption remain matters of concern (84).

## References and Notes

1. L. V. Berkner and L. S. Marshall, *Advan. Geophys.* **12**, 309 (1967); S. I. Rasool, *Science* **157**, 1466 (1967).
2. The climatic consequences of an original single continent, continental drift, changing ocean size, and changing positions of the continents with respect to the poles are not discussed here.
3. M. I. Budyko, *Sov. Geogr.: Rev. Transl.* **10**, 429 (1969); *J. Appl. Meteorol.* **9**, 310 (1970). For a discussion and extension of Budyko's models, see W. D. Sellers, ibid. **8**, 392 (1969); ibid. **9**, 311 (1970).

4. For a recent review of the principal thoughts in this area, based primarily on work by C. E. P. Broooks (1951), W. D. Sellers (1965), and M. I. Budyko (1968), see H. L. Ferguson, *Atmosphere* **6**, 133 (1968); ibid., p. 145; ibid., p. 151.
5. M. Milankovitch, "Canon of Insolation and the Ice-Age Problem," translation of *Kgl. Serbische Akad. Spec. Publ.* **132** (1941) by *Israel Program Sci. Transl.* (1969), *U.S. Dep. Comm. Clearing House Fed. Sci. Tech. Inform.*
6. C. Emiliani and J. Geiss, *Geol. Rundschau* **46**, 576 (1957); C. Emiliani, *J. Geol.* **66**, 264 (1958); ibid. **74**, 109 (1966); *Science* **154**, 851 (1966); W. S. Broecker, D. L. Thurber, J. Goddard, T. L. Ku, R. K. Matthews, K. J. Mesolella, ibid. **159**, 297 (1968).
7. M. I. Budyko, *Tellus* **21**, 611 (1969): D. M. Shaw and W. L. Donn, *Science* **162**, 1270 (1968).
8. J. R. Bray, *Science* **168**, 571 (1970).
9. F. Baur, *Meteorol. Abhandl.* **50**, No. 4 (1967).
10. For a recent review of the many factors causing climatic changes, see *Meteorol. Monogr.* **8**, No. 30 (1968); for a divergent view on the problem, see L. R. Curry, *Ann. Ass. Amer. Geogr.* **52**, 21 (1962); for factors involved in artificially induced changes, see H. Flohn, *Bonner Meteorol. Abhandl. No. 2* (1963).
11. H. C. Fritts, *Mon. Weather Rev.* **93**, 421 (1965).
12. G. Manley, *Quart. J. Roy. Meteorol. Soc.* **79**, 242 (1953); F. Baur, in Linke's *Meteorologisches Taschenbuch, Neue Ausgabe*, F. Baur, Ed. (Akademische Verlagsgesellschaft Geest und Portig, Leipzig, 1962), vol. 1, p. 710; Y. S. Rubinstein and L. G. Polozova, *Sovremennoe Izmenenie Klimata* (Gidrometeorolgicheskoe Izdatelstvo, Leningrad, 1966); H. H. Lamb, *The Changing Climate* (Methuen, London, 1966); H. von Rudloff, in *Europa seit dem Beginn der regelmässigen Instrumentenbeobachtungen (1670)* (Vieweg, Brunswick, 1967); H. E. Landsberg, *Weatherwise* **20**, 52 (1967); M. Konĉek and K. Cehak, *Arch. Meteorol. Geophys. Bioklimatol. Ser. B. Allg. Biol. Klimatol.* **16**, 1 (1968); T. Anderson, "Swedish Temperature and Precipitation Records since the Middle of the 19th Century," *National Institute of Building Research, Stockholm, Document D4* (1970); for the Far East a particularly pertinent paper is H. Arakawa, *Arch. Meteorol. Geophys. Bicklimatol. Ser. B. Allg. Biol. Klimatol.* **6**, 152 (1964).
13. H. E. Landsberg, C. S. Yu, L. Huang, "Preliminary Reconstruction of a Long Time Series of Climatic Data for the Eastern United States," *Univ. Md. Inst. Fluid Dyn. Appl. Math. Tech. Note BN-571* (1968); for other assessments of climatic fluctuations in the United States, see also E. W. Wahl, *Mon. Weather Rev.* **96**, 73 (1968); D. G. Baker, *Bull. Amer. Meteorol. Soc.* **41**, 18 (1960).
14. C. W. Hurley, Jr., *Mt. Washington News Bull.* **10**, No. 3, 13 (1969).
15. W. S. Carlson, *Science* **168**, 396 (1970).
16. J. Bjerknes, *Advan. Geophys.* **10**, 1 (1964); S. I. Rasool and J. S. Hogan, *Bull. Amer. Meteorol. Soc.* **50**, 130 (1969); N. I. Yakovleva, *Izv. Acad. Sci. USSR, Atm. Ocean. Phys. Ser.* (American Geophysical Union translation) **5**, 699 (1969).
17. J. Namias, in *Proc. Amer. Water Resources Conf. 4th* (1968), p. 852; *J. Geophys. Res.* **75**, 565 (1970).
18. The term *greenhouse effect*, which has been commonly accepted for spectral absorption by atmospheric gases of long-wave radiation emitted by the earth, is actually a misnomer. Although the opaqueness of the glass in a greenhouse for long-wave radiation keeps part of the absorbed or generated heat inside, the seclusion of the interior space from advective and convective air flow is a very essential part of the functioning of a greenhouse. In the free atmosphere such flow is, of course, always present.
19. S. Arrhenius, *Worlds in the Making* (Harper, New York, 1908), pp. 51–54.
20. G. S. Callendar, *Quart. J. Roy. Meteorol. Soc.* **64**, 223 (1938).
21. G. N. Plass, *Amer. J. Phys.* **24**, 376 (1956).
22. W. Bischof and B. Bolin, *Tellus* **18**, 155 (1966); K. W. Brown and N. J. Rosenberg, *Mon. Weather Rev.* **98**, 75 (1970).

23. G. S. Callendar, *Tellus* **10**, 253 (1958); B. Bolin and C. D. Keeling, *J. Geophys. Res.* **68**, 3899 (1963); T. B. Harris, *Bull. Amer. Meteorol. Soc.* **51**, 101 (1970); *ESSA* [*Environ. Sci. Serv. Admin.*] *Pam. ERLTM-APCL9* (series 33, 1970).
24. H. Lieth, *J. Geophys. Res.* **68**, 3887 (1963).
25. E. K. Peterson, *Environ. Sci. Technol.* **3**, 1162 (1969).
26. R. Revelle and H. E. Suess, *Tellus* **9**, 18 (1957); H. E. Suess, *Science* **163**, 1405 (1969); R. Berger and W. F. Libby, ibid. **164**, 1395 (1969).
27. S. Manabe and R. T. Wetherald, *J. Atmos. Sci.* **24**, 241 (1967).
28. G. J. Symons, Ed., *The Eruption of Krakatoa and Subsequent Phenomena* (Royal Society, London, 1888).
29. W. J. Humphreys, *Physics of the Air* (McGraw-Hill, New York, ed. 3, 1940), pp. 587–618.
30. R. A. Ebdon, *Weather* **22**, 245 (1967); J. M. Mitchell, Jr. [personal communication and presentation in December 1969 at the Boston meeting of the AAAS] attributes about two-thirds of recent hemispheric cooling to volcanic eruptions.
31. A. B. Meinel and M. P. Meinel, *Science* **155**, 189 (1967); F. E. Volz, *J. Geophys. Res.* **75**, 1641 (1970).
32. In the 1930's I made a large number of counts of Aitken condensation nuclei [see H. Landsberg, *Mon. Weather Rev.* **62**, 442 (1934); *Ergeb. Kosm. Phys.* **3**, 155 (1938)]. These gave a background of ~ 100 to 200 nuclei per centimeter. Measurements made in the last decade indicate an approximate doubling of this number [see C. E. Junge, in *Atmosphärische Spurenstoffe und ihre Bedeutung für den Menschen* (1966 symposium, St. Moritz) (Birkhäuser, Basel, 1967)].
33. R. A. McCormick and J. H. Ludwig, *Science* **156**, 1358 (1967).
34. J. T. Peterson and R. A. Bryson. ibid. **162**, 120 (1968).
35. R. A. Bryson advocates this hypothesis. He states, in *Weatherwise* **21**, 56 (1968): "All other factors being constant, an increase of atmospheric turbidity will make the earth cooler by scattering away more incoming sunlight. A decrease of dust should make it warmer." This remains a very simplified model, because "all other factors" never stay constant. See also W. M. Wendland and R. A. Bryson, *Biol. Conserv.* **2**, 127 (1970). E. W. Barret in "Depletion of total short-wave irradiance at the ground by suspended particulates," a paper presented at the 1970 International Solar Energy Conference, Melbourne, Australia, calculates for various latitudes the depletion of radiation received at the ground because of dust. For geometrical reasons this is a more pronounced effect at higher than at lower latitudes. He therefore postulates that an order-of-magnitude increase in the amount of dust will redistribute the energy balance at the surface sufficiently to cause changes in the general circulation of the atmosphere.
36. W. T. Roach, *Quart. J. Roy. Meteorol. Soc.* **87**, 346 (1961); K. Bullrich, *Advan. Geophys.* **10**, 101 (1964); H. Quenzel, *Pure Appl. Geophys.* **71**, 149 (1968); R. J. Charlson and M. J. Pilat, *J. Appl. Meteorol.* **8**, 1001 (1969).
37. H. E. Landsberg, *Bull. Amer. Meteorol. Soc.* **21**, 102 (1940); N. Georgievskii, *Sev. Morskoi Put. No. 13* (1939), p. 29; A. Titlianov, *Dokl. Vses. (Ordena Lenina) Akad. Sel'skokhoz. Nauk Imeni V. I. Lenina* **6**, No. 8, 8 (1941); A. I. Kolchin, *Les. Khoziaistvo* **3**, 69 (1950); *Les i Step* **3**, 77 (1951); G. A. Ausiuk, *Priroda (Moskva)* **43**, No. 3, 82 (1954).
38. F. F. Davitaya, *Trans. Soviet Acad. Sci. Geogr. Ser. 1965 No. 2* (English translation) (1966), p. 3.
39. M. R. Block, *Paleogeogr. Paleoclimatol. Paleoecol.* **1**, 127 (1965).
40. J. O. Fletcher, "The Polar Ocean and World Climate," *Rand Corp., Santa Monica, Calif., Publ. P-3801* (1968); "Managing Climatic Resources," *Rand Corp., Santa Monica, Calif., Publ. P-4000-1* (1969).
41. T. Jefferson, letter written from Monticello to his correspondent Dr. Lewis Beck of Albany, dated July 16, 1824.

42. R. Geiger, *Das Klima der bodennahen Luftschicht* (Vieweg, Brunswick, 1961), p. 503.
43. J. van Eimern, L. R. Razumova, G. W. Robinson, "Windbreaks and Shelterbelts," *World Meteorological Organ., Geneva, Tech. Note No. 59* (1964); J. M. Caborn, *Shelterbelts and Windbreaks* (Faber and Faber, London, 1965).
44. I. A. Campbell, according to a news item in *Arid Land Research Newsletter No. 33* (1970), p. 10, studied the Shonto Plateau in northern Arizona, where he found that all gullies were stabilized, remaining just as they were 30 years ago. Yet there are now far more sheep in the area. He concluded that accelerated erosion there was caused by climatic variations and not by overgrazing.
45. C. W. Thornthwaite, in *Man's Role in Changing the Face of the Earth*, W. L. Thomas, Jr., Ed. (Univ. of Chicago Press, Chicago, 1956), p. 567.
46. L. A. Joos, "Recent rainfall patterns in the Great Plains," paper presented 21 October 1969 before the American Meteorological Society; F. Begemann and W. F. Libby, *Geochim. Cosmochim. Acta* **12**, 277 (1957).
47. In this context it is important to stress again the inadequacy of the ordinary rain gage as a sampling device. With about one gage per 75 square kilometers, we are actually sampling $5 \times 10^{-10}$ of the area in question. But precipitation is usually unevenly distributed, especially when rain occurs in the form of showers. Then the sampling errors become very high. Even gages close to each other often show 10 percent differences in monthly totals. It takes, therefore, a long time to determine whether differences are significant or trends are real. This same caveat applies to analyses of rainmaking or to changes induced by effects of cities. This problem is often conveniently overlooked by statisticians unfamiliar with meteorological instruments and by enthusiasts with favorite hypotheses [see H. E. Landsberg, *Physical Climatology* (Gray, Dubois, Pa., ed. 2, 1966), p. 324; G. E. Stout, *Trans. Ill. Acad. Sci.* **153**, 11 (1960)].
48. S. Zych and H. Dubaniewicz, *Zesz. Nauk. Univ. Lodz Riego Ser. II* **32**, 3 (1969); S. Gregory and K. Smith, *Weather* **22**, 497 (1967).
49. V. F. Pushkarev and G. P. Leochenko, *Sov. Hydrol. Select. Pap.* **3**, 253 (1967); M. Gangopadhyaya and S. Venkataraman, *Agr. Meteorol.* **6**, 339 (1969); R. Kapesser, R. Greif, I. Cornet, *Science* **166**, 403 (1969).
50. W. B. Beckwith, in "Human Dimensions of Weather Modification," *Univ. Chicago, Dep. Geogr. Res. Pap. No. 150* (1966), p. 195; B. A. Silverman, *Bull. Amer. Meteorol. Soc.* **51**, 420 (1970).
51. "Weather and Climate Modification, Problems and Prospects," *Nat. Acad. Sci. Nat. Res. Counc. Publ. No. 1350* (1966); M. Neiburger, "Artificial Modification of Clouds and Precipitation," *World Meteorol. Organ., Geneva, Tech. Note No. 105* (1969); "Weather Modification, a Survey of the Present Status with Respect to Agriculture," *Res. Branch, Can. Dep. Agr., Ottawa, Publ.* (1970); M. Tribus, *Science* **168**, 201 (1970).
52. L. Le Cam and J. Neyman, Eds., *Weather Modification Experiments* (Proceedings of the 5th Berkeley Symposium on Mathematical Statistics and Probability (Univ. of California Press, Berkeley, 1967).
53. J. R. Stinson, in *Water Supplies for Arid Regions*, F. L. Gardner and L. E. Myers, Eds. (Univ. of Arizona Press, Tucson, 1967), p. 10; U.S. Department of the Interior, Office of Atmospheric Water Resources, Project Skywater 1969 Annual Report, Denver (1970).
54. R. A. Schleusner, *J. Appl. Meteorol.* **7**, 1004 (1968); "Metody vozdeistviia na gradovye protsessy," in *Vysokogornyi Geofiz. Trudy 11* (Gidrometeorologicheskoe Izdatelstvo, Leningrad, 1968).
55. R. C. Gentry, *Science* **168**, 473 (1970).

56. G. W. Cry, "Effects of Tropical Cyclone Rainfall on the Distribution of Precipitation over the Eastern and Southern United States," *ESSA [Environ. Sci. Serv. Admin.] Prof. Pap. No. 1* (1967); A. L. Sugg. *J. Appl. Meteorol.* **7**, 39 (1968).
57. H. E. Landsberg, in *Man's Role in Changing the Face of the Earth*, W. L. Thomas, Jr., Ed. (Univ. of Chicago Press, Chicago, 1956), p. 584; A. Kratzer, *Das Stadtklima*, vol. 90 of *Die Wissenshaft* (Vieweg, Brunswick, 1956); H. E. Landsberg, in "Air over Cities," *U.S. Pub. Health Serv. R. A. Taft Sanit. Eng. Center, Cincinnati, Tech. Rep. A 62-5* (1962); J. L. Peterson, "The Climate of Cities: A Survey of Recent Literature," *Nat. Air Pollut. Contr. Admin., Raleigh, N.C., Publ. No. AP-59* (1969).
58. P. M. Tag, in "Atmospheric Modification by Surface Influences," *Dep. Meteorol., Penn. State Univ., Rep. No. 15* (1969), pp. 1–71; M. A. Estoque, "A Numerical Model of the Atmospheric Boundary Layer," *Air Force Cambridge Res. Center, GRD Sci. Rep.* (1962); L. O. Myrup, *J. Appl. Meteorol.* **8**, 908 (1969).
59. H. E. Landsberg, in "Urban Climates," *World Meteorol. Organ., Geneva, Tech. Note No. 108* (1970), p. 129.
60. T. R. Oke and F. G. Harnall, ibid., p. 113.
61. J. Dettwiller, *J. Appl. Meteorol.* **9**, 178 (1970).
62. R. T. Jaske, J. F. Fletcher, K. R. Wise, "A national estimate of public and industrial heat rejection requirements by decades through the year 2000 A.D.," paper presented before the American Institute of Chemical Engineers at its 67th National Meeting, Atlanta, 1970).
63. J. Coulomb, *News Report, Nat. Acad. Sci. Nat. Res. Counc.* **20**, No. 3, 6 (1970).
64. W. A. Hass, W. H. Hoecker, D. H. Pack, J. K. Angell, *Quart. J. Roy. Meteorol. Soc.* **93**, 483 (1967).
65. F. Pooler, *J. Appl. Meteorol.* **2**, 446 (1963); R. E. Munn, in "Urban Climates" *World Meteorol. Organ., Geneva, Tech. Note No. 108* (1970), p. 15.
66. W. H. K. Espey, C. W. Morgan, F. D. Marsh, "Study of Some Effects of Urbanization on Storm Run-off from Small Watersheds," *Texas Water Develop. Board Rep. No. 23* (1966); L. A. Martens, "Flood Inudation and Effects of Urbanization in Metropolitan Charlotte, North Carolina," *U.S. Geol. Surv. Water Supply Pap. 1591-C* (1968); G. E. Seaburn, "Effects of Urban Development on Direct Run-off to East Meadow Brook, Nassau County, Long Island, N.Y.," *U.S. Geol. Surv. Prof. Pap. 627-B* (1969).
67. J. Dettwiller, "Levent a sommet de la tour Eiffel," *Monogr. Meteorol. Nat. No. 64* (1969).
68. See, for example, *Air Pollution*, A. C. Stern, Ed. (Academic Press, New York, ed. 2, 1968).
69. A. J. Hagen-Smit, C. E. Bradley, M. M. Fox, *Ind. Eng. Chem.* **45**, 2086 (1953); J. K. Angell, D. H. Pack, G. C. Holzworth, C. R. Dickson, *J. Appl. Meteorol.* **5**, 565 (1966); M. Neiburger, *Bull. Amer. Meteorol. Soc.* **50**, 957 (1969); in "Urban Climates," *World Meteorol. Organ., Geneva, Tech. Note No. 108* (1970), p. 248.
70. F. Lauscher and F. Steinhauser, *Sitzungsber. Wiener Akad. Wiss. Math. Naturw. Kl. Abt. 2a* **141**, 15 (1932); *ibid.* **143**, 175 (1934).
71. R. P. McNulty, *Atmos. Environ.* **2**, 625 (1968); R. S. Charlson, *Environ. Sci. Technol.* **3**, 913 (1969); R. O. McCaldin, L. W. Johnson, N. T. Stephens, *Science* **166**, 381 (1969); C. G. Collier, *Weather* **25**, 25 (1970); London Borough Association press release, quoted from UPI report of 14 Jan. 1970.
72. H. E. Landsberg, *Bull. Amer. Meteorol. Soc.* **18**, 172 (1937).
73. B. W. Atkinson, "A Further Examination of the Urban Maximum of Thunder Rainfall in London, 1951–60," *Trans. Pap. Inst. Brit. Geogr. Publ. No. 48* (1969), p. 97.
74. J. von Kienle, *Meteorol. Rundschau* **5**, 132 (1952); W. M. Culkowski, *Mon. Weather Rev.* **90**, 194 (1962).

75. R. H. Frederick, *Bull. Amer. Meteorol. Soc.* **51**, 100 (1970).
76. S. A. Changnon, in "Urban Climates," *World Meteorol. Organ., Geneva, Tech. Note 108* (1970), p. 325; B. G. Holzman and H. C. S. Thom, *Bull. Amer. Meteorol. Soc.* **51**, 335 (1970); S. A. Changnon, *ibid.*, p. 337.
77. G. Langer, in *Proc. 1st Nat. Conf. Weather Modification, Amer. Meteorol. Soc.* (1968), p. 220; P. V. Hobbs and L. F. Radke, *J. Atmos. Sci.* **27**, 81 (1970); *Bull. Amer. Meteorol. Soc.* **51**, 101 (1970).
78. G. Band, "Der Einfluss der Siedlung auf das Freilandklima," *Mitt. Inst. Geophys. Meteorol. Univ. Köln* (1969), vol. 9.
79. V. J. Schaefer, Science **1154**, 1555 (1966); A. W. Hogan, *ibid.* **158**, 800 (1967); V. J. Schaefer, *Bull. Amer. Meteorol. Soc.* **50**, 199 (1969); State University of New York at Albany, *Atmospheric Sciences Research Center, Annual Report* 1969; J. P. Lodge, Jr., *Bull. Amer. Meteorol. Soc.* **50**, 530 (1969 ); G. Langer, ibid. **51**, 102 (1970).
80. R. Gunn and B. B. Phillips, *J. Meteorol.* **14**, 272 (1957).
81. P. A. Allee, *Bull. Amer. Meteorol. Soc.* **51**, 102 (1970).
82. G. N. Chatham, *Mt. Washington Observ. News Bull.* **11**, No. 1, 18 (1970): P. M. Kuhn, *Bull. Amer. Meteorol. Soc.* **51**, 101 (1970); F. F. Hall, Jr., ibid., p. 101; V. D. Nuessle and R. W. Holcomb, *Science* **168**, 1562 (1970).
83. P. Dansereau, *BioScience* **14**, No. 7, 20 (1964); in *Future Environments of North America*, S. F. Darling and J. P. Milton, Eds. (Natural History Press, Garden City, N.Y., 1966), p. 425; R. Dubos, "A theology of the earth," lecture presented before the Smithsonian Institution, 1969; M. Bundy, "Managing knowledge to save the environment," address delivered 27 Jan. 1970 before the 11th Annual Meeting of the Advisory Panel to the House Committee on Science and Astronautics.
84. The work discussed here has been supported in part by NSF grants GA-1104 and GA-13353.

# 38

# The world's water resources, present and future

G. P. Kalinin
V. D. Bykov

## The Hydrologic Cycle and the World Water Balance

The world's water resources form a single entity. Nothing illustrates this more clearly than water in the atmosphere-hydrosphere-lithosphere cycle, basic mechanism of the relatively stable distribution of water as between land, sea and atmosphere. This cycle, the hydrologic cycle (Fig. 38–1), is influenced by solar activity and the latter's general effect on the circulation of air, by processes taking place in the sea and on land, reflecting the correlation of the heat budgets and water balances of land and sea, and, to some extent, by factors of cosmic origin.

The influence of the oceans, which occupy two-thirds of the earth's surface, on the natural hydrologic cycle must obviously be considerable but, although occupying a smaller area, the land mass is more varied, more complex and less stable in its properties and structure.

Considerable changes can be brought about in the natural conditions on land by human activity as well as by the forces of nature. The total balance of free water in historical times and even back into the prehistoric era may be regarded as constant. Aristotle was the first, in his *Meteorology*, to express the view that the earth's water balance is constant. The same conclusion was reached, on the basis of calculations, by V. I. Vernadsky, P. Queney, Penck and others.

---

Reprinted with permission from *Impact of Science on Society*, 19(2):135–150(1969).

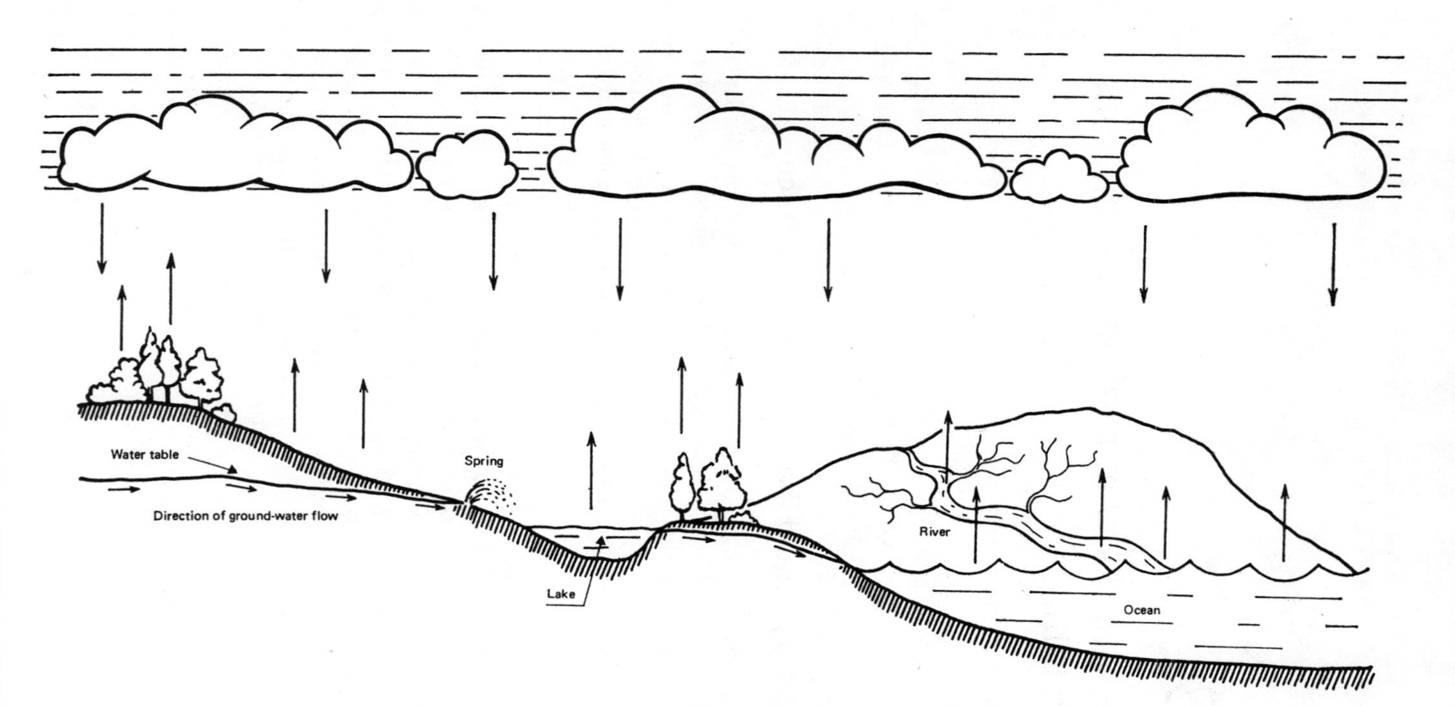


FIGURE 38–1. Simplified depiction of the hydrologic cycle. Annual evaporation exactly equals annual precipitation. Rain that falls on land which does not evaporate back to the atmosphere descends into the ground water or moves into rivers and streams, the ground-water flow and the stream flow eventually returning this water into the oceans.

The water balances of the land and sea taken separately are a different matter. The distribution of water between the oceans and land can only remain stable when the 'credit' and 'debit' sides of the water account balance out for land and sea individually. Such stability can only occur if the natural hydrologic cycle is stable. But it is a dynamic process, governed by a number of variable factors which operate in uncoordinated fashion. For this reason, at various geological epochs and during historical times, the balance between the waters on land and in the ocean has been upset, resulting in a change in the level of the oceans and consequently a change in the relative proportions between sea and land areas.

By analysing variations in surface waters and groundwater, variations of glaciation and changes in plant communities, A. V. Shnitnikov[1] has shown that these variations are governed by general laws and, more especially, that there is a 1,800–2,000 year cycle in terrestrial humidity. G. K. Tushinsky[2] arrived at much the same result by studying the rhythms of snow cover and glaciation.

According to Shnitnikov, we are at present passing through a period of transition from a humid phase to a dry one in which the continents are losing and the oceans are gaining water. Analysis of observations over the last sixty to eighty years has shown that the level of the ocean is rising by 0.05 inch (1.2 mm) a year, on the average. To put it another way, this means that about 105 cubic miles of water (430 cubic kilometres) are being lost yearly from the resources of the land. This yearly addition to the 'credit' side of the ocean's balance is greater than the annual discharge of all rivers flowing into the Caspian Sea. If this situation continues, we can expect a further rise in the level of the seas and a corresponding decrease in the area of land and in its water resources.

It must also be recognized that over long periods of time—several thousands of years—there is a possibility of considerable changes in the world's water resources. The nature of these changes can be determined only by studying the hydrologic balance of the land and the oceans separately.

On the assumption that no changes occur in the earth's resources of moisture (as by loss into space) during the period under consideration, water gains and losses on land and sea balance out, as can be shown by a simple mathematical analysis. Hence, the total quantity of precipitation that falls on the earth is exactly equal to the quantity that evaporates from its surface. Table 38–1 (slightly adapted from M. I. Lvovich)

[1]A. V. Shnitnikov, 'Cycles in Stream Flow and in Variations of the Level of Lakes in Northern Europe and Solar Activity', *Proceedings of the All-Union Geographical Society*, Vol. 93, No. 1, 1961. (In Russian.)

[2]G. K. Tushinsky, *Outer Space and the Natural Rhythms of the Earth*, Moscow, Prosveščenie Publishing House, 1966. (In Russian.)

shows the magnitudes of the component elements of the earth's water balance.

(In land areas where there is vegetation, a part of the return of moisture to the atmosphere results from the transpiration by plants of water which has been absorbed from the ground through their roots or which is a product of metabolism. 'Evapotranspiration' is the term generally applied to the combined return of moisture by both evaporation and transpiration. It shall be understood that in Table 38–1 and elsewhere in this article when evaporation from land areas is referred to this also includes transpiration.)

TABLE 38–1

*World's annual water balance*

| Area and process | Gain | | Loss | |
|---|---|---|---|---|
| | Volume (cu km) | Depth[1] (mm) | Volume (cu km) | Depth[1] (mm) |
| *Peripheral areas of continents (117,504,000 sq km)* | | | | |
| Precipitation | 101 000 | 860 | | |
| River discharge[2] | | | 37 300 | 310 |
| Evaporation | | | 63 700 | 550 |
| *Landlocked areas of continents (31,124,000 sq km)* | | | | |
| Precipitation | 7 400 | 240 | | |
| Evaporation | | | 7 400 | 240 |
| *Continents as a whole (148,628,000 sq km)* | | | | |
| Precipitation | 108 400 | 720 | | |
| River discharge[2] | | | 37 300 | 250 |
| Evaporation | | | 71 100 | 470 |
| *The oceans (361,455,000 sq km)* | | | | |
| Precipitation | 411 600 | 1 140 | | |
| Discharge from rivers[2] | 37 300 | 100 | | |
| Evaporation | | | 448 900 | 1 240 |
| *The earth as a whole (510,083,000 sq km)* | | | | |
| Precipitation | 520 000 | 1 020 | | |
| Evaporation | | | 520 000 | 1 020 |

1. The depth of water if the volume involved in the process were distributed uniformly over the area in question.
2. Including water originating from the melting of Arctic and Antarctic glaciers.

It would be well now to examine the world's water resources so we can see what proportion of it is disposable for man's use.

As Table 38–2 shows, over 90 percent of the earth's water is found in the oceans. At present this water is not directly available for use, with insignificant exceptions. However, it is the evaporation of this water, which then falls as rain, which may be taken as the initiating step of the hydrologic cycle.

The rainwater is directly used, in the first instance, as it falls from the skies. Much of it, however, percolates through the surface and subsoil to replenish the ground-water supply, from which some of it is taken for use before it is eventually discharged, *via* rivers, into the oceans.

The ground water is that water which lies in the saturated region of the ground; the upper surface of this saturated region is the water table. As Table 38–2 shows, the total ground water (free gravitational waters) equals approximately 60 million cubic kilometres. However, most of this water lies so deeply down that for all practical purposes it does not participate in the hydrologic cycle and is not available for use. It

TABLE 38–2

*World's water resources*

| Resource | Volume ($W$) (in thousands of cu km) | Annual rate of removal ($Q$) (in thousands of cu km) and process | Renewal period $\left(T = \frac{W}{Q}\right)$ |
|---|---|---|---|
| Total water on earth | 1 460 000 | 520, evaporation | 2 800 years |
| Total water in the oceans | 1 370 000 | 449, evaporation | 3 100 years |
| | | 37, difference between precipitation and evaporation | 37 000 years |
| Free gravitational waters in the earth's crust (to a depth of 5 km) | 60 000 | 13, underground run-off | 4 600 years |
| (Of which, in the zone of active water exchange | 4 000 | 13, underground run-off | 300 years) |
| Lakes | 750 | — | |
| Glaciers and permanent snow | 29 000 | 1.8, run-off | 16 000 years |
| Soil and subsoil moisture | 65 | 85, evaporation and underground run-off | 280 days |
| Atmospheric moisture | 14 | 520, precipitation | 9 days |
| River waters | 1.2 | 36.3,[1] run-off | 12(20) days |

1. Not counting the melting of Antarctic and Arctic glaciers.

is the 4 million cubic kilometres in the upper region of the crust which participates in active exchanges—into which rainwater percolates and which runs off into lakes, wells, irrigation networks, rivers and the ocean—which is of direct interest to us when we talk of directly usable water resources (though the deeper waters may be increasingly tapped and brought into the hydrologic cycle in the future).

It is in the estimation of ground water that the greatest possibility of error occurs, though there are doubtless inaccuracies in the precipitation and evaporation figures. The average error in the figures shown in Table 38–2 is probably of the order of 10–15 percent.

We might mention that the total volume of water in the earth's crust between subsoil and mantle is actually far larger than the figure shown. Most of this water, is, however, in various states. physically or chemically bound to minerals. Taking into account water in these forms, Vernadsky[3] estimates that the total volume of water in the upper 20 to 25 kilometres of crust is of the order of 1,300 million cubic kilometres, approximately equalling the volume of the oceans.

Table 38–2 shows that the last two items, atmospheric moisture and river waters, represent the two most active agents of water exchange in the hydrologic cycle. River flow also helps to intensify the rate of groundwater exchange. (Two renewal periods have been shown for river waters. The twelve-day period is that for river systems having small catchment areas measuring only a few tens of thousands of square miles. Twenty days is the mean renewal period for major rivers which empty directly into the sea.)

Clearly, the exchange of water between atmosphere and earth's surface is of considerable practical interest. Let us look at some modern views on this point. It is recognized that a portion of the evaporated ocean moisture which falls as rain on land evaporates—the rest eventually making its way back to the ocean—and some of this reevaporated moisture falls as rain on the land again. The relative total quantity of rain which results from an initial precipitated quantity of ocean moisture before all this moisture gets back to the sea is expressed as the coefficient for the hydrologic cycle.

Holzman and Thorntwaite[4] using aerological observations, established that the coefficient for the hydrologic cycle over the territory of the United States was 1.25. Similar calculations by K. I. Kashin and K. P. Pogosyan and later by M. I. Budyko and O. A. Drozdov[5] using different methods, showed coefficients for the territory of the European U.S.S.R. of 1.13 and 1.14 respectively.

Drozdov[6] points out that the coefficient increases somewhat more rapidly than the area of the territory covered, but comparison of coeffi-

[3]V. I. Vernadsky, *History of Minerals in the Earth's Crust. Vol. 2: History of Natural Waters*, Part 1, Sections I-III, 1933, 1934, 1936. (In Russian.)

[4]B. Holzman and C. Thorntwaite, 'A New Interpretation of the Hydrologic Cycle', *Trans. Amer. Geog. Union*, Vol. 19, 1938, p. 11.

[5]K. I. Kashin and K. P. Pogosyan, 'The Atmospheric Water Cycle', *The Hydrometeorological Effect of Growing Trees as Wind Breaks.* Leningrad, Gidrometeoizdat, 1950. (In Russian); M. I. Budyko and O. A. Drozdov, 'General Laws Governing the Atmospheric Water Cycle', *Proceedings of the Academy of Sciences of the U.S.S.R.*, Geography Series, No. 4, 1953. (In Russian.)

[6]O. A. Drozdov, 'Data on the Water Cycle in the European Territory (of the U.S.S.R.) and Central Asia', *Journal of the Main Geophysical Observatory*, No. 45 (107), 1954. (In Russian.)

cients with areas showed a fairly close connexion between them (Fig. 38–2). In Table 38–3 below is an approximate calculation of the

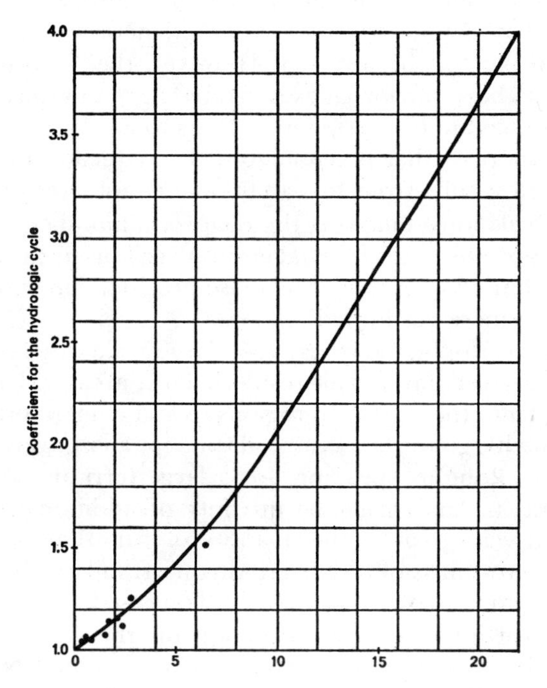


FIGURE 38–2. Relationship between land area and the coefficient for the hydrologic cycle.

TABLE 38–3

*Calculated hydrologic cycle coefficients for the continents*

| Territory | Area (in millions of sq km) | Coefficient for the hydrologic cycle |
|---|---|---|
| Australia | 7.96 | 1.15 |
| Europe | 9.67 | 1.20 |
| South America | 17.98 | 1.30 |
| North America | 20.44 | 1.35 |
| Africa | 29.81 | 1.45 |
| Asia | 42.28 | 1.55 |
| Eurasia | 51.95 | 1.65 |
| Africa and Eurasia | 81.76 | 1.90 |

coefficient of the hydrologic cycle for various continents, worked out from the graph of Figure 38–2. Interesting results are obtained by applying Drozdov's observation on the coefficient/area relationship to measuring the influence of land-improvement schemes in which water normally lost as run-off is diverted to use. If by virtue of such schemes the evaporation from the southern part of the European U.S.S.R. were to be increased by 15 mm, precipitation would increase by 20 mm, mainly in the eastern part of the territory. However, the importance of local evaporation has been somewhat exaggerated in recent calculations. This is because an increase in precipitation leads to a substantial increase in the rate of runoff of that precipitation; the reduction in atmospheric moisture which results from this additional runoff was not taken into account in calculations. Likewise, the reciprocal links between humidity and precipitation were not fully taken into account in calculations: not only does the humidity of the air increase precipitation, but an increase in precipitation increases humidity.

There is no doubt, however, that large-scale land improvement schemes produce very favourable results in large areas with river systems that discharge into the ocean. In regions (continental interiors) without runoff and discharge to the oceans, all precipitation is given off again as evaporation. Roughly speaking, for a large territory of the scale of the Afro-Eurasian land mass the quantity of additional precipitation gained is nearly as great as the quantity of runoff which is diverted to use. About two-thirds of the diverted run-off falls again as precipitation, returning to the rivers where it can be reused.

The present increase in precipitation for the Afro-Eurasian land mass due to the evaporation of diverted runoff cannot be very great, but as the demand for water is developing very rapidly and, according to calculations, is doubling every 10 to 15 years, it is to be expected that in 35 to 50 years the diversion of runoff, which subsequently evaporates, will reach 15–20 per cent. This will produce a noticeable increase in precipitation—of the order of 30 to 40 mm—over this territory, which in turn will lead to a return runoff of the order of 20 to 30 mm, so a significant portion of the diverted runoff will be regained.

We should not discount the possibility that in future man will learn to control the water cycle and will be able to accelerate changes in the humidity regimen of the continents, especially as it is theoretically possible to produce enormous increases in precipitation (hundreds of millimeters) by diverting run-off (including return run-off) and by increasing the evaporation of effluents.

## Short-Term Variations in the Water Content of the Earth's Rivers

The spread of stream-flow gauging stations and modern methods of computing has permitted large-scale investigations into cyclic variations of stream flow. However, the observation periods being too short (at most 150 years), it is impossible to establish long-term cycles with sufficient reliability.

Processing of readings covering many years for sixty major rivers of the northern hemisphere by the spectral functions method has shown that cycles recur most frequently with periods of 2 to 3, 5 to 7, 11 to 13 and 22 to 28 years. The 11 to 13 year cycles are normally regarded as being governed by solar activity. The nature of the shorter cycles has not as yet been adequately explained. The 2 to 3 year cycle is observable not only in variations of stream flow but also in the level of the oceans. Similar cycles occur in a number of other geophysical phenomena.

Periods of greater and lesser humidity affect vast areas of the earth but never the earth as a whole. A period of increased humidity in one part of the world is simultaneously accompanied by reduced humidity elsewhere. Research has shown that, between 1881 and 1950, there was both a synchronous and an asynchronous variation in runoff in the drainage basins of the major rivers of the northern hemisphere. The zones affected by increased or reduced runoff (as compared with the mean) cover considerable areas, the boundaries of which are fairly constant, e.g. runoff was abundant in Europe and North America between 1926 and 1930, reduced in the rest of the northern hemisphere. For 1931–40, this relationship has reversed.

Synchronous and asynchronous variations of run-off are thus geographically localized, a fact of great importance for the calculation of water power and the water balance. There are important possibilities of economic gains by planning hydroelectrical installations not as individual units but as part of a broad, integrated geographical scheme taking into account the synchronous and compensating asynchronous variations in stream flow in different parts of the world, including the U.S.S.R.

One curious fact is that the peaks and lows of the cyclical variations of runoff (2 to 3 years, 5 to 7 years, etc.) occur at different times, in different well-defined, fairly large zones of the earth. This fact can be made use of for forecasting purposes.

Research has been done on this point by A. I. Davydova[7] under the supervision of G. P. Kalinin. By correlating time-graphs depicting runoff and various atmospheric factors, calculations were made of the future annual stream flow for a number of rivers in the northern hemisphere. These investigations proved that it is indeed possible to forecast the water content of rivers for a long time in advance (4 to 7 years, depending on the cycle).

Despite the possibilities for forecasting stream flow, considerable difficulties face us before water resources can be fully utilized. These arise

[7]G. P. Kalinin and A. I. Davydova, 'Cyclical Variations in the Stream Flow of Rivers in the Northern Hemisphere', in *Problems of Stream Flow*, Moscow, Moscow University Publishing House, 1968. (In Russian.)

from the way in which water is distributed in nature as compared to human needs for it. The latter are in direct ratio to solar energy received (because of the consequent potential evaporation), but the water resources annually renewed (surface and underground runoff) are equal to the difference between precipitation and evaporation, i.e. they are in inverse ratio to potential evaporation. (Exceptions to this general rule about water distribution *versus* need may be caused by local topographic or geographic conditions or peculiarities in the atmospheric circulation.) In the particular case of the U.S.S.R., 80 percent of the 1,000 cubic kilometres of regular runoff are concentrated in its most sparsely inhabited parts, and only 20 percent in those where the demand for water is highest.

Another problem in the utilization of water is its uneven seasonal distribution in time, particularly in the case of rivers of the plain. The stream flow of mountain rivers is greatest during the warm part of the year, which is good for agriculture in the southern regions, but even in this case considerable variations from one year to the next create economic difficulties.

## Forecasting Water Demand

To evaluate future water requirements, we need to know about future economic and population growth. According to United Nations experts, the world population will reach 7,000 million by the year 2000, over 50 percent living in towns. Apart from the quantity of drinking water this implies, all branches of the economy will require water in various quantities and qualities. The total is difficult to assess. A very approximate estimate (not counting water for power generation and transport, which do not actually consume water) is given in Table 38–4.

TABLE 38–4

*Total annual world water requirements by year 2000*[1]

| Usage | Water required (cu km) | |
|---|---|---|
| | Total | Irretrievable (lost in evaporation) |
| Irrigation | 7 000 | 4 800 |
| Domestic | 600 | 100 |
| Industrial | 1 700 | 170 |
| Dilution of effluents and waste | 9 000 | |
| Other | 400 | 400 |
| TOTAL | 18 700 | 5 470 |

1. From G. P. Kalinin, *Problems of Global Hydrology*, Leningrad, Gidrometeoizdat, 1968.

The figure for the dilution of industrial and domestic effluent wastes is an expert evaluation assuming dilution with four times their volume of clean water after a preliminary treatment. This is undoubtedly a considerable underestimate of the ratio of clean water to effluents.

It can be seen in Table 38–4 that, by the year 2000, half of all the earth's annually renewed waters—the 37,000 cubic kilometres of evaporated ocean water (Table 38–2) which precipitates on land rather than back into the ocean—will be in use by man. With so high a demand, and the uneven distribution in space and time, work will soon have to be undertaken on an unprecedented scale to regulate runoff and divert the waters of rivers into areas affected by drought. The problem is particularly acute, since demand is growing several times faster than the population. When the world population reaches 20,000 million in the twenty-first century, the demand will be several times greater than in the year 2000.

These water-demand forecasts are rough approximations only, because data on growth of population, irrigation, industry and consumption are also only approximate, and estimates of pollution are either too high or too low. Very likely, the future demand for clean, fresh water is exaggerated because: (a) as demand grows, means will be sought to reduce consumption per unit of industrial and agricultural production; (b) some branches of industry could use salt water or brackish water, as is already being done in some cases; (c) desalination of seawater and brackish ground water will increase; and (d) the purifying of polluted water will be improved and accelerated.

Nevertheless, the figures must be regarded as approximately correct for the next thirty-five to fifty years, i.e. until A.D. 2000–15.

## The Conservation of Continental Waters

Until recently, the utilization of water was mainly regarded in local terms, e.g. a hydraulic scheme for a particular river, or reservoirs for a limited area. At best, the planning of the installations on a river or rivers might be coordinated with other objectives, such as obtaining the maximum hydroelectric yield, or ensuring free passage for shipping, and so on. The same limited approach was also true about fighting pollution, the factory effluents being to some extent purified upon the responsibility of the enterprise itself.

The inevitable consequence of this attitude is that as the demand for water has increased, so natural water resources are coming nearer and nearer to exhaustion, and with progressive deterioration in the quality of the water. In the areas most important to man, nature's own renewal of the quantity and quality of water has not compensated for the deterioration man has caused.

Clearly, the present approach to water utilization must be fundamentally changed, and science must provide the practical means for reversing this degrading of freshwater resources. To do this: (a) man must put back at least as much water as he takes out; (b) the water he returns to nature must be of at least as high a quality as the water he takes out. It is only in this way that we can improve continental waters while still making active use of them.

To meet this objective, localized methods are obviously inadequate. Only large-scale integrated approaches will work.

Water use and conservation can be considered to be done at seven levels:

1. Local (watercourse and ground water).
2. A complete watercourse.
3. A hydrographic system.
4. An entire hydrographic basin (surface water and ground water).
5. Multiple basin.
6. Intercontinental.
7. Global.

As industry develops, the scale of water use gradually moves from the local level upwards.

At the fourth level, the main methods of increasing quantity are: (a) carry over storage: superfluous flood waters can be held over for periods of drought, and part of the runoff in plentiful years can be held over for low years; (b) transfer to regions of shortage within the same hydrographic system; (c) drainage of excessively wet land under agricultural and woodland improvement schemes; (d) agricultural engineering measures to reduce unproductive evaporation; (e) reduction of evaporation by deepening shallows and by chemical methods; and (f) increasing the recharge of groundwater by diverting surface water underground (afforestation, reservoirs and so on).

Runoff control at the fifth level would also make it possible (a) to transfer water from high rainfall areas to low rainfall areas and (b) to increase evaporation of diverted runoff, thus increasing the coefficient for the hydrologic cycle.

Levels (6) and (7) involve modifying the hydrologic cycle in neighbouring continents or in all the continents by planned water management, and even possibly directly using the waters and ice of the world's oceans.

By 2000–15, the loss of water through evaporation may represent one-sixth of total runoff, and be highest in Afro-Eurasia (one quarter to one-fifth of total runoff). Precipitation will increase (because of the higher moisture content of the atmosphere) by close to the quantity of normal runoff which is diverted to use, i.e. 40–50 mm. The increased precipitation which results will in turn cause a return run-off equivalent to about two-thirds of the original diverted runoff (25–35 mm). Thus, despite the high level of water consumption, total runoff is unlikely to be reduced by more than 5–8 percent—probably even less, as some of the demand will by that time be met from new sources (primarily seawater). This latter will introduce further moisture into the hydrologic cycle, thus increasing the moisture content of the atmosphere.

## Fully Exploiting the Earth's Heat

A number of major desalination plants are now in operation—in Kuwait, for instance. Here the development of atomic energy is very pertinent, as atomic energy and desalination at a single large plant make a very efficient combination.

Using new methods, it may prove possible, if only in the distant future, to carry out existing plans for transferring water from the Black Sea to the Caspian. The utilization of the Dead Sea, the Qattara Depression, and other enclosed basins in Africa and Eurasia could draw a considerable additional quantity of water from the oceans into the hydrologic cycle by increased evaporation, and this advantage could be further exploited by integrating several activities, e.g. production of fresh water, chemical industries, fisheries, and recreation areas.

The earth has the enormous heat resources necessary to evaporate extra moisture, but in certain cases they might prove inadequate. In fact, if we take the maximum possible evaporation (fully utilizing all the heat presently available, combined with unlimited irrigation of the continents) as being equivalent to 740 mm, and subtract from this the present evaporation from continents (470 mm, cf. Table 38–1), the remainder—the maximum additional evaporation we could get—is 270 mm. This is virtually equivalent to the present continental runoff figure (river discharge of 250 mm in Table 38–1).

Runoff redistributed to correspond to the distribution of the earth's heat resources would be the ideal solution, and would ensure maximum possible evaporation over the whole surface of the globe, but this is unattainable, even in the very distant future. This is why the argument advanced by many scientists, that by the end of the twenty-first century we will have been able to put all our water resources fully to use, has no adequate physical foundation.

To evaluate the actual potential for consumption of water by evaporation, the following factors must be taken into account: (a) water resources and the possibility of their redistribution; (b) heat resources and their distribution; (c) availability of land suitable for irrigation or the creation of reservoirs; (d) industrial consumption of water. In the long run, then, the diversion of runoff to use will be limited by the amounts of water, heat and land resources available. If the heat and land resources are limited, this will, of course, mean that by no means all the water resources can be used.

## The Planet's Refrigerant

So far we have not considered the changes in the earth's heat resources caused by power generation. Power is now generated and consumed by industry at such an accelerating rate that, even in the near future, this will seriously affect the earth's heat budget.

M. I. Budyko[8] has shown that the radiation balance at the earth's surface and the power generated by industry stand in the ratio of 49 : 0.02. (The radiation balance is the difference between the radiation arriving from the sun and that radiated back to space.) If power generated increases annually by 10 percent, in 100 years it will be greater than the radiation balance and will then have an effect comparable to that of solar radiation. There is thus clearly a real danger that the earth may become overheated. Water resources accordingly become particularly important, since a large increase in water consumption (thus, evaporation) would lead to a corresponding cooling effect.

Here we see water resources in an entirely new light: as one of the main means of preventing the overheating of the earth. In the future world, power production and water consumption will have to be combined in such a way as to ensure optimum moisture and heat conditions.

In sum, it appears highly probable to us that, against the general background of cyclical and incidental variations of climate, there will be a general tendency towards greater humidity (greater quantity of precipitation) of the continents as a result of increased water consumption.

In the U.S.S.R. there are a number of projects for diverting northern Soviet rivers southwards to get greater water utilization. Some (e.g. the scheme for diverting the waters of northern rivers into the Volga-Kama chain of reservoirs and the Caspian) are relatively near to becoming operational. Others (e.g. the diversion of Siberian rivers into Central Asia) are unlikely in the next ten to fifteen years, not only because of the cost, but also because of the problem of the side effects of flooding. The difficulties will eventually be overcome, however, since the problem of water supply cannot be solved without using the runoff of the northern and Siberian rivers.

The measures indicated above for increasing the quantity of water and converting it in the final stage of consumption into evaporation must be taken together with the 'debit' items in the water management balance to attain the optimum utilization.

[8]M. I. Budyko, *The Heat Balance of the Earth's Surface,* Leningrad, Gidrometeoizdat, 1956. (In Russian.)

## The Question of Pollution

The quantitative and qualitative aspects of water utilization cannot be separated. This brings up the matter of pollution.

The main forms of pollution are: (a) basin pollution caused by sediments produced by erosion, poisonous chemicals washed out of the soil and off roads, minerals leached out of the soil by irrigation, etc.; (b) watercourse pollution—the dumping of organic and inorganic waste compounds by sewers and factories, pollutants produced by shipping, minerals leached out of the river bed, etc.; (c) heat pollution—the dumping of water at high temperature by thermal power stations and factories; this increasingly common form of contamination causes a drop in the oxygen content of water which harms fish life and encourages excessive growth of aquatic plants; (d) hydrobiological contamination from a variety of factors, such as run-off control producing shallows with abundant vegetation, the dumping of organic compounds whose residual decomposition products (nitrogen, phosphorus, carbon) encourage the growth of vegetation, etc.

As a result of the increasing use of chemicals in agriculture and of the development of irrigated farming, increasing quantities of chemicals which may be classified as pollutants (weed killers, pesticides, etc.) are bound to enter rivers. Also, large quantities of minerals in solution are washed into rivers during reclamation work on saline soils.

At the basin level, methods of controlling pollution include: (a) on-the-spot processing of sewage and industrial effluents; (b) dilution of effluents, making use of opportunities to control the flow; (c) improvement of the sanitary condition of natural water by purification, by eliminating shallows, and by measures to reduce the quantity of organic and inorganic compounds washed out of agricultural land and woodlands into rivers; (d) improvement of the quality of brackish natural water by diverting part of the purer surface runoff underground; (e) recycling of water used in industry to prevent its being dumped into rivers; (f) putting waste-carrying effluents to use in agriculture.

At the multiple basin, intercontinental and global levels, another pollution-combating measure is to supply additional fresh water from diverted rivers (and perhaps from the sea) while increasing the return of water to the hydrologic cycle by diverting part of used river water for evaporation.

The methods of controlling the quantity and quality of natural waters can and must be harmoniously combined.

## Conclusions

The following conclusions may be drawn.

The development of society inevitably involves increased consumption of water and increased return of used water into the hydrographic system. Since water consumption and the deterioration in the quality of available natural water will inevitably continue, it is imperative that, to allow society to develop normally, the replenishment and qualitative improvement of natural waters must proceed faster than their consumption and deterioration.

As all natural waters form a single entity, within which there is constant interchange, control and conservation must be organized at the level of river basins, continents, even groups of continents.

The time is coming when mankind will have to face up to the problem of reorganizing the hydrographic system in such a way as to divert water in the necessary quantities from areas where there is a surplus to areas where there is a shortage. The creation of irrigated zones in what are now deserts will undoubtedly become extremely urgent in the very near future.

Reorganization of the hydrographic system and control of the hydrologic cycle will become a matter of great importance in the next decades because of the enormously high level which water consumption will reach by the beginning of the twenty-first century. There is growing recognition of the importance of research on these subjects.

The particular feature of water resources which offers prospects for overcoming the water shortage is that not only do they constantly renew themselves but that the intensity with which they renew themselves is proportional to that with which they are used. However, water withdrawn from the hydrologic cycle in one place returns only in minute quantities to the same place. Renewal of resources takes place by: (a) increase in underground run-off; (b) increase in precipitation due to increased humidity of the atmosphere as a result of extraction of water. A study of the mechanisms involved in these processes provides important information for the scientific control of water resources.

Localized water management schemes touch off processes which affect all continents and the planet as a whole. Since the waters of the continents are in constant interaction with the physical and geographical environment, investigation and exploitation must be integrated. Water management is already on such a scale as to have a direct effect on physical, geographical and geological processes as a whole. Hence, hydrological factors are of the greatest importance in working out the theoretical foundations for forecasting changes in the totality of the environment.

We already possess considerable knowledge enabling us to predict the environmental effects of our various interventions in our water resources. We have: (a) the principles and methods for calculating the deformation of the beds of streams and reservoirs and forecasting the state of these beds for many years ahead; (b) a theory of interaction between surface water and ground water—this can be used to assess changes in the level of ground water and the exchanges of water between elements of the hydrographic system and the surrounding territory; (c) experimental data on and theories relating to the movement of masses of air over open water—these can be used to evaluate meteorological changes immediately above the water and its adjacent areas; (d) general laws showing how changes in the elements of the water balance bring about related changes in soil-forming processes; (e) principles and methods for calculating the transport of suspended sediments both in surface run-off and in streams; these enable the determination of changes in the processes of denudation of the earth's crust which are taking place or which will take place with the reorganization of the hydrographic system (these will be on such a large scale that they will have a considerable influence on the future face of the planet).

As far as man is concerned, rivers have two functions: (a) to supply fresh water: (b) to channel off the polluted waste created by domestic and industrial activities. There is such a sharp contradiction between these two functions that in the not too distant future their head-on collision will have grave consequences.

The organic world as we know it today is the end product of an evolutionary process whereby the waters of the continents and the rest of the geographical environment have reached a certain equilibrium. However, the quality of these waters is now being changed by human activity at such an ever-increasing rate that the organic world (and man himself) can hardly adapt itself undamaged to these changes. This situation cannot continue forever. The scale of human activities is becoming such that man cannot go on simply helping himself recklessly to what nature has provided. There is only one way out. Man cannot do without fresh water, therefore he must organize the environment at a higher level. As in the living organism, in addition to the arteries—in this case rivers—he must also have veins to carry away the industrial and domestic waste. This is an enormous challenge, but one which must sooner or later be met.

To create such a purification system, adequate to the needs, he will have to take advantage of all the possibilities. In addition to the methods previously mentioned, these possibilities include: (a) developing supplementary artificial surface and underground hydrographic systems into which industrial and domestic effluents can be dumped (using part of the existing system, suitably modified); (b) dumping waste directly into deep seas; (c) underground dumping of industrial waste, taking care not to contaminate usable ground water.

The question of dumping into the oceans must be handled with extreme care. The zone of active and direct interaction between rivers and seas is the relatively small continental shelf, with depths of up to 200 metres, which makes up about 8 percent of the area of the oceans. It is the part of the ocean richest in vegetable and animal life; this falls off by factors of tens and even thousands as one goes away from the coasts and towards the deep ocean. The sea coasts have played and have an extremely important role to play in the evolution of life on earth—an intermediate zone enriching both seas and land.

Contamination of the ocean by industrial effluents, oil and petroleum products is greatest in coastal waters and in the surface layers of the deep ocean—precisely in those places where biological productivity is greatest. With increasing use of the oceans, care must be taken both to protect them and to assure them an inflow of needed mineral and organic substances. Thus, even when the hydrographic system is reorganized to keep clean and polluted water separate, the problem of purifying effluents will not lose its importance. We shall always have to dump waste into the ocean but it must be so purified and dumped at such a depth that it does not represent a threat to marine life; it must enter into natural processes without lessening the biological productivity of the seas.

Runoff causes both mechanical denudation of land surfaces and chemical denudation—that is, by taking ground minerals into solution. Most of the substances carried away in suspension as a result of mechanical denudation and part of those carried away in solution are deposited on the continental shelf at a rate of 30 metres of sediment in 100,000 years. This is undoubtedly a major reason for the continental shelf's biological productivity. Therefore, when managing run-off we must be careful to take into account the possible consequences of altering the existing conditions of denudation of the earth's surface.

In view of all the above considerations, it is clear that our problem today is to secure an optimum level of control of the run-off and of the exchange of matter for each physical and geographical zone. To do this, we must simultaneously take into consideration water consumption requirements, changes induced in the physical and geographical environment by the retention of run-off, and changes induced in the hydrologic cycle and in the circulation of materials in the earth's natural processes.

# 39

# Cycles of elements

*Task Group: Frederick Smith, Chairman*
*Deborah Fairbanks, Reporter*

*Ronald Atlas*
*C. C. Delwiche*
*Douglas Gordon*
*William Hazen*
*Dian Hitchcock*
*David Pramer*
*John Skujins*
*Minze Stuiver*

## Introduction

Of the many elements necessary for life, six rank as the most important: carbon, hydrogen, oxygen, nitrogen, sulfur, and phosphorus. The first three are combined in energy-rich materials such as carbohydrates and oils. These, together with nitrogen and sulfur, are essential ingredients in all proteins. The sixth, phosphorus, is needed for the transfer of chemical energy within protoplasm, whether this energy is used for activity (respiration) or growth.

Life flourishes only where all of these necessary elements are available. For example, a lack of water (the major source of hydrogen) prevents the development of life in deserts, and a lack of minerals (especially phosphates and nitrates) similarly depresses biological activity in vast areas of the open oceans. On the other hand, these essential materials are relatively abundant in the coastal oceans and vegetated land areas of the planet. Only in these regions can the energy in sunlight and carbon dioxide in air or water become significant factors limiting biological productivity.

---

Reprinted with permission from *Man in the Living Environment* (Madison: © 1972 by the Regents of the University of Wisconsin), pp. 41–89.

Each of these six elements circulates through air, land, sea and living systems in a vast biogeochemical cycle. The circulation of water (hydrologic cycle), the very slow erosion and uplift of continents (geologic cycle) and the opposing processes of photosynthesis and respiration (ecologic cycle) are all involved. Human activities and demands have become measurable on the scale of these global cycles, and, locally, they are sometimes overwhelming. Man's substantive effect on carbon dioxide and his trivial influence on global supplies of oxygen, for example, are documented in the SCEP report (1). In this report, therefore, our attention is restricted to the cycles of phosphorus, sulfur and nitrogen, except where they interact with the cycles of carbon and oxygen.

This focus on phosphorus, sulfur and nitrogen emphasizes the dependence of life on more than the flows of energy and carbon, two aspects of biological cycles most often studied. It emphasizes the roles of protein and protoplasm in the nutritional needs of man and affirms that human activities have a much more profound influence on the cycles of these materials than they do on those of carbon dioxide, oxygen or water.

## Human Activity

Three kinds of human activities may affect the cycles of phosphorus, sulphur and nitrogen:

1) Production for industrial or agricultural use
2) Inadvertent release from other activities
3) Secondary concentration following man's use of these elements.

Phosphorus is produced primarily from the mining of phosphatic rock, and most of it is marketed as phosphate fertilizers.

Sulfur is also produced primarily from mining elemental sulfur or pyrites which contain elemental sulfur. Increasingly, sulfur is being recovered as a by-product in the fossil fuel industries. Most sulfur is converted to sulfuric acid for use—the largest of which is in the extraction of phosphates from phosphatic rock. This, together with its use in sulfate fertilizers, accounts for half of the production. The rest is used in a variety of industrial processes.

Nitrogen is produced primarily by industrial fixation from nitrogen gas in the atmosphere, yielding ammonia and various nitrate compounds. Its primary use is in fertilizers. Table 39–1 shows the commercial production and consumption of these materials for the last 15 years (an annual production of 3.6 million metric tons is equivalent in 1970 to a world average of one kilogram per person).

Such production represents a mobilization of these elements from inactive to active forms in the biosphere. Additional sources arise from other activities, associated primarily with the combustion of fossil fuel.

TABLE 39–1

*Annual production and consumption of phosphates, sulfuric acid, and nitrogenous fertilizers. Units are millions of metric tons of the elements P, S, N (2).*

| | | 1953 | 1963 | 1968 |
|---|---|---|---|---|
| Production: | Phosphate Rock (estimated 12% P) | 3.1 | 5.9 | 10.2 |
| | Sulfuric Acid | 10.3 | 18.8 | 26.3 |
| | Nitrogenous Fertilizer | 6.6 (1954) | 14.9 | 26.6 |
| | | 1954 | 1963 | 1968 |
| Consumption: | Phosphate Fertilizer | 3.3 | 5.4 | 7.6 |
| | Nitrogenous Fertilizer | 6.3 | 14.0 | 24.5. |

The amounts of phosphorus mobilized in this fashion are negligible. Sulfur oxides, however, are released to the atmosphere from the oxidation of sulfur present in the fuel, and nitrogen oxides are released from the high-temperature fixation of gaseous nitrogen as the air passes through the combustion chamber. The amounts of both that are produced over time have tended to follow the increase in consumption of fossil fuel (about 5% per year). Estimates of release from various sources in 1968 are shown on Table 39–2.

These materials mobilized by man are directly or ultimately dispersed to the environment. Along the way, however, they may be reconcentrated to a significant degree, effectively resulting in secondary sources of release. Both primary and secondary sources offer opportunities for pollution control and recycling.

TABLE 39–2

*Atmospheric release in 1968 of oxides of sulfur and nitrogen due to human activities. Units are millions of metric tons of the elements S and N. U.S. estimates from SCEP (1), extrapolated to world estimates using Table 7.3 of SCEP (1).*

| | $SO_x$ | | $NO_x$ | |
|---|---|---|---|---|
| **Source** | U.S. | World | U.S. | World |
| **Coal** | 10.1 | 50.0 | 1.2 | 5.9 |
| **Oil** | 2.6 | 7.3 | 2.7 | 7.6 |
| **Gas** | - | - | 1.5 | 2.3 |
| **Other** | 3.9 | 11.2 | 0.9 | 2.6 |
| **Total** | 16.6 | 68.5 | 6.3 | 18.4 |

Phosphorus and nitrogen, and organic matter in general, are increasingly reconcentrated by three kinds of activities: 1) sewerage systems and treatment plants, 2) food-processing industries and 3) feedlots. All three are growing considerably faster than the population and apparently as fast as the general rate of industrial growth.

## Cycles of Phosphorus, Sulfur and Nitrogen

The effect of man upon global cycles of these elements can be evaluated only by considering the workings of these cycles, our understanding of which is qualitatively fair and quantitatively poor. Only the major components will be considered here, and quantitative analysis is restricted to those areas in which human activity intervenes. More detailed quantitative analyses are given at the end of the chapter.

The cycles differ in both major and minor respects. The phosphorus cycle is the simplest and that of nitrogen the most complex. The major pool of phosphorus is found in rocks (lithosphere), while the major pool of sulfur is in the oceans (hydrosphere) and nitrogen is pooled in the atmosphere. The effects of these differences on the dynamics of the cycles will become evident.

## Phosphorus Cycle

Phosphorus exists in the biosphere almost exclusively as phosphate. Due to the abundance of calcium, aluminum and iron, all of which form phosphate salts with very low solubilities in water, most of the earth's phosphorus is immobilized in rock, soil or sediment.

The natural movement of phosphorus is slow. Phosphates which are leached (in dissolved form) or eroded (in particulate form) from the land find their way to streams and lakes (see Figure 39–1 and Table 39–3). Some of them are precipitated in lake sediments, but the rest enter the ocean were they, too, are precipitated.

As the bodies of animals and plants that have accumulated phosphorus fall to lower levels, surface waters of the oceans become depleted of phosphate supplies. But, deep water tends to be nearly saturated (calcium is abundant), so that the additions from above are precipitated to the sediment. Upwelling of deep water returns some phosphorus to the surface, but this amount is always limited by the relative insolubility of calcium phosphate.

Return of phosphorus to the land depends then almost entirely upon the geological uplift of sediments. It is the formation of new land, and not a return to old land, that closes the cycle; and the rate at which this takes place is so low that, on the human scale, it is virtually zero. For man's purposes, phosphorus is essentially a "nonrenewable" resource.

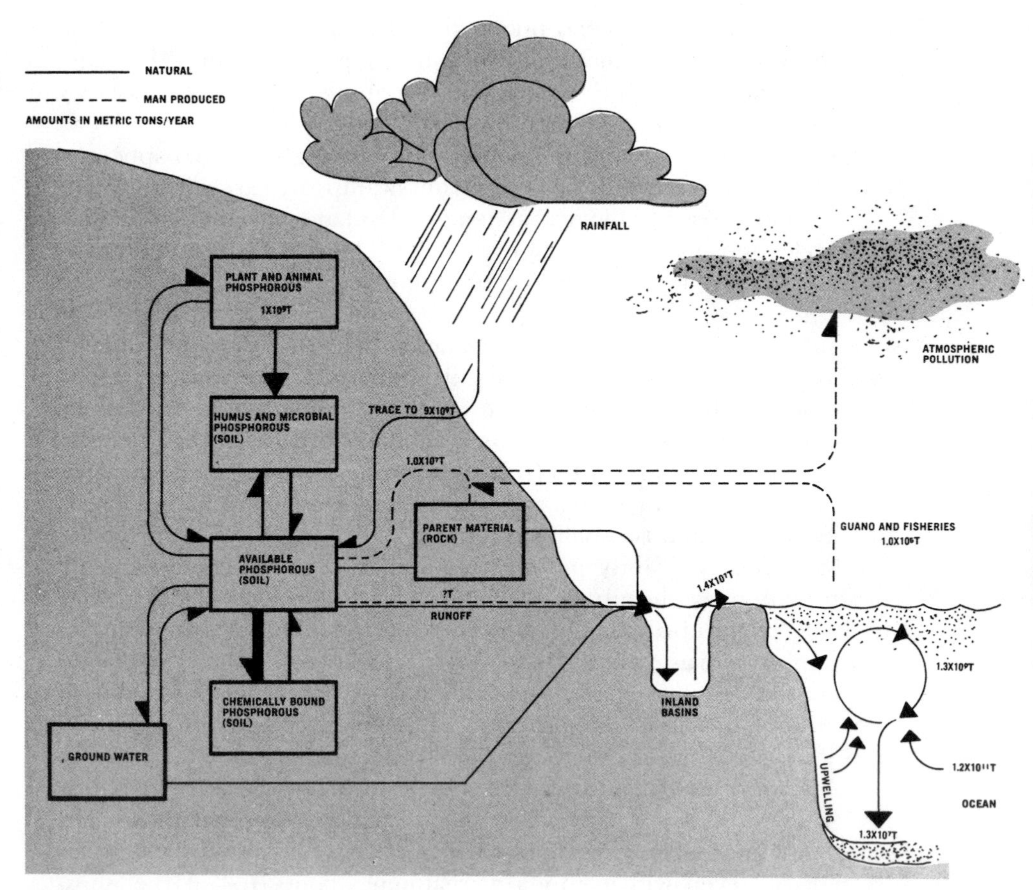


FIGURE 39–1. The Phosphorus Cycle

TABLE 39–3

*Movement of Phosphorus in the Biosphere (3)*

| | Millions of Metric Tons |
|---|---|
| Annual land to sea | 14 |
| Loss to sediments | 13 |
| Phytoplankton cycling | 1,300 |
| Ocean reserve | 120,000 |
| Return to land | 0.1 |

The only significant exceptions to geological rates of return to land are the land deposits of manure from fish-eating birds (guano) and man's fish harvests. Although these are small amounts on a global scale, they are important to man as sources of fertilizer and food.

The amounts of phosphorus that move through the atmosphere as dust or are emitted as phosphene gas from swamps are exceedingly small.

On a local scale, ecological systems on land and in water accumulate and cycle phosphorus (see Figure 39–1). These systems are adapted to environments low in available phosphorus and respond strongly to enrichment.

Subsequent cropping of the vegetation depletes this ecological pool of phosphorus, most rapidly where soil organic matter is low (wet tropics) and least rapidly where it is high (temperate grasslands). Thereafter, unless phosphate fertilizers are used, yields are limited to the rate at which phosphorus is released from its insoluble salts. It is estimated that these natural rates of mobilization would support a world population of between one and two billion people (4).

Man has significantly increased the natural rate of phosphate mobilization by mining phosphatic rock and extracting phosphate fertilizers. In 1968, fertilizer applications exceeded 50% of the estimate of total global phosphorus runoff to the oceans (compare Tables 39–1 and 39–3). The major part of phosphorus added as fertilizer is immobilized in the soil because of abundant supplies of calcium, aluminum and iron which bind with it in the presence of oxygen.

A second part of phosphate fertilizer is removed in crops. This is released later as waste when the crops are processed or consumed. Urbanization and sewerage systems concentrate wastes so that they can be treated, but even with secondary treatment about 70% of the phosphorus passes through into the effluent (5). Other concentrations of phosphorus occur at food processing factories and at feedlots, often contributing to the enrichment of waterways. Although only a minor part of phosphate fertilizer is leached or eroded from the land, drainage systems collect from large areas; and the total downstream effect can be considerable. Table 39–4 shows an analysis of the sources of phosphorus to a lake. At least three-fourths of this phosphorus input is due to human activities, even though they cannot be separated one from another.

Phosphorus in waterways may cycle rapidly through aquatic systems, greatly increasing their biomass, but it is soon lost to the sediment and must be replaced continuously by new additions.

Thus, all phosphorus mobilized by man is eventually immobilized in soil or sediment; and once in the soil, it is slowly transferred by leaching and erosion to sediment. The loop is not closed in less than geologic time. Phosphatic rock containing 12% phosphorus is mined, processed,

TABLE 39–4

*Sources of Phosphorus Entering Lake Mendota (6)*

| Source | % of Total P. |
|---|---|
| Precipitation | 2 |
| Ground Water | 2 |
| Runoff | |
| un-manured rural | 12 |
| manured rural | 30 |
| urban | 17 |
| Waste | 36 |

used and lost in materials which contain less than 0.1% phosphorus (Table 39–5). Maximum recycling from secondary sources of concentration (i.e., sewage) could not reclaim more than 20% of the phosphorus used on the fields (8).

The rapid immobilization of phosphorus in soil precludes an efficient use of added amounts. An empirical study of the relation between fertilizer input and agricultural yield shows that fertilizer use must increase 2.7 times more rapidly than the increase in yield (9). Assuming that the present level of food per person is a minimal goal for the future, we can conclude from this that the use of fertilizers must increase at least 2.7 times faster than the population.

Phosphatic rock reserves and potential supplies are variously estimated to contain between 11,000 and 22,000 million tons of phosphorus (8,10). This includes rock with a phosphorus content as low as 8%, which, though not now economically useful, is expected to become so in the future (10). The U.S. Bureau of Mines estimates the potential yield to be 19,800 million tons of phosphorus (10).

The total production of phosphate in 1968 is estimated at 11.3 million tons of phosphorus (2). Thus, at present rates of use, known reserves and supplies will last 1750 years. It is expected, however, that rates will increase as world populations increase and higher standards of living are achieved.

At present, 39% of the world's population (in the more developed

TABLE 39–5

*Concentration of ppm of P in Natural Material (7)*

| | |
|---|---|
| Igneous rock | 1,050 |
| Shales | 700 |
| Sandstone | 170 |
| Limestone | 400 |
| Phosphate rock | 120,000 (est. ave. 12%) |
| Soil | 650 |
| Fresh water | 0.005 |
| Ocean | 0.07 |
| Oven-dry plant material | 2,000 |
| Dry invertebrate tissue | 4,000 - 9,000 |
| Dry mammal tissue | 43,000 |
| Lake sediments | 150 |

countries) uses 86% of the world's fertilizers (9). Bringing all people to this level of consumption would increase the use of fertilizers by a factor of 2.6. If this were done immediately, the lifetime of known supplies would be only 675 years.

The effect of population growth is much more severe. If present average nutritional levels are maintained, the fertilizer requirements for a population of any size can be estimated by using the 2.7 growth ratio of fertilizers/yield described above. Such estimates are given on Table 39–6, together with the lifetime of known reserves if such populations existed now. A doubling of the present population is expected very early in the 21st century, and levels of 12 to 20 billion people have been suggested as future "steady states" which the world can support (53). This analysis does not support such predictions. By contrast, if phosphorus use is estimated for half the present world population, the lifetime of reserves would be increased many thousands of years.

Another approach is based upon the current rates of increase of phosphorus and of populations and their expected rates in the future. Recently the population has grown at an annual rate of 1.9% (2), and the use of phosphate fertilizers at 5.25% (10). The second is 2.76 times higher than the first, conforming well with the fertilizer/yield analysis given above. If these growth rates are projected into the future, known reserves and supplies will be used up in 90 years—stranding a

TABLE 39–6

*Estimated use of phosphorus fertilizer to feed world populations of various sizes at the present average nutritional level combining data from the UN (2) and the phosphorus-to-yield function of PSAC (9)*

| Population (thousand millions) | Phosphorus Use/Year (millions metric tons) | Lifetime of known reserves (years) |
|---|---|---|
| 1.8 | 1.7 | 11,700 |
| 3.6 | 11.3 | 1,750 |
| 7.2 | 73 | 271 |
| 12 | 291 | 68 |
| 20 | 1170 | 17 |

population of 20 billion. At somewhat lower rates of growth, reserves may last for 130 years.

New reserves will certainly be found. The total phosphorus supply, however, is not limitless. An upper boundary can be set by the accumulated effects of erosion and deposition on this planet over the last half billion years. The total amount of primary (igneous) rock eroded has mobilized an estimated 1,600,000 million tons of phosphorus. Sedimentary rocks account for the loss (at concentrations of less than 0.1% phosphorus) of about 1,000,000 million tons (11). Thus, about 600,000 million tons may exist in materials with higher concentrations. The great bulk of this is not likely to contain more than one or two percent phosphorus (see reference 12 for a discussion of the origin of phosphatic rock). If as much as 5% is sufficiently rich in phosphorus to be potentially usable (8% or more), then no more than 30,000 million tons of usable phosphorus exist on the planet. Supplies are limited.

This being so, what must we do to insure a supply of phosphorus for the longest possible time? While none of the following proposals alone will increase the supply by more than a factor of two, if used in unison, they might extend the planet's phosphorus supply by a total factor of 5 to 10:

1) We should increase recovery from phosphate rock presently being lost in washing and flotation operations. We should protect and rework phosphate rock tailing dumps before they are lost by natural erosion and leaching processes.

Losses in washer and flotation operations, which follow removal of phosphate rock from the ground, range from 40% of the phosphorus in the Florida land-pebble fields to more than 50% in some Tennessee areas (10). Similar beneficiation processes (which separate out that portion of ore containing more phosphate) are used in the U.S.S.R. deposits. The higher-grade ores of the Western United States, however, need little or no washing or beneficiation (8). Discarded material from washing and flotation contains 5.5 to 7.5% phosphorus and not only entails the loss of a valuable resource but is expensive to dispose of as well. The U.S. Bureau of Mines is presently conducting research to develop alternate beneficiation methods.

Underground mining of Western U.S. and North African phosphorus deposits (8) is designed to uncover high-grade beds. But, because it is often necessary to strip low-grade phosphate beds (4.5 to 8% P) to recover the underlying high-grade ores, a large amount of phosphate rock is lost to tailing dumps. These dumps are analogous in concentration of phosphorus to the ores which will have to be mined in less than 100 years. Therefore, they should be safeguarded either by being covered over or by other means, so that they will be available in the future. To neglect them will result in a large loss of materials to wind and rain.

2) Contaminants of phosphate rock should be salvaged as by-products. This recovery, in the future, could defray part of the increasing cost of mining poorer grade phosphate rock and will extend our dwindling resources of uranium and fluoride. Furthermore, the recovery of these elements will prevent additional chemical and radioactive pollution of the land, soil and water.

The uranium content of phosphate rock varies from 45.4 g to 181.4 g of uranium per ton. It can be recovered by the wet process method, although the cost of this process is higher than that of mining and processing uranium ores. In 1968, more than 454 tons of uranium could have been recovered in the United States (10) from phosphate rock. Since contaminant levels will increase as lower-grade ores begin to be used, recovering them could become more profitable.

3) Nations should decrease the amount of phosphorus fertilizers they use to approach the recommended 2/1 ratio. This will conserve phosphorus and still produce desirable, high yields.

Agronomists indicate that a 2/1 ratio of nitrogen to phosphate ($N/P_2O_5$) fertilizers will best meet the overall requirements of the soils and crops of the developing nations (9). However, the ratio of nitrogenous to phosphatic fertilizers actually used by different nations is highly variable and appears to be dictated more by fertilizer availability and economics.

4) We should curtail uses of phosphates for purposes other than fertilizers.

In the U.S., 80% of the phosphate supplies are used in fertilizer manufacture, 15% in phosphate detergents and 5% in producing phosphorus compounds for industry (12). In the rest of the world, the proportion used for fertilizers is higher (85% to 90%). Thus, eliminating non-fertilizer uses will increase the ultimate amount available for fertilizer by 20 to 25% at the most. Fortunately, detergent manufacturers, because of tightening controls on release of phosphorus to the environment, are endeavoring to develop more suitable substitutes. The use of phosphorus in industry is proportionately so small that it almost can be considered inconsequential.

5) We should develop economic methods to recover phosphates released to the environment in effluents.

The likelihood of reclaiming unused phosphate fertilizer and detergents from streams and estuaries is relatively poor because of their great dilution and mixture with other materials. By the time such material reaches the sea, its concentration is no greater than 0.1%, essentially the same percent as occurs naturally. Likewise, recovery from seawater will probably forever remain economically infeasible because of the tremendous pumping costs required. But, recovery from tertiary sewage treatment plants, though costly, may prove a source of reclaimed phosphorus.

6) We should intensify the search for additional phosphate raw materials.

At present, apatite or natural phosphate rock (calcium and/or aluminum hydroxy-fluorophosphate) is the major source of commercial phosphates. As resources dwindle over the next century, alternative sources should be sought. Phosphorite deposits (protophosphate rock) have been fairly well delineated off the Georgia-South Carolina coast of the U.S.A.; similar deposits on continental shelves should be sought and mapped in other parts of the world as future resources.

Implementation of all these proposals suggests the need for an institution of adequate prestige to manage the world use of phosphorus and to work toward its conservation. Phosphate supplies are in short supply and may run out in the next century. As a result, phosphate rocks should be treated as a common resource in order to conserve the world supply. Most nations must import phosphates from the few that produce them (2).

## Sulfur Cycle

Sulfates, the oxidized forms of sulfur, are more soluble than phosphates and move easily from the land to the oceans, where they accumulate (Figure 39–2). Sulfur is 240 times more abundant in seawater than in fresh water, and sulfate deposits occur when shallow seas evaporate and calcium sulfate crystals (gypsum) are formed. This process of removal from the oceans is not taking place in this millennium, but a reduced form of sulfur is deposited continuously in ocean sediments. When sulfate is thrown into the atmosphere in sea spray, the spray evaporates and the particles formed fall back into the sea.

A second source of atmospheric sulfur is in the mud flats bordering seawater. Unlike phosphates, sulfates can be reduced biologically in the absence of oxygen. Various microbes are able to use the oxygen released in this process in respiration, producing hydrogen sulfide. In the deep waters of lakes or seas, this hydrogen sulfide either combines with iron to precipitate as iron pyrites or drifts upward into oxygenated water where it oxidizes spontaneously back to sulfate. When this process occurs in sediments, the hydrogen sulfide tends to precipitate as iron pyrites.

Along the coast where the supply of sulfate from seawater is large, hydrogen sulfide gas escapes to the air when mud, rich in organic matter, is exposed at low tide. The global amount of $H_2S$ produced in this manner is entirely unknown. In the air, it oxidizes rapidly and forms very small particles that ultimately fall as sulfate in rainwater. Pollution with organic matter can increase the rate of sulfate reduction greatly (13).

Sulfur is essential to living systems. It is used in the folding of amino acid chains to form protein molecules in protoplasm. The sulfur, in a reduced form, is attached to several kinds of amino acids. Unlike plants and microbes, which can reduce sulfate for use in protein synthesis, animals depend upon adequate supplies of sulfur-bearing amino acids in their diet.

Under natural conditions, sulfate is abundant in the environment. It tends to be leached from soils but is replenished slowly in rain, and natural vegetations are usually able to meet their needs. However, continual cropping can deplete sulfate supplies, making it advantageous to add moderate amounts in fertilizer.

Man, himself, has created a third source of atmospheric sulfur. Sulfur exists as an impurity in fossil fuel and is converted to sulfur oxides when man burns these fuels. In the presence of atmospheric water, these sulfur oxides become sulfate and mix to some degree with naturally produced sulfate. It is estimated that in 1968, 68 million metric tons of sulfur were released to the atmosphere in this way (see Table 39–2).

The importance of this source depends upon the magnitude of natural sources, and the natural biological production of hydrogen sulfide has not yet been measured. Sulfate in rainfall is estimated to be 165 million metric tons per year for the entire planet (14). If this figure were used to estimate what went up, pollution emissions would account for 43% of the atmospheric circulation of sulfur.

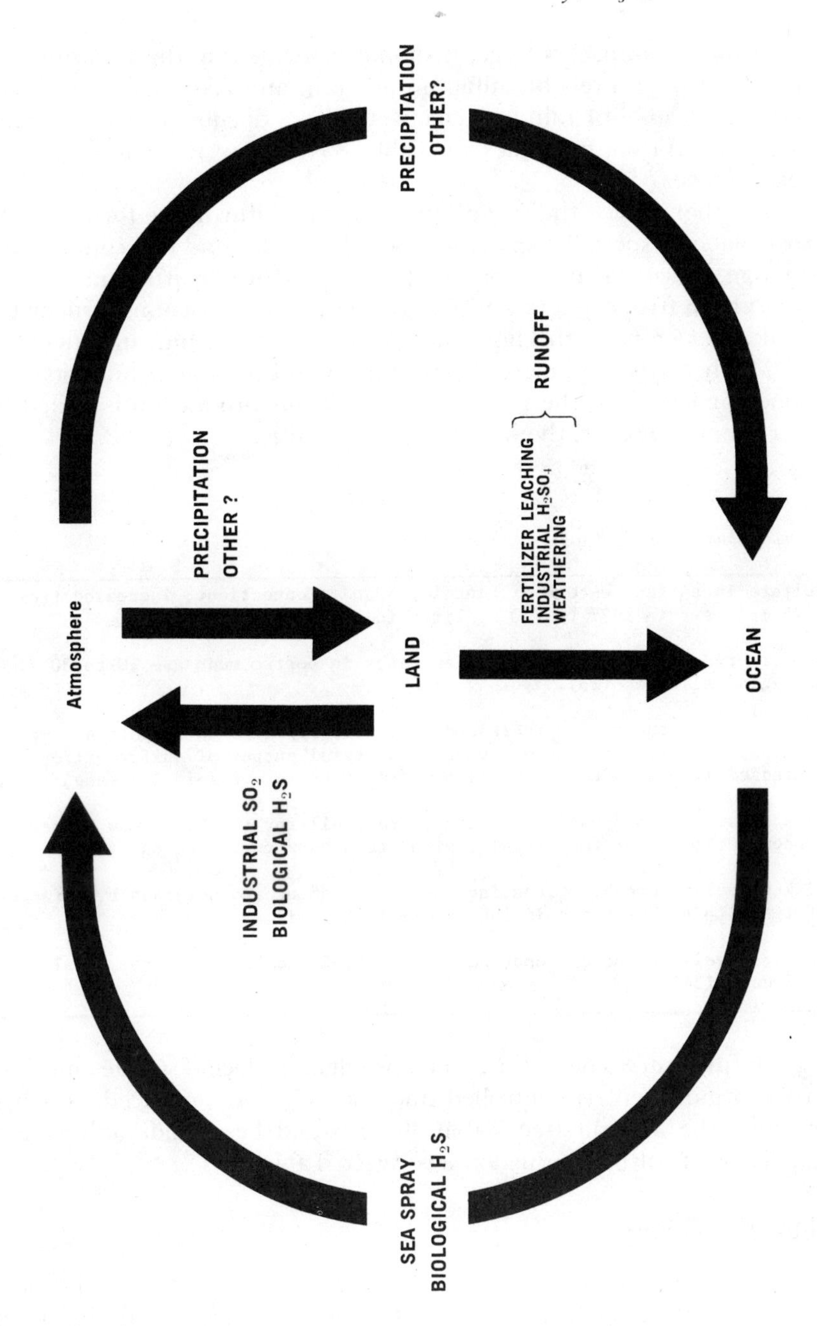


FIGURE 39–2. The Sulfur Cycle.

If human impact is large, it should be evident in the environment and it is. Rising levels of sulfate in air, rain and lakes, along with the increasing acidity of rain and lake waters, are documented locally and regionally. A few examples in Table 39–7 show the magnitude of these changes.

Another aspect, the role of atmospheric Sulfur in the formation of particulates—especially very small particles—is beyond our competence to judge. Since sulfur is a dominant component of particulates in the atmosphere (troposphere and stratosphere), specialists on particulates should reexamine in this light the possible effects of human activity.

Much of the commercially produced sulfur in acid finds its way sooner or later into the rivers. Even if all this production is added to river water, however, the increment is not large.

TABLE 39–7

*Examples showing trends of sulfur in the environment*

1. Sulfate in surface waters of Linsdley Pond, Connecticut, increased from 0.25 mg/liter in 1937 to 7.0 mg/liter in 1963 (15).
2. Sulfate reduction and precipitation rates in bottom muds are 10 to 30 times post-glacial rates (15).
3. Sulfate in Japanese rain increased from 1.13 mg/liter in 1946 to 4.5 mg/liter in 1959, (a 4-fold increase) while industrial output of sulfur oxides increased from 0.26 to 1.2 million metric tons (a 4.6 fold increase) (16).
4. Maps of average non-marine sulfate in rainfall for the U.S. show higher concentrations over industrial regions than over rural regions (17).
5. Aitken nuclei concentrations increased in Wyoming and Colorado by a factor of ten within five years in the middle 1960's (1).
6. Sulfur levels in the air doubled between 1952 and 1962 in three rural areas of Sweden (54).

Sulfur reserves do not present a problem. In fact, if sulfate emission to the atmosphere is controlled and the sulfur is produced as a by-product, the amount (see Table 39–2) would be considerably larger than current sulfur production levels (see Table 39–1).

## Nitrogen Cycle

The global cycling of nitrogen is biologically and chemically the most complex of the biogeochemical cycles. Its major processes are shown on Figure 39–3, a chemical flow chart for nitrogen.

The largest reservoir of nitrogen is found in the atmosphere, which

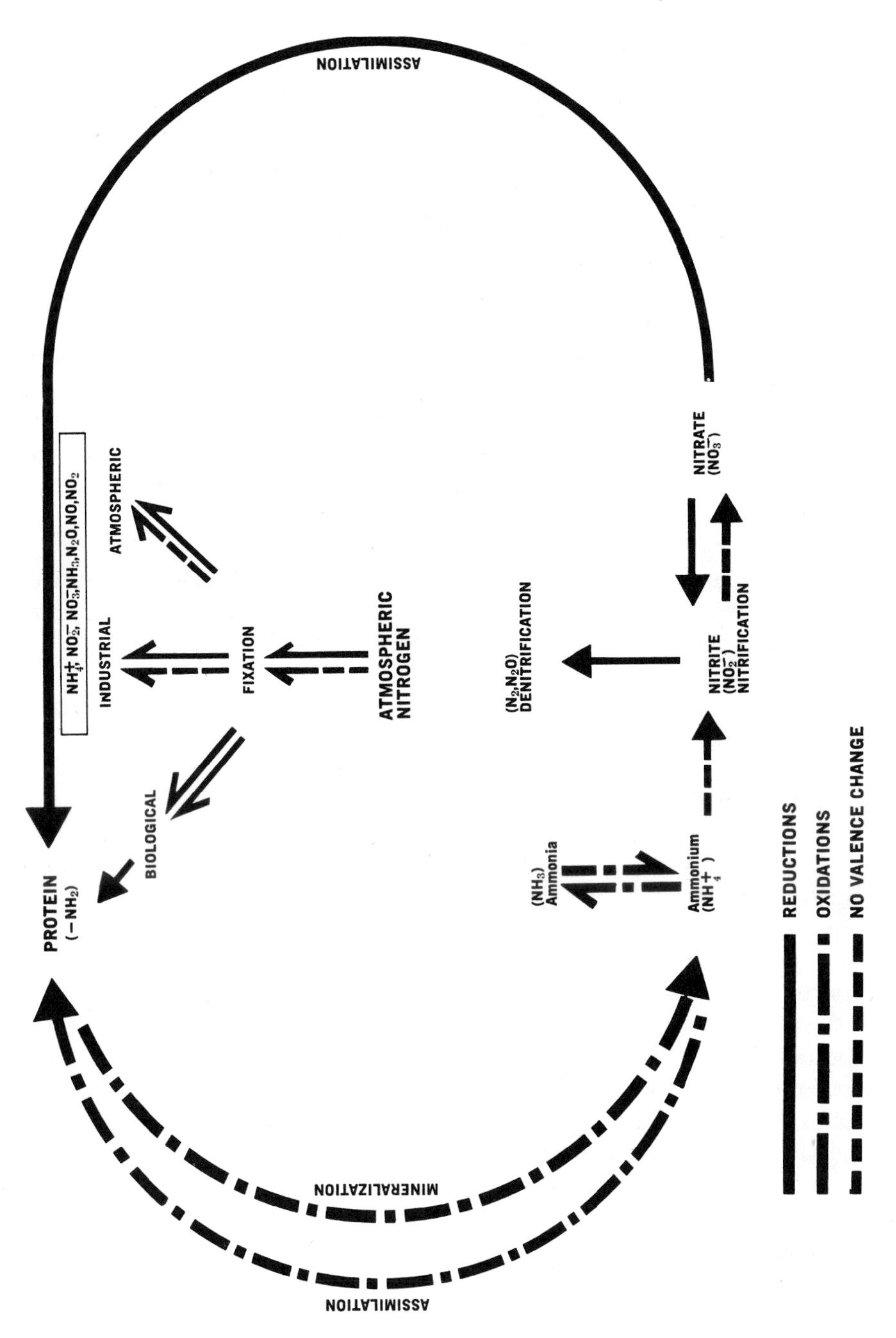


FIGURE 39–3. The Nitrogen Cycle.

is 78% nitrogen gas. A portion of atmospheric nitrogen is removed by microbial action and incorporated into living tissue by a variety of biochemical pathways. Such a process is called nitrogen fixation.

The fixed nitrogen can then be taken up by plants (including microbes) and assimilated, a complex process by which fixed nitrogen is incorporated into protein. Later, this material is either returned directly to the environment or consumed by animals (including man) before it is ultimately released.

When fixed nitrogen is returned to the environment, a process of degradation takes place during which ammonia is released. Ammonia may be reassimilated by plants, evaporated into the air or changed by microorganisms into nitrate.

This latter process, known as nitrification, is an extensive one in nature. It changes the ammonia, which is retained by soils, to nitrate, which leaches into groundwaters. Plants can assimilate only a part of those naturally formed nitrates present in soils.

A natural microbial process, denitrification, removes most of the nitrate by changing it to nitrogen or to nitrous oxide, a gas which requires reasonably large amounts of organic matter as food for the microorganisms and an absence of oxygen. Thus, it has its limitations of effectiveness.

The atmospheric and ecologic cycles of nitrogen are so strong and rapid that its geologic aspects are trivial.

### Nitrogen Fixation

Man has roughly doubled the rate at which fixation occurs. In the next 30 years, a further increase to four or five times the "natural" fixation rate is expected (9). Though this will have no effect on the vast amount of atmospheric nitrogen, it may produce changes in other parts of the nitrogen cycle.

Man's fixed nitrogen tends to be transformed to nitrate which, not bound by soils, is leached into groundwater, rivers and estuaries and eventually reaches the ocean. Some of this nitrate is converted to nitrogen gas or nitrous oxide and returned to the atmosphere; but the capacity of this denitrification process and, hence, its capability of coping with increased amounts of nitrates is unknown.

### Nitrates

A trend of increased nitrate concentrations in natural waters has followed the increased use of industrially fixed nitrogen fertilizers. Trends over time are shown for four Illinois rivers on Figure 39–4

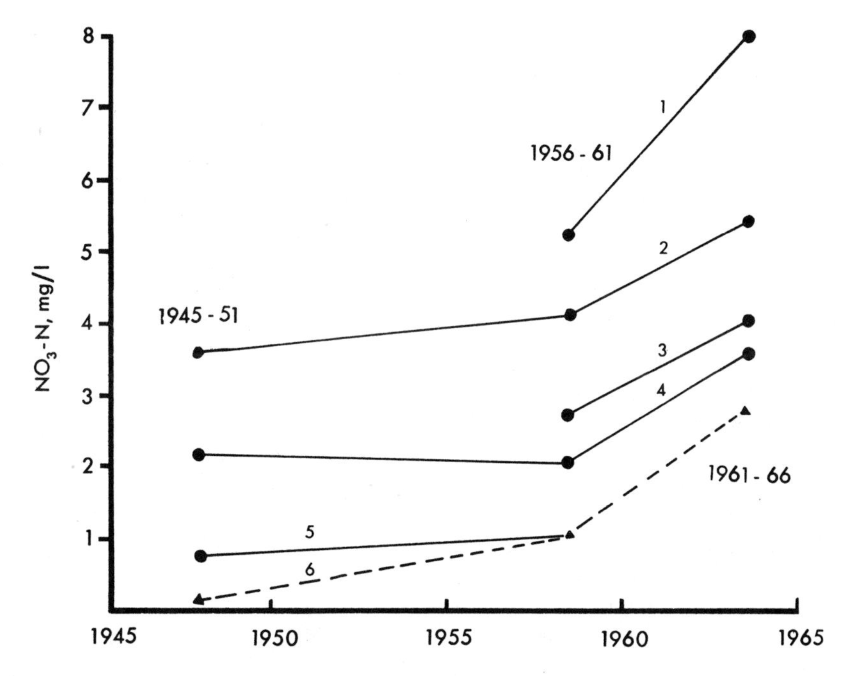


FIGURE 39–4. Nitrate levels in surface waters of Illinois rivers. 1. Kaskaskia River (Shelbyville). 2. Illinois River (Peoria). 3. Rock River (Como). 4. Kaskaskia River (New Athens). 5. Skillet Fork (Wayne City). 6. Nitrogen fertilizer used in Illinois, 100,000 metric tons of N. (Ordinate scale the same as that used for mg./l. of $NO_3$–N in river water.) River data are 90th percentiles for the period indicated, fertilizer data are averages. (18, 55).

(18). When very high levels occur, they tend to coincide in time with the spring crest of high water flow. A curve of fertilizer use has been added for comparison (55). Between 1955–61 and 1961–66 fertilizer use increased 179 percent, while the average increase in the four sets of river data was 54%. Although fertilizer use increased seven-fold during the earlier period, levels were apparently low enough to have little effect on natural waters.

A similar increase of nitrate in sewage releases has been noted for the last twenty years. Another possible source of nitrate in natural waters is increased irrigation practices which can leach naturally accumulated nitrates from soils, particularly in arid or semiarid areas.

The main effect of increased nitrate concentrations in natural waters is eutrophication or over-fertilization. Decay of the excess plant growth requires a lot of oxygen, depleting the water's oxygen supply and resulting in a loss of fish life and damage to the water's economic and recreational value (19).

Increased nitrate concentrations in groundwaters used for drinking are also undesirable. Under anoxic conditions in the digestive tract, this nitrate is reduced to nitrite. When this happens in infants, it produces a "blue baby" condition called methemoglobinemia. Hogs and cattle are also affected adversely by nitrates and grow poorly. Nitrates can also cause gastroenteritis and diarrhea and in some cases can be lethal. Toxic human doses of nitrate have been reported as 0.56 g nitrate-nitrogen per day (20) and 1.04 g nitrate-nitrogen/kg body weight (21). The U.S. limit for approved drinking water is 45 mg/l nitrate (2).

A further concern with nitrate buildup is its influence on the sulfur cycle. Nitrate is known to decrease the reduction of sulfate to sulfide (3), and this could result in a lowering in the pH of soils and natural waters.

Methods have been proposed for removing nitrate-rich waters locally by flushing them into economically unimportant areas. But such proposals do not remedy the problem of nitrate in usable drinking water sources; and, generally, these methods do not remove excess fixed nitrogen from the biosphere. Other methods proposed by Wuhrmann (22), Ludzak and Ettinger (23) and Johnson and Schoepfer (24) are based on controlled microbial denitrification processes which would recycle the nitrate into atmospheric nitrogen gas. A final method which might become essential if the nitrate in natural water continues to rise would be instituting desalinization processes to produce suitable nitrate-free drinking water.

### *Nitrogen Dioxide*

A second focal point of concern with the nitrogen cycle is with levels of nitric oxide (NO) and nitrogen dioxide ($NO_2$). These two oxides are the products of high temperature combustion. Nitric oxide is rapidly transformed to nitrogen dioxide by atmospheric ozone. Though there are natural sources of nitrogen dioxide in the atmosphere which have not yet been elucidated, man puts about 20 million tons of $NO_2$ into the atmosphere yearly. The present pool of nitrogen dioxide and nitric oxide in the atmosphere is estimated at about one million metric tons (25). The present yearly input of nitrogen dioxide from man-made sources is about 20 million tons. The amount of nitrogen dioxide that, in turn, precipitates out of the atmosphere has not been clearly established; but by assuming that equal amounts are continuously added and removed from the atmosphere and that other sources are small, the residence time of nitrogen dioxide is less than three weeks. Nevertheless, a continuous increase in the rate of nitrogen dioxide input into the atmosphere will undoubtedly increase its pool size in the atmosphere.

Increased nitrogen dioxide levels in the atmosphere have many unpleasant side effects for man. Nitrogen dioxide can be toxic in high concentrations and forms corrosive acids when hydrated. Beyond that, it is a key compound in the photochemical production of smog. Even though nitrogen dioxide is present in the atmosphere from natural sources, the concentrations necessary for smog production are of man-made origin. This is exemplified by the fact that the nitrogen dioxide concentrations in cities is about ten times that found in rural areas (25). Continued fuel-burning practices will only further increase this pollution of city air. And, as yet, no methods of $NO_2$ removal appear feasible. It is now removed naturally from the air by rainwater and is dissipated in soil as nitrate and nitrite ions.

Because nitrogen dioxide can be found in well-isolated areas, seemingly beyond the range of contamination from polluted city air, several possible explanations have been proposed in the literature for the origin of natural nitric oxide and hence natural nitrogen dioxide emissions. These range from biological means (26) to the oxidation of ammonia (27) to the transformation of $N_2O_2$ (27). None of these proposed origins seems satisfactory. There is no convincing evidence that these oxides can be generated biologically. In certain acid soils, which also contain high nitrate concentrations, nitric oxide will evolve; but such soils are not common. We are left with an enigma and are unable to determine whether any human practices other than fuel combustion add to atmospheric nitrogen dioxide.

## Interrelations of Biogeochemical Cycles

The diverse cycles of hydrogen, oxygen, carbon, nitrogen, sulfur and phosphorus ultimately intersect in living material, where they show fairly constant ratios of abundance to each other. The cycles also interact in other places, with the result that human influence on one cycle can affect another. Although many of these interactions have not been well studied, enough is known about some of them to permit discussion.

### *Organic Synthesis*

Organisms require various forms of carbon, nitrogen, sulfur and phosphorus and are not always able to make them. Generally, organic carbon (carbohydrate) is produced by green plants through photosynthesis and then distributed to animals and decomposers. The organic nitrogen and sulfur in proteins are produced by green plants and decomposers and must be supplied to animals in their food. Phosphates can be acquired from inorganic sources by all organisms.

Organic carbon and organic nitrogen production need not proceed at the same rate in the biosphere. Land plants are relatively rich in carbon and poor in nitrogen, due primarily to the extensive production of cellulose. Aquatic algae are relatively rich in protein and, hence, have relatively more nitrogen than land plants have. Since the global rates of organic carbon production are reasonably well known (1, 28) and the carbon to nitrogen ratios are known for a variety of materials (11, 28), the global rates of organic nitrogen production by green plants can be estimated. Nitrogen production by decomposers can then be estimated by considering bacterial and fungal growth in relation to plant and animal litter (29). The results are shown on Table 39–8. Even though they are only approximate, they suggest that the ocean is the most important source of protein and that ocean food chains could be larger producers of meat than terrestrial systems. A major problem is the length of the food chain, which is about twice as long in fisheries as in agriculture. The table also suggests that a large potential source of protein exists in the soil.

TABLE 39–8

*Preliminary estimates of rates of production of organic nitrogen (protein) in the biosphere (1,11,28). Units are thousand millions of metric tons per year. Production rates of carbon (1) are shown for comparison.*

| Source | Green Plants | | Bacteria & Fungi | | Total | |
|---|---|---|---|---|---|---|
| | N | C | N | C | N | C |
| Land | 1.6 | 56 | 1.7 | - | 3.3 | 56 |
| Oceans | 3.7 | 22 | 0.2 | - | 3.9 | 22 |
| Total | 5.3 | 78 | 1.9 | - | 7.2 | 78 |

### Alternative Sources of Oxygen

Living in a world rich in oxygen, we are accustomed to the use of molecular oxygen for respiration. In soil, mud, sediments and in the deep waters of many lakes and seas, however, oxygen is depleted; and organic matter accumulates there faster than oxygen can enter by diffusion.

Microbes therefore utilize several other sources of oxygen, two of which are described here. In the absence of oxygen, some organisms are able to use the oxygen in nitrates, releasing nitrogen to the atmosphere in the process. In the absence of both oxygen and nitrates, microbes use the oxygen in sulfates, producing hydrogen sulfide in the process. Whether first the oxygen, then the nitrate and finally the sulfate are depleted depends upon the amount of organic matter (carbohydrate) present. Thus, all four cycles interact in this portion of the biosphere.

Man affects this system in several ways, two of which may be important. The flow of nitrates to groundwater and then to drinking water is prevented primarily by denitrification. This in turn depends on the presence of organic material and an absence of oxygen, both of which are affected by intensive agricultural practices. In general, then, agricultural practices tend to increase oxygen content and deplete organic matter which could depress denitrification and lead to an increase in the amount of nitrates in drinking water. However, the magnitude of this effect is not known.

A second area of man's influence is on the coastal tidal flats. In these muds, the rate of sulfate reduction depends upon the amount of organic matter present. Non-polluted mud produces relatively little hydrogen sulfide, most of which is precipitated with iron. Polluted muds, however, can produce 10 to 20 times as much hydrogen sulfide, a large part of which may escape to the atmosphere (13).

### *Atmospheric Interactions*

Atmospheric sulfur is most rapidly removed from the air as a complex salt of ammonia and sulfate. In polluted urban air, however, removal of man's added sulfur oxides tends to be slowed by a relative scarcity of ammonium.

Nitrogen oxides added to urban air tend to hasten the formation of particulates from hydrocarbons that have also been emitted, but it is scant comfort that the processes leading to photochemical smog production also hasten the rate of removal of hydrocarbons from the air.

### *Eutrophication of Water*

Runoff from agricultural lands is relatively rich in nitrates, since these are more soluble than phosphates. By contrast, effluent from sewage treatment plants is relatively poor in nitrates and rich in phosphates. Not only is much of the original nitrate denitrified under the anoxic conditions of sewage treatment, but the amount of phosphate is increased by its use in laundry and dishwater detergents.

The ratio of phosphate to nitrate influences the kinds of algae that flourish in lakes. Green algae, which are more edible and rapidly enter the food chain, require nitrate or ammonia in the water; whereas many blue-green algae, which are often inedible and, so, tend to accumulate, are able to fix their own nitrogen. Thus, pollution with sewage effluents favors the growth of undesirable algae and large aquatic plants.

Algal overgrowth can occur only if all of the essential elements are abundant. Some lakes, in basins lacking limestone, are deficient in carbonate. Others are acid, making the biological fixation of nitrogen difficult or slow. Most water bodies are relatively deficient in phosphorus, for the reasons described in the preceding sections. Furthermore, phosphorus is the least renewable resource, and its absence precludes any possibility of algae overgrowth.

### *Eutrophication of Land*

Human activity is "stirring up" the biogeochemical cycles of many materials. To the degree that atmospheric levels of nutrient materials are increased, the possibility exists that man is also enriching the land.

Carbon dioxide levels in the atmosphere are rising (1). Since laboratory studies show a linear relation between photosynthetic rates and carbon dioxide levels (1), enhancement of photosynthesis by man-made increases in $CO_2$ is expected in terrestrial vegetation.

The same processes that produce carbon dioxide also add oxides of sulfur and nitrogen in the atmosphere. In addition, man's pollution of mud flats may increase atmospheric sulfates; and where he uses ammonium fertilizers there may be increases in atmospheric ammonia (30). Phosphatic dust from phosphate industries and additional traces of phosphate from other industries also find their way into the atmosphere, although the total amount is small.

Once the acid effect of oxides is neutralized, all of these materials become nutrients and many of them come down in rain. As a result, the possibility of a small but worldwide fertilization of vegetation should not be overlooked.

Since land systems have problems in retaining nutrient material, as opposed to aquatic systems in which these materials accumulate, this kind of eutrophication of land should be generally useful and may in general increase land productivity.

## Quantitative Estimates

Although this section may seem to be of interest primarily to the scientific community, it provides an objective means of evaluating the accuracy of the foregoing analyses and concludes with a major recommendation regarding world monitoring programs.

Some quantitative problems are common to all these cycles: estimates of river runoff, precipitation values, rates of biological activities, etc. These are discussed here at length for the sulfur cycle but are meant to apply as well to the cycles of phosphorus and nitrogen.

The phosphorus cycle is relatively simple, and major rates of transfer have already been presented (Table 39–3). These are considered to be accurate enough for the use that is made of them, even though they have probable errors of ±40% (if the estimate is 20, the odds are one-half that the true value lies between 12 and 28). But, if changes in the phosphorus cycle are to be detected, better estimates will be necessary.

TABLE 39–9

*Budget for the nitrogen cycle. All numbers are millions of metric tons. The error columns list plus-or-minus probable errors as a percentage of the estimate.*

| | Land | | Sea | | Atmosphere | |
|---|---|---|---|---|---|---|
| | rate/yr | % error | rate/yr | % error | rate/yr | % error |
| **Input** | | | | | | |
| Biological nitrogen fixation | -- | -- | 10 | 50 | -- | -- |
| Symbiotic (31) | 14 | 25 | -- | -- | -- | -- |
| Non-symbiotic (31) | 30 | 50 | -- | -- | -- | -- |
| Atmospheric nitrogen fixation (31) | 4 | 100 | 4 | 100 | -- | -- |
| Industrially fixed nitrogen fertilizer (31) | 30 | 5 | -- | -- | -- | -- |
| N-oxides from combustion * | 14 | 25 | 6 | 25 | 20 | 25 |
| Return of volatile nitrogen compounds in rain | ? | -- | ? | -- | -- | -- |
| River influx (31) | -- | -- | 30 | 50 | -- | -- |
| $N_2$ from biological denitrification (31) | -- | -- | -- | -- | 83 | 100 |
| Natural $NO_2$ | -- | -- | -- | -- | ? | -- |
| Volatilization ($HN_3$) | -- | -- | -- | -- | ? | -- |
| Total Input | 92+ | | 50 | | 103+ | |
| **Storage** | | | | | | |
| Plants (31) | 12,000 | 30 | 800 | 50 | -- | -- |
| Animals (31) | 200 | 30 | 170 | 50 | -- | -- |
| Dead organic matter (31) | 760,000 | 50 | 900,000 | 100 | -- | -- |
| Inorganic nitrogen (31) | 140,000 | 50 | 100,000 | 50 | -- | -- |
| Dissolved nitrogen (31) | -- | -- | 20,000,000 | 10 | -- | -- |
| Nitrogen gas (31) | -- | -- | -- | -- | 3,800,000,000 | 3 |
| $NO + NH_4$ (25) | -- | -- | -- | -- | Less than 1 | 50 |
| $NH_3 + NH_4$ (17) | -- | -- | -- | -- | 12 | 50 |
| $N_2O$ (33) | -- | -- | -- | -- | 1,000 | 50 |
| Total Storage | 912,200 | | 21,000,970 | | 3,800,001,013 | |
| **Loss** | | | | | | |
| Denitrification (31) | 43 | -- | 40 | 100 | -- | -- |
| Volatilization | ? | -- | ? | -- | -- | -- |
| River runoff (31, 32)(includes enrichment from fertilizers) | 30 | 50 | -- | -- | -- | -- |
| Sedimentation (31) | -- | -- | 0.2 | 50 | -- | -- |
| $N_2$ in all fixation processes | -- | -- | -- | -- | 92 | 50 |
| $NH_3$ in rain (17) | -- | -- | -- | -- | Less than 40 | 50 |
| $NO_2$ in rain | -- | -- | -- | -- | ? | -- |
| $N_2O$ in rain | -- | -- | -- | -- | ? | -- |
| Total Loss | 73 | | 40.2 | | 132+ | |

* See Table 3.2.

The nitrogen cycle is so complex and has so many exchange rates, that time did not allow a critical evaluation of their accuracy or of the widely different values used by different authors. Table 39–9 lists the rates that were used for the analysis of the nitrogen cycle. Also given is the group's best guess on the probable errors of these estimates. These ranges do not include published estimates that have been rejected, which in some cases differ by factors of 10 to 100.

### Sulfur

The distribution of sulfur on our planet is shown in Table 39–10. Most of the sulfur is involved in a very slow geologic cycle between the land and the oceans, while a much smaller amount is involved in the much more rapid atmospheric and ecologic cycles. Only the latter are of interest on the human time scale.

TABLE 39–10

*Sulfur reservoirs*

| | | Metric Tons |
|---|---|---|
| Land | Inorganic S in igneous rock (11) | 5,700,000,000 |
| | Inorganic S in sedimentary rock (11) | 3,600,000,000 |
| | Organic and inorganic S in soils (11) | 28,000 |
| | Land biomass (46) | 510 |
| Ocean | S as $SO_4^{--}$ | 1,215,000,000 |
| | Organic S in ocean biomass (11,47) | 23 |
| Atmosphere (Troposphere) | $H_2S$ (40) | 10 to 48 |
| | $SO_2$ (40) | 5 to 50 |
| | Aerosol and particulate $SO_4^{--}$ | variable |
| | Sea salt particles | variable |

On the land, the sulfur reservoir is comprised of igneous rock which contains 260 ppm of sulfur and sedimentary rocks, in which it is present primarily as iron sulfides and calcium sulfate. Most weathered sulfur is derived from sedimentary rocks which represent a sulfur reservoir of 3.6 billion million tons (11, 13, 34).

Sulfur is also abundant in the ocean as sulfate ions and is present in much smaller quantities as organic sulfur in the biomass.

Atmospheric sulfur occurs in gaseous form as $H_2S$ and $SO_2$, both of which are short lived, and as sulfate or "ammoniated sulfate" in aerosol or particulate form. In addition, some sulfur is present in sea salt particles formed by evaporation of sea spray. The concentrations, lifetimes, sources and fates of these atmospheric forms of sulfur have been the subject of extensive experimental and theoretical investigation, but the phenomena are so complex that no clear picture has yet emerged. (See References 14, 17, 26, 35–95.)

Most of the sulfur acquired by plants is taken up by the roots from sulfate in soil water, but they may also take in gaseous $SO_2$ directly (26). Some sulfur, in turn, is liberated to the air when vegetable matter decomposes.

The most significant biological role in the sulfur cycle is the reduction of sulfate to hydrogen sulfide in aquatic or mud environments rich in organic matter and depleted of oxygen and nitrate. Under these anoxic conditions, bacteria (generally of the Desulphovibrio type) utilize dissolved sulfate ions as hydrogen acceptors and oxidize organic matter according to the following overall reactions:

$$2CH_2O + SO_4^{--} \rightarrow H_2S + 2HCO_3^{-}$$

$$4CH_2NH_2COOH + 4H_2O + 3SO_4^{--} \rightarrow$$

$$H_2S + 2HS^{-} + 8HCO_3^{-} + 4NH_4 +$$

There are two important sulfur cycles which must be considered in the construction of a sulfur budget: the transfer of weathered sulfur from the land to the ocean by means of continental runoff and the exchange of sulfur between the atmosphere and the surface. Estimates of exchange rates for these cycles are given in Table 39–11. These two cycles interact in that some of the sulfur in river runoff is derived from the atmosphere-surface cycle.

The surface sources of atmospheric sulfur are: volcanic emissions of $SO_2$ and $H_2S$, industrial emissions of $SO_2$ (mostly from fossil fuel combustion), sulfate in sea salt derived from the evaporation of ocean spray, and biogenic hydrogen sulfide produced in anoxic marine or freshwater environments.

Atmospheric sulfur returns to the surface in precipitation and by means of "dry deposition," the sedimentation of atmospheric particles or their scavenging by vegetation, and the direct gaseous absorption of $SO_2$ by water and land surfaces.

The agricultural use of sulfate fertilizer and sulfur-containing industrial effluents added to streams also contribute to the continental river runoff.

Human influences can be estimated from production and consump-

TABLE 39–11

*Annual Rates of Sulfur Exchange*

| | Sulfur (millions metric tons) |
|---|---|
| Land surface to ocean (river runoff) (48) | 123 |
| Rock weathering (13,14) | 15-43 |
| Atmosphere to surface | |
| Sulfate in precipitation over land*(14) | 60 |
| Sulfate in precipitation over ocean*(14) | 60 |
| Dry deposition over land and ocean*** | 0.5-100 |
| Surface to atmosphere | |
| Volcanic emissions ($SO_4^{--}$ and $H_2S$) (16) | 7-12 |
| Industrial emissions ($SO_2$) | 68 |
| Industrial effluents on land**(2) | 27 |
| Ocean to atmosphere | |
| Sea salt (precipitated over land) (14,51) | 5-25 |
| Sea salt (precipitated over ocean) (14,51) | 39-195 |
| $H_2S$ | unknown |
| Ocean to ocean sediments (iron pyrites) (34) | 7 |

* Excludes S originating from sea salt.
** Industrial sulfuric acid including fertilizers.
*** Calculated from dry deposition rates given in (26) and (14).

tion statistics. The sulfur in river runoff and precipitation can be monitored. Most of the remaining exchanges of sulfur between air, land, water and living systems are difficult to measure.

Global estimates of river runoff for various elements have been compiled by Livingstone (48). When his estimates are used, it is worthwhile to examine his sources. In the case of sulfur, many of the data for European rivers were recorded in 1848 and most of it before 1900. Most of the values for the United States wwere recorded in the 1940's and early 1950's. Beyond that, differences in techniques for chemical analysis and the lack of multiple sampling or sampling in all seasons make them unreliable for present use, as it has been shown that ion concentrations vary during the year with varying precipitation (48). Finally, the distribution of samples does not represent the different land areas of the globe equally.

Precipitation data appear to be scanty and difficult to interpret because the ion content of the rain depends upon the rain's intensity, duration and frequency and also upon the gaseous and particulate contents of the atmosphere, which may vary widely with locality, season and meteorological conditions. Except for very isolated locations, we have found few useful sources of bulk precipitation data from which mean annual atmosphere-to-surface transport rates may be calculated (49, 50), even though such data are easy to collect. Most atmospheric composition and precipitation chemistry studies have been concentrated in a few regions of relatively industrialized temperate zone countries and should not be extrapolated to the globe as a whole.

Recent studies of remote, nonurban areas (Panama, Antarctica and the Amazon rain forest) suggest that their atmospheric chemistry may differ appreciably from that of temperate continental zones (40, 42, 43). The introduction of new measurement techniques and their use in remote areas have resulted in revisions of estimates of probable global mean concentrations of trace elements in the atmosphere. Table 39–12 reflects the magnitude of the revisions, contrasting widely used estimates of global tropospheric trace element composition, compiled in 1963, with more recent estimates. The more recent estimates, themselves, should be considered tentative (Cadle, personal communication).

Tables 39–10 and 39–11 indicate the possible sulfur reservoir sizes and exchange rates. For the most part, they are based on figures frequently quoted in the literature. Where possible, we have attempted

TABLE 39–12

*Trace Gases in the Atmosphere, PPB (40)*

| | Polluted | Background | |
|---|---|---|---|
| | | Published Means | Recent |
| Oxygen | | 209,000,000 | |
| Carbon Dioxide | | 300,000 | 320,000 |
| Methane | | 1,400 | 1,250 |
| Hydrogen | | 700 | 500 |
| Nitrous Oxide | 2,000 | 420 | 240 |
| Hydrogen Sulfide | 500 | 10 | 2 |
| Sulfur Dioxide | 2,000 | 10 | 1 |
| Ammonia | 2,000 | 10 | 15 |
| Formaldehyde | 1,000 | 5 | 4 |
| Nitrogen Dioxide | 2,000 | 2 | 0.5 |
| Nitric Oxide | | | 0.5 |

to indicate the uncertainty in these data by showing a range of estimates presented by different authors. With few exceptions, the figures may be seriously questioned. Taken together, they do not provide an adequate basis for an evaluation of the impact of human activities on the sulfur cycle; and, since this impact may be large, we urge action to improve this information base.

These problems are common to analyses of many biogeochemical cycles. At the present time, efforts are being made to design global monitoring networks which will include measures of various rates of flow in element cycles. For those elements essential to life, this may give rise to several compatible and consistent sources of information. This will provide not only a check on the validity of existing data but will also add insight into man's influence on the environment.

## References


1. Study of Critical Environmental Problems (SCEP). 1970. *Man's impact on the global environment*. The MIT Press, Cambridge. 319 p.
2. United Nations. 1969. *United Nations statistical yearbook*. Statistical Office of the United Nations, New York. 770 p.
3. Alexander, M. 1971. *Microbial ecology*. John Wiley and Sons, New York. 511 p.
4. Goeller, H. E. 1971. *The ultimate mineral resource situations*. ORNL DWG 71-3865 (Oak Ridge National Laboratory, Tenn.).
5. American Chemical Society, Committee on Chemistry and Public Affairs. 1969. *Cleaning our environment; the chemical basis for action*. American Chemical Society, Washington, D.C. 249 p.
6. Biggar, J. W., and R. B. Corey. 1967. "Agricultural drainage and eutrophication." In *National Academy of Sciences, Eutrophication: causes, consequences, correctives. NAS, Proc. of Symposium*. Washington, D.C. 661 p.
7. Fortescue, J. A. C., and G. G. Marten. 1970. "Micronutrients: forest ecology and systems analysis, p. 173–198. In D. E. Reichle (ed.), *Analysis of temperate forest ecosystems*. Springer-Verlag, New York. 304 p.
8. Sauchelli, V. 1965. *Phosphate in agriculture*. Reinhold Publ. Co., New York. 277 p.
9. President's Science Advisory Committee, Panel on the World Food Supply. 1967. *The world food problem*. Vol. III. U.S. Govt. Printing Office, Washington, D.C. 332 p.
10. U.S. Bureau of Mines. 1970. Mineral facts and problems. *U.S. Dept. Int., Bull.* 650. Washington, D.C. 1291 p.
11. Bowen, H. J. M. 1966. *Trace elements in biochemistry*. Academic Press, New York. 241 p.
12. *Chemical Economics Handbook*. 1969. Stanford Research Institute, Menlo Park, Calif.
13. Berner, R. A. 1971a. Sulfate reduction, pyrite formation, and the oceanic sulfur budget. *Paper presented at Nobel Symposium*, Stockholm, 1971.
14. Eriksson, E. 1963. The yearly circulation of sulfur in nature. *J. Geophys. Res.* 60:4001–4008.
15. Stuiver, M. 1967. The sulfur cycle in lake waters during thermal stratification. *Geochimica st. Cosmochimica Acta* 31:2151–2167.
16. Koyama, T., N. Nakai, and E. Kamata. 1965. Possible discharge rate of hydrogen sulfide from polluted coastal belts in Japan. *J. of Earth Science* (Nagoya Univ.) 13:1–11.

17. Junge, C. E. 1963. *Air chemistry and radioactivity*. Academic Press, New York and London. 382 p.
18. Larson, T. E., and B. D. Larson. 1968. Quality of surface waters in Illinois. In 1957 *Interim report on the presence of nitrate in Illinois surface waters*. Illinois State Water Survey, Urbana.
19. Hasler, A. D. 1970. Man-induced eutrophication of lakes, p. 110–125. In S. F. Singer (ed.), *Global effects of environmental pollution*. Springer-Verlag, New York. 218 p.
20. Gilbert, C.S., H. F. Eppson, W. B. Bradley, and O. A. Beath. 1946. Nitrate accumulation in cultivated plants and weeds. *Wyoming Agr. Sta. Bull.* p. 227.
21. Whitehead, E. I., and A. L. Moxon. 1952. Nitrate poisoning. *S. Dakota Agr. Exp. Sta. Bull.* 424 p.
22. Wuhrmann, K. 1964. Nitrogen removal in sewage treatment processes. *Verh. Int. Ver. Limnol.* 15:580–596.
23. Ludzak, F. J., and M. C. Ettinger. 1961. Controlling operation to minimize activated sludge effluent nitrogen. *J. Water Pollut. Control Fed.* 34:920–931.
24. Johnson, W. K., and G. J. Schroepfer. 1964. Nitrogen removal by nitrification and denitrification. *J. Water Pollut. Control Fed.* 36:1015–1036.
25. Ripperton, L. A., L. Kornreich, and J. J. B. Worth. 1970. Nitrogen dioxide and nitric oxide in non-urban air. *J. Air Pollut. Center Assoc.* 20:589–592.
26. Robinson, E., and R. C. Robbins. 1970. "Gaseous atmospheric pollutants from urban and natural sources," p. 51–64. In S. F. Singer (ed.), *Global effects of environmental pollution*. Springer-Verlag, New York. 218 p.
27. Georgii, R. W. 1963. Oxides of nitrogen and ammonia in the atmosphere. *J. Geophys. Res.* 68:3933–3970.
28. Olson, J. S. 1970. "Models of the hydrologic cycles," p. 268–285. In D. E. Reichle (ed.), *Analysis of temperate forest ecosystems*. Springer-Verlag, New York. 304 p.
29. McLaren, A. D., and J. Skujins (eds.). 1971. *Soil biochemistry*. Vol. 2. Marcel Dekker, Inc., New York. 527 p.
30. Loewenstein, H., L. E. Engelbert, O. J. Attoe, and O. N. Allen. 1957. Nitrogen loss in gaseous form from soils as influenced by fertilizers and management. *Soil Science Society, Proc.* 21:397–400.
31. Delwiche, C. C. 1970. The nitrogen cycle. *Sci. American* 223:137–158.
32. Clarke, F. W. 1924. The date of geochemistry. 5th ed. *U.S. Geo. Survey Bull.* 770. 841 p.
33. Schutz, K., C. E. Junge, R. Beck, and B. Albrecht. 1970. Studies of atmospheric $N_2O$. *J. Geophys. Res.* 75:2230–2246.
34. Berner, R. A. 1971b. Worldwide sulfur pollution of rivers. *J. Geophys. Res.* (Under review).
35. Eriksson, E. 1959. The yearly circulation of chloride and sulfur in nature; meteorological, geochemical, and pedological implications, 1. *Tellus* 11:375–403.
36. Martell, E. A. 1966. The size distribution and interaction of radioactive and natural aerosols in the atmosphere. *Tellus* 18:486–498.
37. Friend, J. P. 1966. Properties of the stratospheric aerosol. *Tellus* 18:465–473.
38. Junge, C. E., E. Robinson, and F. L. Ludwig. 1969. A study of aerosols in Pacific air masses. *J. Applied Meteorology* 8:340–347.
39. Georgii, H. W. 1970. Contribution to the atmospheric sulfur budget. *J. Geophys. Res.* 75:2365–2371.
40. Pate, J. B., J. P. Lodge, Jr., D. C. Sheesley, and A. F. Wartburg. 1970. "Atmospheric chemistry of the tropics." In *Symp. Proc. on Environment in Amazonia*; 24 April 1970. Part I. Instit. Nacio. Pesq. Amazon, Manaus, Brazil. p. 43.
41. Beilke, S., and H. W. Georgii. 1968. Investigation on the incorporation of sulfur-dioxide into fog and rain droplets. *Tellus* 20:435–442.
42. Lodge, J. P., Jr., and J. B. Pate. 1966. Atmospheric gases and particulates in Panama. *Science* 153:408–410.

43. Fischer, W. H., J. P. Lodge, Jr., J. B. Pate, and R. D. Cadle. 1969. Antarctic atmospheric chemistry: preliminary exploration. *Science* 167:66–67.
44. Cadle, R. D., and E. R. Allen. 1970. Atmospheric photochemistry. *Science* 167:243–249.
45. Cadle, R. D. 1971. Formation and chemical reactions of atmospheric particles. *J. of Colloid and Interface Science*. (In press).
46. Deevey, E. S., Jr. 1970. Mineral cycles. *Sci. Amer.* 223:148–159.
47. Whittaker, R. H., and G. E. Likens. 1961. Woodland Forest Working Group of International Biological Program. (Unpublished).
48. Livingstone, D. A. 1963. "Chemical composition of rivers and lakes," Chapt. G. In M. Fleischer (ed.), *Data of geochemistry. 6th ed. Geolog. Survey Prof. Paper* 440–G. U.S. Govt. Printing Office, Washington, D.C. 64 p.
49. Likens, G. F., F. H. Bormann, N. M. Johnson, D. W. Fisher, and R. S. Pierce. 1970. Effects of forest cutting and herbicidal treatment on nutrient budgets in the Hubbard Brook Watershed-Ecosystem. *Ecol. Monogr.* 40:23–47.
50. Feth, J. H. 1967. Chemical characteristics of bulk precipitation in the Mojave Desert region, California. *U.S. Geol. Survey Prof. Paper* 575-C:222–227.
51. Blanchard, D. C. 1963. "The electrification of the atmosphere by particles from bubbles in the sea," p. 71–202. In M. Sears (ed.), *Progress in oceanography.* Vol. 1. Pergamon Press, New York.
52. Duvigneaud, P., and S. Denaeyer-DeSmet. 1970. "Biological cycling of minerals in temperate deciduous forests," p. 199–225. In D. E. Reichle (ed.), *Analysis of temperate forest ecosystems*. Springer-Verlag, New York. 304 p.
53. Keyfitz, N. 1971. On the momentum of population growth. *Demography* 8:71–81.
54. Royal Commission on Natural Resources. 1968. *Environmental research*. Report of the Royal Comm. on Nat. Resour., Stockholm.

# IX

# Outlook for the Future

# 40

# The conservation ethic

*Aldo Leopold*

When god-like Odysseus returned from the wars in Troy, he hanged all on one rope some dozen slave-girls of his household whom he suspected of misbehavior during his absence.

This hanging involved no question of propriety, much less of justice. The girls were property. The disposal of property was then, as now, a matter of expediency, not of right and wrong.

Criteria of right and wrong were not lacking from Odysseus' Greece: witness the fidelity of his wife through the long years before at last his black-prowed galleys clove the wine-dark seas for home. The ethical structure of that day covered wives, but had not yet been extended to human chattels. During the three thousand years which have since elapsed, ethical criteria have been extended to many fields of conduct, with corresponding shrinkages in those judged by expediency only.

This extension of ethics, so far studied only by philosophers, is actually a process in ecological evolution. Its sequences may be described in biological as well as philosophical terms. An ethic, biologically, is a limitation on freedom of action in the struggle for existence. An ethic, philosophically, is a differentiation of social from antisocial conduct. These are two definitions of one thing. The thing has its origin in the tendency of interdependent individuals or societies to evolve modes of

---

Reprinted with special permission from the *Journal of Forestry*, 31:634–643 (October, 1933).

coöperation. The biologist calls these symbioses. Man elaborated certain advanced symbioses called politics and economics. Like their simpler biological antecedents, they enable individuals or groups to exploit each other in an orderly way. Their first yardstick was expediency.

The complexity of coöperative mechanisms increased with population density, and with the efficiency of tools. It was simpler, for example, to define the antisocial uses of sticks and stones in the days of the mastodons than of bullets and billboards in the age of motors.

At a certain stage of complexity, the human community found expediency yardsticks no longer sufficient. One by one it has evolved and superimposed upon them a set of ethical yardsticks. The first ethics dealt with the relationship between individuals. The Mosaic Decalogue is an example. Later accretions dealt with the relationship between the individual and society. Christianity tries to integrate the individual to society, Democracy to integrate social organization to the individual.

There is as yet no ethic dealing with man's relationship to land and to the nonhuman animals and plants which grow upon it. Land, like Odysseus slave-girls, is still property. The land relation is still strictly economic, entailing privileges but not obligations.

The extension of ethics to this third element in human environment is, if we read evolution correctly, an ecological possibility. It is the third step in a sequence. The first two have already been taken. Civilized man exhibits in his own mind evidence that the third is needed. For example, his sense of right and wrong may be aroused quite as strongly by the desecration of a nearby woodlot as by a famine in China, a near-pogrom in Germany, or the murder of the slave-girls in ancient Greece. Individual thinkers since the days of Ezekiel and Isaiah have asserted that the despoliation of land is not only inexpedient but wrong. Society, however, has not yet affirmed their belief. I regard the present conservation movement as the embryo of such an affirmation. I here discuss why this is, or should be, so.

Some scientists will dismiss this matter forthwith, on the ground that ecology has no relation to right and wrong. To such I reply that science, if not philosophy, should by now have made us cautious about dismissals. An ethic may be regarded as a mode of guidance for meeting ecological situations so new or intricate, or involving such deferred reactions, that the path of social expediency is not discernible to the average individual. Animal instincts are just this. Ethics are possibly a kind of advanced social instinct in the making.

Whatever the merits of this analogy, no ecologist can deny that our land-relation involves penalties and rewards which the individual does not see, and needs modes of guidance which do not yet exist. Call these what you will, science cannot escape its part in forming them.

## Ecology—its Role in History

A harmonious relation to land is more intricate, and of more consequence to civilization, than the historians of its progress seem to realize. Civilization is not, as they often assume, the enslavement of a stable and constant earth. It is a state of *mutual and interdependent coöperation* between human animals, other animals, plants, and soils, which may be disrupted at any moment by the failure of any of them. Land-despoliation has evicted nations, and can on occasion do it again. As long as six virgin continents awaited the plow, this was perhaps no tragic matter—eviction from one piece of soil could be recouped by despoiling another. But there are now wars and rumors of wars which foretell the impending saturation of the earth's best soils and climates. It thus becomes a matter of some importance, at least to ourselves, that our dominion, once gained, be self-perpetuating rather than self-destructive.

This instability of our land-relation calls for example. I will sketch a single aspect of it: the plant succession as a factor in history.

In the years following the Revolution, three groups were contending for control of the Mississippi valley: the native Indians, the French and English traders, and American settlers. Historians wonder what would have happened if the English at Detroit had thrown a little more weight into the Indian side of those tipsy scales which decided the outcome of the Colonial migration into the cane-lands of Kentucky. Yet who ever wondered why the cane-lands, when subjected to the particular mixture of forces represented by the cow, plow, fire, and axe of the pioneer, became bluegrass? What if the plant succession inherent in this "dark and bloody ground" had, under the impact of these forces, given us some worthless sedge, shrub, or weed? Would Boone and Kenton have held out? Would there have been any overflow into Ohio? Any Louisiana Purchase? Any transcontinental union of new states? Any Civil War? Any machine age? Any depression? The subsequent drama of American history, here and elsewhere, hung in large degree on the reaction of particular soils to the impact of particular forces exerted by a particular kind and degree of human occupation. No statesman-biologist selected those forces, nor foresaw their effects. That chain of events which on the Fourth of July we call our National Destiny hung on a "fortuitous concourse of elements," the interplay of which we now dimly decipher *by hindsight only.*

Contrast Kentucky with what hindsight tells us about the Southwest. The impact of occupancy here brought no bluegrass, nor other plant fitted to withstand the bumps and buffetings of misuse. Most of these soils, when grazed, reverted through a successive series of more and more worthless grasses, shrubs, and weeds to a condition of unstable equilibrium. Each recession of plant types bred erosion; each increment to erosion bred a further recession of plants. The result today is a progressive and mutual deterioration, not only of plants and soils, but of the animal community subsisting thereon. The early settlers did not expect this; on the cienegas of central New Mexico some even cut artificial gullies to hasten it. So subtle has been its progress that few people know anything about it. It is not discussed at polite tea-tables or go-getting luncheon clubs, but only in the arid halls of science.

All civilization seem to have been conditioned upon whether the plant successon, under the impact of occupancy, gave a stable and habitable assortment of vegetative types, or an unstable and uninhabitable assortment. The swampy forests of Caesar's Gaul were utterly changed by human use—for the better. Moses' land of milk and honey was utterly changed—for the worse. Both changes are the unpremeditated resultant of the impact between ecological and economic forces. We now decipher these reactions retrospectively. What could possibly be more important than to foresee and control them?

We of the machine age admire ourselves for our mechanical ingenuity; we harness cars to the solar energy impounded in carboniferous forests; we fly in mechanical birds; we make the ether carry our words or even our pictures. But are these not in one sense mere parlor tricks compared with our utter ineptitude in keeping land fit to live upon? Our engineering has attained the pearly gates of a near-millennium, but our applied biology still lives in nomad's tents of the stone age. If our system of land-use happens to be self-perpetuating, we stay. If it happens to be self-destructive we move, like Abraham, to pastures new.

Do I overdraw this paradox? I think not. Consider the transcontinental airmail which plies the skyways of the Southwest—a symbol of its final conquest. What does it see? A score of mountain valleys which were green gems of fertility when first described by Coronado, Espejo, Pattie, Abert, Sitgreaves, and Couzens. What are they now? Sandbars, wastes of cobbles and burroweed, a path for torrents. Rivers which Pattie says were clear, now muddy sewers for the wasting fertility of an empire. A "Public Domain," once a velvet carpet of rich buffalo-grass and grama, now an illimitable waste of rattlesnake-bush and tumbleweed, too impoverished to be accepted as a gift by the states within which it lies. Why? Because the ecology of this Southwest happened to be set on a hair-trigger. Because cows eat brush when the grass is gone, and thus postpone the penalties of over-utilization. Because certain grasses, when grazed too closely to bear seed-stalks, are weakened and give way to inferior grasses, and these to inferior shrubs, and these to weeds, and these to naked earth. Because rain which spatters upon vegetated soil stays clear and sinks, while rain which spatters upon devegetated soil seals its interstices with colloidal mud and hence must run away as floods, cutting the heart out of country as it goes. Are these phenomena any more difficult to foresee than the paths of stars which science deciphers without the error of a single second? Which is the more important to the permanence and welfare of civilization?

I do not here berate the astronomer for his precocity, but rather the ecologist for his lack of it. The days of his cloistered sequestration are over:

"Whether you will or not,
You are a king, Tristram, for you are one
Of the time-tested few that leave the world,
When they are gone, not the same place it was.
Mark what you leave."

Unforeseen ecological reactions not only make or break history in a few exceptional enterprises—they condition, circumscribe, delimit, and warp all enterprises, both economic and cultural, that pertain to land. In the cornbelt, after grazing and plowing out all the cover in the interests of "clean farming," we grew tearful about wild-life, and spent several decades passing laws for its restoration. We were like Canute commanding the tide. Only recently has research made it clear that the implements for restoration lie not in the legislature, but in the farmer's toolshed. Barbed wire and brains are doing what laws alone failed to do.

In other instances we take credit for shaking down apples which were, in all probability, ecological windfalls. In the Lake States and the Northeast lumbering, pulping, and fire accidentally created some scores of millions of acres of new second-growth. At the proper stage we find these thickets full of deer. For this we naively thank the wisdom of our game laws.

In short, the reaction of land to occupancy determines the nature and duration of civilization. In arid climates the land may be destroyed. In all climates the plant succession determines what economic activities can be supported. Their nature and intensity in turn determine not only the domestic but also the wild plant and animal life, the scenery, and the whole face of nature. We inherit the earth, but within the limits of the soil and the plant succession we also *rebuild* the earth—without plan, without knowledge of its properties, and without understanding of the increasingly coarse and powerful tools which science has placed at our disposal. We are remodelling the Alhambra with a steam-shovel.

## Ecology and Economics

The conservation movement is, at the very least, an assertion that these interactions between man and land are too important to be left to chance, even that sacred variety of chance known as economic law.

We have three possible controls: Legislation, self-interest, and ethics.

Before we can know where and how they will work, we must first understand the reactions. Such understanding arises only from research. At the present moment research, inadequate as it is, has nevertheless piled up a large store of facts which our land using industries are unwilling, or (they claim) unable, to apply. Why? A review of three sample fields will be attempted.

Soil science has so far relied on self-interest as the motive for conservation. The landholder is told that it pays to conserve his soil and its fertility. On good farms this economic formula has improved land-practice, but on poorer soils vast abuses still proceed unchecked. Public acquisition of submarginal soils is being urged as a remedy for their misuse. It has been applied to some extent, but it often comes too late to check erosion, and can hardly hope more than to ameliorate a phenomenon involving in some degree *every square foot* on the continent. Legislative compulsion might work on the best soils where it is least needed, but it seems hopeless on poor soils where the existing economic set-up hardly permits even uncontrolled private enterprise to make a profit. We must face the fact that, by and large, no defensible relationship between man and the soil of his nativity is as yet in sight.

Forestry exhibits another tragedy—or comedy—of *Homo sapiens*, astride the runaway Juggernaut of his own building, trying to be decent to his environment. A new profession was trained in the confident expectation that the shrinkage in virgin timber would, as a matter of self-interest, bring an expansion of timber-cropping. Foresters are cropping timber on certain parcels of poor land which happen to be public, but on the great bulk of private holdings they have accomplished little. Economics won't let them. Why? He would be bold indeed who claimed to know the whole answer, but these parts of it seem agreed upon: modern transport prevents profitable tree-cropping in cut-out regions until virgin stands in all others are first exhausted; substitutes for lumber have undermined confidence in the future need for it; carrying charges on stumpage reserves are so high as to force perennial liquidation, overproduction, depressed prices, and an appalling wastage of unmarketable grades which must be cut to get the higher grades; the mind of the forest owner lacks the point-of-view underlying sustained yield; the low wage-standards on which European forestry rests do not obtain in America.

A few tentative gropings toward industrial forestry were visible before 1929, but these have been mostly swept away by the depression, with the net result that forty years of "campaigning" have left us only such actual tree-cropping as is under-written by public treasuries. Only a blind man could see in this the beginnings of an orderly and harmonious use of the forest resource.

There are those who would remedy this failure by legislative compulsion of private owners. Can a landholder be successfully compelled to raise any crop, let alone a complex long-time crop like a forest, on land the private possession of which is, for the moment at least, a liability? Compulsion would merely hasten that avalanche of tax-delinquent land-titles now being dumped into the public lap.

Another and larger group seeks a remedy in more public ownership. Doubtless we need it—we are getting it whether we need it or not—but how far can it go? We cannot dodge the fact that the forest problem, like the soil problem, *is coextensive with the map of the United States*. How far can we tax other lands and industries to maintain forest lands and industries artificially? How confidently can we set out to run a hundred-yard dash with a twenty foot rope tying our ankle to the starting point? Well, we are bravely "getting set," anyhow.

The trend in wildlife conservation is possibly more encouraging than in either soils or forests. It has suddenly become apparent that farmers, out of self-interest, can be induced to crop game. Game crops are in demand, staple crops are not. For farm-species, therefore, the immediate future is relatively bright. Forest game has profited to some extent by the accidental establishment of new habitat following the decline of forest industries. Migratory game, on the other hand, has lost heavily through drainage and over-shooting; its future is black because motives of self-interest do not apply to the private cropping of birds so mobile that they "belong" to everybody, and hence to nobody. Only governments have interests coextensive with their annual movements, and the divided counsels of conservationists give governments ample alibi for doing little. Governments could crop migratory birds because their marshy habitat is cheap and concentrated, but we get only an annual crop of new hearings on how to divide the fast-dwindling remnant.

These three fields of conservation, while but fractions of the whole, suffice to illustrate the welter of conflicting forces, facts, and opinions which so far comprise the result of the effort to harmonize our machine civilization with the land whence comes its sustenance. We have accomplished little, but we should have learned much. What?

I can see clearly only two things:

First, that the economic cards are stacked against some of the most important reforms in land-use.

Second, that the scheme to circumvent this obstacle by public ownership, while highly desirable and good as far as it goes, can never go far enough. Many will take issue on this, but the issue is between two conflicting conceptions of the end towards which we are working.

One regards conservation as a kind of sacrificial offering, made for us vicariously by bureaus, on lands nobody wants for other purposes, in propitiation for the atrocities which still prevail everywhere else. We have made a real start on this kind of conservation, and we can carry it as far as the tax-string on our leg will reach. Obviously, though it conserves our self-respect better than our land. Many excellent people accept it, either because they despair of anything better, or because they failed to see the *universality of the reactions needing control*. That is to say their ecological education is not yet sufficient.

The other concept supports the public program, but regards it as merely extension, teaching, demonstration, an initial nucleus, a means to an end, but not the end itself. The real end is a *universal symbiosis with land*, economic and esthetic, public and private. To this school of thought public ownership is a patch but not a program.

Are we, then, limited to patchwork until such time as Mr. Babbitt has taken his Ph. D. in ecology and esthetics? Or do the new economic formulae offer a short-cut to harmony with our environment?

## The Economic Isms

As nearly as I can see, all the new isms—Socialism, Communism, Fascism, and especially the late but not lamented Technocracy—outdo even Capitalism itself in their preoccupation with one thing: The distribution of more machine-made commodities to more people. They all proceed on the theory that if we can all keep warm and full, and all own a Ford and a radio, the good life will follow. Their programs differ only in ways to mobilize machines to this end. Though they despise each other, they are all, in respect of this objective, as identically alike as peas in a pod. They are competitive apostles of a single creed: *salvation by machinery*.

We are here concerned, not with their proposals for adjusting men and machinery to goods, but rather with their lack of any vital proposal for adjusting men and machines to land. To conservationists they offer only the old familiar palliatives: Public ownership and private compulsion. If these are insufficient now, by what magic are they to become sufficient after we change our collective label?

Let us apply economic reasoning to a sample problem and see where it takes us. As already pointed out, there is a huge area which the economist calls sub-marginal, because it has a minus value for exploitation. In its once-virgin condition, however, it could be "skinned" at a profit. It has been, and as a result erosion is washing it away. What shall we do about it?

By all the accepted tenets of current economics and science we ought to say "let her wash." Why? Because staple land-crops are over-produced, our population curve is flattening out, science is still raising the yields from better lands, we are spending millions from the public treasury to retire unneeded acreage, and here is nature offering to do the same thing free of charge; why not let her do it? This, I say, is economic reasoning. *Yet no man has so spoken.* I cannot help reading a meaning into this fact. To me it means that the average citizen shares in some degree the intuitive and instantaneous contempt with which the conservationist would regard such an attitude. We can, it seems, stomach the burning or plowing-under of over-produced cotton, coffee, or corn, but the destruction of mother-earth, however "sub-marginal," touches something deeper, some sub-economic stratum of the human intelligence wherein lies that something—perhaps the essence of civilization—which Wilson called "the decent opinion of mankind."

## The Conservation Movement

We are confronted, then, by a contradiction. To build a better motor we tap the uttermost powers of the human brain; to build a better countryside we throw dice. Political systems take no cognizance of this disparity, offer no sufficient remedy. There is, however, a dormant but widespread consciousness that the destruction of land, and of the living things upon it, is wrong. A new minority have espoused an idea called conservation which tends to assert this as a positive principle. Does it contain seeds which are likely to grow?

Its own devotees, I confess, often give apparent grounds for skepticism. We have, as an extreme example, the cult of the barbless hook, which acquires self-esteem by a self-imposed limitation of armaments in catching fish. The limitation is commendable, but the illusion that it has something to do with salvation is as naive as some of the primitive taboos and mortifications which still adhere to religious sects. Such excrescences seem to indicate the whereabouts of a moral problem, however irrelevant they be in either defining or solving it.

Then there is the conservation-booster, who of late has been rewriting the conservation ticket in terms of "tourist-bait." He exhorts us to "conserve outdoor Wisconsin" because if we don't the motorist-on-vacation will streak through to Michigan, leaving us only a cloud of dust. Is Mr. Babbitt trumping up hard-boiled reasons to serve as a screen for doing what he thinks is right? His tenacity suggests that he is after something more than tourists. Have he and other thousands of "conservation workers" labored through all these barren decades fired by a dream of augmenting the sales of sandwiches and gasoline? I think not. Some of these people have hitched their wagon to a star—and that is something.

Any wagon so hitched offers the discerning politician a quick ride to glory. His agility in hopping up and seizing the reins adds little dignity to the cause, but it does add the testimony of his political nose to an important question: is this conservation something people really want? The political objective, to be sure, is often some trivial tinkering with the laws, some useless appropriation, or some pasting of pretty labels on ugly realities. How often, though, does any political action portray the real depth of the idea behind it? For political consumption a new thought must always be reduced to a posture or a phrase. It has happened before that great ideas were heralded by growing-pains in the body politic, semicomic to those onlookers not yet infected them. The insignificance of what we conservationists, in our political capacity, say and do, does not detract from the significance of our persistent desire to do something. To turn this desire into productive channels is the task of time, and ecology.

The recent trend in wildlife conservation shows the direction in which ideas are evolving. At the inception of the movement fifty years ago, its underlying thesis was to save species from extermination. The means to this end were a series of restrictive enactments. The duty of the individual was to cherish and extend these enactments, and to see that his neighbor obeyed them. The whole structure was negative and prohibitory. It assumed land to be a constant in the ecological equation. Gunpowder and blood-lust were the variables needing control.

There is now being superimposed on this a positive and affirmatory ideology, the thesis of which is to prevent the deterioration of environment. The means to this end is research. The duty of the individual is to apply its findings to land, and to encourage his neighbor to do likewise. The soil and the plant succession are recognized as the basic variables which determine plant and animal life, both wild and domesticated, and likewise the quality and quantity of human satisfactions to be derived. Gun-powder is relegated to the status of a tool for harvesting one of these satisfactions. Blood-lust is a source of motive-power, like sex in social organization. Only one constant is assumed, and that is common to both equations: the love of nature.

This new idea is so far regarded as merely a new and promising means to better hunting and fishing, but its potential uses are much larger. To explain this, let us go back to the basic thesis—the preservation of fauna and flora.

Why do species become extinct? Because they first become rare. Why do they become rare? Because of shrinkage in the particular environments which their particular adaptations enable them to inhabit. Can such shrinkage be controlled? Yes, once the specifications are known. How known? Through ecological research. How controlled? By modifying the environment with those same tools and skills already used in agriculture and forestry.

Given, then, the knowledge and the desire, this idea of controlled wild culture or "management" can be applied not only to quail and trout, but to *any living thing* from bloodroots to Bell's vireos. Within the limits imposed by the plant succession, the soil, the size of the property, and the gamut of the seasons, the landholder can "raise" any wild plant, fish, bird, or mammal he wants to. A rare bird or flower need remain no rarer than the people willing to venture their skill in *building it a habitat.* Nor need we visualize this as a new diversion for the idle rich. The average dolled-up estate merely proves what we will some day learn to acknowledge: that bread and beauty grow best together. Their harmonious integration can make farming not only a business but an art; the land not only a food-factory but an instrument for self-expression, on which each can play music of his own choosing.

It is well to ponder the sweep of this thing. It offers us nothing less than a renaissance—a new creative stage—in the oldest, and potentially the most universal, of all the fine arts. "Landscaping," for ages dissociated from economic land-use, has suffered that dwarfing and distortion which always attends the relegation of esthetic or spiritual functions to parks and parlors. Hence it is hard for us to visualize a creative art of land-beauty which is the prerogative, not of esthetic priests but of dirt farmers, which deals not with plants but with biota, and which wields not only spade and pruning shears, but also draws rein on those invisible forces which determine the presence or absence of plants and animals. Yet such is this thing which lies to hand, if we want it. In it are the seeds of change, including, perhaps, a rebirth of that social dignity which ought to inhere in land-ownership, but which, for the moment, has passed to inferior professions, and which the current processes of land-skinning hardly deserve. In it, too, are perhaps the seeds of a new fellowship in land, a new solidarity in all men privileged to plow, a realization of Whitman's dream to "*plant companionship as thick as trees along all the rivers of America.*" What bitter parody of such companionship, and trees, and rivers, is offered to this our generation!

I will not belabor the pipe-dream. It is no prediction, but merely an assertion that the idea of controlled environment contains colors and brushes wherewith society may some day paint a new and possibly a better picture of itself. Granted a community in which the combined beauty and utility of land determines the social status of its owner, and we will see a speedy dissolution of the economic obstacles which now beset conservation. Economic laws may be permanent, but their impact reflects what people want, which in turn reflects what they know and what they are. The economic set-up at any moment is in some measure the result, as well as the cause, of the then prevailing standard of living. Such standards change. For example: some people discriminate against manufactured goods produced by child-labor or other anti-social processes. They have learned some of the abuses of machinery, and are willing to use their custom as a leverage for betterment. Social pressures have also been exerted to modify ecological processes which happened to be simple enough for people to understand—witness the very effective boycott of birdskins for millinery ornament. We need postulate only a little further advance in ecological education to visualize the application of like pressures to other conservation problems.

For example: the lumberman who is now unable to practice forestry because the public is turning to synthetic boards may be able to sell man-grown lumber "to keep the mountains green." Again: certain wools are produced by gutting the public domain; couldn't their competitors, who lead their sheep in greener pastures, so label their product? Must we view forever the irony of educating our sons with paper, the offal of which pollutes the rivers which they need quite as badly as books? Would not many people pay an extra penny for a "clean" newspaper? Government may some day busy itself with the legitimacy of labels used by land-industries to distinguish conservation products, rather than with the attempt to operate their lands for them.

I neither predict nor advocate these particular pressures—their wisdom or unwisdom is beyond my knowledge. I do assert that these abuses are just as real, and their correction every whit as urgent, as was the killing of egrets for hats. *They differ only in the number of links composing the ecological chain of cause and effect.* In egrets there were one or two links, which the mass-mind saw, believed, and acted upon. In these others there are many links; people do not see them, nor believe us who do. The ultimate issue, in conservation as in other social problems, is whether the mass-mind *wants to* extend its powers of comprehending the world in which it lives, or, granted the desire, *has the capacity to do so.* Ortega, in his "Revolt of the Masses," has pointed the first question with devastating lucidity. The geneticists are gradually, with trepidations, coming to grips with the second. I do not know the answer to either. I simply affirm that a sufficiently enlightened society, by changing its wants and tolerances, can change the economic factors bearing on land. It can be said of nations, as of individuals: "as a man thinketh, so is he."

It may seem idle to project such imaginary elaborations of culture at a time when millions lack even the means of physical existence. Some may feel for it the same honest horror as the Senator from Michigan who lately arraigned Congress for protecting migratory birds at a time when fellow-humans lacked bread. The trouble with such deadly parallels is we can never be sure which is cause and which is effect. It is not inconceivable that the wave phenomena which have lately upset everything from banks to crime-rates might be less troublesome if the human medium in which they run *readjusted its tensions.* The stampede is an attribute of animals interested solely in grass.